# Benchmark Papers in Genetics

Series Editor: David L. Jameson
University of Houston

**Published Volumes and Volumes in Preparation**

GENETICS AND SOCIAL STRUCTURE / Paul Ballonoff
POPULATION GENETICS / James F. Crow and Carter Denniston
MEDICAL GENETICS / William Jack Schull
ECOLOGICAL GENETICS / W. W. Anderson
METABOLIC GENETICS / R. P. Wagner
QUANTITATIVE GENETICS / R. E. Comstock
GENETIC RECOMBINATION / Rollin Hotchkiss
REGULATION GENETICS / Werner K. Maas
ANIMAL BREEDING / Robert C. Carter
PLANT BREEDING / D. F. Matzinger
DEVELOPMENTAL GENETICS / Antonie W. Blackler and Richard Hallberg
CYTOGENETICS / Ronald L. Phillips and Charles H. Burnham
EUGENICS / Carl Bajema

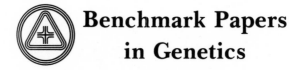

Benchmark Papers in Genetics

A *BENCHMARK* ® Books Series

# GENETICS AND SOCIAL STRUCTURE:

## Mathematical Structuralism in Population Genetics and Social Theory

Edited by
**PAUL BALLONOFF**
*The University of Texas, Houston*

Dowden, Hutchinson & Ross, Inc.
Stroudsburg, Pennsylvania

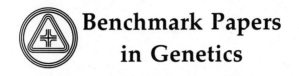

Benchmark Papers in Genetics, Volume 1
Library of Congress Catalog Card Number: 73–20412
ISBN: 0–87933–067–8



Manufactured in the United States of America.

Exclusive distributor outside the United States and Canada: John Wiley & Sons, Inc.

74 75 76 5 4 3 2 1

**Library of Congress Cataloging in Publication Data**

Ballonoff, Paul A comp.
Genetics and social structure.

(Benchmark papers in genetics, v. 1)
CONTENTS: Macfarlane, A. Analysis of relationships of consanguinity and affinity.--Kroeber, A. L. Classificatory systems of relationship.--Pearl, R. The measurement of the intensity of inbreeding. [etc.]
1. Population genetics--Addresses, essays, lectures. 2. Kinship--Addresses, essays, lectures. I. Title. [DNLM: 1. Genetics, Population. 2. Mathematics. W1 BE516 v.1 1974 / QH431 G3285 1974]
GN289.B34 301.42'1'0157321 73-20412
ISBN 0-87933-067-8

# Acknowledgments and Permissions

**ACKNOWLEDGMENTS**

Genetics Society of America—*Genetics*
"The Probability of Consanguineous Marriages"

Life Sciences and Agricultural Experiment Station of the University of Maine at Orono—*Bulletin 218*
"The Measurement of the Intensity of Inbreeding"

Mathematical Societies of Denmark, Iceland, Norway, Finland, and Sweden—*Mathematica Scandinavica*
"Genetic Algebras Studied Recursively and by Means of Differential Operators"

Society for the Study of Evolution—*Evolution*
"The Interpretation of Population Structure by F-Statistics with Special Regard to Systems of Mating"

**PERMISSIONS**

The following papers have been reprinted with the permission of the authors and copyright holders.

The Biometric Society—*Biometrics*
"Genetic Information Given by a Relative"

Cambridge University Press—*Annals of Human Genetics*
"Analysis of Population Structure: II. Two-dimensional Stepping Stone Models of Finite Length and Other Geographically Structured Populations"

Department of Anthropology, The University of New Mexico—*Southwestern Journal of Anthropology*
"A Formal Analysis of Prescriptive Marriage Systems Among the Australian Aborigines"

Gauthier-Villars—*Comptus Rendus de l'Académie des Sciences Mathématiques*
"Démonstration mathématique de la loi d'hérédité de Mendel"
"Principe de stationarité et généralisations de la loi de Mendel"

Mrs. Helen Haldane
"An Enumeration of Some Human Relationships"

Harper & Row, Publishers—*The Effects of Inbreeding on Japanese Children*
"The Interpretation of the Effects of Inbreeding"

Prentice-Hall, Inc.—*An Anatomy of Kinship: Mathematical Models for Structures of Cumulated Roles*
"Models of Kinship Systems with Prescribed Marriage"

The Royal Anthropological Institute of Great Britain and Ireland—*Man*
"Classificatory Systems of Relationship"

The Royal Society of Edinburgh—*Proceedings of the Royal Society of Edinburgh*

"On Non-associative Combinations"
"Genetic Algebras"
"Non-associative Algebra and the Symbolism of Genetics"

Science Council of Japan—*VIIIth Congress of Anthropological and Ethnological Sciences, Volume II: Ethnology*
"Formal Analysis of Prescriptive Marriage System: The Murngin Case"

Scottish Academic Press Ltd.—*Proceedings of the Edinburgh Mathematical Society*
"Duplication of Linear Algebras"

# Series Editor's Preface

The study of any discipline assumes the mastery of the literature of the subject. In many branches of science, even one as new as genetics, the expansion of knowledge has been so rapid that there is little hope of learning of the development of all phases of the subject. The student has difficulty mastering the textbook, the young scholar must tend to the literature near his own research, the young instructor barely finds time to expand his horizons to meet his class preparation requirements, the monographer copes with a wider literature but usually from a specialized viewpoint, and the textbook author is forced to cover much the same material as previous and competing texts to respond to the user's needs and abilities.

Few publishers have the dedication to scholarship to serve primarily the limited market of advanced studies. The opportunity to assist professionals at all stages of their careers has been recognized by the publishers and by a distinguished group of editors knowledgeable in specific aspects of the literature of genetics. Some have contributed heavily to the development of that literature, some have studied with the early scholars, and some have developed and are in the process of developing entirely new fields of genetic knowledge. In many cases, the judgments of the editors become a historical document recording their opinion of the important steps in the development of the subject. These editors have selected papers and portions of papers that demonstrate both the development of knowledge and the atmosphere in which that knowledge was developed. There is no substitute for reading great papers. Here you can learn how questions are asked, how they are approached, and how difficult and essential it is to obtain definitive answers and clear writing.

My own pleasure in working with this distinguished panel is exceeded only by the considerable pleasure of reading their remarks and their selections. Their dedication and wisdom are impressive.

This volume presents a selection of literature constituting the Benchmark papers in the intersect between the social structure of populations and the classical literature of mathematical population genetics. The volume is representative of both the trend to interdisciplinary work and the movement to relevance. A new field and a young editor with a good start and considerable promise provide us with challenging, thought provoking, and meaningful papers.

D. L. Jameson

# Preface

## Basic Purpose

Most readers of this volume will undoubtedly find its contents different from what the title might on first sight suggest. Therefore, the first paragraphs of the introduction are devoted to a specific definition—for present purposes—of the terms "genetics" and "social structure." The remainder of the introduction is a discussion of the primary topic and the interrelationships of papers in the volume.

The present purpose is of course to collect into a single volume historic or current, yet fundamental, papers on the topic "genetics and social structure," or more precisely, mathematical/structural foundations of population genetic theory and similar foundations of social theory where there are relationships between them. In several places, there was great temptation to divide the volume into two: one on the structural foundation of genetics, which would have contained Parts II, III, IV, and VI; and one on the foundations of social theory, which would have included Parts I and V. This would have allowed the inclusion of many papers whose importance earned them a place here, but which were excluded for one or more of the following reasons: they were easily available elsewhere; they were overly long; their sensible interpretation depended on papers included; and a balance of topics was needed.

## Uses of the Volume

However unusual the present juxtaposition of materials may appear to a reader, I know of no other place where papers on either of the two leading topics may be found together. For that matter, there is no place where collections of these mathematical/structural papers on either subtopic could be found separately. I hope that the contrast and interrelationships will be intriguing.

In constructing the general introduction, I presumed that my principal audience has some interest in, if not knowledge of, one or the other of the leading topics in some depth, and a very minimal knowledge of vocabulary in both. I presumed a knowledge of mathematics that is at about the junior level of college algebra. Certainly it requires at least this level to read intelligently most of the papers selected. However, all my introductory materials are in narrative style, and I have tried to minimize use of the normal jargon from either field.

Uses of this volume include the following: as source material for researchers or courses in structural and/or mathematical anthropology, or similarly, in mathematical sociology; as background and historic materials in the structural theory of population genetics; as sources of insight for mathematical geneticists of some large holes in the existing theory; and finally, as an unusual perspective on the history of population genetics. I must admit that when I began research on the volume, I expected that the first two purposes would predominate. Now that I have looked at the existing literature in more detail, I strongly believe that the most important effect of this material *ought to be* the finding and filling of holes—I am now quite convinced that a very large and generally unsuspected gap most certainly exists in population genetic theory. I hope that after reviewing this material, the reader may agree and offer his or her contribution to the work that must be done.

## Organization of Papers

In the introduction, papers are treated in historical order, but for ease in comparing papers, and to simplify and condense introductory materials, papers in the volume itself are grouped into the following parts:

I. History of Social Theory
II. History of Genetic Theory
III. Foundations of Genetic Algebra
IV. Calculus for Statistico-genetics
V. New Departures in Structures
VI. The Current State of Unsolved Problems

In each part, a common introduction is provided, and in some cases, additional introductions are provided for particular papers. Thus, Part III, which contains five papers by only two authors, needs only one introduction discussing the selection of these papers and the related literature, while Part VI contains papers so diverse that discussion of a common literature would be farcical. Such bibliography as is provided is intended to be a selective survey of the most important related publications. A complete survey would of course be a massive volume in itself. I have thus principally referred to three kinds of papers: those which might well have been included here, but were not for reasons of space or balance; those with good bibliographies of the topic; and those of primary historic significance.

## Sources of Support

This volume would have been impossible both in its basic form and final content without the generous assistance of two individuals. William J. Schull was very generous in his suggestions of possible historic sources, in discussions of areas of the history, and in editorial comment on the general introduction. Similarly, I. M. H. Etherington freely offered suggestions of background literature on genetic algebras. Naturally, neither of these individuals is in any way responsible for the final selection of papers; nor can either of them be held responsible for some of the possibly highly speculative statements made in the various introductions. My mistakes remain my own responsibility.

This work was supported under N.I.G.M.S. Grant 19513.

Paul L. Ballonoff

# Contents

# Contents by Author

# Introduction

The primary purpose of this volume is to expose, by way of a series of early and recent papers, the structural foundations of population genetics and of social theory as it bears on these foundations. Not surprising to workers in both fields of study, there are important areas of complementation. This is especially true if "social structure" is viewed as the study of genealogical interconnectedness and shared genealogical history of individuals, while "genetics" is taken as the study of gene frequency and of the distribution of genotypes in a particular genealogical structure. "Structuralism" as it will be used here is the same topic that others have called "axiomatization," "logistic," "formalization," "algebraic approaches," and related terms. It is related to the philosophical/literary structuralism now popular in certain circles, but it is not my purpose to explore this relationship.

The papers selected come primarily from two sources: from a century of attempts by anthropologists and sociologists to describe and enumerate kin-based systems of reckoning social relatedness; and from a tradition of structuralism in population genetics which appears in this decade to again be bearing fruit. As there are a large number of possible papers that could fit even into our slightly narrowed definitions, the reader will find that two kinds of introductory materials are provided. At the beginning of each paper is a summary discussion of related literature. Since the field of study of interest here has not been widely recognized in the form in which it will be discussed, these short surveys will provide historical and complementary references that might otherwise occur in a general introduction. This will leave the latter open for discussion of the papers selected and their interrelationships.

It is useful to recognize that the conceptual problems which have characterized social and genetic theories have been quite different and that this difference perhaps has been one reason for the historical separation of the fields. In particular, where social theory has been concerned with underlying finite geometries and algebras which describe genealogical histories, genetics has, at least recently, been concerned mainly with probabilistic and stochastic models for gene frequencies, and has treated genealogical structures as of secondary interest even where they determine the form of a particular expression. The great importance of algebraic models (as is so beautifully developed in Cotterman's work, Paper 11) is that they provide direct means of coupling statistical to structural arguments, just as the matrix mechanics and Dirac notation of the 1920s and 1930s in physics enabled the body of working physicists to more easily grasp this same relationship.

It is useful to begin by discussing a number of topics from both source fields for these papers, to demonstrate that there is indeed a correspondence in subject matter. Perhaps the most obvious comparison is between what geneticists call "regular systems of mating" and what social theorists variously call "marriage theory," "elementary structures," or "minimal structures." In each of these areas, the primary concern is an idealized version of a closed (to migration) population that follows rigorously a specified rule or pattern in selecting partners for "marriage." Although it is true that "natural" populations of humans or other organisms may not actually follow such rules, it is surely legitimate to study the idealized case for its genetic or demographic consequences. Such systems are also interesting since they have quite noticeable mathematical properties, including a group theoretical symmetry that has yet to be properly exploited for theory construction.

Beyond these most basic aspects, a large number of other areas of similarity may be discovered. I shall identify a few of these here, since it is not possible to explore all of them in this collection of papers. One omission may be contrary to the intuition of some readers: assortative mating. In the minds of many, a principal relationship between genetics and social structure is the way in which class structure, achievement, intelligence, and other apparently quantitative characters affect and are affected by gene frequencies or "social values." [See Halsey (1972) for a good summary.] These have been omitted for several reasons. First, they are typically considered in the literature relating to specific problems or philosophies, and the concern here is on a more abstract level of theory construction. Second, one hope of the present volume is to present, with coherent perspective, a number of related papers. This perspective includes belief that a study of the relationship of assortative mating to social organization requires prior understanding of nonassortative mating, and that therefore a more fundamental theory lies in this realm.

Another topic *not* covered here is that of "altruistic evolution" and the related topic "kin selection." While the reader should certainly read the innovative literature on this topic [see Hamilton (1972)], the evolution of altruism per se is a false issue, for as is clearly pointed out by Monod (1971, p. 119), evolutionary theory does not concern itself with anthropomorphic struggles for existence but with differential reproduction. Thus the essential question in "altruistic" species is under what condition the population that contains "altruistic" individuals reproduces more offspring than populations which do not. The altruism of an individual wasp, say, is but one phenotypic expression of a potential genotype of a particular gene pool. Its existence is no more (or less) mysterious than the evolution of tails on monkeys or hands on men. [Note that Haldane (1931, p. 208) shares this opinion.] On the other hand, construction of a priori theories that can predict a phenotype which might occur under a particular condition will provide evolutionary theory with powerful techniques. Study of altruism more properly belongs in a volume dedicated to that more general question. As with assortative mating, it can be confused with social theory as defined here only on specious grounds and because there has not been given sufficient, nor sufficiently careful, thought to description and interpretation of organismic behavior.

## Selection of Papers

Selection of papers for this volume was difficult for two reasons. In social theory there has only been a small number of papers published whose structural value might be clearly recognized by geneticists. (This is largely because the orientation of anthropologists, for example, has been toward psychological more than structural questions.) However, this small number is highly original. The first difficult choice was therefore to minimize inclusion of social theoretical materials to these few mathematical papers, and include only two "empirical" papers that give the reader at least a flavor of the context in which work was originally done.

The second difficulty was in selecting papers from the huge body of mathematical genetics without the selection being simply a survey of theoretical genetics. However, once one begins to focus on *foundations* of theory rather than on the theory itself, the availability of papers again quickly diminishes. In addition, the best papers were also the longest, and since it is difficult in mathematical work to edit sections from a discussion, the choices were inclusion or exclusion of whole papers.

Papers in the volume are approximately ordered by date, the oldest first. The oldest selected paper is that by Macfarlane, who was evidently a Scottish mathematician with a concern for anthropology. This paper is remarkable first because of its age. It represents a tradition of work (essentially the use of semigroups) that, except for this paper, flowered hardly at all until the middle of the twentieth century. Second, it is and remains an innovative technique for enumeration and description of kin relations, and for describing the different types of marriage rules and their consequences for social organization. These consequences are principally the equating of several different types of kin into a single category. It is interesting that no one, to my knowledge, has ever studied what the average correlation of uniting gametes might be in each system of the types generated by such enumeration. (However, see Haldane and Jayakar, Paper 17.)

Some reasons why social theorists have not been interested in this question are documented by Kroeber in Paper 2. After outlining a number of conceptual questions on relations between social organization and demography (which have still not been adequately answered!), Kroeber argues that one should not confuse social categories with biological ones or, for that matter, cognitive categories with social ones. The principal task of social theory in the last century has been in discovering and answering this kind of question. For a theoretical geneticist, however, the important questions are conditional ones: *If* a system ever followed such a rule, what would be the consequences? As I shall argue later, there are some potentially useful answers for genetics in the literature generated by cognitive questions.

The third paper, which is the first in this volume by a geneticist, represents a tradition of work that flourished in 1910–1920 and was concerned, as in social theory, with description of genealogical systems and their consequences. Much more than the newer work (by Haldane, Fisher, and Wright), which seems to have displaced these papers by the early 1920s, the early tradition was structuralist. And although the genetic interpretation given by the authors of these papers was quite often wrong, it appears to have been a necessary early stage. Indeed, there is an open historical question [see Provine (1971, pp. 136–137)] as to how directly this material stimulated the work of Sewall Wright. I have included this paper both because it demonstrates an important historic period and because it proposes an interesting measure on genealogical history. Although genetically not useful in the form proposed, the utility of measures for reducing complicated histories to fairly simple and calculable numbers will eventually have to be treated in a structural theory, especially if it is to have demographic importance. The various coefficients (of kinship, etc.) used in inbreeding studies are such examples in genetics.

Were it not for their wide availability in other forms, the next paper or papers in this collection would undoubtedly have been by Sewall Wright (1921). The most interesting fact about these historic papers for our purposes is the explicit dependence on genetic results on the particular genealogic structures [in fact, "minimal" structures per Ballonoff (1973)] or regular system of mating which generated them. Although it is true that Fisher (1918) had already published one treatment of this problem based on recurrence equation methods and that Haldane's major papers used similar methods, the structural foundation of the technique and its explicit connection to these techniques was, in 1921, yet to be elaborated. It is, after all, a principal aim of structural or axiomatic approaches to remove the implicit from scientific insight, and to make comprehensible, or better yet, imperative, the reasons for choice or exclusion of one or another mathematical form.

The next two papers are clearly in this structural tradition. Although the Hardy–Weinberg law relating genotypic to gene frequencies at equilibrium was certainly known before these papers by Bernstein, he has presented an early rationalization not only that it works but that it *ought to work* under some specified conditions. Without in any sense demeaning more empirical material (which is the whole point of scientific work in the first place), it must be recognized that the step from noticing a relationship between particular observations and particular models, to creating a framework in which this correspondence is a necessary consequence, is the most fundamental step in theory construction. With Bernstein's papers we can begin to see most clearly the distinction between population genetics and the *foundations* of population genetics. Making another comparison to theoretical physics, we must recognize the difference between study of physics per se (the properties of particular atoms, for example) and the study of Hilbert space and operators on Hilbert space (from which one hopes to derive the forms used to describe these properties, a priori). To take the most philosophically extreme position, one must even say that the existence of empirical constants in the resulting theory is simply an indication that the foundations are not yet adequately known.

The next important work is one that is omitted, as it is an entire book—J. H. Woodger's *The Axiomatic Method in Biology*, published in 1937. The text was highly influenced by the work of Bertrand Russell and A. N. Whitehead and contains an appendix by Alfred Tarski. Inspired by the logical ideas that were later to revolutionize linguistic theory and provide a foundation for theories of computer languages, Woodger set himself the goal of axiomatizing biology, especially population genetics. That he at least made a major contribution is without doubt, but the perversity of his notation system made the work unaccessible to all but the most dedicated biologists. Indeed, it is safe to argue that one reason for the apparent retreat of the positivists from dominance in philosophy is the cumbersomeness of their techniques. Although the goals of complete axiomatization must still be sought, other papers in the volume demonstrate that algebraic and set theoretic rather than logical techniques are required, or are at least more productive. It may be true that all of any one technique may be derived from any other, but it is also true that many mathematical results become much more apparent, if not obvious, when seen in their proper framework.

It is certainly untrue, however, to call the logistic approach unproductive; it lead to the first formalization of the probability approach to consanguinity in sketchy form by Woodger, and more fully by Charles Cotterman in his doctoral dissertation. The complete dissertation was never published and appears here for the first time. It predates, by eight years, publication of remarkably similar techniques by Malécot, and Cotterman surely deserves recognition along with Malécot as founder of the approach. Although the text does show the effects of a cumbersome notation on accessibility of a result to readers, it also shows the workability of the approach for one with much imagination. Not the least of Cotterman's contributions was the invention of a form of bayesian statistics with which to handle conditional probabilities.

The period of World War II produced a number of important but unrecognized works. Besides that of Cotterman, certain papers by I. M. H. Etherington have only recently begun to receive the attention they deserve. These papers have several highly important characteristics. First, they provide (to my knowledge) the first *formal* demonstration that the recurrence method was the correct one. Second, they provide a similarly rare demonstration that the recurrence method is appropriate to large classes of genetic systems. Third, they provide the earliest treatment of the *foundations* of population genetics, which was at once comprehensive and well founded in techniques easily extended to broader applications or "hooked up" to other mathematical apparatus. It is very useful to compare Woodger's later work (1952) to the papers of Etherington to recognize that they are after similar objectives but that the real power of the techniques is only apparent when the coefficients generated by "logical" analysis can be put to some use. Fourth, these papers created a tradition of work on algebraic foundations of genetics, of which one example is included here (by Reiersöl) and to which other references are provided in the paper's introduction.

By the end of the 1940s Malécot had published his probabilistic approach, and developments had also taken place in social theory. They were essentially recognition that, first, there are common features to marriage systems throughout the world and that this was much more complete than had been previously believed. Second, at least some of these systems had regular properties amenable to treatment as mathematical groups. Because the volume is dedicated to generally unavailable mathematical materials, I have chosen to not include a selection from Claude Lévi-Strauss's *The Elementary Structures of Kinship,* and I mention it here only for historical continuity. Mathematically, the work was important for its appendix by André Weil, which provoked several of the papers in the present volume—those by Courrège, White, and Liu. However, to represent the types of structures that were being recognized and discussed, I have selected the paper of Frank Livingstone as being the treatment that is clearest and most compact, yet still empirical. Other references are provided in the introduction to Livingstone's paper.

The 1950s were basically years of clarification of ideas suggested earlier in both fields, but the 1960s brought new work in a form that suggests a strength both in numbers and in intellectual depth and potential. In 1962, two critical papers were published. One of these followed in the path of Etherington. Olav Reiersöl proposed a treatment of chromosome structure which has since received elaboration in the paper by Lyubich to be discussed below. Reiersöl's is the most clearly genetic paper under my previous definition. Although this is slightly untrue when the papers of Bernstein and portions of Cotterman's work are being considered, it is still useful to remember this when comparing the paper to that of Haldane and Jayakar, which appeared in the same year. Although the enumeration techniques and notation differ, there is remarkable similarity between the kinship algebra used here and that of Macfarlane, the first paper in the volume. Likewise, the genetic measures used would have been traceable to Cotterman had only the work been published.

I shall summarize briefly the developments so far. First, semigroups that describe kin systems have been experimented with and their application to genetic systems begun. Second, axiomatic foundations of both genealogical and purely genetic theories have been explored. Third, group theory techniques have begun to be tapped. In this is another parallel and also perhaps a lesson to be gained from physical theory. In its early phases, physical theory was largely a statistical theory. By the 1920s, structural approaches in the form of algebraic foundations under the rubric "quantum theory" had already proved critical. (Was it an accident that the early training of Etherington was in relativity theory?) Thus it was for physicists in the decades up to the present to explore the effects of different systems of symmetry on various physical statistics. This is precisely the position of social and genetic theory at present, and the next few papers were chosen to make this point.

The first such paper is that by Harrison White, whose first doctorate was in solid-state physics and his second in sociology. The book that is summarized in this article was one of the more brilliant theoretical excursions of the decade; it elaborated the group theoretical structure of a number of systems, argued about how many such were possible, and suggested ideas on relational structures that have contributed to more than one doctoral dissertation and to a major work in French soon to be published (Lorrain, 1974). This work is also strongly based upon the companion paper by Philippe Courrège (translated for its first English publication in this volume), which takes a somewhat different group theoretical approach to the same problem. The paper is also significant because it is a mathematical treatment of the ideas proposed by Lévi-Strauss, mentioned above. (I break my historical order to note that the paper by Liu is another tack on the same course.)

What is missing from all these papers, but is so tantalizingly close, is a direct statement of the relationship of such structures to the genetic systems they could theoretically determine. The next papers should clear up any doubts that the topic is worth studying, or that there are new results to be gained by approaching it. Sewall Wright was the pioneer in direct use of elementary structures in genetic studies. This article also shows that he is pioneer in exploring the subtle but significant differences in empirical prediction that result from differences in theoretical framework. While Wright is exploring the effects of the whole structure (i.e., assuming a closed, endogamous system), the paper by Cavalli-Sforza et al. has a more practical aim. They are concerned with geographic diffusion and the relation of this to kin-based structures. Although the decomposition techniques used here are perhaps insightful, in the present context the most important aspect of the paper is its use of operator algebras for the study of diffusion. If the suggestions made in this introduction about other techniques prove correct, then a very natural development will be the extension of the endogamous system theory to open or exogamous systems. The fact that matrix techniques appear useful in this context is at least strongly suggestive that this is worth exploring.

The paper by Maruyama has been selected for the same reason. Although not much is known mathematically about the eigenvalues of a graph, there are to my knowledge two suggestions that it is important to study this topic for its utility in theory construction. One of these is Chapter 8 in Ballonoff (1974b), where it is argued that these eigenvalues may be used to construct a visual representation of a regular mating system which has the same periodicities as the group structures of Courrège's representation. In this paper Maruyama shows how the eigenvalues of the incidence graph of interbreeding populations interact directly with the statistics of the distribution of gene frequencies.

The final papers are selected to show that the theoretical and empirical problems of the interactions of genealogical (social) and genetic structures are by no means solved. The paper by Jacquard is selected because it appears to simultaneously use ideas from some literature referenced in my introduction to the paper, because it has strong relations to other papers in this volume (compare the enumeration technique to that of Cotterman), and because it demonstrates the use of the trimat notation developed by Jacquard for these problems. The paper by Schull shows that there is yet a gap from the theoretical structures to empirical ones, but that the social theory literature discussed here has already proved valuable in treating these.

If the reader is now looking for a comprehensive survey of the theory proposed here, I strongly suggest the 1971 article by Lyubich (which was brought to my attention by Etherington). Although the article very clearly does not deal with social structures, it also very correctly deals with the most fundamental aspects of genetic structures and is strongly based on the genetic literature summarized here. It is not included only because of its size.

## Needed Lines of Research

Rather than write a conclusion, it seems more necessary to indicate what directions of research are mandated by the present state of structural theory (the order of presentation does not reflect importance):

1. The cardinalities of minimal systems (i.e., the actual size of minimal populations for the regular mating systems, which is also usually a simple multiple of the order of the mathematical group that represents it) undoubtedly have some relationship to the genetic (and demographic!) statistics of any real population using such a system.

2. The study of semigroups and groups of operators over descent structures seems well in order. At the least, we can expect this to produce a neat rack on which to hang the remainder of the theory.

3. On the same lines, the perspective of nonassociative and genetic algebras should, can, and undoubtedly will continue under the strong momentum it appears to have gained.

4. The theory of free populations (i.e., populations without selection) as defined by Lyubich must certainly be extended to cases that include selection and social structure.

5. Diffusion or stochastic process models need to be related to the structural models surveyed here. This is a topic in its own right but also will be a necessary step in the study of cardinalities.

6. The use of various representational techniques of modern algebra (such as commutative diagrams to summarize aspects of theory) needs to be continued beyond the brief start given it by Woodger (1952) and Lyubich.

7. The enumeration of various regular systems of mating and study of their consequences has hardly begun. Enumeration will likely be done on structural grounds (such as group theory or the theory of minimal structures), and almost certainly contains more surprises, such as those discussed by Wright (Paper 18).

8. The application of symmetry group theory to the study of regular mating will undoubtedly simplify treatment of complex mating systems, group inbreeding, and regular mating. It should be pursued independently of studies under point 7.

9. An "updated" version of the work begun by Woodger is in order, although perhaps the greater power of algebraic methods will make this a point of only historic and philosophical interest.

10. Once point 4 has been carried out, the possibility of studying adaptive surfaces between different social systems will present itself. In such a study, the statistics (genetic *and* demographic) implied by a particular system will be jointly minimized, much as in thermodynamics. It should be possible to theoretically classify systems into

states according to which sets of statistics are jointly minimized by a particular structure, and to predict the occurrence of these states in natural populations.

11. A historian of science should meticulously study the interrelations of quantum theory and the body of literature surveyed here. In particular, works by the mathematician A. A. Albert and their influences should be noted and weighed against the philosophical outlook of the day. This should include a searching examination of why it appears that statistical theories have preceded structural ones in both physical and social-genetic studies.

I conclude this section with a deliberately provocative, if arguable statement: The more-fundamental theories are always structural. If we are confused theoretically (as in the hidden-variable argument in modern theoretical physics), it is quite likely because we have lost track of, or never saw, the fundamental structures.

There are undoubtedly other topics worthy of note. These are the ones that appear most significant and most productive to this writer. I hope that presentation of these papers may lead to further progress.

# I

# History of Social Theory

# Editor's Comments on Papers 1 and 2

1 **Macfarlane:** Analysis of Relationships of Consanguinity and Affinity
*J. Roy. Anthropol. Soc.*, **12**, 46–63 (1882)

2 **Kroeber:** Classificatory Systems of Relationship
*Man*, **39**, 77–84 (1909)

Social theory as discussed by its modern practitioners has a short but complicated history, not at all simplified by the cleavages of language and continent, academic department, and ideology. However, it is fair to say that the primary concerns of the early authors was not structuralism per se, nor was most early work mathematical. The works of Henry Sumner Maine (1861) on foundations of marriage and descent practices in ancient law, of Lewis Henry Morgan (1871, 1877) on the variety and evolutionary significance of various types of human organization, or of Ernest Crawley (1927) on customs associated with marriage exemplify the concerns of theorists well into the twentieth century. Nonetheless, work such as that by C. Staniland Wake (1967, but originally 1889) demonstrated that by the 1880s even nonmathematical theorists were using structural techniques and recognizing empirically observed structures.

Thus the appearance in 1882 of a paper such as that by Macfarlane is not completely surprising but is at least mildly so. What is more surprising is that so little resulted immediately from publication of this work. A recent survey article by Needham (1971) helps in understanding the historical context, but Needham's arguments are not completely satisfactory largely because they do not account for developments discussed in the introduction to Part V. His survey is, however, an intriguing entry into the schizophrenic social theory literature.

Another of the many personalities of social theory is demonstrated by the paper by Kroeber, which is still a classic work of anthropological thought. (The reader will easily find references to many works by A. L. Kroeber in the card file in his library.) The implications of this article should not be ignored. If Kroeber is even partly correct, the completion of the structural theory argued for in this volume will not be enough to always predict the occurrence of the full variety of social systems nor all the variations within a single system. However, as I argued in the introduction, the conditional question of what would happen *if* the varieties of possible structures were followed as rigorous mating patterns is valid and useful. Also, Hajnal (1963) provided a mathematical treatment of many questions related to the demographic implications of these same questions which should be read by every serious social theorist of whatever persuasion.

# 1

Reprinted from *J. Roy. Anthropol. Soc.*, **12**, 46–63 (1882)

## ANALYSIS *of* RELATIONSHIPS *of* CONSANGUINITY *and* AFFINITY.

By A. MACFARLANE, M.A., D.Sc., F.R.S.E., Examiner in Mathematics in the University of Edinburgh.

[WITH PLATES II TO V.]

THE problem we have to consider may be described as how to develop a systematic notation capable of denoting any relationship of consanguinity or affinity. Such a notation, it is evident, will be able to serve as an instrument in further inquiries, and will bear a relation to the ordinary system of terms, the same as that which the notation of chemistry bears to the arbitrarily chosen names of substances. Like the chemist, we first analyse as much as is possible, then choose symbols for the elements resulting from our analysis, and express the compound ideas in terms of these fundamental symbols. Further, a graphic method can be developed analogous to that used by the chemist.

In several papers recently published,[1] I have considered the problem from the purely mathematical point of view; at present, I wish to present the method, and some applications, in a simple, self-contained form. I was invited to undertake this task by the distinguished anthropologist, Dr. E. B. Tylor, in the hope that the method may prove of service in investigating certain problems of comparative jurisprudence.

I have found from my own course of study, and also from the nature of other notations which I have met with, that there is a tendency to stop the analysis before pushing it far enough. I refer specially to the ingenious notation of Mr. Francis Galton, as used by him in his work on "Hereditary Genius." For example, with a single symbol to denote such an idea as brother, it is impossible to build up a scientific notation; the idea must be resolved into its constituent ideas. At first,[2] I took for a basis the four ideas of son of a man, son of a woman, daughter of a man, daughter of a woman; next,[3] I found it more convenient to proceed with symbols denoting child of a man, and child of a woman; and, finally,[4] I found what I believe is the proper basis, namely, the separation of the idea of sex from the idea of descent.

There are two fundamental relationships of the highest generality, namely, *child* and *parent*, the one relationship being the reciprocal of the other. These can be combined so as to express any of the complex relationships; thus, grandchild is expressed by *child of child;* grand-parent by *parent of parent;* brother or sister by *child of parent;* and consort by *parent of child.* The two latter expressions are taken subject to a certain condition (*see* p. 48). In the same way, great grandchild is expressed by *child of child of child,* nephew or niece by *child of child of parent,* and so on.

For the sake of shortness, let $c$ be used to denote child, $p$ to denote parent, and let "of" be expressed by juxtaposition, then grandchild will be denoted by $c\,c$, brother or sister by $c\,p$, consort by $p\,c$, grandparent by $p\,p$. This method leads to an exhaustive and orderly notation for relationships, as will be seen by turning to Table I. It contains what may be called the general relationships of the first five orders. The order of a relationship is defined as depending upon the number

[1] "Proc. Roy. Soc., Edinb.," vol. x, p. 224; vol. xi, pp. 5 and 162. "Phil. Mag," June 1881. "Educational Times," reprint vol. xxxvi.
[2] "Proc. Roy. Soc., Edinb.," vol. x, p. 224.
[3] Ibid., vol. xi, p. 6.
[4] Ibid., vol. xi, p. 162.

lines which can collapse may collapse, and in the other case, by supposing that such lines may not collapse. Figs. 15 to 18, Plate II, indicate graphically the examples considered above. It is evident that a relationship of an odd order can reduce only to one of an odd order, and a relationship of an even order only to one of an even order.

Regarding the use of the terms in the fourth column, it is necessary to make the following observations. By *brother* is meant what is usually denoted by half-brother, that is, son of the same father *or* son of the same mother. In accordance with this system, son of the same father *and* son of the same mother is considered as *two-fold* brother. To develop a complete scientific notation demands this view of the subject; for, consider the relationship of *first cousin.* In this country it may exist singly, or two-fold, or three-fold, or four-fold. We should then require to speak of cousin, three-quarters cousin, half cousin, quarter cousin. But, in addition to the awkwardness of employing fractions, there is this defect, that the four-fold limit depends, not upon biological but upon moral law. Hence for the purpose of an exact investigation, it is preferable to say cousin, two-fold cousin, three-fold cousin, four-fold cousin.

The expression *consort* may be taken in three different senses, according to the nature of the investigation; first, in the simple sense of co-parent of a child; secondly, in the sense of legitimate co-parent of a child; thirdly, in the sense of husband or wife, that is, legitimate, actual or potential, co-parent of a child. In what follows, the term is generally used in the last signification, but it may be used in either of the other significations should a particular investigation demand it.

The term *step-child* is used in a sense which is probably more general than the sense ordinarily attached. Suppose that $A$ marries $B$, and that they have a child $X$, and that $B$ afterwards marries $C$, and that they have a child $Y$, then $X$ would, in the ordinary acceptation of the term, be a step-child of $C$; but in a systematic nomenclature, it is convenient to extend the meaning of the term, so that it may apply equally to the relationship of $Y$ to $A$. I use the term *step* in this extended sense throughout.

In the case of certain irreducible relationships, equivalent terms are, so far as I know, wanting in the English language. For example, *p c p c*, which from its analogy to *c p c p* (step-brother or step-sister) I have ventured to express as *step-consort;* also *c p c p c*, which I have expressed as *step-step-child.* It will be observed that a special irreducible term is required for, and only for, each genus which has its letters arranged alternately.

Column fifth contains a classification of the genus relationships proceeding upon their characteristic parts. Suppose that from each relationship which has a combination of *c's* or of *p's* at its front, or at its end, all the letters of the combination are cut off excepting one, then, those relationships which leave the same remainder may be said to have the same *characteristic*. Such a group of relationships fall naturally into a class. The several characteristics to be met with in relationships occurring within the first five orders are exhibited on Plate III. Words are in common use to express the classes determined by the first three characteristics, namely, 1st descending lineal or descendant, 2nd ascending lineal or ancestor, and 3rd collateral; but there is, so far as I am aware, no single term to denote the fourth. It embraces all the ancestors of any consort of any descendant of self (including consort of self). As this group embraces the relationships by affinity in the strictest sense of the phrase, it may, for the sake of shortness, and to provide a means of developing a nomenclature for the more complex classes, be denoted by *affinal*.

Each class comprises a number of sub-classes (col. 6th), determined by the number of letters in the combination of *c's* (or *p's*) at the end of the relationship. If, further, the number of letters in the combination at the front of the relationship be specified (col. 7th), the genus is then wholly determined. This last entry has the best title to the denomination of the degree, but to avoid the use of that ambiguous word, I shall call it the *Number*. Not only is it only relationships of the same class, but it is only relationships of the same sub-class which can properly be compared as to degree. As it is, the degree is reckoned by different authorities in different ways. In the case of the first two classes, the lineal ascending and the lineal descending, there is no ambiguity; the degree coincides with the number of the table. In the case of the third class—the collateral—the degree of the civilians is equal to the sum of the sub-class and number, while that of the canonists is the greater of the two. In the case of the fourth class, there is room for still greater ambiguity, owing to the difficulty of reckoning the degree of *c p*, that is, of consort. The only unambiguous and perfectly general method, is first to specify the class, then the sub-class, and then the number.

In the eighth column I have entered the *Index* of the Relationship. It is obtained from the notation in the second column by counting the number of *c's* or the number of *p's* following one another, and writing the sum of the *c's* with a + sign before it, and the sum of the *p's* with a − sign before it. When the relationship is given to be irreducible, the

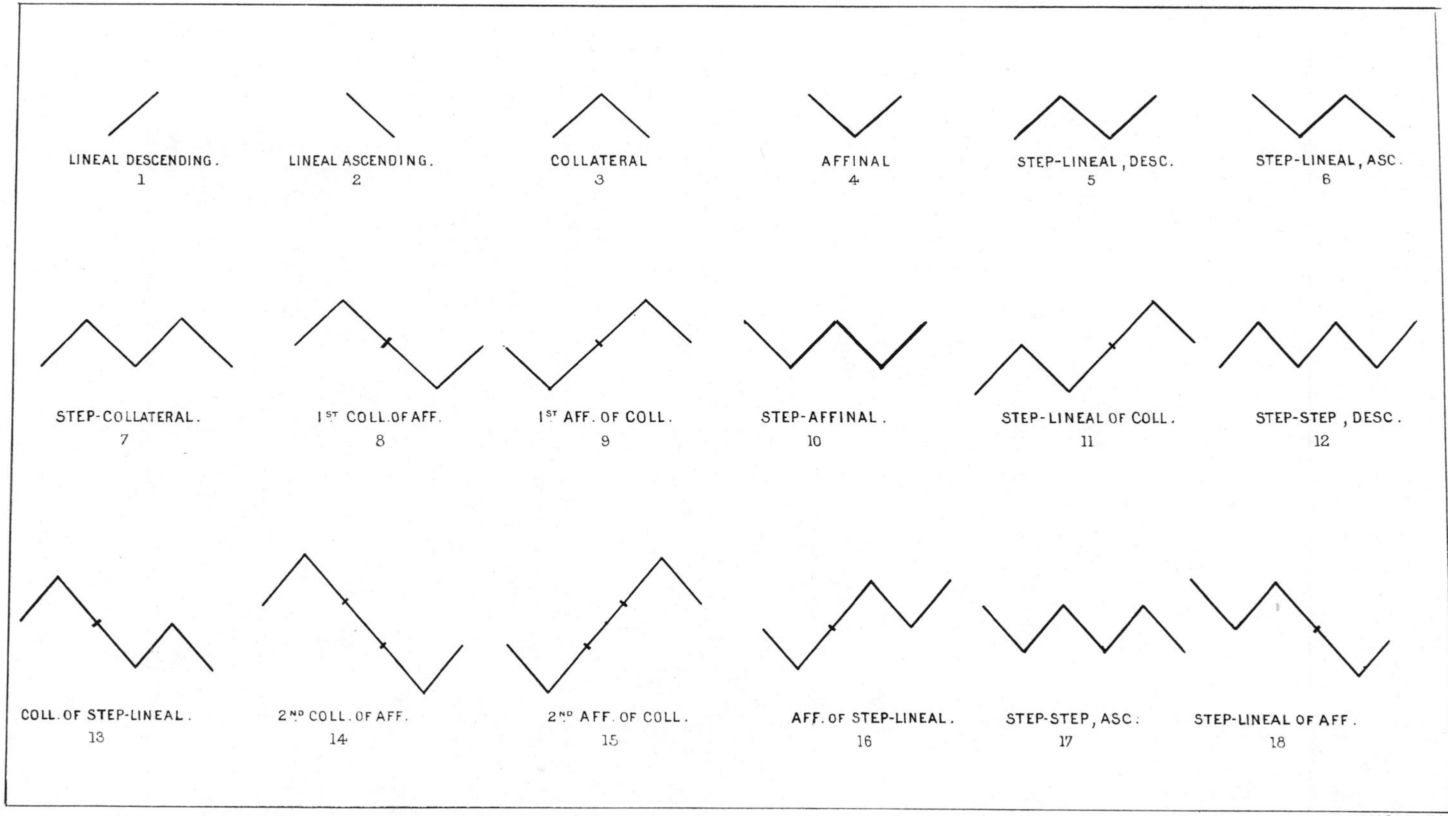


J. & W. Emslie, lith.

PL. III. CHARACTERISTICS OF THE CLASSES OCCURRING IN THE FIRST FIVE ORDERS.

numbers in the index cannot in any case destroy one another. But, if the relationship is reducible, the reduction may take place in any way by which a positive 1 can destroy a neighbouring negative 1. A relationship, and the forms to which it can reduce, compose a class of relationships which naturally group together. It is evident that the class, sub-class, and number, may be read off from the index; and, were it not for the distinction of sex, which still requires to be symbolised, the index might be a sufficient notation.

The ninth column contains another classification—proceeding upon what may be called the *sign* of the relationships. The sign is determined by the first direction of the line, and by the subsequent number of changes; hence, it may be deduced from the index, by neglecting the numbers, and retaining only the several signs. The common property denoted by the sign + is descendant; by − ancestor; by + − descendant of ancestor; by − + ancestor of descendant. The irreducible meanings of the two latter are *collateral* and *affinal* respectively. So far this classification agrees with that in column fifth, but when we proceed to the next class + − + that is, descendant of ancestor of descendant, we find that it embraces several of the Classes, namely, step-lineal descending, 1st collateral of affinal, and 2nd collateral of affinal. Its irreducible meaning is any descendant of an affinal, not being, as such, a descendant of self. Similarly, − + − means ancestor of descendant of ancestor, its irreducible meaning being any ancestor of a collateral, not being, as such, an ancestor of self. The other signs may be read off in a similar manner.

In the last column there is entered the *interval*, by which is meant the number of generations separating the two extremes of the relationship. A cipher indicates that they are of the same generation; a number without a sign that the relation is younger than the origin by the given number of generations; and a number with the − sign, that the relation is older than the origin by the given number of generations. The value of the interval is deduced from the notation by summing up all the *c's*, and all the *p's*, and subtracting the latter sum from the former.

A very natural classification of the general relationships is formed by grouping together those having the same interval. The result is the systematic development of the idea involved in the Chinese grades. "All men who are born into the world," says a Chinese author, "have nine ranks of relations. My own generation is one grade, my father's is one, my grandfather's is one, that of my grandfather's father is one, and that of my grandfather's grandfather is one; thus above me are four grades: my

E 2

son's generation is one grade, my grandson's is one, that of my grandson's son is one, and that of my grandson's grandson is one; thus below me are four grades of relations: including myself in the estimate, there are in all nine grades. These are brethren, and though each grade belongs to a different house or family, yet they are all my relations, and these are called the nine grades of relations."[1] The relationships of the first five orders fall within the first eleven grades.

The classification by grade is capable of serving as a basis for a nomenclature. A common term is required to denote any general relationship falling into a given grade, and qualifying words or phrases to denote the several ways in which the relationship may pass from *self* or the grade 0, to the given grades. The nature of the connecting line corresponds to the "different house or family" mentioned above. For example, the first four general relationships ending in the grade 1 are child proper, nephew or niece, step-child, child-in-law. Here the idea of the grade, namely, child, enters into three of these English terms, and the genera are separated by adding on qualifying phrases. What more reasonable to expect, than that the second *genus* should also in some languages be named on the same principle? If we examine the terms for the relationships ending in grade 0, we shall find that they exhibit a similar tendency to group under a generalised idea of *brother* or *sister*, the principal exception being *consort*. In gesture language, however, consort is represented by the same sign as brother or sister, namely, by the two forefingers placed close to one another.[2] Any nomenclature built upon this basis is called by Morgan *classificatory*; but the distinction is very rough, for there is more or less of this kind of classification in every nomenclature. It is so natural that I had drawn it out before hearing of Morgan's classificatory systems.

Having classified the general relationships in various ways, I now proceed to divide them into species by the introduction of a notation for sex. Let $m$ be used to denote *male*, and $f$ to denote *female*; then as the adjective male or female may apply to each of the nouns child or parent, we may attach an $m$ or an $f$ to any letter in a general relationship. It is convenient to place the symbol of the adjective before the symbol of the noun to which it refers: thus $m\,c$ denotes son, $m\,c\,m\,c$ son of son, $m\,p$ father, and so on. Also as the origin of a relationship may be either man or woman, we may have an $m$ or an $f$ after the last $c$ or $p$ of the relationship; for example $m\,c\,m$ denotes son of a man, and $m\,c\,f$ son of a woman. A relationship which has

[1] Morgan's "Systems of Consanguinity and Affinity," p. 415.
[2] Tylor's "Early History of Mankind," p. 37.

neither *m* nor *f* at its end is applicable to any person independently of sex.

The symbols *m* and *f* are conveniently represented on a diagram by the marks × and o respectively. I find these marks so used in genealogical tables by Mr. Galton.[1] I used to employ a short transverse stroke, instead of the cross, but it is better to reserve the stroke for indicating the position of an intermediate person of either, or of indeterminate sex, in cases where it is not necessarily indicated by a corner (p. 48). This notation is exemplified in figs 19–26, Plate II, where we have the different species of brother or sister relationships indicated.

A general relationship is specialised as much as is possible with respect to sex, when it has a sex-symbol for either extreme, and for each of the intermediates. In Table II the general relationships of the first two orders are broken up into species of the kind referred to. The permutations of the sex-symbols *m* and *f* are formed in the same manner as those of the descent symbols *c* and *p* (p. 48), that is, by first taking *m* and *f*, then prefixing *m* before each of these, and also prefixing *f*, then by prefixing *m* and *f* severally before each of these four results, and so on. The manner in which the sex-symbols follow one another gives us the idea of *line*. To find the species into which the general relationships of a given order break up, all that we have to do is to write, as in Table II, the permutations of *c* and *p*, in a vertical column, and those of *m* and *f* in a horizontal row; then the result to be entered in a given place is determined by the row and the column which intersect in that place. The species in the second row of the second order are those represented graphically in figs. 19–26, Plate I. I use the term *brother german*, to denote brother on the father's side, following McLennan[2]; Sir H. Maine[3] uses the longer term *brother consanguineous*. In the case of the third genus of the same order, we have several remarkable species. The sixth and the eighth species necessarily reduce to simple forms—a mother of the son of a woman is necessarily the woman referred to, and a mother of a daughter of a woman is necessarily the woman referred to. The two corresponding male species,—the first and the third—are not so necessarily reducible; they are so only in countries where monandry is established. Hence the rule is, that *f p c f* always reduces to *f*; and *m p c m* to *m* where monandry is established. On the other hand *m p c f* and *f p c m* are necessarily irreducible, owing to the fact that sex in mankind is diœcious. Hence of the

[1] Galton's "Hereditary Genius," p. 93.
[2] McLennan's "Studies in Ancient History," p. 176.
[3] Sir H. Maine's "Ancient Law," p. 152.

four lines (figs. 27–30, Plate II) the first necessarily collapses where monandry is established, the second and third cannot collapse, and the fourth necessarily collapses.

I may observe here that the expression in words entered below a notation is always intended to be the exact equivalent of the notation, so far as existing English words can convey the meaning. An entry of this sort, of course, differs from one which means that the relationship denoted belongs to the class described; for then it is the sum of all the relationships, which are said to belong to the given class, that is the equivalent of the class. For instance, brother or sister-in-law is not the equivalent of *cppc*, but of *cppc* and *pccp* taken together. The relationship to which a given relationship reduces may not be the exact equivalent of the relationship; it is one which necessarily follows from the given relationship, as such.

Another important system of relationships (Table III) is obtained by supposing the sex of the extremes to be given; that is, by specifying *m* or *f* at the front and at the end. When the relationships are considered to be irreducible, the specification of the sex of the relation may determine the sex of some of the intermediates, or of the origin. This depends on the Laws of Reduction stated on page 53. The rule for putting in the consequent specifications of sex is as follows:—When a relationship begins with *p c*, the sex-symbol after the *p c* is the opposite of that in front, and should this *p c* be followed by another, the sex-symbol following the latter will be the same as that in front. In the same way the sex-symbol at the end, when immediately preceded by *p c*, requires the sex-symbol before the *p c* to be its opposite, and so on.

In the table referred to, I have developed the general relationship *first* for the relation being male, and the origin female; and *secondly*, for the relation being female, and the origin male. The first series fully developed gives all the possible relationships of a man to a woman, the second series all the possible relationships of a woman to a man. Corresponding to any relationship in the one series, there is a relationship in the other series which is its reciprocal. Two relationships may be said to be reciprocal to one another, if when one denotes the relationship of *R* to *A*, the other denotes the consequent relationship of *A* to *R*. Hence the rule for deducing the reciprocal of a relationship is—Write the given relationship backwards, at the same time changing each *c* into *p*, and each *p* into *c*. For example, the reciprocal of *m c c p f* is *f c p p m*; if *R* is the nephew of the woman *A*, then *A* is the aunt of the man *R*.

The deducing of the reciprocal relationship is a special case of the problem—Given a proposition stating a relationship between two persons, into how many equivalent forms can the statement be put? The solution will be best explained by means of an example. Suppose the given statement to be that represented by fig. 31, Plate II, namely, $R$ is a son of a sister of the father of the woman $A$. This is expressed in the analytical notation by—

$$R = m\,c\,f\,c\,p\,m\,p\,f\,A, \qquad (1)$$

It follows that

$$p\,m\,R = f\,c\,p\,m\,p\,f\,A \qquad (2)$$

A parent of the man $R$ is a sister of the father of the woman $A$.

$$p\,f\,p\,m\,R = p\,m\,p\,f\,A, \qquad (3)$$

A parent of the mother of the man $R$ is a parent of the father of the woman $A$;

$$c\,p\,f\,p\,m\,R = m\,p\,f\,A, \qquad (4)$$

A brother of the mother of the man $R$ is the father of the woman $A$.

$$\text{and } f\,c\,m\,c\,p\,f\,p\,m\,R = A, \qquad (5)$$

A daughter of a brother of the mother of the man $R$ is $A$.

Thus the statement can be thrown into as many forms as there are persons involved in the relationship, each successive form being derived by taking away a $c$ or a $p$, from the front of the right hand side, and putting a $p$ or a $c$ at the front of the left hand side. The final form is the reciprocal of the original form.

A statement of the laws of marriage of a country is obtained by marking those relationships of the first series, which are inconsistent with the relationship of husband, or those of the second series, which are inconsistent with the relationship of wife. I have marked with an asterisk the relationships explicitly excluded by the English Table of Degrees. Theoretically, no doubt, all the relationships of the lineal classes are excluded, those only being stated which are not rendered impossible by difference of grade. By the law of the Greek Church, all the relationships of this table, with the necessary exception of wife, and the impossible exception of wife of husband, are excluded. Not only so—to form a table exhibiting all the excluded relationships would require one embracing the first nine orders.

Table IV exhibits an important mode of developing the relationships of consanguinity. These embrace the general relationships of the lineal and collateral classes only; and they coincide with the *cognates* of the Romans, provided we generalise the meaning of $c$ so as to denote not only actual child, but also child by adoption. The principle by which the division into species is effected, is by writing $m$ or $f$ after each $c$ and before each $p$ of the general relationship. By grouping together the relationships at the beginning of the several rows, that is, all those traced exclusively through males, we obtain the agnatic system of the Roman law[1]; and by grouping together the relationships at the end of the several rows we obtain the uterine system, that is, the system resulting from tracing kinship through females only[2]. We can also obtain by separating out from this table, the system resulting from any other law of tracing kinships, as, for example, by tracing alternately through a male and a female.

It will be observed that to express fully the different specific relationships we require four and only four irreducible terms, namely, brother-german, brother-uterine, sister-german, sister-uterine, the reason being that the only change of letter that we can have is that from $c$ to $p$. This is what Morgan calls a purely descriptive system. But other irreducible terms, though not required, might be introduced, and their introduction would not make the system less descriptive. On the other hand, if a language does not provide simple terms for the four collateral relationships mentioned, it is needless to expect that it will provide simple terms for the more complex collateral relationships.

It is now necessary to consider the proper mode of denoting compound relationships. An elementary relationship is one which denotes a single line of connection between the extremes; a compound relationship is one which denotes the simultaneous existence of several such lines. The simplest example is in the case of full brother or full sister. To denote that $R$ is the full brother of $A$, we may write

$$R = m \begin{Bmatrix} c\,m\,p \\ c\,f\,p \end{Bmatrix} A$$

using a bracket to embrace the two members of the bifurcation. When the bifurcation does not commence with the relation or terminate in the origin, the common part may be written outside the bracket. For example, the statement that $R$ is a child of a full brother of a grandparent of $A$ may be written

$$R = c\,m \begin{Bmatrix} c\,m\,p \\ c\,f\,p \end{Bmatrix} p\,p\,A.$$

Figs. 32, 33, Plate II, show how the above statements are expressed by the graphic notation; and other examples are to

[1] Sir H. Maine's "Ancient Law," p. 146.
[2] McLennan's "Studies in Ancient History," p. 124.

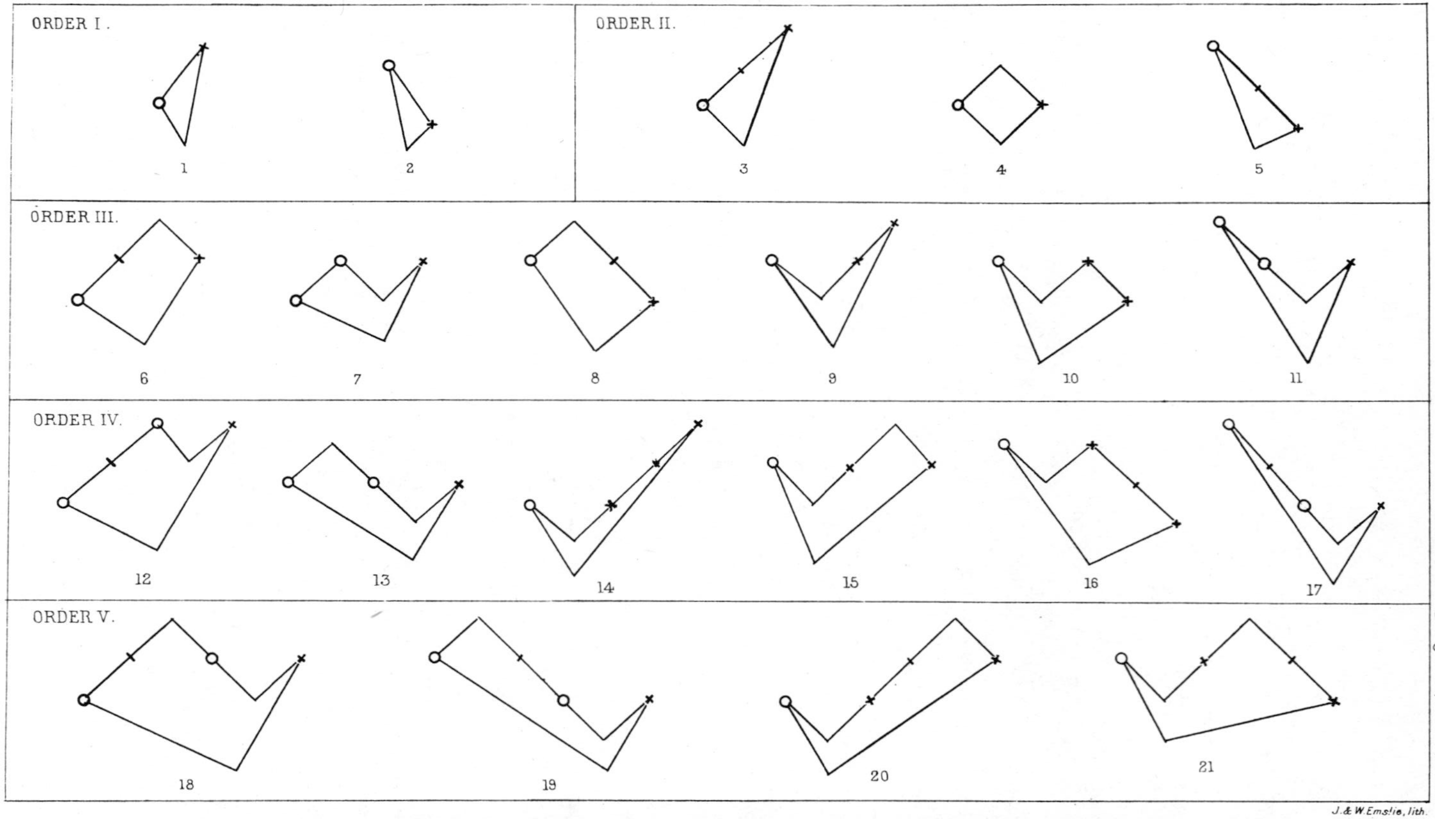


PL. IV. GRAPHICAL STATEMENT OF THE ENGLISH LAWS OF MARRIAGE AND THEIR CONSEQUENCES.

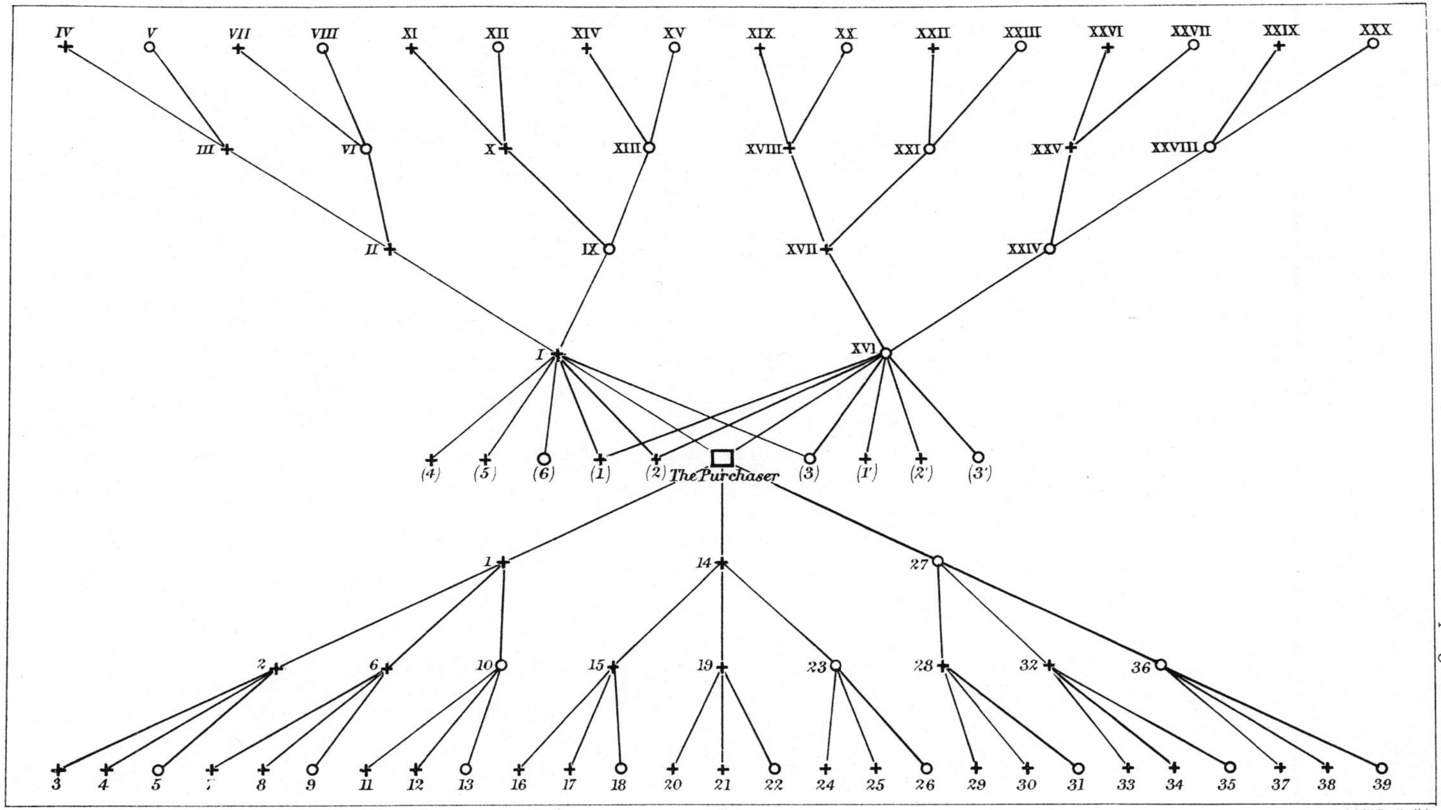


J. & W. Emslie, lith.

PL. V. DESCENT OF PROPERTY ACCORDING TO THE ENGLISH LAW.

be found on Plates IV and V. The two branches of a bifurcation may or may not be of the same genus relationship.

It is best to consider compound relationships as embracing any combination of elementary relationships, and then to classify them into the possible and the impossible. The latter may be further classified according to the law or laws which render them impossible. Several of these laws have already been referred to, namely—(1) the diœcious nature of sex; (2) the definiteness of mother; (3) the definiteness of father where monandry prevails. Other laws are, (4) the continuity of a person's life, which prevents any ancestor of a person from also being a descendant of that person; (5) the maximum length of human life compared with the minimum length of a generation, which renders impossible the marriage of parties separated by a certain number of generations; (6) the marriage laws of a country preventing marriage between parties already nearly related.

On Plate IV I have exhibited the combinations rendered impossible by the English Laws of Marriage (following the Table of Degrees, p. 55). In each case we have a cyclic relationship, and the impossibility of the existence of this cycle may be expressed in various ways. We can take each person in turn as being both relation and origin of the relationship, and then transform each of these statements in accordance with the rule (p. 55). For example, take the fourth impossible cycle, the primary meaning of which is that a man cannot be the husband of a sister of himself. This is the reading obtained by taking No. 1 (*see* fig. 34, Plate II) as both relation and origin of the supposed relationship. By taking No. 2 we obtain—A person cannot be the child of a sister of the father of him or herself. By taking No. 3 we obtain—A woman cannot be the sister of the husband of herself. Finally by taking No. 4—A person cannot be the parent of the husband of the daughter of him or herself. To show how any one of these statements may be further transformed, in accordance with the rule on p. 55, take the first—

| | | | | |
|---|---|---|---|---|
| | $m\,A$ | cannot be | $m\,p\,c\,f\,c\,p\,m\,A.$ | (1) |
| Then | $c\,m\,A$ | cannot be | $c\,f\,c\,p\,m\,A;$ | (2) |
| and | $p\,c\,m\,A$ | cannot be | $f\,c\,p\,m\,A;$ | (3) |
| and | $p\,f\,p\,c\,m\,A$ | cannot be | $p\,m\,A;$ | (4) |
| and | $c\,p\,f\,p\,c\,m\,A$ | cannot be | $m\,A.$ | (5) |

The meanings of these several transformations are:

A child of the man $A$ cannot be a son of a sister of $A$; (2)
A wife of $A$ cannot be a sister of $A$; (3)

A parent-in-law of the man $A$ cannot be a parent of $A$; (4)

A brother of the wife of $A$ cannot be $A$ himself. (5)

The last form is the reciprocal of the first; it is obtained by going in the opposite direction round the cycle. It may be shown in a similar manner that each of the other three principal statements has five forms. Let the number of persons involved in such a cycle be $n$; then the number of principal statements is $n$, and the number of forms for each of these is $n + 1$; hence the total number of forms is $n (n + 1)$.

The above is the only impossible cycle which occurs in the combination of two cousin relationships.

By means of this notation we can easily calculate the amount of consanguinity existing between two persons connected by a given relationship, provided we can first settle two principles, namely, the relative parts to be ascribed to father and mother, and secondly, how far the consanguinity derived by one child from a parent is equivalent to the consanguinity derived by another child from the same parent. Suppose that the answers to these questions respectively are that the parts are equal, and that the consanguinities are wholly equivalent; then in the case of any lineal relationship the consanguinity will be measured by a product of as many halves as there are letters in the relationship; and in the case of any collateral relationship the number of times half is repeated in the product will be less than the number of letters in the relationship by one. In the case of a compound relationship the total value of the consanguinity is the sum of the consanguinities of the elements. The value for single first cousin is one-eighth; hence for two-fold first cousin it must be a quarter; for three-fold first cousin three-eighths; and for four-fold first cousin one-half.

I propose to accept the late Dr. Morgan's invitation[1] to criticise the data furnished in his tables. He took for the basis of his schedule of questions the Roman method of denoting relationships. That method was no doubt sufficient for the purpose for which it was intended; but for the purpose of a scientific inquiry, which to be useful must involve the discrimination of very nice differences, we require a more exact analysis. In Morgan's Tables we nowhere find the distinction between an elementary and a compound relationship: thus, for example, *brother* may mean brother with respect to father, or brother with respect to mother, or brother with respect to both father and mother. The questions of the schedule are not test questions, but aim at being exhaustive. They amount to 268—a number

[1] "Systems of Consanguinity and Affinity," p. 9.

sufficiently great to make it difficult to keep an American Indian to the task of answering, and to cause the filling up of a schedule by another person to occupy two or three years, but still a number very far from being exhaustive, when we consider that without going beyond the fifth order there are more than 27,000 elementary relationships. These 268 questions are not distributed so well as they might have been, for 228 are devoted to the first two classes—the lineal and the collateral, while only 37 are devoted to the remaining classes. This disproportion becomes all the more striking, when we bear in mind that the principal application to which Morgan attempted to put the data was to determine the forms through which the institution of marriage is supposed to have passed.

When we examine his tables we find that the specification of sex in a relationship is introduced or omitted in a very arbitrary manner. For example, in his first Table, question 4, we have given as the equivalents of mother of great grandfather terms which are really equivalent to (1) grandmother of grandmother, (2) great grandmother of father, (3) grandmother of grandfather, (4) great great grandmother, while we have, in addition, terms which are really equivalent to the heading. Question 13 is "grandson" (common term), and question 14 is "grandson" (descriptive phrase). Under the former heading, besides proper equivalents of grandson, I find some terms which are equivalent to grandchild, others to son of son, and one to son of daughter; and there is no difference in the nature of the entries under the other heading, excepting that son of daughter is more frequently introduced.

The tables have three columns, one of which is devoted to the description of a relationship in English, the second to the corresponding relationship in the foreign language, and the third to a translation of the entry in the second. Now if the second entry is the precise equivalent of the first, then the first is the proper translation of the second, and accordingly we find that the entries in the first and third columns are frequently the same. There is room for a third column, when and only when the question is understood to be what is the idiomatic expression in the foreign language of the given relationship, and what is a literal translation of that expression into English. This is the case with the Chinese method. But in the case of the American Indian methods this cannot be said to be the meaning of the entry of the third column. It is not co-extensive with, but includes the entry of the second column.

The analysis of this paper suggests two methods of dealing with Morgan's data or of recording more exact ones. First, the general relationships of Table I, broken up, if necessary, into species, may be taken as the schedule of questions, and their equivalents in the particular language entered in another column; second, the principal relationship words or phrases in the particular language may be made the argument, and their equivalent in the scientific notation be made the entry. The first method tests whether the data are complete within the orders of relationship considered; and the mere arrangement of the data, when so stated, is sufficient to show the principal characteristics of the particular system examined.

In Table V I have given an example of how the second method may be employed. The relationship words and phrases in the English language are defined not in terms of one another but in terms of an exact scientific notation. Mr. Francis Galton suggested to me that I should show that this could be done. The relationship terms of any language may be exactly defined in this manner.

I have supplied (Plate V) the graphic notation to the problem of giving a complete representation of the descent of property according to the English law. The purchaser is the origin of the scheme; three generations of lineal descendants and four generations of lineal ascendants are taken into account. A family is sufficiently represented by two sons and one daughter, because the elder son succeeds before the younger, and the youngest son before any daughter, but all the daughters together. The order of succession among the lineal descendants is indicated by the numbers. Suppose the issue of the purchaser exhausted, then the inheritance goes back to the lineal ascendants or their issue in the order indicated by the Roman numerals. Each lineal ancestor forms a stock and his family breaks up into sub-stocks, which succeed in the manner indicated by the numbers enclosed within the brackets. The issue of each sub-stock succeeds in the same order as the issue of the purchaser. The sub-stocks 1, 2, 3, 4, 5, 6, succeed after the father, while (1′), (2′), (3′), succeed after the mother. The diagram supposes sub-stocks attached to each pair of stocks, and issue to each of the sub-stocks.

*Appendix.*—After I read the above paper Mr. Francis Galton suggested to me that the notation would be improved were the symbols so taken that the expressions could be spoken. The simplest way of carrying out this idea seems to me to be to use the vowels *a* and *o* instead of the consonants *c* and *p*; to employ *m* and *f* as before to denote male and female, while *mf* may be taken to denote both; and to introduce *y* as a consonant between two vowels not separated by *m*, *f*, or *mf*. On Table V will be found the vocalised equivalents of the ordinary terms of relationship formed in accordance with these principles. After further study of this matter I may be able to make improvements; but the scheme given is so far a construction of a small portion of the scientific language discussed by Professor Max Müller in his "Lectures on the Science of Language."[1]

*Explanation of Plates II to V.*

PLATE II.

Figures illustrative of the text of the paper.

PLATE III.

Diagram showing characteristics of the classes occurring in the first five orders of the author's system.

PLATE IV.

Graphical statement of the English Laws of Marriage and their consequences.

PLATE V.

Diagram showing Descent of Property according to the English Law.

DISCUSSION.

Mr. GALTON said that the attempt to express relationship was essentially a difficult task, not to be got through by any Royal road; it was like attempting to define the position of a large number of draughtsmen on a board, which could not be done without a great deal of detailed description. We were apt to underrate the difficulty of expressing relationship owing to the imperfect nomenclature to which habit had accustomed us, but as soon as we found it necessary to define a relationship accurately, the imperfection of our language and the vagueness of our ordinary conceptions became manifest. There was an especial source of verbal confusion in the way in which the same relationship was sometimes singly and sometimes doubly expressed. We say, for example, on the one hand, that A is father of B, or conversely that B is son of A, and on the other hand that the relationship between A and B is that of father and son. There was an incongruity in using the two phrases as equivalent. "Father and son" in the single sense means the father and the son of a third person, and refers to three generations, viz.: to the father of A, to A. and to A's son, whereas in the double sense it refers to two generations only.

He thought that Dr. Macfarlane had attacked the problem of relationship with thoroughness, ability, and success, and that he had done a very acceptable work for all who concerned themselves with genealogies of the complicated descriptions referred to by

[1] Max Müller's "Lectures on the Science of Language," vol. ii, p. 48.

Dr. Macfarlane. The diagrammatic form seemed to himself the most distinctive and self-explanatory. Some few, however, of the series of letters were perhaps a little too long and cumbrous compared with the simplicity of the relationship they conveyed, as, for example, the formula by which a husband's sister was expressed. He should like to receive an assurance from the author that he was able himself readily to decipher his own formulæ, after he had laid the subject by for a time and had temporarily ceased to be familiar with it.

Mr. PARK HARRISON, the Rev. Professor HARLEY, and the CHAIRMAN also took part in the discussion.

Dr. MACFARLANE, in reply to questions asked, stated that a little practice was sufficient to enable one to use either the analytical or graphical notation, while in reading off the notation to others the difficulty consisted in framing an expression in ordinary words having a meaning exactly equivalent to that concisely and precisely expressed by the notation; that the expression of the complex relationships in terms of the fundamental symbols $c$, $p$, $m$, $f$, while a principle of the analysis, did not preclude the introduction of single letters to denote the more frequently occurring complex ideas, just as the chemist, while expressing the composition of every substance in terms of the elementary substances, introduced special symbols to denote frequently occurring combinations; and that he wrote $m$ and $f$ not as suffixes but in the same letter as $c$ and $p$, though they were symbols of a different kind, because the expressions were then more easily written and printed, and besides, for some applications numerical suffixes had to be introduced to distinguish the different children, or the different sons, or the different daughters.

TABLE I.—GENERAL RELATIONSHIPS OF THE FIRST

| Order. | Genus. | General Meaning. | Irreducible Meaning. |
|---|---|---|---|
| I. .... | *c* .. | child .. .. .. .. .. .. | child .. .. .. .. .. |
| | *p* .. | parent .. .. .. .. .. .. | parent .. .. .. .. .. |
| II. .. | *cc* .. | grandchild .. .. .. .. .. | grandchild.. .. .. .. .. |
| | *cp* .. | child of parent.. .. .. .. .. | brother or sister .. .. .. .. |
| | *pc* .. | parent of child.. .. .. .. .. | consort .. .. .. .. .. |
| | *pp* .. | grandparent .. .. .. .. .. | grandparent .. .. .. .. |
| III. .. | *ccc* .. | great grandchild .. .. .. .. | great grandchild .. .. .. .. |
| | *ccp* .. | grandchild of parent .. .. .. .. | nephew or niece .. .. .. .. |
| | *cpc* .. | child of parent of child .. .. .. | step-child .. .. .. .. .. |
| | *cpp* .. | child of grandparent .. .. .. .. | uncle or aunt .. .. .. .. |
| | *pcc* .. | parent of grandchild .. .. .. .. | child-in-law .. .. .. .. |
| | *pcp* .. | parent of child of parent .. .. .. | step-parent .. .. .. .. |
| | *ppc* .. | grandparent of child .. .. .. .. | parent-in-law .. .. .. .. |
| | *ppp* .. | great grandparent .. .. .. .. | great grandparent .. .. .. |
| IV. .. | *cccc* .. | great great grandchild .. .. .. | great great grandchild .. .. .. |
| | *cccp* .. | great grandchild of parent .. .. .. | grandnephew or grandniece .. .. |
| | *ccpc* .. | grandchild of parent of child .. .. | child of step-child .. .. .. |
| | *ccpp* .. | grandchild of grandparent .. .. .. | first cousin.. .. .. .. .. |
| | *cpcc* .. | child of parent of grandchild .. .. | step-child of child .. .. .. |
| | *cpcp* .. | child of parent of child of parent .. .. | step-brother or step-sister .. .. |
| | *cppc* .. | child of grandparent of child.. .. .. | brother or sister of consort .. .. |
| | *cppp* .. | child of great grandparent .. .. .. | granduncle or grandaunt.. .. .. |
| | *pccc* .. | parent of great grandchild .. .. .. | consort of grandchild .. .. .. |
| | *pccp* .. | parent of grandchild of parent .. .. | consort of brother or sister .. .. |
| | *pcpc* .. | parent of child of parent of child .. .. | step-consort .. .. .. .. |
| | *pcpp* .. | parent of child of grandparent .. .. | step-parent of parent .. .. .. |
| | *ppcc* .. | grandparent of grandchild .. .. .. | parent-in-law of child .. .. .. |
| | *ppcp* .. | grandparent of child of parent .. .. | parent of step-parent .. .. .. |
| | *pppc* .. | great grandparent of child .. .. .. | grandparent of consort .. .. .. |
| | *pppp* . | great great grandparent .. .. .. | great great grandparent .. .. .. |
| V. .. | *ccccc* . | great great great grandchild .. .. .. | great great great grandchild .. .. |
| | *ccccp* . | great great grandchild of parent .. .. | great grandnephew or niece .. .. |
| | *cccpc* . | great grandchild of parent of child .. .. | grandchild of step-child .. .. .. |
| | *cccpp* | great grandchild of grandparent .. .. | child of first cousin .. .. .. |
| | *ccpcc* . | grandchild of parent of grandchild .. | child of step-child of child .. .. |
| | *ccpcp* . | grandchild of parent of child of parent .. | child of step-brother or sister .. .. |
| | *ccppc* | grandchild of grandparent of child .. .. | nephew or niece of consort .. .. |
| | *ccppp* | grandchild of great grandparent .. .. | child of granduncle or grandaunt .. |
| | *cpccc*. | child of parent of great grandchild .. .. | step-child of grandchild .. .. .. |
| | *cpccp*. | child of parent of grandchild of parent .. | step-child of brother or sister .. .. |
| | *cpcpc*. | child of parent of child of parent of child .. | step-step-child .. .. .. .. |
| | *cpcpp* | child of parent of child of grandparent .. | step-brother or step-sister of parent .. |
| | *cppcc*. | child of grandparent of grandchild .. .. | brother or sister of children-in-law .. |
| | *cppcp* | child of grandparent of child of parent .. | brother of sister of step-parent .. .. |
| | *cpppc* | child of great grandparent of child .. .. | uncle or aunt of consort .. .. .. |
| | *cpppp* | child of great great grandparent .. .. | great granduncle or aunt .. .. |
| | *pcccc*. | parent of great great grandchild .. .. | consort of great grandchild .. .. |
| | *pcccp*. | parent of great grandchild of parent.. .. | consort of nephew or niece .. .. |
| | *pccpc* | parent of grandchild of parent of child .. | consort of step-child .. .. .. |
| | *pccpp* | parent of grandchild of grandparent.. .. | consort of uncle or aunt .. .. .. |
| | *pcpcc* | parent of child of parent of grandchild .. | step-consort of child .. .. .. |
| | *pcpcp* | parent of child of parent of child of parent.. | step-step parent .. .. .. .. |
| | *pcppc* | parent of child of grandparent of child .. | step-parent of consort .. .. .. |
| | *pcppp* | parent of child of great grandparent .. | step-parent of grandparent .. .. |
| | *ppccc* | grandparent of great grandchild .. .. | parent-in-law of grandchild .. .. |
| | *ppccp* | grandparent of grandchild of parent .. | parent-in-law of brother or sister .. |
| | *ppcpc* | grandparent of child of parent of child .. | parent of step-consort .. .. .. |
| | *ppcpp* | grandparent of child of grandparent .. | parent of step-parent of parent.. .. |
| | *pppcc* | great grandparent of grandchild .. .. | parent of parent-in-law of child .. |
| | *pppcp* | great grandparent of child of parent .. | grandparent of step-parent .. .. |
| | *ppppc* | great great grandparent of child .. .. | grandparent of parent-in-law .. .. |
| | *ppppp* | great great great grandparent .. .. | great great great grandparent .. .. |

| Class. | Sub-class. | Number. | Index. | Sign. | Grade. |
|---|---|---|---|---|---|
| lineal, descending | .. | first | 1 | + | 1 |
| lineal, ascending | .. | first | −1 | − | −1 |
| lineal, descending | .. | second | 2 | + | 2 |
| collateral | first | first | 1−1 | +− | 0 |
| affinal | first | first | −1+1 | −+ | 0 |
| lineal, ascending | .. | second | −2 | − | −2 |
| lineal, ascending | .. | third | 3 | + | 3 |
| collateral, | first | second | 2−1 | +− | 1 |
| step-lineal, descending | first | first | 1−1+1 | +−+ | 1 |
| collateral | second | first | 1−2 | +− | −1 |
| affinal | second | first | −1+2 | −+ | 1 |
| step-lineal, ascending | first | first | −1+1−1 | −+− | −1 |
| affinal | first | second | −2+1 | −+ | −1 |
| lineal, ascending | .. | third | −3 | − | −3 |
| lineal, descending | .. | fourth | 4 | + | 4 |
| collateral | first | third | 3−1 | +− | 2 |
| step-lineal, descending | first | second | 2−1+1 | +−+ | 2 |
| collateral | second | second | 2−2 | +− | 0 |
| step-lineal, descending | second | first | 1−1+2 | +−+ | 2 |
| step-collateral | first | first | 1−1+1−1 | +−+− | 0 |
| first collateral of affinal | first | first | 1−2+1 | +−+ | 0 |
| collateral | third | first | 1−3 | +− | −2 |
| affinal | third | first | −1+3 | −+ | 2 |
| first affinal of collateral | first | first | −1+2−1 | −+− | 0 |
| step-affinal | first | first | −1+1−1+1 | −+−+ | 0 |
| step-lineal, ascending | second | first | −1+1−2 | −+− | −2 |
| affinal | second | second | −2+2 | −+ | 0 |
| step-lineal, ascending | first | second | −2+1−1 | −+− | −2 |
| affinal | first | third | −3+1 | −+ | −2 |
| lineal, ascending | .. | fourth | −4 | − | −4 |
| lineal, descending | .. | fifth | 5 | + | 5 |
| collateral | first | fourth | 4−1 | +− | 3 |
| step-lineal, descending | first | third | 3−1+1 | +−+ | 3 |
| collateral | second | third | 3−2 | +− | 1 |
| step-lineal, descending | second | second | 2−1+2 | +−+ | 3 |
| step-collateral | first | second | 2−1+1−1 | +−+− | 1 |
| first collateral of affinal | first | second | 2−2+1 | +−+ | 1 |
| collateral | third | second | 2−3 | +− | −1 |
| step-lineal, descending | third | first | 1−1+3 | +−+ | 3 |
| first step-lineal of collateral | first | first | 1−1+2−1 | +−+− | 1 |
| step-step-lineal, descending | first | first | 1−1+1−1+1 | +−+−+ | 1 |
| step-collateral | second | first | 1−1+1−2 | +−+− | −1 |
| first collateral of affinal | second | first | 1−2+2 | +−+ | 1 |
| first collateral of step-lineal | first | first | 1−2+1−1 | +−+− | −1 |
| second collateral of affinal | first | first | 1−3+1 | +−+ | −1 |
| collateral | fourth | first | 1−4 | +− | −3 |
| affinal | fourth | first | −1+4 | −+ | 3 |
| second affinal of collateral | first | first | −1+3−1 | −+− | 1 |
| first affinal of step lineal | first | first | −1+2−1+1 | −+−+ | 1 |
| first affinal of collateral | second | first | −1+2−2 | −+− | −1 |
| step-affinal | second | first | −1+1−1+2 | −+−+ | 1 |
| step-step-lineal, ascending | first | first | −1+1−1+1−1 | −+−+− | −1 |
| first step-lineal of affinal | first | first | −1+1−2+1 | −+−+ | −1 |
| step-lineal, ascending | third | first | −1+1−3 | −+− | −3 |
| affinal | third | second | −2+3 | −+ | 1 |
| first affinal of collateral | first | second | −2+2−1 | −+− | −1 |
| step-affinal | first | second | −2+1−1+1 | −+−+ | −1 |
| step-lineal, ascending | second | second | −2+1−2 | −+− | −3 |
| affinal | second | third | −3+2 | −+ | −1 |
| step-lineal, ascending | first | third | −3+1−1 | −+− | −3 |
| affinal | first | fourth | −4+1 | −+ | −3 |
| lineal, ascending | .. | fifth | −5 | | −5 |

## TABLE II.—GENERAL RELATIONSHIPS OF THE FIRST

### ORDER I.

Genus.

| Line. | *m m* | *m f* | *f m* | *f f* |
|---|---|---|---|---|
| | *m c m*<br>son of man | *m c f*<br>son of woman | *f c m*<br>daughter of man | *f c f*<br>daughter of woman |
| *p* | *m p m*<br>father of man | *m p f*<br>father of woman | *f p m*<br>daughter of man | *f p f*<br>daughter of woman |

Genus.

| Line. | *m m m* | *m m f* | *m f m* | *m f f* |
|---|---|---|---|---|
| *c c* | *m c m c m*<br>son of son of man | *m c m c f*<br>son of son of woman | *m c f c m*<br>son of daughter of man | *m c f c f*<br>son of daughter of woman |
| *c p* | *m c m p m*<br>brother-german of man | *m c m p f*<br>brother-german of woman | *m c f p m*<br>brother-uterine of man | *m c f p f*<br>brother-uterine of woman |
| *p c* | *m p m c m*<br>father of son of man (man) | *m p m c f*<br>father of son of woman | *m p f c m*<br>father of daughter of man (man) | *m p f c f*<br>father of daughter of woman |
| *p p* | *m p m p m*<br>father of father of man | *m p m p f*<br>father of father of woman | *m p f p m*<br>father of mother of man | *m p f p f*<br>father of mother of woman |

TWO ORDERS DIVIDED INTO THEIR ULTIMATE SPECIES.

ORDER II.

| *fmm* | *fmf* | *ffm* | *fff* |
|---|---|---|---|
| *fcmcm*<br>daughter of son of man | *fcmcf*<br>daughter of son of woman | *fcfcm*<br>daughter of daughter of man | *fcfcf*<br>daughter of daughter of woman |
| *fcmpm*<br>sister-german of man | *fcmpf*<br>sister-german of man | *fcfpm*<br>sister-uterine of man | *fcfpf*<br>sister-uterine of woman |
| *fpmcm*<br>mother of son of man | *fpmcf*<br>mother of son of woman (woman) | *fpfcm*<br>mother of daughter of man | *fpfcf*<br>mother of daughter of woman (woman) |
| *fpmpm*<br>mother of father of man | *fpmpf*<br>mother of father of woman. | *fpfpm*<br>mother of mother of man | *fpfpf*<br>mother of mother of woman |

TABLE III.—POSSIBLE RELATIONSHIPS OF A MAN TO A WOMAN,

| | | Man to Woman. | |
|---|---|---|---|
| Order. | Genus. | Notation. | Meaning. |
| I. .. | *c* .. .. | *m c f* .. .. | *son .. .. .. .. .. |
| | *p* .. .. | *m p f* .. .. | *father .. .. .. .. |
| II. .. | *c c* .. .. | *m c c f*.. .. | *grandson .. .. .. .. |
| | *c p* .. .. | *m c p f* .. | *brother .. .. .. .. |
| | *p c* .. .. | *m p c f* .. | husband .. .. .. .. |
| | *p p* .. .. | *m p p f* .. | *grandfather .. .. .. |
| III. .. | *c c c*.. .. | *m c c c f* .. | great grandson .. .. . |
| | *c c p*.. .. | *m c c p f* .. | *nephew .. .. .. .. |
| | *c p c* .. | *m c m p c f* .. | *son of husband .. .. .. |
| | *c p p* .. | *m c p p f* .. | *uncle .. .. .. .. |
| | *p c c*.. .. | *m p c f c f* .. | *husband of daughter .. .. |
| | *p c p* .. | *m p c f p f* .. | *stepfather .. .. .. .. |
| | *p p c* .. | *m p m p c f* .. | *father of husband .. .. |
| | *p p p* .. | *m p p p f* .. | great grandfather.. .. .. |
| IV. .. | *c c c c* .. | *m c c c c f* .. | great grandson .. .. .. |
| | *c c c p* .. | *m c c c p f* .. | grandnephew .. .. .. |
| | *c c p c* .. | *m c c m p c f* .. | *son of step-child .. .. .. |
| | *c c p p* .. | *m c c p p f* .. | first cousin (male) .. .. |
| | *c p c c* .. | *m c p c c f* .. | stepson of child .. .. .. |
| | *c p c p* .. | *m c p c p f* .. | step-brother .. .. .. |
| | *c p p c* .. | *m c p m p c f* .. | *brother of husband .. .. |
| | *c p p p* .. | *m c p p p f* .. | granduncle .. .. .. |
| | *p c c c* .. | *m p c f c c f* .. | *husband of granddaughter .. |
| | *p c c p* .. | *m p c f c p f* .. | *husband of sister.. .. .. |
| | *p c p c* .. | [*m p c f p c m*]. | [husband of wife] .. .. |
| | *p c p p* .. | *m p c f p p f* .. | *stepfather of parent .. .. |
| | *p p c c* .. | *m p p c c f* .. | father-in-law of child .. .. |
| | *p p c p* .. | *m p p c p f* .. | father of step-parent .. .. |
| | *p p p c* .. | *m p p m p c f* .. | *grandfather of husband .. .. |
| | *p p p p* .. | *m p p p p f* .. | great great grandfather .. .. |
| V. .. | *c c c c c* .. | *m c c c c c f* .. | great great great grandson .. |
| | *c c c c p* .. | *m c c c c p f* .. | great grandnephew .. .. |
| | *c c c p c* .. | *m c c c m p c f*.. | grandson of step-child .. .. |
| | *c c c p p* .. | *m c c c p p f* .. | son of first cousin .. .. |
| | *c c p c c* .. | *m c c p c c f* .. | son of step-child of child .. |
| | *c c p c p* .. | *m c c p c p f* .. | son of step-brother or step-sister |
| | *c c p p c* .. | *m c c p m p c f* . | *nephew of husband .. .. |
| | *c c p p p* .. | *m c c p p p f* .. | son of granduncle or grandaunt |
| | *c p c c c* .. | *m c p c c c f* .. | step-son of grandchild .. .. |
| | *c p c c p* .. | *m c p c c p f* .. | step-son of brother or sister .. |
| | *c p c p c* .. | *m c f p c m p c f* | step-step-son .. .. .. |
| | *c p c p p* .. | *m c p c p p f* .. | step-brother of parent .. .. |
| | *c p p c c* .. | *m c p p c c f* .. | brother of child-in-law .. .. |
| | *c p p c p* .. | *m c p p c p f* .. | brother of step-parent .. .. |
| | *c p p p c* .. | *m c p p m p c f*. | *uncle of husband.. .. .. |
| | *c p p p p* .. | *m c p p p p f*.. | great granduncle .. .. .. |
| | *p c c c c* .. | *m p c f c c c f*.. | husband of great granddaughter |
| | *p c c c p* .. | *m p c f c c p f*.. | *husband of niece .. .. .. |
| | *p c c p c* .. | *m p c f c m p c f* | husband of step-daughter .. |
| | *p c c p p* .. | *m p c f c p p f*.. | *husband of aunt .. .. .. |
| | *p c p c c* . | *m p c f p c m c f* | other husband of daughter-in-law |
| | *p c p c p* .. | *m p c f p c m p f* | step-step-father .. .. .. |
| | *p c p p c* .. | *m p c f p m p c f* | step-father of husband .. .. |
| | *p c p p p* .. | *m p c f p p p f*. | step-father of grandparent .. |
| | *p p c c c* .. | *m p p c c c f* .. | father-in-law of grandchild .. |
| | *p p c c p* .. | *m p p c c p f* .. | father-in-law of brother or sister |
| | *p p c p c* .. | *m p f p c m p c f* | father of another wife of husband |
| | *p p c p p* .. | *m p p c p p f*.. | father of step-parent of parent .. |
| | *p p p c c* .. | *m p p p c c f* .. | father of parent-in-law of child.. |
| | *p p p c p* .. | *m p p p c p f*.. | grandfather of step-parent .. |
| | *p p p p c* .. | *m p p p m p c f*. | great grandfather-in-law.. .. |
| | *p p p p p* .. | *m p p p p p f*.. | great great great grandfather .. |

| | Woman to Man. | |
|---|---|---|
| | Notation. | Meaning. |
| .. | *fcm* | *daughter. |
| .. | *fpm* | *mother. |
| .. | *fccm* | *granddaughter. |
| .. | *fcpm* | *sister. |
| .. | *fpcm* | wife. |
| .. | *fppm* | *grandmother. |
| .. | *fcccm* | great granddaughter. |
| .. | *fccpm* | *niece. |
| .. | *fcfpcm* | *daughter of wife. |
| .. | *fcppm* | *aunt. |
| .. | *fpcmcm* | *wife of son. |
| .. | *fpcmpm* | *step-mother. |
| .. | *fpfpcm* | *mother. |
| .. | *fpppm* | great grandmother. |
| .. | *fccccm* | great great granddaughter |
| .. | *fcccpm* | grandniece. |
| .. | *fccfpcm* | *daughter of step-child. |
| .. | *fccppm* | first cousin (female). |
| .. | *fcpccm* | step-daughter of child. |
| .. | *fcpcpm* | step-sister. |
| .. | *fcpfpcm* | *sister of wife. |
| .. | *fcpppm* | grandaunt. |
| .. | *fpcmccm* | *wife of grandson. |
| .. | *fpcmcpm* | *wife of brother. |
| .. | [*fpcmpcf*] | [wife of husband]. |
| .. | *fpcmppm* | *step-mother of parent. |
| .. | *fppccm* | mother-in-law of child. |
| .. | *fppcpm* | mother of step-parent. |
| .. | *fppfpcm* | *grandmother of wife. |
| .. | *fppppm* | great great grandmother. |
| .. | *fcccccm* | great great great granddaughter. |
| .. | *fccccpm* | great grandniece. |
| .. | *fcccfpcm* | granddaughter of step-child. |
| .. | *fcccppm* | daughter of first cousin. |
| .. | *fccpccm* | daughter of step-child of child. |
| .. | *fccpcpm* | daughter of step-brother or step-sister. |
| .. | *fccpfpcm* | *niece of wife. |
| .. | *fccpppm* | daughter of granduncle or grandaunt. |
| .. | *fcpcccm* | step-daughter of grandchild. |
| .. | *fcpccpm* | step-daughter of brother or sister. |
| .. | *fcmpcfpcm* | step-step-daughter. |
| .. | *fcpcppm* | step-sister of parent. |
| .. | *fcppccm* | sister of child in-law. |
| .. | *fcppcpm* | sister of step-parent. |
| .. | *fcppfpcm* | *aunt of wife. |
| .. | *fcppppm* | great grandaunt. |
| .. | *fpcmcccm* | wife of great grandson. |
| .. | *fpcmccpm* | *wife of nephew. |
| .. | *fpcmcfpcm* | wife of step-son. |
| .. | *fpcmcppm* | *wife of uncle. |
| .. | *fpcmpcfcm* | other wife of son-in-law. |
| .. | *fpcmpcfpm* | step-step-mother. |
| .. | *fpcmpfpcm* | step-mother of wife. |
| .. | *fpcmpppm* | step-mother of grandparent. |
| .. | *fppcccm* | mother-in-law of grandchild. |
| .. | *fppccpm* | mother-in-law of brother or sister. |
| .. | *fpmpcfpcm* | mother of ano her husband of wife |
| .. | *fppcppm* | mother of step-parent of parent. |
| .. | *fpppccm* | grandmother-in-law of child. |
| .. | *fpppcpm* | grandmother of step-parent. |
| .. | *fpppfpcm* | great grandmother-in-law. |
| .. | *fpppppm* | great great great grandmother. |

Table IV.—CONSANGUINEOUS RELATIONSHIPS OF THE

AGNATIC SYSTEM FORMED BY THE EXTREME TERMS ON THE LEFT;

| Order. | Genus. | | | | | | | | |
|---|---|---|---|---|---|---|---|---|---|
| I. | *c* | *c m* child of man. | | | | | | | |
| | *p* | *m p* father. | | | | | | | |
| II. | *cc* | *c m c m* child of son of man. | *c m c f* child of son of woman. | | | | | | |
| | *c p* | *c m p* brother or sister-german. | | | | | | | |
| | *p p* | *m p m p* father of father. | *m p f p* father of mother. | | | | | | |
| III. | *c c c* | *c m c m c m* child of son of son of man. | *c m c m c f* child of son of son of woman. | *c m c f c m* child of son of daughter of man. | *c m c f c f* child of son of daughter of woman. | | | | |
| | *c c p* | *c m c m p* child of brother-german. | *c m c f p* child of brother uterine. | | | | | | |
| | *c p p* | *c m p m p* brother or sister-german of father. | *c m p f p* brother or sister-german of mother. | | | | | | |
| | *p p p* | *m p m p m p* father of father of father. | *m p m p f p* father of father of mother. | *m p f p m p* father of mother of father. | *m p f p f p* father of mother of mother. | | | | |
| IV. | *c c c c* | *c m c m c m c m* child of son of son of son of man. | *c m c m c m c f* child of son of son of woman. | *c m c m c f c m* child of son of son of daughter of man. | *c m c m c f c f* child of son of son of daughter of woman. | *c m c f c m c m* child of son of daughter of son of man. | *c m c f c m c f* child of son of daughter of son of man. | *c m c f c f c m* child of son of daughter of daughter of man. | *c m c f c f c f* child of son of daughter of daughter of woman. |
| | *c c c p* | *c m c m c m p* child of son of brother-german. | *c m c m c f p* child of son of brother-uterine. | *c m c f c m p* child of son of sister-german. | *c m c f c f p* child of son of sister-uterine. | | | | |
| | *c c p p* | *c m c m p m p* child of brother-german of father. | *c m c m p f p* child of brother-german of mother. | *c m c f p m p* child of brother-uterine of father. | *c m c f p f p* child of brother-uterine of mother. | | | | |
| | *c p p p* | *c m p m p m p* brother or sister-german of father of father. | *c m p m p f p* brother or sister-german of father of mother. | *c m p f m p* brother or sister-german of mother of father. | *c m p f p f p* brother or sister-german of mother of mother. | | | | |
| | *p p p p* | *m p m p m p m p* father of father of father of father. | *m p m p m p f p* father of father of father of mother. | *m p m p f p m p* father of father of mother of father. | *m p m p f p f p* father of father of mother of mother. | *m p f p m p m p* father of mother of father of father. | *m p f p m p f p* father of mother of father of mother. | *m p f p f p m p* father of mother of mother of father. | *m p f p f p f p* father of mother of mother of mother. |

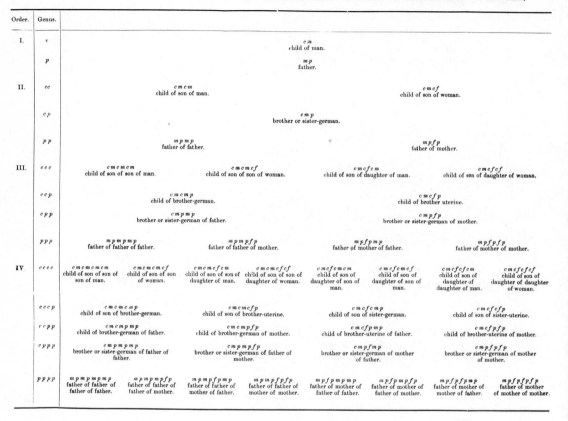

FIRST FOUR ORDERS GROUPED IN LINES AND SPECIES.

UTERINE SYSTEM FORMED BY THE EXTREME TERMS ON THE RIGHT.

*c f*
child of woman.

*f p*
mother.

*c f c m*
child of daughter of man.

*c f c f*
child of daughter of woman.

*c f p*
brother or sister-uterine.

*f p m p*
mother of father.

*f p f p*
mother of mother.

*c f c m c m*
child of daughter of son of man.

*c f c m c f*
child of daughter of son of woman.

*c f c f c m*
child of daughter of daughter of man.

*c f c f c f*
child of daughter of daughter of woman.

*c f c m p*
child of sister-german.

*c f c f p*
child of sister-uterine.

*c f p m p*
brother or sister-uterine of father.

*c f p f p*
brother or sister-uterine of mother.

*f p m p m p*
mother of father of father.

*f p m p f p*
mother of father of mother.

*f p f p m p*
mother of mother of father.

*f p f p f p*
mother of mother of mother.

*c f c m c m c m*
child of daughter of son of son of man.

*c f c m c m c f*
child of daughter of son of son of woman.

*c f c m c f c m f*
child of daughter of son of daughter of man.

*c f c m c f c f*
child of daughter of son of daughter of woman.

*c f c f c m c m*
child of daughter of daughter of son of man.

*c f c f c m c f*
child of daughter of daughter of son of woman.

*c f c f c f c m*
child of daughter of daughter of daughter of man.

*c f c f c f c f*
child of daughter of daughter of daughter of woman.

*c f c m c m p*
child of daughter of brother-german.

*c f c m c f p*
child of daughter of brother-uterine.

*c f c f c m p*
child of daughter of sister-german.

*c f c f c f p*
child of daughter of sister-uterine.

*c f c m p m p*
child of sister-german of father.

*c f c m p f p*
child of sister-german of mother.

*c f c f p m p*
child of sister-uterine of father.

*c f c f p f p*
child of sister-uterine of mother.

*c f p m p m p*
brother or sister-uterine of father of father.

*c f p m p f p*
brother or sister-uterine of father of mother.

*c f p f p m p*
brother or sister-uterine of mother of father.

*c f p f p f p*
brother or sister-uterine of mother of mother.

*f p m p m p m p*
mother of father of father of father.

*f p m p m p f p*
mother of father of father of mother.

*f p m p p m p*
mother of father of mother of father.

*f p m p f p f p*
mother of father of mother of mother.

*f p f p m p m p*
mother of mother of father of father.

*f p f p m p f p*
mother of mother of father of mother.

*f p f p f p m p*
mother of mother of mother of father.

*f p f p f p f p*
mother of mother of mother of mother.

## Table V.—DEFINITION OF THE ENGLISH TERMS OF RELATIONSHIP.

| Term or Phrase. | Equivalent. | Vocalised Equivalent. |
|---|---|---|
| Aunt, half blood | $fcpp$ | fayoyo. |
| ,, full blood | $fc^{m}_{f}pp$ | famfoyo. |
| ,, half blood, paternal | $fcpmp$ | fayomo. |
| ,, full blood, ,, | $fc^{m}_{f}pmp$ | famfofo. |
| ,, half blood, maternal | $fcpfp$ | fayofo. |
| ,, full blood, ,, | $fc^{m}_{f}pfp$ | famfofo. |
| Brother, half blood | $mcp$ | mayo. |
| ,, full blood | $mc^{m}_{f}p$ | mamfo. |
| ,, german | $mcmp$ | mamo. |
| Brother uterine | $mcfp$ | mafo. |
| Child | $c$ | ya. |
| Consort | $pc$ | yoya. |
| Cousin, first | $ccpp$ | yayayoyo. |
| ,, second | $cccppp$ | yayayayoyoyo. |
| Daughter | $fc$ | fa. |
| Daughter-in-law | $fpcmc$ | foyama. |
| Father | $mp$ | mo. |
| Father-in-law | $mppc$ | moyoya. |
| Grandchild | $cc$ | yaya. |
| Granddaughter | $fcc$ | faya. |
| Grandson | $mcc$ | maya. |
| Husband | $mpcf$ | moyaf. |
| Mother | $fp$ | fo. |
| Mother-in-law | $fppc$ | foyoya. |
| Nephew, half blood | $mccp$ | mayayo. |
| ,, full blood | $mcc^{m}_{f}p$ | mayamfo. |
| Niece, half blood | $fccp$ | fayayo. |
| ,, full blood | $fcc^{m}_{f}p$ | fayamfo. |
| Parent | $p$ | yo. |
| Sister, half blood | $fcp$ | fayo. |
| ,, full blood | $fc^{m}_{f}p$ | famfo. |
| ,, german | $fcmp$ | famo. |
| ,, uterine | $fcfp$ | fafo. |
| Son | $mc$ | ma. |
| Son-in-law | $mpcfc$ | moyafa. |
| Step-brother | $mcpcp$ | mayoyayo. |
| Step-child | $cpc$ | yayoya. |
| Step-daughter | $fcpc$ | fayoya. |
| Step-father | $mpcfp$ | moyafo. |
| Step-mother | $fpcmp$ | foyamo. |
| Step-parent | $pcp$ | yoyayo. |
| Step-sister | $fcpcp$ | fayoyayo. |
| Step-son | $mcpc$ | mayoya. |
| Uncle, half blood | $mcpp$ | mayoyo. |
| ,, full blood | $mc^{m}_{f}pp$ | mamfoyo. |
| ,, half blood, paternal | $mcpmp$ | mayomo. |
| ,, full blood, ,, | $mc^{m}_{f}pmp$ | mamfomo. |
| ,, half blood, maternal | $mcpfp$ | mayofo. |
| ,, full blood ,, | $mc^{m}_{f}pfp$ | mamfofo. |
| Wife | $fpcm$ | foyam. |

# 2



Reprinted from *Man* (Journal of the Royal Anthropological Society), **39**, 77–84 (1909)

## CLASSIFICATORY SYSTEMS OF RELATIONSHIP.

BY A. L. KROEBER.

THE distinction between classificatory and descriptive systems of relationship has been widely accepted, and has found its way into handbooks and general literature. According to the prevalent belief the systems of certain nations or languages group together distinct relationships and call them by one name, and are therefore classifying. Other systems of consanguinity are said to indicate secondary differences of relationship by descriptive epithets added to their primary terms, and to be therefore descriptive.

Nothing can be more fallacious than this common view. A moment's reflection is sufficient to show that every language groups together under single designations many distinct degrees and kinds of relationship. Our word brother includes both the older and the younger brother and the brother of a man and of a woman. It therefore embraces or classifies four relationships. The English word cousin denotes both men and women cousins; cousins on the father's or on the mother's side; cousins descended from the parent's brother or the parent's sister; cousins respectively older or younger than one's self, or whose parents are respectively older or younger than the speaker's parents; and cousins of men or women. Thirty-two different relationships are therefore denoted by this one English word. If the term is not strictly limited to the significance of first cousin, the number of distinct ideas that it is capable of expressing is many times thirty-two. Since then it is not only primitive people that classify or fail to distinguish relationships, the suspicion is justified that the current distinction between the two classes or systems of indicating relationship is subjective, and has its origin in the point of view of investigators, who, on approaching foreign languages, have been impressed with their failure to discriminate certain relationships between which the languages of civilized Europe distinguish, and who, in the enthusiasm of formulating general theories from such facts, have forgotten that their own languages are filled with entirely analogous groupings or classifications which custom has made so familiar and natural that they are not felt as such.

The total number of different relationships which can be distinguished is very large, and reaches at least many hundred. No language possesses different terms for all of these or even for any considerable proportion of them. In one sense it is obvious that a language must be more classificatory as the number of its terms of relationship is smaller. The number of theoretically possible relationships remaining constant, there must be more ideas grouped under one term in proportion as the number of terms is less. Following the accepted understanding of what constitutes classificatory consanguinity, English, with its twenty terms of relationship, must be not less but more classificatory than the languages of all primitive people who happen to possess twenty-five, thirty, or more terms.

It is clear that if the phrase classificatory consanguinity is to have any meaning it must be sought in some more discriminating way. The single fact that another people group together various relationships which our language distinguishes does not make their system classificatory. If there is a general and fundamental difference between the systems of relationship of civilized and uncivilized people, its basis must be looked for in something more exact than the rough and ready expressions of subjective point of view that have been customary.

It is apparent that what we should try to deal with is not the hundreds or thousands of slightly varying relationships that are expressed or can be expressed by the various languages of man, but the principles or categories of relationship which underlie these. Eight such categories are discernible.

1. *The difference between persons of the same and of separate generations.*—The distinctions between father and grandfather, between uncle and cousin, and between a person and his father, involve the recognition of this category.

2. *The difference between lineal and collateral relationship.*—When the father and the father's brother are distinguished, this category is operative. When only one term is employed for brother and cousin, it is inoperative.

3. *Difference of age within one generation.*—The frequent distinction between the older and the younger brother is an instance. In English this category is not operative.

4. *The sex of the relative.*—This distinction is carried out so consistently by English, the one exception being the foreign word cousin, that the discrimination is likely to appear self-evident. By many people, however, many relationships are not distinguished for sex. Grandfather and grandmother, brother-in-law and sister-in-law, father-in-law and mother-in-law, and even such close relationships as son and daughter, are expressed respectively by single words.

5. *The sex of the speaker.*—Unrepresented in English and most European languages, this category is well known to be of importance in many other languages. The father, mother, brother, sister, and more distant relatives may receive one designation from a man and another from his sister.

6. *The sex of the person through whom relationship exists.*—English does not express this category. In consequence we frequently find it necessary to explain whether an uncle is a father's or a mother's brother, and whether a grandmother is paternal or maternal.

7. *The distinction of blood relatives from connections by marriage.*—While this distinction is commonly expressed by most languages, there are occasional lapses; just as in familiar English speech the father-in-law is often spoken of as father. Not strictly within the domain of relationship, but analogous to the occasional

failure to express this category, is the frequent ignoring on the part of primitive people of the difference between actual relatives and fictitious clan or tribal relatives.

8. *The condition of life of the person through whom relationship exists.*—The relationship may be either of blood or by marriage; the person serving as the bond of relationship may be alive or dead, married or no longer married. Many North American Indians refrain from using such terms as father-in-law and mother-in-law after the wife's death or separation. Some go so far as to possess terms restricted to such severed relationship. It is natural that the uncle's relation to his orphaned nephew should tend to be somewhat different from his relation to the same boy while his natural protector, his father, was living. Distinct terms are therefore sometimes found for relatives of the uncle and aunt group after the death of a parent.

The subjoined table indicates the representation of the eight categories, and the degree to which they find expression, respectively in English and in several of the Indian languages of North America.

| | | N.A. Indian. | | | | | California Indian. | | | | | | |
|---|---|---|---|---|---|---|---|---|---|---|---|---|---|
| | English. | Arapaho. | Dakota. | Pawnee. | Skokomish. | Chinook. | Yuki. | Pomo. | Washo. | Miwok. | Yokuts. | Luiseño. | Mohave. |
| No. of terms ... ... ... | 21[1] | 20 | 31 | 19 | 18 | 28 | 24 | 27 | 28 | 24 | 28 | 34 | 35 |
| Generation ... ... ... | 21 | 20 | 31 | 11 | 13 | 23 | 24 | 21 | 27 | 24 | 22 | 30 | 26 |
| Blood or marriage ... ... | 21 | 19 | 31 | 17 | 18 | 26 | 24 | 27 | 28 | 24 | 28 | 32 | 34 |
| Lineal or collateral ... ... | 21 | 10 | 20 | 5 | 11 | 25 | 24 | 21 | 28 | 18 | 26 | 34 | 28 |
| Sex of relative ... ... ... | 20 | 18 | 29 | 17 | 2 | 12 | 16 | 21 | 20 | 20 | 17 | 18 | 22 |
| Sex of connecting relative ... | 0 | 6 | 6 | 2 | 0 | 20 | 13 | 13 | 14 | 10 | 14 | 19 | 21 |
| Sex of speaker ... ... ... | 0 | 3 | 18 | 4 | 0 | 15 | 3 | 3 | 10 | 2 | 12 | 10 | 14 |
| Age in generation ... ... | 0 | 3 | 7 | 2 | 2 | 2 | 3 | 4 | 4 | 4 | 4 | 12 | 8 |
| Condition of connecting relative | 0 | 0 | 0 | 0 | 8 | 1 | 0 | 0 | 0 | 0 | [2] | 0 | 1 |

It appears that English gives expression to only four categories. With the exception, however, of the one and foreign word cousin, every term in English involves the recognition of each of these four categories. All the Indian languages express from six to eight categories. Almost all of them recognize

[1] All terms are omitted, such as great grandfather, great-uncle, and second-cousin, which are not generally used in ordinary speech and exist principally as a reserve available for specific discrimination on occasion.

[2] Terms denoting relatives by marriage undergo a vocalic change to indicate the death of the connecting relative.

seven. But in all the Indian languages the majority of the categories occurring are expressed in only part of the terms of relationship found in the language. There are even Indian languages, such as Pawnee and Mohave, in which not a single one of the seven or eight categories finds expression in every term. While in English the degree of recognition which is accorded the represented categories is indicable by a percentage of 100 in all cases but one, when it is 95, in Pawnee corresponding percentages range variously from about 10 to 90, and in Mohave from 5 to 95. All the other Indian languages, as compared with English, closely approach the condition of Pawnee and Mohave.

It is clear that this difference is real and fundamental. English is simple, consistent, and, so far as it goes, complete. The Indian systems of relationship all start from a more elaborate basis, but carry out their scheme less completely. This is inevitable from the fact that the total number of terms of relationship employed by them is approximately the same as in English. The addition of only one category to those found in English normally doubles the number of terms required to give full expression to the system; and the presence of three additional categories multiplies the possible total by about eight. As the number of terms occurring in any of the Indian languages under consideration is not much more than half greater than in English, and sometimes is not greater at all, it is clear that at least some of their categories must find only very partial expression.

In short, as far as the expression of possible categories is concerned, English is less complete than any of the Indian languages; but as regards the giving of expression to the categories which it recognizes, English is more complete. In potentiality, the English scheme is poorer and simpler; but from its own point of view it is both more complete and more consistent. As English may evidently be taken as representative of European languages, it is in this point that the real difference is to be found between the systems that have been called classificatory and those that have been called descriptive.

The so-called descriptive systems express a small number of categories of relationship completely; the wrongly-named classificatory systems express a larger number of categories with less regularity. Judged from its own point of view, English is the less classificatory; looked at from the Indian point of view it is the more classificatory, inasmuch as in every one of its terms it fails to recognize certain distinctions often made in other languages; regarded from a general and comparative point of view, neither system is more or less classificatory.

In short, the prevalent idea of the classificatory system breaks down entirely under analysis. And in so far as there is a fundamental difference between the languages of European and of less civilized peoples in the method of denoting relationship, the difference can be determined only on the basis of the categories described and can be best expressed in terms of the categories.[1]

[1] A tendency toward reciprocal expression is sometimes of importance and may influence the degree to which categories are given expression. Reciprocal terms are such that all the persons included in the relationship expressed by one term call by one name all the persons

The categories serve also to indicate the leading characteristics of systems of the same general order. It is obvious, for instance, that the most important difference between Dakota and Arapaho is the strong tendency of the former to recognize the sex of the speaker. Chinook is notable for laying more stress on the sex of the speaker and of the connecting relation than on the sex of the relative.[1] General differences such as naturally occur between the languages of one region and of another can also be expressed in terms of the categories. All the California systems, for instance, lay much more stress upon the sex of the connecting relative than do any of the Plains languages examined. The Plains systems are conspicuous for their weak development of the distinction between lineal and collateral relationship, this finding expression in two-thirds of all cases in Dakota, half in Arapaho, one-fourth in Pawnee. In seven California languages the corresponding values lie between three-fourths and complete expression. The method can be applied successfully even in the case of smaller and contiguous geographical areas. Of the seven California languages Luiseño and Mohave are spoken in southern California. Their systems show a unity as compared with the systems of the five languages from northern and central California. Both the southern California languages have a greater number of terms; both are stronger in the expression of the categories of the sex of the connecting relative and of age within the same generation; and both are weaker in the category of sex of the relative, than the others. Again, Chinook and Skokomish, both of the North Pacific Coast, are alike in indicating the condition of the connecting relative and in failing, on account of the possession of grammatical sex gender, to distinguish the sex of relatives themselves in many terms of relationship. There is a very deep-going difference between them, however, in the fact that Skokomish

who apply this term to them. In the most extreme form of reciprocity the two groups of relatives use the same term. The paternal grandparents call their sons' children, whether boys or girls, by the same term which these children, both boys and girls, apply to their fathers' parents. Nevertheless, the reciprocal relation is just as clear, though less strikingly expressed, when each of the groups uses a different term for the other. Our English words father and child, or brother and sister, are not reciprocal, for the term child is employed also by the mother, and brother is used by the brother as well as by the sister. In fact the only reciprocal term in English is cousin. The tendency toward reciprocal expression is developed in many Indian languages. It is particularly strong in California. In some languages this tendency has brought it about that different categories are involved in the terms applied to a pair of mutual relationships. The term father's sister indicates the sex of the relative but not of the speaker. The exact reciprocal of father's sister is woman's brother's child. This term, however, does not recognize the sex of the relative indicated, but does imply the sex of the speaker. The two reciprocal terms therefore each involve a category which the other does not express. If the same categories were represented in the two terms, brother's daughter would correspond to father's sister and exact reciprocity would be impossible. When, therefore, the terms found are father's sister and woman's brother's child, it is clear that the tendency toward the establishment of exactly reciprocal terms has been stronger than the feeling favoring the consistent use or neglect of certain categories; in other words, the extent to which certain categories are expressed has been determined by the vigor of the reciprocal tendency.

[1] No doubt, as has been pointed out, owing to the fact that the sex of the relative is indicable by purely grammatical means in this and certain other languages.

is as free as English from recognizing the sex of the speaker and of connecting relatives, while Chinook generally expresses both categories. In short, the categories present a means of comparing systems of terms of relationship along the basic lines of their structure and of expressing their similarities and differences without reference to individual terms or details.

The reason why the vague and unsatisfactory idea of a classificatory system of consanguinity has found such wide acceptance is not to be sought in any primary interest in designations of relationship as such, but in the fact that terms of relationship have usually been regarded principally as material from which conclusions as to the organization of society and conditions of marriage could be inferred. If it had been more clearly recognized that terms of relationship are determined primarily by linguistic factors, and are only occasionally, and then indirectly, affected by social circumstances, it would probably long ago have been generally realized that the difference between descriptive and classificatory systems is subjective and superficial. Nothing is more precarious than the common method of deducing the recent existence of social or marital institutions from a designation of relationship. Even when the social condition agrees perfectly with expressions of relationship, it is unsafe to conclude without corroborative evidence that these expressions are a direct reflection or result of the condition.

In the Dakota language, according to Riggs, there is only one word for grandfather and father-in-law. Following the mode of reasoning sometimes employed, it might be deduced from this that these two relationships were once identical. Worked out to its implications, the absurd conclusion would be that marriage with the mother was once customary among the Sioux.

In the same language the words for woman's male cousin and for woman's brother-in-law have the same radical, differing only in a suffix. Similar reasoning would induce in this case that marriage of cousins was or had been the rule among the Sioux, a social condition utterly opposed to the basic principles of almost all Indian society.

The use of such identical or similar terms for distinct relationships is due to a considerable similarity between the relationships. A woman's male cousin and her brother-in-law are alike in sex, are both of opposite sex from the speaker, are of the same generation as herself, and are both collateral, so that they are similar under four categories. In view of the comparative paucity of terms as compared with possible relationships, it is entirely natural that the same word, or the same stem, should at times be used to denote two relationships having as much in common as these two.

No one would assume that the colloquial habit in modern English of speaking of the brother-in-law as brother implies anything as to form of marriage, for logically the use of the term could only be an indication of sister marriage. It is easily conceivable that in the future development of English the more cumbersome of these two terms might come into complete disuse in daily life and the shorter take its place, without the least change in social or marital conditions.

The causes which determine the formation, choice, and similarities of terms of relationship are primarily linguistic. Whenever it is desired to regard terms of relationship as due to sociological causes and as indicative of social conditions, the burden of proof must be entirely with the propounder of such views.

Even the circumstances that the father's brother is frequently called father is not necessarily due to or connected with the custom of the Levirate; nor can group marriage be inferred from the circumstance that there is frequently no other term for mother's sister than mother. A woman and her sister are more alike than a woman and her brother, but the difference is conceptual, in other words linguistic, as well as sociological. It is true that a woman's sister can take her place in innumerable functions and relations in which a brother cannot; and yet a woman and her sister, being of the same sex, agree in one more category of relationship than the same woman and her brother, and are therefore more similar in relationship and more naturally denoted by the same term. There are so many cases where the expression of relationship cannot have been determined by sociological factors and must be purely psychological, as in the instances just discussed, that it is fair to require that the preference be given to the psychological cause, or that this be admitted as of at least equal probability, even in cases where either explanation is theoretically possible and supporting evidence is absent.

On the whole it is inherently very unlikely in any particular case that the use of identical terms for similar relationships can ever be connected with such special customs as the Levirate or group marriage. It is a much more conservative view to hold that such forms of linguistic expression and such conditions are both the outcome of the unalterable fact that certain relationships are more similar to one another than others. On the one hand this fact has led to certain sociological institutions; on the other hand, to psychological recognitions and their expression in language. To connect the institutions and the terms causally can rarely be anything but hazardous. It has been an unfortunate characteristic of the anthropology of recent years to seek in a great measure specific causes for specific events, connection between which can be established only through evidence that is subjectively selected. On wider knowledge and freedom from motive it is becoming increasingly apparent that causal explanations of detached anthropological phenomena can be but rarely found in other detached phenomena, and that it is even difficult to specify the most general tendencies that actuate the forms taken by culture, as the immediate causes of particular phenomena.

The following conclusions may be drawn:—

1. The generally accepted distinction between descriptive and classificatory systems of terms of relationship cannot be supported.
2. Systems of terms of relationship can be properly compared through an examination of the categories of relationship which they involve and of the degree to which they give expression to these categories.
3. The fundamental difference between systems of terms of relationship

of Europeans and of American Indians is that the former express a smaller number of categories of relationship than the latter and express them more completely.

4. Terms of relationship reflect psychology, not sociology. They are determined primarily by language and can be utilized for sociological inferences only with extreme caution.

# II

# History of Genetic Theory

# Editor's Comments on Papers 3, 4, and 5

**3 Pearl:** The Measurement of the Intensity of Inbreeding
*Life Sciences and Agricultural Experiment Station, University of Maine, Bulletin 215,* 123–138 (1913)

**4 Bernstein:** Démonstration mathématique de la loi d'hérédité de Mendel
*Compt. Rend. Sci. Math.,* Pt. 1, **177**, 528–531 (Sept. 1923)

**5 Bernstein:** Principe de stationarité et généralisations de la loi de Mendel
*Compt. Rend. Sci. Math.,* Pt. 1, **177**, 581–584 (Oct. 1923)

The selected articles in this section show both the potential and error of early works in population genetics. Pearl, in particular, was quite active in 1910–1920 in writing a series of papers analyzing inbreeding (1913a, 1914a, 1914b, 1914c, 1917a, 1917b; Pearl and Miner, 1913). The earliest of these created an argument, in which Pearl was wrong, on the effect of various systems of inbreeding on the eventual homozygosity of a population. [Jennings (1914) and Fish (1914) are the principal corrections; Wright (1922) compares to path analysis.] However, Pearl was creative in his approach; he dealt with the population as a unit rather than with particular individuals and was concerned with the time path of evolution of a population. After the above mentioned correcting articles were published, Pearl's work improved greatly and he used techniques similar to those elaborated by Karlin (1968) for determination of coefficients in genotypic distributions.

Largely because of recognition of the utility of Fibonacci series to determine values of coefficients of genotypic proportions in regular systems of breeding, the most recognized early author is Jennings (1914, 1916). The 1916 article, in particular, was a rather elaborate treatment of the use of series in finding coefficients in numerous cases. An unfortunately clumsy technique was devised by Robbins (1917, 1918a, 1918b) for determining the general term in a series, but the work of Wright (1921) eventually proved to be the most tractable approach.

Should the reader be curious about what population genetics might be like had Pearl's early paper been correct, see Mycielski and Ulam (1969), which was evidently written without knowledge of the existing population genetic literature or attention to the small bibliography they provide. On the other hand, there is still need for a genetic theory that directly depends on the graph theoretic structure of a population, and were it not for the good papers of Lyubich or Maruyama, this poor reference would be the best available place to start.

3

Reprinted from *Life Sciences and Agricultural Experiment Station, University of Maine, Bulletin* 215, 123–138 (1913)

## BULLETIN No. 215.

# THE MEASUREMENT OF THE INTENSITY OF INBREEDING.[1]

RAYMOND PEARL.

The effect of inbreeding on the progeny is a much discussed problem of theoretical biology and of practical breeding. It has been alternately maintained, on the one hand, that inbreeding is the most pernicious and destructive procedure which could be followed by the breeder, and on the other hand, that without its powerful aid most of what the breeder has accomplished in the past could not have been gained and that it offers the chief hope for further advancement in the future. While there is now, among animal breeders at least, a more widespread tendency than was formerly the case towards the opinion that inbreeding *per se* is not a surely harmful thing, nevertheless this opinion is by no means universally held and in any case does not rest upon a definite and well-organized body of evidence. Aside from a relatively small amount of definite experimental data one's judgment in the matter (so far as it is not wholly speculative) is finally formed on the basis of his *interpretation* of the vast accumulation of material comprised in the recorded experience of the breeders of registered (pedigreed) livestock.

[1]This bulletin is essentially an abstract (with some new material—see p. 134) of a more extended technical discussion of the subject published, with the title "A Contribution towards an Analysis of the Problem of Inbreeding," in the American Naturalist (Vol. XLVII, 1913). The complete paper contains illustrative pedigrees and examples of the calculation of coefficients of inbreeding, together with a more detailed discussion of the theoretical significance of these coefficients. Anyone wishing to make use of these coefficients of inbreeding should consult the original paper, in addition to this abstract. The paper referred to is *not* available for general distribution by the Maine Agricultural Experiment Station. Recourse must be had by those wishing to examine it to the files of the American Naturalist, a journal available in many of the larger public libraries, or by purchase at a nominal cost from the publishers.

2

In order that progress may be made in the analysis of this important problem of inbreeding there is a fundamental need which must first be met. This is the need for an appropriate and valid method of pedigree analysis, which possesses generality, and can on that account be depended on to give comparable results when applied to two (or more) different pedigrees. Specifically, there seems not to have been worked out any *adequate general method of measuring quantitatively the degree of inbreeding which is exhibited in a particular pedigree.* Without such a measure it is clearly impossible to proceed far in the analysis of inbreeding.

It is the purpose of the paper of which this is an abstract to present a method for measuring, and expressing numerically in the form of a coefficient, the degree of inbreeding which exists in any particular case, and to show by illustrations the manner in which these coefficients may be computed. It is shown that the method is (a) *unique,* in the sense that the value obtained in any particular instance can only be affected by the degree or amount of inbreeding which has been practiced in the line of descent under consideration, and (b) *general,* in the sense that it is equally applicable to all pedigrees and to all degrees and types of inbreeding.

## Preliminary Definitions.

In attempting any general analysis of the problem of inbreeding from the theoretical standpoint one is confronted with the necessity for a definition of inbreeding, which shall be at once precise and general, that is, such as to include all of the most diverse ways in which this sort of breeding may be practised.

Leaving aside for the moment all consideration of details as to *how* a particular piece of inbreeding may be brought about is to be found the concept of a *narrowing* of the network of descent as a result of mating together individuals genetically *related* to one another in some degree. Let us take this as our basic concept of inbreeding. It means that the number of potentially *different* germ-to-germ lines, or "blood-lines" concentrated in a given individual animal is *fewer* if the individual is inbred than if it is not. In other words, *the inbred individual possesses fewer different ancestors in some particular generation or generations than the maximum possible number for that generation or generations.* This appears to be the most general form in which the concept of inbreeding may be expressed. In whatever way the mating of relatives is accomplished, or whatever the degree of relationship of the individuals mated together, the case in last analysis comes back to the above statement, namely that there are actually in the pedigree of the inbred individual fewer *different* ancestors in some particular generation or generations than the maximum possible number.

## The Measurement of the Degree of Inbreeding.

This brings us to a consideration of a practical and general measure of the degree of inbreeding exhibited in a particular pedigree. This problem has been attacked by a number of other investigators, but so far as I have been able to learn, all previous measures have been modifications in one form or another of the scheme of Lehndorff. All systems based on the number of "free generations" alone, as in Lehndorff's, do not furnish a precise or reliable measure of the real intensity of inbreeding. The essential reason for this failure, stated baldly, is that they do not take account of the composition of the generation to which the "common ancestor" of an inbred pair belongs.

In developing a general measure of the intensity of inbreeding we may well start from the conception set forth in the preceding section, namely that the inbred individual possesses fewer different ancestors than the maximum possible number. Besides this factor account must be taken of the generation or generations in which the reduced number of different ancestors is found, and the extent to which these generations are removed (in the sense of Lehndorff mentioned above) from the individual or generation under consideration. In other words the two factors which must be included in a general measure of the intensity of inbreeding are (a) the *amount* of ancestral reduction in successively earlier generations, and (b) the *rate* of this reduction ever any specified number of generations.

Both of these demands are met by taking as a measure of the intensity of inbreeding in any generation the proportionate degree to which the actually existent number of different ancestral individuals fails to reach the maximum possible number, and by specifying the location in the series of the generation under discussion.

This statement is amplified and made more precise in the following propositions.

1. The production of the individual must be the point of departure in any analytical consideration of inbreeding, leading towards its measurement. That is, the question to which one wants an answer is: What degree of inbreeding was involved in the production of this particular animal?

2. It is therefore necessary practically to *start* with the individual and work *backwards* into the ancestry in measuring inbreeding, rather than to start back in the ancestry and work down towards the individual.

3. In the genetic passage from the $n+1$'th generation to the $n$'th, or in other words the contribution of the matings of the $n+1$'th generation to the total amount of inbreeding involved in the production of an individual, the degree of inbreeding involved will be measured by the expression

$$Z_n = \frac{100\,(p_{n+1} - q_{n+1})}{p_{n+1}} \qquad \text{(i)}$$

where $p_{n+1}$ denotes the maximum possible number of different individuals involved in the matings of the $n+1$ generation, $q_{n+1}$ the *actual* number of different individuals involved in these matings. $Zn$ may be called a *coefficient of inbreeding.* If the value of $Z$ for successive generations in the ancestral series be plotted to the generation number as a base, the points so obtained will form a curve which may be designated as the *curve of inbreeding.*

It will be noted that the coefficient of inbreeding $Z$ is the percentage of the difference between the maximum possible number of ancestors in a given generation and the actual number realized in the former. The coefficient may have any value between 0 and 100. When there is no breeding of relatives whatever (that is, in the entire absence of inbreeding) its value for each generation is 0. As the intensity of the inbreeding increases the value of the coefficient rises.

4. The above measure of inbreeding has to do solely with the *relationship* aspect of the problem. It has nothing whatever

to do *directly* with the gametic or zygotic constitution of individuals.

5. Since the only possible infallible criterion of relationship between individuals is common ancestry in some *earlier* generation, we are led to the practical rule, in measuring the degree of inbreeding in a pedigree, *to regard all different individuals as entirely unrelated until the contrary is proved by the finding of a common ancestor.* This no doubt appears at this stage of the discussion as an exceedingly obvious truism. The reader is urged to accept it as such, and hold fast to it, because it will help him over some apparent paradoxes later.

The method of calculating coefficients of inbreeding, and their real significance will be made much clearer by the consideration of illustrative examples of their application. To these we may therefore turn.

## The Calculation of Coefficients of Inbreeding.

We may first consider some simple hypothetical pedigrees,

### Pedigree Table I. (Hypothetical).

*To illustrate the Breeding of Brother × Sister, out of Brother × Sister, Continued for a Series of Generations.*

| | | | | |
|---|---|---|---|---|
| *x* | *a* | *c* | *e* | *g* |
| | | | | *h* |
| | | | *f* | *g* |
| | | | | *h* |
| | | *d* | *e* | *g* |
| | | | | *h* |
| | | | *f* | *g* |
| | | | | *h* |
| | *b* | *c* | *e* | *g* |
| | | | | *h* |
| | | | *f* | *g* |
| | | | | *h* |
| | | *d* | *e* | *g* |
| | | | | *h* |
| | | | *f* | *g* |
| | | | | *h* |

before attacking the more complicated ones actually realized in stock-breeding.

ILLUSTRATION I. CONTINUED BROTHER × SISTER BREEDING.

Let us begin with the most extreme type of inbreeding possible, namely the mating of brother with sister for a series of generations. Pedigree Table I gives the pedigree of an individual so bred.

Let us now proceed to calculation of the coefficients of inbreeding $Z_0$, $Z_1$, $Z_2$, and $Z_3$. For $Z_0$ we have

$$p=2,$$
$$q=2,$$

whence $$Z_0=\frac{100\ (0)}{2}=0$$

In the same way $$Z_1=\frac{100\ (4-2)}{2}=50$$

$$Z_2=\frac{100\ (8-2)}{8}=75$$

$$Z_3=\frac{100\ (16-2)}{16}=87.5$$

These results may be expressed verbally in the following way: In the last two ancestral generations $x$ is 50 per cent inbred; in the last three generations it is 75 percent inbred; and in the last four generations it is 87.5 per cent inbred.

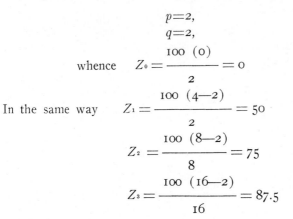

This pedigree table and the constants will repay further consideration, since the case is a limiting one. With the table at hand it is possible to grasp a little more clearly the precise biological meaning of the coefficients of inbreeding. Thus it is seen that what the value of $Z_1=50$ really signifies is that because the individual $a$ and $b$ were brother and sister the number of different ancestors which $x$ can possibly have in any ancestral generation cannot be *more* than 50 percent of the total number theoretically possible for the generation. That is, $x$'s sire and dam having been brother and sister means that $x$ cannot have more than 2049 different great-great-great-great-great-great-great-great-great-great-great-great-great-grandparents, instead of the possible 4096. He may have had fewer than 2049, but $Z_1=50$ tells us that he could not have had more. Similarly $Z_2=75$ indicates that since *c* and *d,* the grand-sire and grand-dam of *x* were brother and sister, *x* cannot have in any ancestral generation more than 25 percent of the theoretically possible number of ancestors for that generation. And so on for the other values of *Z*.

In the limiting case of the closest inbreeding possible the successive Z's will have the values given in the following table.

TABLE I.

*Values of the Successive Coefficients of Inbreeding ($Z_0$ to $Z_{15}$) in the Case of the Most Intense Inbreeding Possible (Brother × Sister Out of Brother × Sister Continued).*

| COEFFICIENT OF INBREEDING. | Ancestral Generations Included. | Numerical Value of Coefficient. |
|---|---|---|
| $Z_0$ | 1 | 0 |
| $Z_1$ | 2 | 50 |
| $Z_2$ | 3 | 75 |
| $Z_3$ | 4 | 87.5 |
| $Z_4$ | 5 | 93.8 |
| $Z_5$ | 6 | 96.9 |
| $Z_6$ | 7 | 98.4 |
| $Z_7$ | 8 | 99.2 |
| $Z_8$ | 9 | 99.6 |
| $Z_9$ | 10 | 99.8 |
| $Z_{10}$ | 11 | 99.9 |
| $Z_{11}$ | 12 | 99.95 |
| $Z_{12}$ | 13 | 99.98 |
| $Z_{13}$ | 14 | 99.99 |
| $Z_{14}$ | 15 | 99.994 |
| $Z_{15}$ | 16 | 99.997 |

From this table it is apparent that while the narrowing or exclusion of the possible different source lines of descent proceeds very rapidly in the first few generations of brother × sister breeding, only relatively little change is made by further generations of this sort of breeding. Thus in seven generations of brother × sister breeding all but about 1 1-2 percent of the potentially different ancestral "blood-lines" will have been eliminated.

The values of the *Z*'s in Table 1 are maxima: No particular coefficient of inbreeding can have a higher value than that given in the table. It is not possible, for example, so to breed any domestic animal that its pedigree on analysis will give $Z_3 >$

87.5. If therefore, the coefficients of Table 1 are plotted the result will be the maximum limiting *curve* of inbreeding.

ILLUSTRATION II. PARENT × OFFSPRING BREEDING.

The next illustration of the application of coefficients of inbreeding will be the general case of back-crossing, that is the mating of parent × offspring.

The values of the successive coefficients for parent × offspring breeding for 16 ancestral generations are given in Table 2.

TABLE 2.

*Values of the Successive Coefficients of Inbreeding in the Case of Continued Parent × Offspring Mating.*

| COEFFICIENT OF INBREEDING. | Ancestral Generations Included. | Numerical Value of Coefficient. |
|---|---|---|
| $Z_0$ | 1 | 0 |
| $Z_1$ | 2 | 25 |
| $Z_2$ | 3 | 50 |
| $Z_3$ | 4 | 68.75 |
| $Z_4$ | 5 | 81.25 |
| $Z_5$ | 6 | 89.06 |
| $Z_6$ | 7 | 93.75 |
| $Z_7$ | 8 | 96.48 |
| $Z_8$ | 9 | 98.05 |
| $Z_9$ | 10 | 98.93 |
| $Z_{10}$ | 11 | 99.41 |
| $Z_{11}$ | 12 | 99.68 |
| $Z_{12}$ | 13 | 99.83 |
| $Z_{13}$ | 14 | 99.91 |
| $Z_{14}$ | 15 | 99.95 |
| $Z_{15}$ | 16 | 99.97 |

By comparison of this table with Table 1 it is evident that while the increase in intensity of inbreeding is not so rapid in the first few ancestral generations by this parent × offspring type of breeding as with the brother × sister type, by the time the 10th ancestral generation is reached the values are, for practical purposes, the same.

ILLUSTRATION III. THE PEDIGREE OF THE JERSEY COW, BESS WEAVER (155121).*

Leaving now the hypothetical cases we may consider some pedigrees of actually existing animals. For a first illustration of this sort the Jersey cow Bess Weaver may be taken. Her pedigree through four ancestral generations is shown in pedigree table II.

PEDIGREE TABLE II.

| | | | | |
|---|---|---|---|---|
| No. 155121 Bess Weaver Sex ♀ | No. 53598 Davy Stoke Pogis ♂ | No. 35913 ♂ Sisera's Stoke Pogis | No. 26271 ♂ Juno's Stoke Pogis | No. 14207 ♂ Alphea's Stoke Pogis |
| | | | | No. 14436 ♀ Carlo's Juno |
| | | | No. 37346 ♀ Sisera | No. 18811 ♂ Duchess Stoke Pogis |
| | | | | No. 6246 ♀ Edith Darby |
| | | No. 79860 ♀ Baltimore | No. 19350 ♂ Patrick Fawkes | No. 10469 ♂ Regal Koffee |
| | | | | No. 21574 ♀ Kermesse |
| | | | No. 17900 ♀ Avoca 2nd | No. 3286 ♂ Champion's Son |
| | | | | No. 17769 ♀ Avoca |
| | No. 126629 Peg Weaver ♀ | No. 35913 ♂ ⊕ Sisera's Stoke Pogis | No. 26271 ♂ × Juno's Stoke Pogis | No. 14207 ♂ × Alphea's Stoke Pogis |
| | | | | No. 14436 ♀ × Carlo's Juno |
| | | | No. 87346 ♀ × Sisera | No. 18811 ♂ × Duchess Stoke Pogis |
| | | | | No. 6246 ♀ × Edith Darby |
| | | No. 126626 ♀ Kate Weaver | No. 36382 ♂ General Kelly | No. 19350 ♂ ⊗ Patrick Fawkes |
| | | | | No. 95606 ♀ ⊕ Balm |
| | | | No. 95606 ♀ Balm | No. 7056 ♂ America's Champion |
| | | | | No. 95605 ♀ Maid of Gilead 2nd |

In the twelfth ancestral generation the theoretically possible number of different ancestors is 4096. In a relatively long pedigree, such as arises in dealing with registered cattle, it would obviously be an extremely tedious business to determine the value of $q$ by direct counting, as has been done in the preceding simpler illustrations. The calculation of the coefficients of inbreeding may be greatly simplified in the case of long pedigrees by a system of counting which makes the *line of*

* The illustrations from actual pedigrees used in this abstract are not the same as those used in the complete paper. The two illustrations here given are chosen because of their probably greater interest to the Station's constituency.

*descent* the unit rather than the individual. This system is used in the above pedigree as an illustration of method, although only 4 ancestral generations are here considered. While each individual animal which is eliminated because of previous appearances in a lower ancestral generation is marked with an *X*, those at the apex of a line of descent are marked with a cross within a circle. These latter are all that need to be counted directly. Their elimination automatically eliminates their own ancestors. Thus the bull Sisera's Stoke Pogis first appears in the second ancestral generation as the sire of Davy Stoke Pogis. He next appears (here marked with a cross within a circle) in the same generation as the sire of Peg Weaver. He will, by the general rule for coefficients of inbreeding, not be counted as a "different" ancestor the second time in this generation. But this automatically eliminates his two parents in the third ancestral generation, his four grandparents in the fourth generation, and so on until in the twelfth generation 1024 ancestors of Sisera's Stoke Pogis will be so eliminated. The same consideration applies in every other like case.

Practically then the method of dealing with a pedigree of this sort is first to go through and indicate in a distinctive way every *primary** reappearance of individuals. Then form a table on the plan of Table 3, the character of which is so obvious as not to need detailed explanation.

This table is to be read in the following way: Because of the reappearance of Sisera's Stoke Pogis in the 2nd ancestral generation Bess Weaver has 1 fewer ancestors in that generation than she would have had in the entire absence of inbreeding; 2 fewer in the 3rd generation and so on. The totals of the columns of this table are the values, for each generation, of

$$p_{n+1} - q_{n+1}$$

* By "primary" reappearance in the pedigree is meant a reappearance as the sire or dam of an individual which has not itself appeared before in the lower ancestral generations. Thus Patrick Fawkes makes a *primary* reappearance in the fourth ancestral generation as the sire of General Kelly, a bull which is not found in any generation below the third.

TABLE 3.

*Working Table Used in Calculating the Coefficients of inbreeding for Pedigree Table II.*

| ANIMAL. | ANCESTRAL GENERATION. | | |
|---|---|---|---|
| | 2. | 3. | 4. |
| Sisera's Stoke Pogis............... | 1 | 2 | 4 |
| Patrick Fawkes.................. | – | – | 1 |
| Balm........................... | – | – | 1 |
| Totals...................... | 1 | 2 | 6 |

in (1). These totals, multiplied by 100, have then merely to be divided by $p_{n+1}$ in order to obtain the successive $Z$'s. The whole operation may be very quickly carried out. It is not necessary, in fact, to fill out the whole of the later columns of the table, the entries may be cumulated.

For the present pedigree we have

$$Z_0 = 0, \text{ as always*}$$

$$Z_1 = \frac{100\ (1)}{4} = 25\%$$

$$Z_2 = \frac{100\ (2)}{8} = 25\%$$

$$Z_3 = \frac{100\ (6)}{16} = 37.5\%$$

From these values it is seen that in the first four ancestral generations the cow Bess Weaver is 37.5 percent inbred. This is a perfectly definite figure, directly comparable with similar constants for other animals. Of course, if we were to go back more generations we should find Bess Weaver still more inbred, that is, the coefficients would grow larger with each case of the mating of relatives. Since the case is cited here merely for illustration of method, four generations only are considered.

* The apparent paradox implied in the fact that $Z_0$ must always be zero, or in other words that in the first ancestral generation, considered *alone,* there is no inbreeding will be cleared up, if it strikes the reader as paradoxical, by a reconsideration of the general principle numbered 5 on p. 127. The point of course is that it is impossible to say whether the parents are or are not related to one another until something is known of *their* parentage, or in other words, until a *second* ancestral generation is considered.

## Illustration IV. The Pedigree of the Jersey Cow, Figgis 20th of Hood Farm (190306).

Figgis 20th of Hood Farm (190306) is a cow which, in official test for advanced registry, produced as a two year old 437 lbs. 14.4 oz. butter fat. This is a high record. The complete pedigree of this animal has been worked out for 12 ancestral generations, and the coefficients of inbreeding calculated. Twelve generations cover practically the whole of the known pedigree of this cow. On account of its great length it is impossible in the space here available to reproduce this pedigree. Nor is it necessary. The inbreeding coefficients in a quarter of a page give more clear and definite information regarding the amount of inbreeding practised in the breeding of Figgis 20th than could any visual inspection of the pedigree itself. Furthermore the list of names in Table 4 shows just what animals appear more than once in the pedigree, and how frequent are their reappearances.

Table 4 is precisely the same sort of table for Figgis 20th that Table 3 is for Bess Weaver.

In dealing with this and all other pedigrees it is assumed, in the absence of information on the point and the impossibility of acquiring any, that any imported animal for which there is no further pedigree, was not inbred to any degree whatsoever. This is probably not often strictly true, but, on the other hand, some assumption must be made, and this puts all individuals on an equal footing. It is in accord with the principle laid down earlier (p. 127) that in pedigree analysis all individuals must be considered to be unrelated until the contrary is proved by the evidence of their ancestry. After all, the only thing we can possibly measure is the inbreeding shown in the *recorded* pedigree. All that happened prior to the beginning of the record must be a matter of assumption. The *same* assumption should, however, be made for all cases. What this assumption really means practically is that, in all cases of analysis of actual pedigrees, which are bound after a time to come to an end, the values of the coefficients of inbreeding obtained are *lower limiting values.* They signify that the intensity of inbreeding in a particular case could not have been *less* than that indicated; it may have been more. Whether it was or not is not a question open to scientific determination but only to speculation.

From this table the following coefficients of inbreeding for Figgis 20th are easily calculated.

$Z_0 = 0$
$Z_1 = 0$
$Z_2 = 12.50$ percent
$Z_3 = 12.50$ "
$Z_4 = 21.88$ "
$Z_5 = 25.00$ "
$Z_6 = 32.03$ "
$Z_7 = 35.94$ "
$Z_8 = 37.30$ "
$Z_9 = 37.99$ "
$Z_{10} = 38.48$ "
$Z_{11} = 38.57$ "

These coefficients show that Figgis 20th was at least 38 1-2 percent inbred. That is, she had rather less than two-thirds as many different ancestors as she would have had in the event of no inbreeding whatever. It is further clear that most of this inbreeding took place in the fourth and earlier ancestral generations, chiefly in the fourth, fifth, and sixth generations.

Comparing Figgis 20th with Bess Weaver we see that the latter is practically as much inbred in the first 4 ancestral generations as Figgis 20th is in the first 12. Considering in each case only the first four ancestral generations the figures show that, within these generations, Bess Weaver is exactly 3 times more intensely inbred than Figgis 20th.

## The Relation of Coefficients of Inbreeding to the Hereditary Constitution of the Individual.

What, if any is the relation of coefficients of inbreeding to the zygotic constitution (i. e., the hereditary make-up) of the individual? Do the coefficients tell us anything regarding this matter? A little consideration shows that they do. The successive coefficients of inbreeding indicate the rate and degree to which the possible number of *different* hereditary unit factors present in the ancestry is subsequently reduced as a result of inbreeding. They give no indication of the condition in which the *remaining* factors are present (i. e., whether in homogygous

TABLE 4.

*Working Table for Calculating the Coefficients of Inbreeding of Figgis 20th.*

| Animal. | Ancestral Generations. | | | | | | | | | |
|---|---|---|---|---|---|---|---|---|---|---|
| | 3. | 4. | 5. | 6. | 7. | 8. | 9. | 10. | 11. | 12. |
| Sophie's Tormentor | 1 | 2 | 4 | 8 | 16 | 32 | 64 | 128 | 256 | 512 |
| Ida's Stoke Pogis | – | – | 1 | 2 | 4 | 8 | 16 | 32 | 64 | 128 |
| Catono | – | – | 1 | 2 | 4 | 8 | 16 | 32 | 64 | 128 |
| Rosabelle Hudson | – | – | 1 | 2 | 4 | 8 | 16 | 32 | 64 | 128 |
| Khedive | – | – | – | 1 | 2 | 4 | 8 | 16 | 32 | 64 |
| Young Fancy | – | – | – | 1 | 2 | 4 | 8 | 16 | 32 | 64 |
| Stoke Pogis 3rd | – | – | – | – | 1 | 2 | 4 | 8 | 16 | 32 |
| Marjoram | – | – | – | – | 1 | 2 | 4 | 8 | 16 | 32 |
| Lord Lisgar | – | – | – | – | 1 | 2 | 4 | 8 | 16 | 32 |
| Tormentor | – | – | – | – | 1 | 2 | 4 | 8 | 16 | 32 |
| Oonan | – | – | – | – | 1 | 2 | 4 | 8 | 16 | 32 |
| Landseer | – | – | – | – | 1 | 2 | 4 | 8 | 16 | 32 |
| Optimus | – | – | – | – | 1 | 2 | 4 | 8 | 16 | 32 |
| Jersey Bull of Scituate | – | – | – | – | 1 | 2 | 4 | 8 | 16 | 32 |
| Hebe | – | – | – | – | 1 | 2 | 4 | 8 | 16 | 32 |
| Victor Hugo | – | – | – | – | – | 1 | 2 | 4 | 8 | 16 |
| Lord Lisgar | – | – | – | – | – | 1 | 2 | 4 | 8 | 16 |
| Victor Hugo | – | – | – | – | – | 1 | 2 | 4 | 8 | 16 |
| Roxbury | – | – | – | – | – | 1 | 2 | 4 | 8 | 16 |
| Gen'l Scott | – | – | – | – | – | 1 | 2 | 4 | 8 | 16 |
| Europa | – | – | – | – | – | 1 | 2 | 4 | 8 | 16 |
| Europa | – | – | – | – | – | 1 | 2 | 4 | 8 | 16 |
| Mollie | – | – | – | – | – | 1 | 2 | 4 | 8 | 16 |
| Sir Charles | – | – | – | – | – | 1 | 2 | 4 | 8 | 16 |
| Pharos | – | – | – | – | – | 1 | 2 | 4 | 8 | 16 |
| Clement | – | – | – | – | – | – | 1 | 2 | 4 | 8 |
| Hebe | – | – | – | – | – | – | 1 | 2 | 4 | 8 |
| Lady Mary | – | – | – | – | – | – | 1 | 2 | 4 | 8 |
| Fancy | – | – | – | – | – | – | 1 | 2 | 4 | 8 |
| Sir Charles | – | – | – | – | – | – | 1 | 2 | 4 | 8 |
| Mary Lowndes | – | – | – | – | – | – | 1 | 2 | 4 | 8 |
| Victor | – | – | – | – | – | – | 1 | 2 | 4 | 8 |
| Potomac | – | – | – | – | – | – | – | 1 | 2 | 4 |
| Splendid | – | – | – | – | – | – | – | 1 | 2 | 4 |
| Saturn | – | – | – | – | – | – | – | 1 | 2 | 4 |
| Saturn | – | – | – | – | – | – | – | 1 | 2 | 4 |
| Rhea | – | – | – | – | – | – | – | 1 | 2 | 4 |
| Countess | – | – | – | – | – | – | – | 1 | 2 | 4 |
| Dick Swiveller, Jr | – | – | – | – | – | – | – | 1 | 2 | 4 |
| Countess | – | – | – | – | – | – | – | – | 1 | 2 |
| Taintor | – | – | – | – | – | – | – | – | 1 | 2 |
| Comus | – | – | – | – | – | – | – | – | 1 | 2 |
| Comus | – | – | – | – | – | – | – | – | 1 | 2 |
| St. Clement | – | – | – | – | – | – | – | – | 1 | 2 |
| Major | – | – | – | – | – | – | – | – | 1 | 2 |
| Czar | – | – | – | – | – | – | – | – | 1 | 2 |
| Countess | – | – | – | – | – | – | – | – | 1 | 2 |
| Dick Swiveller | – | – | – | – | – | – | – | – | 1 | 2 |
| Dove | – | – | – | – | – | – | – | – | 1 | 2 |
| Diana | – | – | – | – | – | – | – | – | – | 1 |
| Custard | – | – | – | – | – | – | – | – | – | 1 |
| Prince John | – | – | – | – | – | – | – | – | – | 1 |
| Splendens | – | – | – | – | – | – | – | – | – | 1 |
| Totals | 1 | 2 | 7 | 16 | 41 | 92 | 191 | 389 | 788 | 1,580 |

or heterozygous condition). The meaning here will be clear if a concrete example is considered. When one brother and sister mating is made 50 percent of the maximum possible number of different ancestors is eliminated. It is at least readily conceivable, if indeed it cannot be said to be highly probable, that no two individuals among higher animals and plants are *exactly* alike in zygotic constitution when *all* hereditary characters are taken into account. This means, in last analysis, that each individual must differ from every other by at least one unit factor, possibly more. Once mating of brother and sister will diminish the number of such differences by 50 percent from what it would have been had no such mating occurred. The number of homozygous individuals *with respect to the hereditary differences remaining,* however, will not increase. This is practically equivalent to saying that while self-fertilization increases the proportion of *individuals* homozygous with reference to all characters, the closest inbreeding other than self-fertilization, if continued, increases the proportion of *characters* with respect to which all individuals are homozygous. Thus while both processes tend towards uniformity in the progeny, it is a different kind of uniformity obtained in a different way, in the one case from what it is in the other.

While in the above discussion only brother × sister mating is mentioned it is clear that the same reasoning applies regarding the meaning of the coefficients of inbreeding in all other types of mating.

There are other theoretical relations of inbreeding coefficients which are of interest, but to discuss them in detail here is not possible.

## Concluding Remarks.

In this paper has been presented in abstract a general method of measuring the intensity or degree of the inbreeding practised in any particular case. The method proposed is shown to be perfectly general. It is based on no assumption whatever as to the nature of the hereditary process. On the contrary it is founded on the most completely logical and comprehensive definition of the concept of inbreeding that it seems possible to formulate. This is, in simplest form, that the most fundamental objective criterion which distinguishes an inbred individual from one not inbred is that the former has fewer different ancestors than the latter. It is believed that the proposed coefficients of inbreeding may be made extremely useful in studies of the problem of the effect of inbreeding, whether in relation to its purely theoretical aspects, or in the practical fields of stock-breeding and eugenics.

*Table Showing the Maximum Possible Number of Different Ancestors in the First Twenty Ancestral Generations (Obligate Bisexual Reproduction).*

| ANCESTRAL GENERATION. | Maximum Possible Number of Ancestors. | ANCESTRAL GENERATION. | Maximum Possible Number of Ancestors. |
|---|---|---|---|
| 1 | 2 | 11 | 2,048 |
| 2 | 4 | 12 | 4,096 |
| 3 | 8 | 13 | 8,192 |
| 4 | 16 | 14 | 16,384 |
| 5 | 32 | 15 | 32,768 |
| 6 | 64 | 16 | 65,536 |
| 7 | 128 | 17 | 131,072 |
| 8 | 256 | 18 | 262,144 |
| 9 | 512 | 19 | 524,288 |
| 10 | 1,024 | 20 | 1,048,576 |

# 4



Reprinted from *Compt. Rend. Sci. Math.*, Pt. 1, **177**, 528–531 (Sept. 1923)

STATISTIQUE MATHÉMATIQUE. — *Démonstration mathématique de la loi d'hérédité de Mendel.* Note [1] de M. **Serge Bernstein**, présentée par M. Émile Borel.

Soit H un groupe d'individus contenant trois classes A, B, C différentes. Supposons que les individus de chaque classe puissent appartenir à deux sexes différents et admettons que la probabilité d'appartenir à l'un des deux sexes est la même pour chaque classe. Nous dirons alors que les individus du groupe H sont soumis à une *panmixie normale*, s'il n'y a aucune sélection, c'est-à-dire si tous les croisements sont également probables, également fertiles, et si le taux de mortalité est le même pour toutes les classes. En admettant de plus que tous les croisements conduisent à des individus de même groupe H, nous pouvons déterminer chaque loi permanente d'hérédité

[1] Séance du 10 septembre 1923.

par 18 coefficients non négatifs $E^A_{AA}$, $E^B_{AA}$, ..., $E^A_{CC}$, où $E^C_{AB}$ représente, par exemple, la probabilité pour que le croisement d'un individu de la classe A avec un individu de la classe B donne un individu de la classe C. On aura évidemment six relations de la forme

$$E^A_{AA}+E^B_{AA}+E^C_{AA}=1. \tag{1}$$

Nous dirons qu'une loi d'hérédité satisfait au *principe de stationnarité si, à partir de la seconde génération, la distribution des individus entre les trois classes reste la même dans toutes les générations successives.*

Dans ces conditions nous allons établir le théorème suivant :

*La seule loi d'hérédité, compatible avec le principe de stationnarité, d'après laquelle le croisement d'un individu de la classe* A *avec un individu de la classe* B *donne toujours un individu de la classe* C, *est la loi de Mendel* [les classes A et B seront des lignes pures, c'est-à-dire $E^A_{AA}=E^B_{BB}=1$, et la classe C sera hybride ([1])].

En effet, $\alpha$, $\beta$, $\gamma$ ayant respectivement les probabilités d'appartenir aux classes A, B, C dans la première génération, on aura, en désignant par $\alpha_1$, $\beta_1$, $\gamma_1$ les probabilités d'appartenir aux mêmes classes dans la seconde génération,

$$\alpha_1=f(\alpha,\beta,\gamma),\qquad \beta_1=\varphi(\alpha,\beta,\gamma),\qquad \gamma_1=\psi(\alpha,\beta,\gamma), \tag{2}$$

où $f$, $\varphi$, $\psi$ sont des formes quadratiques à coefficients non négatifs, telles que

$$f+\varphi+\psi\equiv(\alpha+\beta+\gamma)^2, \tag{3}$$

à cause de (1). D'après le principe de stationnarité, on doit avoir identiquement :

$$\begin{cases}(\alpha+\beta+\gamma)^2 f(\alpha,\beta,\gamma)=f[f(\alpha,\beta,\gamma),\varphi(\alpha,\beta,\gamma),\psi(\alpha,\beta,\gamma)],\\(\alpha+\beta+\gamma)^2\varphi(\alpha,\beta,\gamma)=\varphi[f(\alpha,\beta,\gamma),\varphi(\alpha,\beta,\gamma),\psi(\alpha,\beta,\gamma)],\\(\alpha+\beta+\gamma)^3\psi(\alpha,\beta,\gamma)=\psi[f(\alpha,\beta,\gamma),\varphi(\alpha,\beta,\gamma),\psi(\alpha,\beta,\gamma)].\end{cases} \tag{4}$$

La condition, qui résulte de l'hypothèse que le croisement mutuel des individus de la classe A et B conduit nécessairement à des individus de la

---

([1]) Rappelons que, d'après la loi de Mendel,

$$E^A_{CC}=E^B_{CC}=\frac{1}{4},\qquad E^C_{CC}=E^C_{AC}=E^C_{BC}=E^A_{AC}=E^B_{BC}=\frac{1}{2},$$

les autres coefficients étant nuls à cause de (1).

classe C, signifie que dans $f$ et $\varphi$ il n'y a pas de terme en $\alpha\beta$. Or, si, quels que soient $\alpha$, $\beta$, $\gamma$ (avec $\alpha+\beta+\gamma=S=1$), les équations (2) ne donnaient qu'un nombre limité de valeurs à $\alpha_1$, $\beta_1$, $\gamma_1$, les fonctions $f$, $\varphi$ et $\psi$ ne dépendraient que de S, ce qui est incompatible avec la remarque que l'on vient de faire. Donc, on aura nécessairement

$$(5)\quad \begin{cases} f(\alpha,\beta,\gamma)=\alpha S+\ \ F(\alpha,\beta,\gamma),\\ \varphi(\alpha,\beta,\gamma)=\beta S+\ \ F(\alpha,\beta,\gamma),\\ \psi(\alpha,\beta,\gamma)=\gamma S-2F(\alpha,\beta,\gamma),\end{cases}$$

où l'équation du second degré

$$F(\alpha,\beta,\gamma)=0$$

devra être satisfaite par toutes les distributions stationnaires; en remarquant, d'autre part, que le coefficient de $\alpha^2$ dans $f$ et de $\beta^2$ dans $\varphi$ ne saurait dépasser 1, on trouve la forme nécessaire de

$$F(\alpha,\beta,\gamma)=-\alpha\beta+l\alpha\gamma+m\beta\gamma+n\gamma^2.$$

En substituant ensuite les expressions (5) dans (4), on obtient l'identité unique

$$(6)\quad S\equiv 2F'_\gamma-F'_\alpha-F'_\beta,$$

de laquelle on tire immédiatement

$$(7)\quad F(\alpha,\beta,\gamma)=-\alpha\beta+\frac{1}{4}\gamma^2.$$

Par conséquent, nous déduisons de (5) la forme nécessaire des fonctions

$$(8)\quad \begin{cases} f(\alpha,\beta,\gamma)=\alpha S-\ \alpha\beta+\frac{1}{4}\gamma^2=\left(\alpha+\frac{\gamma}{2}\right)^2,\\ \varphi(\alpha,\beta,\gamma)=\beta S-\ \alpha\beta+\frac{1}{4}\gamma^2=\left(\beta+\frac{\gamma}{2}\right)^2,\\ \psi(\alpha,\beta,\gamma)=\gamma S+2\alpha\beta-\frac{1}{2}\gamma^2=2\left(\alpha+\frac{\gamma}{2}\right)\left(\beta+\frac{\gamma}{2}\right),\end{cases}$$

qui exprime précisément la loi d'hérédité de Mendel.

Ce résultat pourrait être appliqué, il me semble, pour reconnaître si la loi d'hérédité dans un groupe H est celle de Mendel ou non; il suffit, après avoir constaté que le croisement des deux classes A et B conduit toujours à la classe unique C, de réaliser *la panmixie normale* (en corrigeant, s'il y a lieu, les inégalités de fertilité et de mortalité), — pour vérifier la loi de Mendel il sera nécessaire et suffisant de constater qu'un régime stationnaire s'est établi dès la seconde génération.

Je ne m'arrêterai pas sur les formules (que j'ai établies ailleurs), par lesquelles il faudrait remplacer les formules (8), si l'on rejetait la condition de Mendel que le croisement de A avec B produit C, et je laisserai également de côté pour le moment le problème de la loi d'hérédité dans le cas où le nombre de classes est supérieur à trois.

5



Reprinted from *Compt. Rend. Sci. Math.*, Pt. 1, **177**, 581–584 (Oct. 1923)

STATISTIQUE MATHÉMATIQUE. — *Principe de stationarité et généralisations de la loi de Mendel.* Note ([1]) de M. **Serge Bernstein**, transmise par M. Émile Borel.

Soit H un groupe contenant $n$ classes $A_1, A_2, \ldots, A_n$ d'individus.

D'après le principe de stationarité énoncé dans ma Communication précédente ([2]), toute loi d'hérédité bisexuelle sera déterminée par $n$ formes quadratiques ([3])

$$f_1(\alpha_1, \alpha_2, \ldots, \alpha_n), \quad \ldots, \quad f_n(\alpha_1, \alpha_2, \ldots, \alpha_n)$$

à $n$ variables, ayant leurs coefficients non négatifs, satisfaisant aux équations fonctionnelles

$$(1) \left\{ \begin{array}{l} (\alpha_1+\alpha_2+\ldots+\alpha_n)^2 f_1(\alpha_1, \alpha_2, \ldots, \alpha_n) = f_1[f_1(\alpha_1, \alpha_2, \ldots, \alpha_n), \ldots, f_n(\alpha_1, \alpha_2, \ldots, \alpha_n)], \\ \ldots\ldots\ldots\ldots\ldots\ldots\ldots\ldots\ldots\ldots\ldots\ldots\ldots\ldots, \\ (\alpha_1+\alpha_2+\ldots+\alpha_n)^2 f_n(\alpha_1, \ldots, \alpha_n) = f_n[f_1(\alpha_1, \ldots, \alpha_n), \ldots, f_n(\alpha_1, \alpha_2, \ldots, \alpha_n)] \end{array} \right.$$

et à la condition

$$f_1 + f_2 + \ldots + f_n = (\alpha_1 + \alpha_2 + \ldots + \alpha_n)^2.$$

Supposons d'abord *qu'aucun des coefficients des formes $f_i$ ne soit nul*, c'est-à-dire *que chaque paire d'individus soit capable de donner naissance à un individu de classe quelconque. Dans ces conditions, on a nécessairement*

$$(2) \qquad f_i = \lambda_i(\alpha_1 + \alpha_2 + \ldots + \alpha_n)^2,$$

([1]) Séance du 10 septembre 1923.

([2]) *Comptes rendus*, t. 177, 1923, p. 528.

([3]) Dans le cas de la reproduction *unisexuelle*, les formes $f_i$ sont *linéaires* et elles sont nécessairement de la forme $\lambda_i(\alpha_1 + \alpha_2 + \ldots + \alpha_n)$; en ce cas, *la distribution permanente réalisée dès la seconde génération est indépendante de la distribution primitive.* La même conclusion s'applique lorsque le nombre des classes est infiniment grand (les formes linéaires sont remplacées par des intégrales).

*où* $\sum_1^n \lambda_i = 1$ : en d'autres termes, *la probabilité pour un individu d'appartenir à une classe déterminée est la même, quels que soient ses parents.* On peut dire que, dans ce cas, les différentes classes ne se distinguent que par leurs propriétés *somatiques* (extérieurement), mais sont identiques au point de vue *génétique*, de sorte que *le groupe* H *ne forme, au fond, qu'une seule race pure polymorphe.*

Admettons à présent qu'au contraire, certains des coefficients des $f_i$ sont nuls; d'une façon plus précise, supposons, premièrement, que $A_1$ et $A_2$ sont deux races pures distinctes (en d'autres termes, le terme en $\alpha_1^2$ ne figure que dans $f_1$ et le terme en $\alpha_2^2$ ne se trouve que dans $f_2$) et, secondement, toutes les autres classes sont des hybrides qui s'obtiennent par le croisement mutuel [1] des classes $A_1$ et $A_2$ (en d'autres termes, le produit $\alpha_1\alpha_2$ n'intervient ni dans $f_1$, ni dans $f_2$, mais entre nécessairement dans toutes les autres formes $f_i$.

Cela étant, *les équations* (1) *n'admettent que deux genres de solutions :*

1° *Genre mendelien.* — Deux individus d'une même classe hybride quelconque peuvent par leur croisement donner naissance à un individu de l'une au moins des deux races pures (en d'autres termes, les coefficients de $\alpha_i^2$, où $i > 2$, sont différents de O au moins dans une des formes $f_1$ ou $f_2$). *Dans ce cas, on doit avoir*

$$f_1 = \left[\alpha_1 + \frac{1}{2}(A_3\alpha_3 + \ldots + A_n\alpha_n)\right]^2, \qquad f_2 = \left[\alpha_2 + \frac{1}{2}(B_3\alpha_3 + \ldots + B_n\alpha_n)\right]^2,$$

et, pour $i > 2$,

$$(3) \qquad f_i = 2c_i\left[\alpha_1 + \frac{1}{2}(A_3\alpha_3 + \ldots + A_n\alpha_n)\right]\left[\alpha_2 + \frac{1}{2}(B_3\alpha_3 + \ldots + B_n\alpha_n)\right],$$

*les coefficients positifs* $c_i$ *et les nombres négatifs* $A_i$ et $B_i$ *satisfaisant aux n conditions*

$$\Sigma c_i A_i = \Sigma c_i B_i = 1, \qquad A_i + B_i = 2.$$

Il y a lieu de remarquer que les formules (3) ne se distinguent pas essentiellement des formules (8) de ma Communication précédente qui correspondent à la loi élémentaire de Mendel (pour $n = 3$), de sorte qu'un observateur qui ne saurait pas distinguer par leurs propriétés exté-

(1) D'après ma Communication précédente cette dernière condition entraînerait à elle seule la loi de Mendel, si la classe hybride était unique.

rieures les différentes classes hybrides pourrait facilement confondre ce cas avec le cas classique de Mendel, surtout si les coefficients $A_i$ et $B_i$ étaient voisins de l'unité.

2, *Genre « en quadrille »*. — Il existe une classe hybride, telle que le croisement de deux individus de cette classe est incapable de reproduire un individu d'une des deux races pures primitives (en d'autres termes, il existe des valeurs de $i$ telles que le terme en $\alpha_i^2$ manque à la fois dans $f_1$ et $f_2$).

*Dans ce cas, on doit avoir*

$$f_1 = (\alpha_1 + \alpha_3 + \ldots + \alpha_k)(\alpha_1 + \alpha_{k+1} + \ldots + \alpha_n),$$
$$f_2 = (\alpha_2 + \alpha_3 + \ldots + \alpha_k)(\alpha_2 + \alpha_{k+1} + \ldots + \alpha_n)$$

et pour $2 < i \leqq k$, où $k$ est un nombre quelconque inférieur à $n$,

$$f_i = c_i(\alpha_1 + \alpha_3 + \ldots + \alpha_k)(\alpha_2 + \alpha_3 + \ldots + \alpha_k),$$

tandis que, pour $k < i \leqq n$,

$$f_i = d_i(\alpha_1 + \alpha_{k+1} + \ldots + \alpha_n)(\alpha_2 + \alpha_{k+1} + \ldots + \alpha_n),$$

*avec*

$$\sum_3^k c_i = \sum_{k+1}^n d_i = 1.$$

J'ai appelé ce genre de loi d'hérédité *en quadrillé*, car pour $n = 4 (k = 3)$, les classes $A_1$ et $A_2$, d'une part, et les classes $A_3$ et $A_4$ d'autre part, forment deux couples jouissant de propriétés mutuelles absolument semblables : ainsi $A_3$ et $A_4$, *qui sont des races provenant du croisement de* $A_1$ *avec* $A_2$, *constituent des races hybrides constantes telles que le croisement de deux individus de la race* $A_3$ *donne toujours naissance à un individu de la même race* (de même pour $A_4$) : *ces deux races sont aussi des races pures, mais le croisement mutuel de* $A_3$ *et* $A_4$ *conduit à des individus* $A_1$ *et* $A_2$ *qui sont aussi des hybrides constants par rapport aux races* $A_3$ *et* $A_4$. Ce genre de loi d'hérédité qui est irréductible au genre mendelien permettrait, peut-être, d'expliquer certaines expériences, incompatibles avec la loi de Mendel, où l'on a vu (M. de Vries) surgir en même temps deux nouvelles races pures par le croisement de deux autres races pures distinctes. Le cas de $n > 4$ se rattache au cas simple de $n = 4$, comme le genre mendelien général se rattache à la loi élémentaire de Mendel pour $n = 3$.

Avant de terminer, remarquons que l'hérédité de propriétés complexes ou de combinaisons de propriétés ne peut satisfaire au principe de stationarité. Ce n'est qu'après plusieurs générations qu'un régime stationnaire tendra à s'établir et sera déterminé par la distribution stationnaire, fixée dès la seconde génération, de chacune des propriétés élémentaires, combinées conformément au théorème de la multiplication des probabilités : ainsi les propriétés dont la distribution dans *une panmixie normale* ne se fixerait pas dès la seconde génération devraient être considérées comme complexes.

# III

# Foundations of Genetic Algebra

# Editor's Comments on Papers 6 Through 10

6 **Etherington:** On Non-associative Combinations
*Proc. Roy. Soc. Edinburgh,* **59**, 153–162 (1939)

7 **Etherington:** Genetic Algebras
*Proc. Roy. Soc. Edinburgh,* **59**, 242–258 (1939)

8 **Etherington:** Duplication of Linear Algebras
*Proc. Edinburgh Math. Soc.,* Ser. 2, **6**, Pt. 4, 222–230 (1941)

9 **Etherington:** Non-associative Algebra and the Symbolism of Genetics
*Proc. Roy. Soc. Edinburgh,* **61**, 24–42, (1941)

10 **Reiersöl:** Genetic Algebras Studied Recursively and by Means of Differential Operators
*Math. Scand.,* **10**, 25–44 (1962)

It would have been sensible to devote an entire volume to the topic for which the following five papers are the most fundamental contributions. It is quite unclear why population genetic theory has had only a small and sporadic romance with structural mathematics, but these papers, together with the referenced materials and the references found in Part IV, most clearly demonstrate that genetic problems yield nicely to such analysis. In this one section selection was simplified by the fact that the important papers were also the most inaccessible, hence the obvious candidates for inclusion. Two works by Etherington (1940, 1941) are the principal omissions.

A highly truncated survey of related materials includes the following. First, for historic and possibly for current research interests, one should see the mathematical papers of A. A. Albert, whose books are accessible in most libraries, and should also see Albert (1942a, 1942b, 1944). Bruck (1944) may also prove useful. Second, there are two survey papers that show the development of the literature, by Schafer (1949) and by Raffin (1951). Third, a surprisingly large literature has arisen directly from Etherington's materials. The excellent paper by Lyubich (1971) also includes the most complete bibliography available; in particular, the papers of Gonshor (1960, 1971), Holgate (1970), and Heuch (1972) are important.

I have summarized the importance of this material in the introduction but want to emphasize that these papers, or at least this approach, is absolutely fundamental to a more organized theory of population genetics.

**ADDED NOTE:** After reviewing a copy of the introductory material for this volume, I. M. H. Etherington volunteered the following comment:

> For the purpose stated [for historic and possibly for current research interests] I think your readers would find Schafer a much kinder author, and I would have liked to see the following references included:
>
> (a) Schafer, R. D. Structure and representation of nonassociative algebras. *Bull. Amer. Math. Soc.,* **61**:469–484 (1955). (This is a historical survey with no proofs, with a bibliography of 112 items.)
>
> (b) Schafer, R. D. *An Introduction to Nonassociative Algebras.* New York: Academic Press, Inc., 1966. (This is written as a textbook for graduate students, and so is at a more elementary level than your references 1, 2, 3, and 10. There is a bibliography of some 200 items.)

Further, $X^{a(b+c)}$ means $(X^a)^{b+c}$, *i.e.* $(X^a)^b(X^a)^c$, which is the same as $X^{ab+ac}$. Hence in the arithmetic of the indices

$$a(b+c)=ab+ac.$$

But in general

$$(b+c)a \neq ba+ca,$$

since $(X^bX^c)^a$ is not the same as $(X^b)^a(X^c)^a$. We may say therefore that in the arithmetic of the indices multiplication is predistributive with addition, but not in general postdistributive.

In these arguments $a$, $b$, $c$ can be any expressions standing for complicated powers: they are not restricted to being simple integers indicating primary powers.

The notation provides an arithmetical method of specifying commutative shapes; for now the *shape* $s$ of any commutative non-associative product can be redefined as the index of the corresponding power obtained by equating all the factors The product $AB.C^2:D$, for instance, has the same shape as the power $(X^2)^2X=X^{2.2+1}$, namely $s=2.2+1$.

Consider what addition and multiplication of shapes mean when we are dealing with products in general instead of powers. Let $\Pi_1$, $\Pi_2$ be any two products with shapes $s_1$, $s_2$. Then $s_1+s_2$ is the shape of the product $\Pi_1\Pi_2$, while $s_1s_2$ is the shape of the product formed by substituting $\Pi_1$ for each of the factor elements of $\Pi_2$.

The procedure of this § may be described as a representation of the set of all commutative non-associative continued products formed from given elements on a non-associative arithmetic, whose integers are commutative shapes $a, b, c, \ldots$ with the rules of combination

$$\left.\begin{array}{lll} a+b=b+a, & ab.c=a.bc, & a(b+c)=ab+ac, \\ ab \neq ba, & (a+b)+c \neq a+(b+c), & (b+c)a \neq ba+ca. \end{array}\right\} \qquad (1)$$

A similar representation is possible when multiplication of the original elements is non-commutative as well as non-associative. It is reflected as non-commutative addition of shapes, the other rules of combination (1) being unchanged. But the numerical specification of non-commutative shapes of increasing complexity rapidly becomes very complicated; to simplify it, some convention is required for distinguishing the $2^{\delta-2}$ distinct primary shapes of any given degree $\delta(>1)$.

## § 3. Classification of Shapes.

Shapes $s$ will be classified by their *degree* $\delta(s)$, *altitude* $\alpha(s)$, and *mutability* $\mu(s)$. Non-commutative shapes will be further classified by the commutative shapes with which they are *conformal.* These terms will now be defined.

The *degree* $\delta$ of a shape $s$ means the number of factor elements in a product having this shape. It may be reckoned by evaluating $s$ as if it were an integer in ordinary arithmetic.

Two non-commutative shapes $s_1$, $s_2$, which become the same shape $s$ when multiplication is regarded as commutative, will be called *conformal* with each other and with $s$. Write $s_1 \sim s_2$ to indicate this. With commutative shapes, $s_1 \sim s_2$ means the same as $s_1 = s_2$, a commutative shape being conformal only with itself. The word is also applicable to products whose shapes are conformal. Thus

$$AB.C:D, \qquad A.BC:D, \qquad A:BC.D, \qquad A:B.CD$$

and their shapes

$$(2+1)+1, \qquad (1+2)+1, \qquad 1+(2+1), \qquad 1+(1+2)$$

are all conformal with the commutative power $A^4$ and its shape 4.

Let shapes be depicted as pedigrees (§ 1). Any non-associative product is then, so to speak, "descended from" its factors. The number of "generations" preceding the product itself is its *altitude* $\alpha$ (Cayley, 1875). At each knot in the pedigree two factors are united; the total number of knots is thus $\delta - 1$. Let a knot be called *balanced* if its two factors are conformal: then the number of unbalanced knots in a pedigree will be called its *mutability* $\mu$. The various terms defined may be applied indiscriminately to the product, shape or pedigree.

If the mutability of any shape $s$ (commutative or not) is $\mu$, then there are evidently just $2^\mu$ distinct non-commutative shapes which will become the same as $s$ when multiplication is commutative. So $\mu$ could be defined alternatively as the logarithm to base 2 of the number of conformal non-commutative shapes.

If $s_1 \sim s_2$, then evidently

$$\delta(s_1) = \delta(s_2), \qquad \alpha(s_1) = \alpha(s_2), \qquad \mu(s_1) = \mu(s_2). \qquad (2)$$

The following formulæ are easily proved, $r$ and $s$ being any shapes, commutative or non-commutative, and $\nu$ an ordinary positive integer:—

$$\delta(r+s) = \delta(r) + \delta(s), \qquad (3)$$

$$\delta(rs) = \delta(r)\delta(s), \qquad (4)$$

$$\delta(s^\nu) = \delta(s)^\nu; \qquad (5)$$

$$\alpha(r+s) = 1 + \alpha(r) \quad \text{or} \quad 1 + \alpha(s) \quad \text{according as} \quad \alpha(r) \geqslant \quad \text{or} \quad \leqslant \alpha(s), \qquad (6)$$

$$\alpha(rs) = \alpha(r) + \alpha(s), \qquad (7)$$

$$\alpha(s^\nu) = \nu\alpha(s); \qquad (8)$$

$$\mu(r+s) = 2\mu(s) \qquad \text{if } r \sim s, \qquad (9)$$

$$= 1 + \mu(r) + \mu(s) \text{ if not}, \qquad (10)$$

$$\mu(rs) = \delta(s)\mu(r) + \mu(s), \qquad (11)$$

$$\mu(s^r) = (1 + \delta + \delta^2 + \ldots + \delta^{r-1})\mu(s), \qquad (12)$$

where

$$\delta = \delta(s).$$

The last result is proved by induction from the preceding one. It may also be written

$$\frac{\mu(s^r)}{\mu(s)} = \frac{\tau(s^r)}{\tau(s)}, \qquad (13)$$

where

$$\tau = \delta - 1.$$

The degree, altitude and mutability can now be readily calculated for any given shape specified numerically. The table below gives all commutative shapes for which $\alpha \leqslant 4$, $\delta \leqslant 6$.

TABLE OF COMMUTATIVE SHAPES.

| $\alpha$. | $\delta$. | $\mu$. | $s$. |
|---|---|---|---|
| 0 | 1 | 0 | 1 |
| 1 | 2 | 0 | 2 |
| 2 | 3 | 1 | 3 |
| 2 | 4 | 0 | 2.2 |
| 3 | 4 | 2 | 4 |
| 3 | 5 | 1 | 2.2 + 1 |
| 3 | 5 | 2 | 3 + 2 |
| 3 | 6 | 1 | 2.3 |
| 3 | 6 | 2 | 3.2 |
| 3 | 7 | 2 | 2.2 + 3 |
| 3 | 8 | 0 | $2^3$ |
| 4 | 5 | 3 | 5 |
| 4 | 6 | 2 | (2.2 + 1) + 1 |
| 4 | 6 | 3 | (3 + 2) + 1, 4 + 2 |
| Etc. | | | |

As the table suggests, we cannot construct a shape with $\alpha$, $\delta$, $\mu$ assigned arbitrarily. Certain relations must be satisfied, namely:

$$2^\alpha \geqslant \delta \geqslant \alpha + 1; \quad \text{i.e.} \quad \delta - 1 \geqslant \alpha \geqslant \log_2 \delta. \qquad (14)$$

$$\delta \geqslant \mu + 2, \quad \textit{except when} \quad \delta = 1. \qquad (15)$$

$$\mu \leqslant 3.2^{\alpha-3} - 1; \quad \text{i.e.} \ \alpha \geqslant 3 + \log_2 \frac{\mu+1}{3}, \quad \textit{except when} \quad \alpha < 3. \qquad (16)$$

*$\delta$ is expressible as the sum of $\mu + 1$ powers of 2, not all alike if $\mu > 0$.* (17)

(14) and (15) are easily proved by consideration of pedigrees. At one extreme, the equality $\delta = \alpha + 1$ holds only when $s$ is primary; and the

same is true of $\delta=\mu+2$. Similarly at the other extreme, $\delta=2^a$, $\mu=0$ occur when and only when $s$ is plenary.

(17) is proved by induction from (3), (5), (9), (10); it being noted that when $\mu=0$, $s$ is of the form $2^a$ (plenary); when $\mu=1$, $s=2^a+2^\beta$ $(a \neq \beta)$; and that two like powers of 2 can be combined if desired into a single power of 2.

To prove (16), let $\mu_a$ be the greatest possible mutability for a shape whose altitude $a$ is given; it will be shown that for $a \geqslant 3$

$$\mu_a = 3.2^{a-3} - 1.$$

In view of (2) it will be sufficient to consider only commutative shapes. By inspection of the table of commutative shapes,

$$\mu_0 = \mu_1 = 0, \qquad \mu_2 = 1, \qquad \mu_3 = 2.$$

Now (see (6)) any shape of altitude $a+1$ is necessarily the sum of two shapes, one of altitude $a$ and one of altitude $\beta \leqslant a$. By (9), (10), $\mu_{a+1}$ must be expressible either as $2\mu_a$ or as $1+\mu_a+\mu_\beta$. Since $\mu_1=0$, $\mu_2=1$, it follows that

$$\mu_{a+1} > \mu_a \quad \text{for} \quad a > 0,$$

so that $\mu_a$ increases monotonically with $a$.

Now let $a$ be any altitude (e.g. $a=3$) for which there exist at least three distinct shapes $s_1$, $s_2$, $s_3$ with the maximum mutability $\mu_a$. Then

$$\mu(s_1) = \mu(s_2) = \mu(s_3) = \mu_a > \mu(s),$$

where $s$ is any shape of lower altitude. Hence for the altitude $a+1$ also there will exist at least three distinct shapes of maximum mutability; namely,

$$s_1+s_2, \qquad s_2+s_3, \qquad s_3+s_1,$$

with the mutability given by (10)

$$\mu_{a+1} = 1 + 2\mu_a. \qquad (a \geqslant 3.)$$

It follows that

$$1+\mu_{a+1} = 2(1+\mu_a).$$

But

$$1+\mu_3 = 3,$$

whence

$$1+\mu_a = 3.2^{a-3},$$

or

$$\mu_a = 3.2^{a-3} - 1 \qquad \text{if} \quad a \geqslant 3.$$

This proves (16).

It will be seen that the equality in (16) is attained by $N_a$ commutative shapes of altitude $a$, where

$$N_{a+1} = \tfrac{1}{2}N_a(N_a - 1), \qquad N_3 = 4. \qquad . \quad . \quad . \quad . \quad (16a)$$

## § 4. Enumeration of Shapes.

Let $a_\delta$, $p_\alpha$ be the numbers of possible shapes of given degree $\delta$ and of given altitude $\alpha$ respectively, when multiplication is non-commutative and non-associative; and let $b_\delta$, $q_\alpha$ be the corresponding numbers when multiplication is commutative and non-associative. Evidently

$$a_1 = b_1 = p_0 = q_0 = 1.$$

Remembering (3) and (6), and considering the different ways in which shapes of given degree or altitude can be formed from those of lower degree or altitude, we obtain the formulæ:

$$a_\delta = a_1 a_{\delta-1} + a_2 a_{\delta-2} + a_3 a_{\delta-3} + \ldots + a_{\delta-1} a_1, \quad (18)$$

$$\left.\begin{aligned} b_{2\delta-1} &= b_1 b_{2\delta-2} + b_2 b_{2\delta-3} + \ldots + b_{\delta-1} b_\delta, \\ b_{2\delta} &= b_1 b_{2\delta-1} + b_2 b_{2\delta-2} + \ldots + b_{\delta-1} b_{\delta+1} + \tfrac{1}{2} b_\delta (b_\delta + 1), \end{aligned}\right\} \quad (19)$$

$$p_{\alpha+1} = 2p_\alpha (p_0 + p_1 + p_2 + \ldots + p_{\alpha-1}) + p_\alpha^2, \quad (20)$$

$$q_{\alpha+1} = q_\alpha (q_0 + q_1 + q_2 + \ldots + q_{\alpha-1}) + \tfrac{1}{2} q_\alpha (q_\alpha + 1). \quad (21)$$

For $\delta = 1, 2, 3, \ldots$ and $\alpha = 0, 1, 2, \ldots$ the sequences start:

$$\begin{aligned} a_\delta &= 1, 1, 2, 5, 14, 42, 132, 429, 1430, 4862, \ldots \\ b_\delta &= 1, 1, 1, 2, 3, 6, 11, 23, 46, 98, \ldots \\ p_\alpha &= 1, 1, 3, 21, 651, 457653, 210065930571, \ldots \\ q_\alpha &= 1, 1, 2, 7, 56, 2212, 2595782, \ldots \end{aligned}$$

Let

$$F(x) = a_1 x + a_2 x^2 + a_3 x^3 + \ldots + a_\delta x^\delta + \ldots \quad (22)$$

and

$$f(x) = -1 + b_1 x + b_2 x^2 + \ldots + b_\delta x^\delta + \ldots \quad (23)$$

The following results are known:—

$$F(x)^2 - F(x) + x = 0, \quad (24)$$

$$f(x)^2 + f(x^2) + 2x = 0; \quad (25)$$

$$F(x) = \tfrac{1}{2} - \tfrac{1}{2}\sqrt{1 - 4x}, \quad (26)$$

$$f(x) = \lim_{n \to \infty} -\sqrt{\vphantom{x}} - 2x + \sqrt{\vphantom{x}} - 2x^2 + \sqrt{\vphantom{x}} - 2x^4 + \ldots + \sqrt{\vphantom{x}} - 2x^{2^n} + 1, \quad (27)$$

where in (27) each $\sqrt{\vphantom{x}}$ covers all that follows it;

$$a_\delta = \frac{(2\delta - 2)!}{(\delta - 1)!\ \delta!} = \frac{1}{\delta} \cdot {}^{2\delta-2}C_{\delta-1}. \quad (28)$$

Of those formulæ, (18), (28) were given by Catalan (1838). (Catalan pointed out that $a_\delta$ is the number of ways in which a convex polygon of $\delta + 1$ sides can be divided up into triangles by diagonals. (28), as a consequence of (18) with $a_1 = 1$, was first established from this point of view, and was known to other writers, apparently first to Euler. Several papers on this topic appear in the *Journ. de Math.*, 1838–39.) Binet (1839) introduced the generating function (22), and deduced (24), (26), (28) from (18). The calculations were repeated by Cayley (1859) from the pedigree point of view; by Schröder (1870); also by Wedderburn (1922), who discussed as well the commutative case, obtaining (19), (25), (27), and made a special study of the functional equation (25) and its more general solutions. (*Cf.* Etherington, 1937.)

It will now be shown that by introducing mutability we can discuss the commutative and non-commutative cases simultaneously and obtain a more general functional equation (33) which includes the two equations (24), (25) as special cases.

Let $c_{\delta\mu}$ be the number of possible commutative shapes of given degree $\delta$ and mutability $\mu$, so that the corresponding number of non-commutative shapes will be

$$n_{\delta\mu} = 2^{\mu} c_{\delta\mu}. \qquad (29)$$

Then $n_{\delta\mu}$, $c_{\delta\mu}$ are defined for all integer values of $\delta$, $\mu$ with $\delta \geqslant 1$, $\mu \geqslant 0$. For all other values of $\delta$ and $\mu$, let $n_{\delta\mu}$, $c_{\delta\mu}$ be defined as zero.

Consider with the aid of (3), (9), (10) the different ways in which a non-commutative shape $s$ of degree $\delta$ and mutability $\mu$ can be formed. Excluding $\delta = 1$, $\mu = 0$, $s = 1$, $s$ must be of the form $s_1 + s_2$ where, by (3),

$$\delta(s_1) + \delta(s_2) = \delta.$$

If (10) held in all cases, we should have

$$1 + \mu(s_1) + \mu(s_2) = \mu,$$

and consequently

$$n_{\delta\mu} = \sum_{i,j,l,m} n_{il} n_{jm} \qquad (i + j = \delta, \quad 1 + l + m = \mu).$$

Subtracting the cases to which (10) does not apply, and adding those to which (9) does, we get as the correct formula

$$\left.\begin{aligned} n_{\delta\mu} &= \sum_{i,j,l,m} n_{il} n_{jm} - 2^{\frac{1}{2}(\mu-1)} n_{\frac{1}{2}\delta, \frac{1}{2}(\mu-1)} + 2^{\frac{1}{2}\mu} n_{\frac{1}{2}\delta, \frac{1}{2}\mu} \\ \text{where} \quad & i + j = \delta, \quad l + m = \mu - 1, \quad \delta \neq 1. \\ \text{Also} \quad & n_{10} = 1. \end{aligned}\right\} \qquad (30)$$

Putting $n_{\delta\mu} = 2^{\mu} c_{\delta\mu}$, and removing the factor $2^{\mu}$,

$$\left.\begin{aligned} c_{\delta\mu} &= \tfrac{1}{2}\Big(\sum_{i,j,l,m} c_{il} c_{jm} - c_{\frac{1}{2}\delta, \frac{1}{2}(\mu-1)}\Big) + c_{\frac{1}{2}\delta, \frac{1}{2}\mu} \\ \text{where} \quad & i + j = \delta, \quad l + m = \mu - 1, \quad \delta \neq 1. \\ \text{Also} \quad & c_{10} = 1. \end{aligned}\right\} \qquad (31)$$

Now let

$$f(x, y) = \sum_{\delta, \mu} c_{\delta\mu} x^\delta y^\mu. \qquad . \quad . \quad . \quad . \quad . \quad (32)$$

Substituting (31) in (32), we obtain the functional equation

$$f(x, y) = x + \tfrac{1}{2}y\{f(x, y)\}^2 + (1 - \tfrac{1}{2}y)f(x^2, y^2). \qquad . \quad . \quad (33)$$

Now, from the definitions of $a_\delta$, $b_\delta$, $c_{\delta\mu}$, $n_{\delta\mu}$,

$$a_\delta = \sum_\mu n_{\delta\mu} = \sum_\mu 2^\mu c_{\delta\mu}, \qquad b_\delta = \sum_\mu c_{\delta\mu}.$$

Consequently, comparing (22), (23), (32),

$$F(x) = f(x, 2), \qquad 1 + f(x) = f(x, 1). \qquad . \quad . \quad . \quad (34)$$

It is readily verified that on putting $y=2$ the equation (33) reduces to (24); and that on putting $y=1$ it reduces to (25), as it should.

If on the right of (33) we substitute the first approximation

$$f(x, y) = x + \ . \ . \ ., \qquad f(x^2, y^2) = x^2 + \ . \ . \ .,$$

we obtain the second approximation

$$f(x, y) = x + x^2 + \ . \ . \ .$$

Similarly the third approximation is

$$\begin{aligned} f(x, y) &= x + \tfrac{1}{2}y(x^2 + 2x^3 + x^4 + \ . \ . \ .) + (1 - \tfrac{1}{2}y)(x^2 + x^4 + \ . \ . \ .) \\ &= x + x^2 + x^4 + x^3y + \ . \ . \ .; \end{aligned}$$

and the process may be repeated to any required extent.

Alternatively, we may proceed in either of the following ways. Write

$$f(x, y) = xf_1(y) + x^2f_2(y) + \ . \ . \ . + x^\delta f_\delta(y) + \ . \ . \ . \qquad . \quad . \quad (35)$$

or

$$f(x, y) = g_0(x) + yg_1(x) + y^2g_2(x) + \ . \ . \ . + y^\mu g_\mu(x) + \ . \ . \ .; \qquad . \quad (36)$$

substitute in the functional equation (33), and equate coefficients. We obtain

$$\left.\begin{aligned} &f_1 = f_2 = 1, \quad f_3 = y, \quad f_4 = 1 + y^2, \\ &f_5 = y + y^2 + y^3, \quad f_6 = y + 2y^2 + 2y^3 + y^4, \ . \ . \ ., \\ &f_{2\delta-1} = y(f_1f_{2\delta-2} + f_2f_{2\delta-3} + \ . \ . \ . + f_{\delta-1}f_\delta), \\ &f_{2\delta} = y(f_1f_{2\delta-1} + f_2f_{2\delta-2} + \ . \ . \ . + f_{\delta-1}f_{\delta+1} + \tfrac{1}{2}f_\delta^2) + (1 - \tfrac{1}{2}y)f_\delta(y^2); \end{aligned}\right\} \quad . \quad (37)$$

$$\left.\begin{aligned} &g_0 = x(1-x)^{-1}, \quad g_1 = x^3(1-x)^{-1}(1-x^2)^{-1}, \\ &g_2 = x^4(1 + x + 2x^2)(1-x)^{-1}(1-x^2)^{-1}(1-x^4)^{-1}, \\ &g_3 = x^5(1 + x + 3x^2)(1-x)^{-2}(1-x^2)^{-1}(1-x^4)^{-1}, \ . \ . \ ., \\ &g_{2\mu-1} = g_0g_{2\mu-2} + g_1g_{2\mu-3} + \ . \ . \ . + g_{\mu-2}g_\mu + \tfrac{1}{2}g_{\mu-1}^2 - \tfrac{1}{2}g_{\mu-1}(x^2), \\ &g_{2\mu} = g_0g_{2\mu-1} + g_1g_{2\mu-2} + \ . \ . \ . + g_{\mu-1}g_\mu + g_\mu(x^2). \end{aligned}\right\} \quad . \quad (38)$$

It will be observed that

$$f_\delta(2) = a_\delta, \qquad f_\delta(1) = b_\delta. \qquad . \quad . \quad . \quad . \quad . \quad (39)$$

The first of these two methods is perhaps the quickest way of calculating many terms of the expansion of $f(x, y)$. By means of the second, we could find explicit formulæ for $c_{\delta 0}, c_{\delta 1}, c_{\delta 2}, c_{\delta 3}, \ldots$

With regard to the convergence of the various generating series, it may be observed that (22), since it is the expansion of (26), is absolutely convergent if $|x| < \frac{1}{4}$. Since $b_\delta \leqslant a_\delta$, it follows that (23) also converges absolutely if $|x| < \frac{1}{4}$; and since $f(x, 2) = \mathrm{F}(x)$, it follows that the double series (32) converges absolutely if $|x| < \frac{1}{4}$, $|y| \leqslant 2$.

SUMMARY.

Non-associative combinations are classified and enumerated with the aid of a representation involving non-associative arithmetic.

---

## REFERENCES TO LITERATURE.

BINET, M. J., 1839. "Réflexions sur le problème de déterminer le nombre de manières dont une figure rectiligne peut être partagée en triangles au moyens de ses diagonales," *Journ. de Math.* (1), vol. iv, pp. 79–90.

CATALAN, E., 1838. "Note sur une équation aux différences finies," *Journ. de Math.*, (1) vol. iii, pp. 508–516. Also 1839, vol. iv, pp. 91–94, 95–99; 1841, vol. vi, p. 74.

CAYLEY, A., 1857, 1859. "On the theory of the analytical forms called trees," *Phil. Mag.*, vol. xiii, pp. 172–176; vol. xviii, pp. 374–378. Also 1875, *Rep. Brit. Assoc. Adv. Sci.*, pp. 257–305; 1881, *Amer. Journ. Math.*, vol. iv, pp. 266–268; 1889, *Quart. Journ. Pure App. Math.*, vol. xxiii, pp. 376–378. (*Collected Math. Papers*, vol. iii, no. 203; iv, 247; ix, 610; xi, 772; xiii, 895.)

ETHERINGTON, I. M. H., 1937. "Non-associate powers and a functional equation," *Math. Gaz.*, vol. xxi, pp. 36–39, 153.

NETTO, E., 1901. *Lehrbuch der Combinatorik*, Leipzig.

RODRIGUES, O., 1838. "Sur le nombre de manières d'effectuer un produit de $n$ facteurs," *Journ. de Math.* (1), vol. iii, p. 549.

SCHRÖDER, E., 1870. "Vier combinatorische Probleme," *Zeits. Math.*, vol. xv, pp. 361–376.

WEDDERBURN, J. H. M., 1922. "The functional equation $g(x^2) = 2ax + [g(x)]^2$," *Ann. Math.* (2), vol. xxiv, pp. 121–140.

(*Issued separately June* 27, 1939.)

# 7



Reprinted from *Proc. Roy. Soc. Edinburgh,* **59**, 242–258 (1939)

## XXIII.—**Genetic Algebras.** By **I. M. H. Etherington,** B.A.(Oxon), Ph.D.(Edin.), Mathematical Institute, University of Edinburgh.

(MS. received May 8, 1939. Revised MS. received September 5, 1939. Read July 3, 1939.)

CONTENTS.



### § 1. Introduction.

Two classes of linear algebras, generally non-associative, are defined in § 3 (*baric algebras*) and § 4 (*train algebras*), and the process of *duplication* of a linear algebra in § 5. These concepts, which will be discussed more fully elsewhere, arise naturally in the symbolism of genetics, as shown in §§ 6–15. Many of their properties express facts well known in genetics; and the processes of calculation which are fundamental in many problems of population genetics can be expressed as manipulations in the genetic algebras. In cases where inheritance is of a simple type (*e.g.* §§ 10–13, 15) this constitutes a new point of view, but perhaps amounts to little more than a change of notation as compared with existing methods. § 14, however, indicates the possibility of generalisations which would seem to be impossible by ordinary methods.

The occurrence of the genetic algebras may be described in general terms as follows. The mechanism of chromosome inheritance, in so far as it determines the probability distributions of genetic types in families and filial generations, and expresses itself through their frequency distributions, may be represented conveniently by algebraic symbols. Such a symbolism is described, for instance, by Jennings (1935, chap. ix);

many applications are given by Geppert and Koller (1938). It is shown in the present paper that the symbolism is equivalent to the use of a system of related linear algebras, in which multiplication (equivalent to the procedure of "chessboard diagrams") is commutative ($PQ = QP$) but non-associative ($PQ . R \neq P . QR$). A population (*i.e.* a distribution of genetic types) is represented by a normalised hypercomplex number in one or other algebra, according to the point of view from which it is specified. If P, Q are populations, the filial generation $P \times Q$ (*i.e.* the statistical population of offspring resulting from the random mating of individuals of P with individuals of Q) is obtained by multiplying two corresponding representations of P and Q; and from this requirement of the symbolism it will be obvious why multiplication must be non-associative. It must be understood that a population may mean a single individual, or rather the information which we may have concerning him in the form of a probability distribution.

Inheritance will be called *symmetrical* if the sex of a parent does not affect the distribution of gametic types produced. Paying attention only to the inheritance of gene differences (not of phenotypes), every regular mode of symmetrical inheritance in theoretical genetics has its fundamental *gametic algebra*, from which other algebras (*zygotic*, etc.) are deduced by duplication. From the nature of the symbolism these are of necessity baric algebras; but it appears on closer examination that they belong in all cases to the narrower category of train algebras.

(The fundamental algebras can be modified to take account of various kinds of selection. They are then no longer train algebras, although the baric property and the relation of duplication sometimes persist.)

Symmetry of inheritance may be disturbed by unequal crossing over in male and female, by sex linkage, or by gametic selection. These cases are not discussed at all in the present paper; but it may be stated briefly that in the absence of selection the corresponding genetic algebras (of order $n$, say) possess train subalgebras (of order $n-1$).

The occurrence of a non-associative linear algebra in the simplest case of Mendelian inheritance was pointed out by Glivenko (1936).

## § 2. NOTATION.

By *principal powers* in a non-associative algebra, I mean powers in which the factors are absorbed one at a time always on the right or always on the left (see (3.6)). Otherwise, for the notation and nomenclature for non-associative products and powers, see my paper "On Non-Associative Combinations" (1939). The word *pedigree* which occurs there can now be interpreted almost in its ordinary biological sense.

Elements of a linear algebra (*i.e.* hypercomplex numbers) will be called *elements* and denoted by Latin letters, generally small ($a$, $b$, . . .); but normalised elements, *i.e.* elements of unit weight (§ 3), will be denoted by Latin capitals (A, B, . . .). The letters $m$, $n$, $r$, however, denote positive integers.

Elements of the field $\mathsf{F}$ over which a linear algebra is defined will be called *numbers* and denoted by small Greek letters ($\alpha$, $\beta$, . . .). Thus, an *element* is determined by its coefficients, which are *numbers*. In the genetical applications, $\mathsf{F}$ may be taken as the field of real numbers. The enumerating indices (subscripts and superscripts) take positive integer values, either 1 to $m$, 1 to $n$, or 1 to $r$, according to the context.

Block capitals ($\mathsf{A}$, $\mathsf{B}$, . . .) denote algebras.

The symbol $\Sigma$ indicates summation with respect to repeated indices, *e.g.* with respect to $\sigma$ in (3.3), with respect to $\sigma$ and $\tau$ in (5.3).

The symbol $1^\mu$ stands for a set of 1's. Thus the formula (6.3) means the same as

$$\sum_{\sigma=1}^{n} \gamma_\sigma^{\mu\nu} = 1.$$

The advantage of this notation is that such formulæ retain their form under linear transformations of the basis of a genetic algebra, $1^\mu$ being replaced by the vector $\xi^\mu$ (*cf.* (6.12)).

## § 3. Baric Algebras.

It is well known that a linear associative algebra possesses a matrix representation. Non-associative algebras in general do not, but may. The simplest such representation would be a scalar representation on the field $\mathsf{F}$ over which the algebra is defined. A linear algebra $\mathsf{X}$, associative or not, which possesses a non-trivial representation of this kind, will be called *baric*.

The definition means that to any element $x$ of $\mathsf{X}$ there corresponds a number $\xi(x)$ of $\mathsf{F}$, not identically zero, such that

$$\xi(x+y)=\xi(x)+\xi(y), \quad \xi(\alpha x)=\alpha\xi(x), \quad \xi(xy)=\xi(x)\xi(y). \quad (x, y<\mathsf{X}, \alpha<\mathsf{F}) \qquad (3.1)$$

$\xi(x)$ will be called the *weight* of $x$, or the *weight function* of $\mathsf{X}$. If $\xi(x)\neq 0$, $x$ can be *normalised*—that is, replaced by the element

$$\mathrm{X}=x/\xi(x) \quad . \quad . \quad . \quad . \quad . \qquad (3.2)$$

of unit weight. Elements of zero weight will be called *nil elements*. The set $\mathsf{U}$ of all nil elements is evidently an invariant subalgebra of $\mathsf{X}$; *i.e.* $\mathsf{XU} \leqslant \mathsf{U}$: it will be called the *nil subalgebra*.

Let the multiplication table of a linear algebra **X** be

$$a^{\mu}a^{\nu} = \Sigma\gamma_{\sigma}^{\mu\nu}a^{\sigma}, \qquad (\mu, \nu, \sigma = 1, \ldots, n) \quad . \quad . \quad . \quad (3.3)$$

and let the general element be denoted

$$x = \Sigma\alpha_{\mu}a^{\mu}. \quad . \quad . \quad . \quad . \quad . \quad . \quad (3.4)$$

*For* **X** *to be a baric algebra, it is necessary and sufficient that the equations* (3.3), *regarded as ordinary simultaneous equations in* **F** *for the unknowns* $a^{\mu}$, *should possess a non-null solution* $a^{\mu} = \xi^{\mu}$. For this is obviously necessary, the $\xi^{\mu}$ being the weights of the basic elements $a^{\mu}$. Conversely, if the condition is satisfied and we take

$$\xi(x) = \Sigma\alpha_{\mu}\xi^{\mu}, \quad . \quad . \quad . \quad . \quad . \quad . \quad (3.5)$$

then (3.1) are at once deducible. The basic weights $\xi^{\mu}$ form the *weight vector* of **X**. In the genetical applications, $\xi^{\mu} = 1^{\mu}$.

Let the right rank equation (Dickson, 1914, § 19), or equation of lowest degree connecting the right principal powers,

$$x, x^2, x^3, \ldots, \qquad x^m = x^{m-1}x, \ldots, \quad . \quad . \quad . \quad . \quad (3.6)$$

be

$$f(x) \equiv x^r + \theta_1 x^{r-1} + \theta_2 x^{r-2} + \ldots + \theta_{r-1}x = 0, \quad . \quad . \quad (3.7)$$

where each coefficient $\theta_m$ is a homogeneous polynomial of degree $m$ in the co-ordinates $\alpha_{\mu}$ of $x$. Then $f(x)$, being zero, is of zero weight. Hence the equation is satisfied when we substitute $\xi(x)$ for $x$; consequently $x - \xi(x)$ must be a factor of $f(x)$. The same is true for the left rank equation. Thus

$\xi(x)$ *is a root of the right and left rank equations.* . . (3.8)

The weight function of an algebra is not necessarily unique. In fact, a commutative associative linear algebra for which the determinant $|\Sigma\gamma_{\sigma}^{\mu\nu}\gamma_{\tau}^{\sigma\tau}|$ does not vanish has $n$ independent weight functions; and its rank equation is hence completely determined by (3.8) (Dickson, 1914, § 55, and the references given there).

## § 4. Train Algebras.

A baric algebra with the weight function $\xi(x)$ and right rank equation (3.7) will be called a *right train algebra* if the coefficients $\theta_m$, in so far as they depend on the element $x$, depend only on $\xi(x)$. A *left train algebra* is defined similarly. For simplicity, suppose multiplication commutative, so that we may drop "left" and "right."

Since $\theta_m$ is homogeneous of degree $m$ in the co-ordinates of $x$, it must in a train algebra be a numerical multiple of $\xi(x)^m$. Hence (if the field **F** be sufficiently extended, *e.g.*, to include complex numbers) the rank equation can be factorised:

$$f(x) \equiv x(x-\xi)(x-\lambda_1\xi)(x-\lambda_2\xi) \ldots = 0. \qquad (4.1)$$

(It is implied that when the left side is expanded, powers of $x$ are interpreted as principal powers.) The numbers 1, $\lambda_1$, $\lambda_2$ . . . are the *principal train roots* of the algebra.

For a normalised element (3.7) becomes

$$f(\mathrm{X}) \equiv \mathrm{X}^r + \theta_1 \mathrm{X}^{r-1} + \theta_2 \mathrm{X}^{r-2} + \ldots + \theta_{r-1}\mathrm{X} = 0, \qquad (4.2)$$

where now the $\theta$'s are constant (*i.e.* independent of X); and (4.1) becomes

$$f(\mathrm{X}) \equiv \mathrm{X}(\mathrm{X}-1)(\mathrm{X}-\lambda_1)(\mathrm{X}-\lambda_2) \ldots = 0. \qquad (4.3)$$

Since (4.2) can be multiplied by X any number of times, it can be regarded as a linear recurrence equation with constant coefficients connecting the principal powers of the general normalised element X. Solving the recurrence relation for $\mathrm{X}^m (m \geqslant r)$ in the usual way, we obtain 1, $\lambda_1$, $\lambda_2$ . . . as the roots of the auxiliary equation; hence a formula for $\mathrm{X}^m$ can be written down in terms of X, $\mathrm{X}^2$, . . . $\mathrm{X}^{r-1}$. Hence also for the general non-nil element $x = \xi\mathrm{X}$, the value of $x^m = \xi^m \mathrm{X}^m$ is known; while for a nil element $u$, $u^m = 0 (m \geqslant r)$.

The properties of train algebras will be studied elsewhere, and the following theorem proved:—

*If* (1) **X** *is a baric algebra*; (2) *its nil subalgebra* **U** *is nilpotent* (Wedderburn, 1908*a*, p. 111); (3) *for* $m = 1, 2, 3, \ldots$, *the subalgebra* $\mathbf{U}^{(m)}$, *consisting of all products of altitude m* (Etherington, 1939, p. 156) *formed from nil elements is an invariant subalgebra of* **X** (*as it necessarily is of* **U**); *then* **X** *is a train algebra.*

For train algebras of rank $r = 2$ or 3, provided that the principal train roots do not include $\frac{1}{2}$, the conditions are necessary as well as sufficient; but I cannot say whether this converse holds more generally or not. I will call **X** a *special train algebra* if it satisfies the conditions (1), (2), (3). In such algebras it can be shown that there are many other sequences which have properties like those of the sequence of principal powers; *i.e.* sequences of elements derived from the general element, which satisfy linear recurrence equations whose coefficients, being functions of the weight only, become constants on normalisation. Such sequences will be called *trains*. For example, the sequence of plenary powers

$$x,\ x^2,\ x^{2\cdot 2},\ x^{2^3}, \ldots, \qquad (4.4)$$

and the sequence of primary products

$$x,\ \mathrm{Y}x,\ \mathrm{Y}\,.\,\mathrm{Y}x,\ \mathrm{Y} : \mathrm{Y}\,.\,\mathrm{Y}x, \ldots, \qquad (4.5)$$

form trains in a special train algebra.

It is convenient to denote the $m$th element of a train as $x^{[m]}$, and to regard it as a symbolic $m$th power of $x$. Let the normalised recurrence equation, or *train equation*, be

$$g[X] \equiv X^{[s]} + \phi_1 X^{[s-1]} + \phi_2 X^{[s-2]} + \ldots + \phi_{s-1} X = 0, \quad . \quad . \quad (4.6)$$

where the $\phi$'s are numerical constants. It is implied that the equation may be symbolically "multiplied all through" by X any number of times. It may also be symbolically factorised:

$$g[X] \equiv X[X-1][X-\mu_1][X-\mu_2] \ldots = 0. \quad . \quad . \quad (4.7)$$

The square brackets indicate that after expansion powers of X are to be interpreted as symbolic powers. The expansion being performed as in ordinary algebra, multiplication of the symbolic factors is commutative and associative. Extra factors may be introduced without destroying the validity of the train equation; but assuming that all superfluous factors have been removed, $s$ is the *rank* of the train, and the numbers $1, \mu_1, \mu_2, \ldots$ are the *train roots*, by means of which a formula for $X^{[m]}$ $(m \geqslant s)$ can be written down.

In the applications to genetics, it will be found that all the fundamental symmetrical genetic algebras are special train algebras. Various trains have genetical significance; the $X^{[m]}$ represent successive discrete generations of an evolving population or breeding experiment, and the train equation is the recurrence equation which connects them.

Thus, for example, plenary powers (4.4) refer to a population with random mating; principal powers (3.6) to a mating system in which each generation is mated back to one original ancestor or ancestral population; and the primary products (4.5) to the descendants of a single individual or subpopulation X mating at random within a population Y. Other mating systems are described by other sequences, and in various well-known cases these have the train property—that is, the determination of the $m$th generation depends ultimately on a linear recurrence equation with constant coefficients. It usually happens that the train roots are real, distinct, and not exceeding unity. Hence it may be shown that $X^{[m]}$ tends to equilibrium with increasing $m$; the rate of approach to equilibrium is ultimately that of a geometrical progression with common ratio equal to the largest train root excluding unity; but it may be some generations (depending on the number of train roots) before this rate of approach is manifest.

Train roots may be described as the eigen-values of the operation of symbolic multiplication by X, or in genetic language, the operation of passing from one generation to the next.

Train algebras of (principal) rank 3, which occur in several contexts in genetics, have certain special properties. For example, *if the train equation for principal powers is* $X(X-1)(X-\lambda)=0$, *then the train equation for plenary powers is* $X[X-1][X-2\lambda]=0$; *and* vice versa. Examples may be seen below in (10.12), (12.4, 5), (15.3), where respectively $\lambda=0$, $\frac{1}{2}(1-\omega)$, $\frac{1}{6}$.

## § 5. Duplication.

Let

$$a^\mu a^\nu = \Sigma \gamma_\sigma^{\mu\nu} a^\sigma \qquad (5.1)$$

be the multiplication table of a linear algebra **X** with basis $a^\mu$ ($\mu=1, \ldots, n$). Then

$$a^\mu a^\nu \,.\, a^\theta a^\phi = \Sigma \gamma_\sigma^{\mu\nu} a^\sigma \,.\, \Sigma \gamma_\tau^{\theta\phi} a^\tau.$$

Writing

$$a^\mu a^\nu = a^{\mu\nu}, \qquad (5.2)$$

this becomes

$$a^{\mu\nu} a^{\theta\phi} = \Sigma \gamma_\sigma^{\mu\nu} \gamma_\tau^{\theta\phi} a^{\sigma\tau}, \qquad (5.3)$$

which may be regarded as the multiplication table of another linear algebra, isomorphic with the totality of quadratic forms in the original algebra. It will be called the *duplicate* of **X**, and denoted **X'**. It is commutative and of order $\frac{1}{2}n(n+1)$ if **X** is commutative; non-commutative and of order $n^2$ if **X** is non-commutative. It is generally non-associative, even if **X** is associative. It is not to be confused with what may be called the *direct square* of **X**, or direct product of two algebras isomorphic with **X**: this would be an algebra of order $n^2$, having the multiplication table

$$a^{\mu\nu} a^{\theta\phi} = \Sigma \gamma_\sigma^{\mu\theta} \gamma_\tau^{\nu\phi} a^{\sigma\tau}, \qquad (5.4)$$

differing from (5.3) in the arrangement of indices.

Some theorems on duplication will be proved elsewhere. It will be shown that the duplicates (i) of a linear transform of an algebra, (ii) of the direct product of two algebras, (iii) of a baric algebra with weight vector $\xi^\mu$, (iv) of a train algebra with principal train roots $1, \lambda, \mu, \ldots$, are respectively (i) a linear transform of the duplicate algebra, (ii) the direct product of the duplicates, (iii) a baric algebra with weight vector $\xi^\mu \xi^\nu$, (iv) a train algebra with principal train roots $1, 0, \lambda, \mu, \ldots$ These theorems are relevant as follows: (iii) in view of §§ 7, 8; (ii) in view of § 9; (i) in connection with the method used in § 14; (iv) in deriving equations such as (10.10), (12.6).

Duplication of an algebra may be compared with the process of forming the second induced matrix of a given matrix (Aitken, 1935; *cf.* also Wedderburn, 1908 *b*).

## § 6. Gametic Algebras.

Consider the inheritance of characters depending on any number of gene differences at any number of loci on any number of chromosomes in a diploid or generally autopolyploid species. Assume that inheritance is symmetrical in the sexes: the sex chromosomes are thus excluded, and crossing over if present must be equal in male and female.

Let $G^1, G^2, \ldots, G^n$ denote the set of gametic types determined by these gene differences. Then there will be

$$m = \tfrac{1}{2}n(n+1) \qquad . \quad . \quad . \quad . \quad . \quad . \quad (6.1)$$

zygotic types $G^\mu G^\nu (= G^\nu G^\mu)$. The formulæ giving the series of gametic types produced by each type of individual, and their relative frequencies, may be written

$$G^\mu G^\nu = \Sigma \gamma_\sigma^{\mu\nu} G^\sigma, \qquad . \quad . \quad . \quad . \quad . \quad . \quad (6.2)$$

with the normalising conditions

$$\Sigma \gamma_\sigma^{\mu\nu} 1^\sigma = 1; \qquad . \quad . \quad . \quad . \quad . \quad . \quad (6.3)$$

$\gamma_\sigma^{\mu\nu}$ is then the probability that an arbitrary gamete produced by an individual of zygotic type $G^\mu G^\nu$ is of type $G^\sigma$.

(I speak of *zygotic types*—individuals distinguished by the gametes from which they were formed—rather than *genotypes*—individuals distinguished by the gametes which they produce—because the $G^\mu G^\nu$ are not all distinct genotypes if more than one chromosome is involved: the zygotic algebra, § 7, will have the same train equation if genotypes are used, but will then not be a duplicate algebra.)

A population P which produces gametes $G^\mu$ in proportions $a_\mu$ may be represented by writing

$$P = \Sigma a_\mu G^\mu. \qquad . \quad . \quad . \quad . \quad . \quad . \quad (6.4)$$

Imposing the normalising condition

$$\Sigma a_\mu 1^\mu = 1, \qquad . \quad . \quad . \quad . \quad . \quad . \quad (6.5)$$

$a_\mu$ denotes the probability that an arbitrary gamete produced by an arbitrary individual of P is of type $G^\mu$.

A population may also be described by the proportions of the zygotic types $G^\mu G^\nu$ which it contains; thus we may write

$$P = \Sigma a_{\mu\nu} G^\mu G^\nu, \qquad . \quad . \quad . \quad . \quad . \quad . \quad (6.6)$$

with the normalising condition

$$\Sigma a_{\mu\nu} 1^\mu 1^\nu = 1, \qquad . \quad . \quad . \quad . \quad . \quad . \quad (6.7)$$

and a similar probability interpretation. We may suppose without loss of generality that $a_{\mu\nu} = a_{\nu\mu}$, so that in (6.6) the coefficient of $G^\mu G^\nu$ is

$2\alpha_{\mu\nu}$ if $\mu \neq \nu$. The two representations are connected by the gametic series formulæ (6.2); that is to say, from the zygotic representation (6.6) follows the gametic representation

$$P = \Sigma \alpha_{\mu\nu} \gamma_{\sigma}^{\mu\nu} G^{\sigma}. \quad \quad (6.8)$$

If two populations P, Q intermate at random, representations of the first filial generation are obtained by multiplying the gametic representations of P and Q; *i.e.* if

$$P = \Sigma \alpha_{\mu} G^{\mu}, \qquad Q = \Sigma \beta_{\mu} G^{\mu}, \quad \quad (6.9)$$

the population of offspring is

$$PQ = \Sigma \alpha_{\mu} \beta_{\nu} G^{\mu} G^{\nu} \quad \quad (6.10)$$

$$= \Sigma \alpha_{\mu} \beta_{\nu} \gamma_{\sigma}^{\mu\nu} G^{\sigma}. \quad \quad (6.11)$$

In particular, the population of offspring of random mating of P within itself is given by $P^2$.

We may now view the situation abstractly. The gametic series (6.2) form the multiplication table of a commutative non-associative linear algebra with basis $G^{\mu} (\mu = 1, \ldots, n)$. It will be called the *gametic algebra* for the type of inheritance considered, and denoted **G**. The equations (6.3) show that **G** is a baric algebra with weight vector

$$\xi^{\mu} = 1^{\mu}. \quad \quad (6.12)$$

With regard to its gametic type frequencies, a population is represented by a normalised element (6.4) of **G**. Multiplication in **G** has the significance described in § 1, and it follows from the multiplicative property of the weight in a baric algebra that PQ will be automatically normalised if P and Q are.

### § 7. Zygotic Algebras.

When individuals of types $G^{\mu}G^{\nu}$, $G^{\theta}G^{\phi}$ mate, the probability distribution of zygotic types in their offspring can be obtained by multiplying the gametic representations (given by (6.2)) together, and leaving the product in quadratic form (as in (6.10)). We obtain

$$G^{\mu}G^{\nu} \,.\, G^{\theta}G^{\phi} = \Sigma \gamma_{\sigma}^{\mu\nu} \gamma_{\tau}^{\theta\phi} G^{\sigma} G^{\tau};$$

or, writing

$$Z^{\mu\nu} = G^{\mu} G^{\nu} \quad \quad (7.1)$$

to emphasise the union of paired gametes into single individuals,

$$Z^{\mu\nu} Z^{\theta\phi} = \Sigma \gamma_{\sigma}^{\mu\nu} \gamma_{\tau}^{\theta\phi} Z^{\sigma\tau}. \quad \quad (7.2)$$

These $\frac{1}{2}m(m+1)$ equations, then, are the formulæ giving the series of zygotic types produced by the mating type or *couple* $Z^{\mu\nu} \times Z^{\theta\phi}$, the probability of $Z^{\sigma\tau}$ being the corresponding coefficient $\gamma_\sigma^{\mu\nu}\gamma_\tau^{\theta\phi}+\gamma_\tau^{\mu\nu}\gamma_\sigma^{\theta\phi}$ (if $\sigma\neq\tau$) or $\gamma_\sigma^{\mu\nu}\gamma_\tau^{\theta\phi}$ (if $\sigma=\tau$).

The linear algebra with basis $Z^{\mu\nu}$ and multiplication table (7.2) will be called the *zygotic algebra* for the type of inheritance considered. It is a baric algebra with weight vector $1^\mu 1^\nu$, the duplicate of the gametic algebra **G**, and will be denoted

$$\mathbf{Z}=\mathbf{G}'. \qquad (7.3)$$

A population, regarded as a distribution of zygotic types, is represented by a normalised element

$$P=\Sigma\alpha_{\mu\nu}Z^{\mu\nu}, \quad \text{where} \quad \Sigma\alpha_{\mu\nu}1^\mu 1^\nu=1;$$

and multiplication in **Z**, as in **G**, has the significance described in § 1. A product left in quadratic form in the Z's gives now the probability distribution of couples $Z^{\mu\nu}Z^{\theta\phi}$ among the parents; or, as I shall call it, the *copular representation* of the population of offspring.

## § 8. Further Duplicate Genetic Algebras.

The process of duplication can be applied repeatedly. Thus the $\frac{1}{2}m(m+1)$ types of paired zygotes, or couples,

$$K^{\mu\nu\cdot\theta\phi}=Z^{\mu\nu}Z^{\theta\phi}, \qquad (8.1)$$

can be taken as the basis of a new linear algebra

$$\mathbf{K}=\mathbf{Z}'=\mathbf{G}''. \qquad (8.2)$$

Call it the *copular algebra.* A normalised element with positive coefficients

$$P=\Sigma\alpha_{\mu\nu\cdot\theta\phi}K^{\mu\nu\cdot\theta\phi}, \quad \text{where} \quad \Sigma\alpha_{\mu\nu\cdot\theta\phi}1^\mu 1^\nu 1^\theta 1^\phi=1,$$

is the copular representation of a population—the probability distribution of couples in the parents of the individuals comprised in the population.

Similarly, in the next duplicate algebra **K'**, the basic symbols would classify tetrads of grandparents.

In all these algebras, multiplication has the significance described in § 1.

## § 9. Combination of Genetic Algebras.

Consider two distinct genetic classifications referring to the same population P, firstly into a set of $m$ genetic types

$$A^1, A^2, \ldots, A^m;$$

secondly into a set of $n$ genetic types

$$B^1, B^2, \ldots, B^n$$

of the same kind (gametic, zygotic, etc.). Let the corresponding genetic algebras be **A**, **B** with multiplication tables

$$A^{\mu}A^{\nu}=\Sigma\gamma_{\sigma}^{\mu\nu}A^{\sigma}, \qquad B^{\theta}B^{\phi}=\Sigma\delta_{\tau}^{\theta\phi}B^{\tau}.$$

By taking account of both classifications at once, we obtain a third classification which may be called their *product*, into *mn* genetic types

$$C^{\mu\theta}=A^{\mu}B^{\theta}.$$

The type $C^{\mu\theta}$ comprises all individuals (gametes, zygotes, etc.) who are of type $A^{\mu}$ in the first classification, $B^{\theta}$ in the second.

If the characters of the two classifications are inherited independently, *i.e.* if they involve two quite distinct sets of chromosomes, then the probabilities $\gamma_{\sigma}^{\mu\nu}$, $\delta_{\tau}^{\theta\phi}$ refer to independent events. Hence the genetic algebra with basis $C^{\mu\theta}$ is the direct product

$$\mathbf{C}=\mathbf{AB};$$

*i.e.* its multiplication table is

$$C^{\mu\theta}C^{\nu\phi}=\Sigma\gamma_{\sigma}^{\mu\nu}\delta_{\tau}^{\theta\phi}C^{\sigma\tau}.$$

It follows that a genetic algebra which depends on several autosomal linkage groups must be a direct product **ABC** . . . of genetic algebras, one factor algebra for each linkage group.

If, however, the A and B classifications are independent but genetically linked, *i.e.* if they involve two quite distinct sets of gene loci but not distinct sets of chromosomes, then the probabilities $\gamma_{\sigma}^{\mu\nu}$, $\delta_{\tau}^{\theta\phi}$ are not independent. Regarded as a linear set, **C** is still the product of the linear sets **A** and **B**; but the algebra **C** will not be the direct product of the algebras **A** and **B** (except in the very exceptional case when all crossing over values between A and B are precisely 50 per cent.). It is, however, still the case that **C** contains subalgebras isomorphic with **A** and **B**. For example, if these algebras are gametic, and if we keep the first index of $C^{\mu\theta}$ constant, we are virtually disregarding all the A-loci, so we obtain a subalgebra isomorphic with **B**; and this can be done in *m* ways.

Hence a genetic algebra based on the allelomorphs of several autosomal loci possesses numerous automorphisms.

It will be shown in § 14 that even when linkage is involved the gametic algebra can be symbolically factorised, and regarded as a symbolic direct product of non-commutative factor algebras, one for each locus (see (14.12)).

## §§ 10–15. Examples of Symmetrical Genetic Algebras.

A more detailed description of practical applications will be given elsewhere. My object here is simply to show that the genetic algebras are train algebras. I give in each case the principal and plenary train equations, *i.e.* the identities of lowest degree connecting respectively the sequences of principal and plenary powers of a normalised element. As explained in § 4, these are really recurrence equations, and have a special significance in genetics.

## § 10. Simple Mendelian Inheritance.

For a single autosomal gene difference (D, R), the gametic multiplication table is

$$DD = D, \qquad DR = \tfrac{1}{2}D + \tfrac{1}{2}R, \qquad RR = R. \qquad (10.1)$$

Writing

$$A = DD, \qquad B = DR, \qquad C = RR, \qquad (10.2)$$

we find, *e.g.*,

$$B^2 = (\tfrac{1}{2}D + \tfrac{1}{2}R)^2 = \tfrac{1}{4}A + \tfrac{1}{2}B + \tfrac{1}{4}C.$$

Hence and similarly the zygotic multiplication table is

$$A^2 = A, \qquad B^2 = \tfrac{1}{4}A + \tfrac{1}{2}B + \tfrac{1}{4}C, \qquad C^2 = C,$$
$$BC = \tfrac{1}{2}B + \tfrac{1}{2}C, \qquad CA = B, \qquad AB = \tfrac{1}{2}A + \tfrac{1}{2}B. \qquad (10.3)$$

Call these two algebras $\mathbf{G}_2$, $\mathbf{Z}_2$ ($\mathbf{Z}_2 = \mathbf{G}_2'$), and denote their general elements

$\mathbf{G}_2$: $$x = \delta D + \rho R, \qquad (10.4)$$

$\mathbf{Z}_2$: $$x = \alpha A + 2\beta B + \gamma C. \qquad (10.5)$$

The principal rank equations are

$\mathbf{G}_2$: $$x^2 - (\delta + \rho)x = 0, \qquad (10.6)$$

$\mathbf{Z}_2$: $$x^3 - (\alpha + 2\beta + \gamma)x^2 = 0; \qquad (10.7)$$

and the plenary rank equations (or identities of lowest degree connecting plenary powers of the general elements) are (10.6) and

$\mathbf{Z}_2$: $$x^{2\cdot 2} - (\alpha + 2\beta + \gamma)^2 x^2 = 0. \qquad (10.8)$$

A population P is represented by an element of unit weight in either algebra, *i.e.* (10.4) or (10.5) with

$$\delta + \rho = 1, \qquad \alpha + 2\beta + \gamma = 1,$$

the ratios $\delta : \rho$, $\alpha : 2\beta : \gamma$ giving the relative frequencies of the gametic types which it produces or genotypes which it contains. In this case (10.6), (10.7), (10.8) become the train equations

$\mathbf{G}_2$: $$P^2 = P, \qquad (10.9)$$

$\mathbf{Z}_2$: $$P^3 = P^2, \qquad P^{2\cdot 2} = P^2, \qquad (10.10)$$

expressing facts well known in genetics. It is convenient to write these equations in the form (*cf.* 4.7)

$\mathbf{G}_2$: $$P(P-1)=0, \qquad (10.11)$$

$\mathbf{Z}_2$: $$P^2(P-1)=0, \qquad P^2[P-1]=0. \qquad (10.12)$$

## § 11. Multiple Allelomorphs.

For $n$ allelomorphs $G^\mu(\mu=1, \ldots, n)$, the gametic and zygotic multiplication tables are

$$G^\mu G^\nu = \tfrac{1}{2}G^\mu + \tfrac{1}{2}G^\nu, \qquad (11.1)$$

$$Z^{\mu\nu}Z^{\theta\phi} = \tfrac{1}{4}Z^{\mu\theta} + \tfrac{1}{4}Z^{\mu\phi} + \tfrac{1}{4}Z^{\nu\theta} + \tfrac{1}{4}Z^{\nu\phi}, \qquad (11.2)$$

where $Z^{\mu\nu}=G^\mu G^\nu$. The algebras $\mathbf{G}_n$, $\mathbf{Z}_n$ so determined reduce to $\mathbf{G}_2$, $\mathbf{Z}_2$ when $n=2$; and they have in general the same train equations (10.11), (10.12).

## § 12. Linked Allelomorphs.

For two linked series of multiple allelomorphs, respectively $m$ and $n$ in number, with crossing over probability $\omega$, the gametic multiplication table is

$$G^{\mu\alpha}G^{\nu\beta} = \tfrac{1}{2}(1-\omega)(G^{\mu\alpha}+G^{\nu\beta}) + \tfrac{1}{2}\omega(G^{\mu\beta}+G^{\nu\alpha}), \qquad (12.1)$$

where $G^{\mu\alpha}(\mu=1, \ldots, m;\ \alpha=1, \ldots, n)$ are the $mn$ gametic types. Denote this gametic algebra $\mathbf{G}_{mn}(\omega)$. The principal and plenary rank equations are

$$x^3 - \tfrac{1}{2}(3-\omega)\xi x^2 + \tfrac{1}{2}(1-\omega)\xi^2 x = 0, \qquad (12.2)$$

$$x^{2\cdot 2} - (2-\omega)\xi^2 x^2 + (1-\omega)\xi^3 x = 0, \qquad (12.3)$$

giving for a normalised element P the train equations

$$P^3 - \tfrac{1}{2}(3-\omega)P^2 + \tfrac{1}{2}(1-\omega)P \equiv P(P-1)\left(P - \frac{1-\omega}{2}\right) = 0, \qquad (12.4)$$

$$P^{2\cdot 2} - (2-\omega)P^2 + (1-\omega)P \equiv P[P-1][P-(1-\omega)] = 0. \qquad (12.5)$$

In the duplicate algebra $\mathbf{Z}_{mn}(\omega)=\mathbf{G}_{mn}(\omega)'$ the corresponding equations are

$$P^2(P-1)\left(P - \frac{1-\omega}{2}\right) = 0, \qquad P^2[P-1][P-(1-\omega)] = 0. \qquad (12.6)$$

## § 13. Independent Allelomorphs.

Consider two series of multiple allelomorphs in separate autosomal linkage groups. This being indistinguishable from the case of § 12 with $\omega=\tfrac{1}{2}$, the gametic algebra is $\mathbf{G}_{mn}(\tfrac{1}{2})$. As in § 9, it may also be expressed as the direct product $\mathbf{G}_m\mathbf{G}_n$.

## § 14. Linkage Group.

I will first rewrite equations (12.1) with a change of notation. I will then write down the analogous equations for the case of three linked loci, and examine the structure of the corresponding algebra. This will be a sufficient indication of the procedure which can be followed out quite generally for a complete linkage group comprising any number of loci on one autosome, with any number of allelomorphs at each locus. The method may be extended to include any number of linkage groups.

Equations (12.1) may be written

$$AB \,.\, A'B' = \tfrac{1}{2}(1-\omega)(AB + A'B') + \tfrac{1}{2}\omega(AB' + A'B). \quad (14.1)$$

Here A and B refer to the two gene loci. $A^{\mu}B^{\alpha}$ would mean the same as $G^{\mu\alpha}$—a gamete with the $\mu$th allelomorph at the A-locus and the $\alpha$th at B; but dropping the indices AB and A′B′ stand for any particular gametic types, the same or different.

(14.1) may again be rewritten

$$AB \,.\, A'B' = \tfrac{1}{2}\varpi(A + \chi A')(B + \chi B'), \quad (14.2)$$

where $\varpi = 1 - \omega$ and $\chi$ is an operator which interchanges $\omega$ and $\varpi$, so that $\chi^2 = 1$ and $\varpi\chi = \omega$.

Now consider the case of three loci A, B, C, having respectively $m$, $n$, $r$ allelomorphs, and crossing over probabilities $\omega_{AB}$, $\omega_{BC}$, $\omega_{AC}$. The gametic algebra may be symbolised conveniently as $\mathbf{G}_{mnr}(\omega)$, where $\omega$ is the symmetrical matrix of the crossing over values, with diagonal zeros. Its multiplication table, comprising $\frac{1}{2}mnr(mnr+1)$ formulæ, is

$$ABC \,.\, A'B'C' = \tfrac{1}{2}\lambda(ABC + A'B'C') + \tfrac{1}{2}\mu(A'BC + AB'C') + \tfrac{1}{2}\nu(AB'C + A'BC') + \tfrac{1}{2}\rho(ABC' + A'B'C), \quad (14.3)$$

where

$$\lambda + \mu + \nu + \rho = 1, \quad (14.4)$$

$$\mu + \nu = \omega_{AB}, \qquad \nu + \rho = \omega_{BC}, \qquad \mu + \rho = \omega_{AC}. \quad (14.5)$$

The $\omega$'s are not independent, but are connected only by an inequality (Haldane, 1918):

$$\omega_{AC} = \omega_{AB} + \omega_{BC} - \kappa\omega_{AB}\omega_{BC}, \quad \text{where} \quad 0 \leqslant \kappa \leqslant 2, \quad (14.6)$$

from which may be deduced

$$\mu\rho \geqslant \nu\lambda. \quad (14.7)$$

Now introduce the following operators:—

$$\left.\begin{array}{llll} \chi_1 \text{ interchanges } \lambda \text{ with } \mu, & \nu \text{ with } \rho, \\ \chi_2 \quad\;\; \text{,,} \quad\;\; \lambda \;\text{,,}\; \nu, & \rho \;\text{,,}\; \mu, \\ \chi_3 \quad\;\; \text{,,} \quad\;\; \lambda \;\text{,,}\; \rho, & \mu \;\text{,,}\; \nu. \end{array}\right\} \quad (14.8)$$

Together with 1, they form an Abelian group, having the relations

$$\left.\begin{array}{c}\chi_2\chi_3=\chi_1, \quad \chi_3\chi_1=\chi_2, \quad \chi_1\chi_2=\chi_3, \\ \chi_1^2=\chi_2^2=\chi_3^2=\chi_1\chi_2\chi_3=1.\end{array}\right\} \qquad (14.9)$$

(14.3) may then be rewritten:

$$ABC \, . \, A'B'C' = \tfrac{1}{2}\lambda(A+\chi_1 A')(B+\chi_2 B')(C+\chi_3 C'). \qquad (14.10)$$

This symbolism can be manipulated with considerable freedom. For example, an expression such as $(\alpha ABC+\beta A'BC)$ can be written $(\alpha A+\beta A')BC$; and when two such expressions are multiplied, the distributive law works. The interchange symbols co-operate in the same way.

(14.10) may again be rewritten

$$ABC \, . \, A'B'C' = (\chi_0 A+\chi_1 A')(\chi_0 B+\chi_2 B')(\chi_0 C+\chi_3 C'), \qquad (14.11)$$

where $\chi_0 = 1$, and the operand $\frac{1}{2}\lambda$ is implied. Finally, (14.11) may be analysed into

$$AA' = \chi_0 A+\chi_1 A', \qquad BB' = \chi_0 B+\chi_2 B', \qquad CC' = \chi_0 C+\chi_3 C'. \qquad (14.12)$$

This separation of the symbols, or factorisation of the algebra (*cf.* end of § 9), will evidently yield valid results, provided that after recombination and application of (14.9), $\chi_0$ is interpreted as $\frac{1}{2}\lambda$, $\chi_1$ as $\frac{1}{2}\mu$, $\chi_2$ as $\frac{1}{2}\nu$, $\chi_3$ as $\frac{1}{2}\rho$. It must be noted that the symbols when separated in this way are non-commutative; *e.g.* $AA' \neq A'A$, since $ABC \, . \, A'B'C' \neq A'BC \, . \, AB'C'$.

Select a particular gametic type $\mathbf{ABC}$, and write

$$\mathbf{A}-A=u, \qquad \mathbf{B}-B=v, \qquad \mathbf{C}-C=w, \qquad (14.13)$$

where $A \neq \mathbf{A}$, $B \neq \mathbf{B}$, $C \neq \mathbf{C}$. Thus the symbols $u$, $v$, $w$ are nil elements having respectively $m-1$, $n-1$, $r-1$ possible values. We have from (14.12):

$$\mathbf{A}^2 = (\chi_0+\chi_1)\mathbf{A},$$
$$\mathbf{A}u = \mathbf{A}^2-\mathbf{A}A = (\chi_0+\chi_1)\mathbf{A}-(\chi_0\mathbf{A}+\chi_1 A) = \chi_1 u,$$
$$u\mathbf{A} = \mathbf{A}^2-A\mathbf{A} = (\chi_0+\chi_1)\mathbf{A}-(\chi_0 A+\chi_1\mathbf{A}) = \chi_0 u,$$
$$u^2 = \mathbf{A}^2-\mathbf{A}A-A\mathbf{A}+A^2 = (\chi_0+\chi_1)\mathbf{A}-(\chi_0\mathbf{A}+\chi_1 A)-(\chi_0 A+\chi_1\mathbf{A})+(\chi_0+\chi_1)A = 0,$$

and eight similar equations.

Now write

$$\left.\begin{array}{llll}\mathbf{ABC}=I, & u\mathbf{BC}=\bar{u}, & \mathbf{A}v\mathbf{C}=\bar{v}, & \mathbf{AB}w=\bar{w}, \\ \mathbf{A}vw=\overline{vw}, & u\mathbf{B}w=\overline{wu}, & uv\mathbf{C}=\overline{uv}, & uvw=\overline{uvw}.\end{array}\right\} \qquad (14.14)$$

The symbols $I$, $\bar{u}$, $\bar{v}$, $\bar{w}$, $\overline{vw}$, $\overline{wu}$, $\overline{uv}$, $\overline{uvw}$ thus introduced are linear and linearly independent in the gametic type symbols; and their number is

$$1+(m-1)+(n-1)+(r-1)+(n-1)(r-1)+(r-1)(m-1)+(m-1)(n-1)$$
$$+(m-1)(n-1)(r-1) = mnr,$$

which is equal to the number of gametic type symbols. They may thus be taken as a new basis for the gametic algebra. The transformed multiplication table is then easily deduced. We find, for example,

$$I^2 = I,$$
$$I\bar{u} = \mathbf{A}u \, . \, \mathbf{B}^2 \, . \, \mathbf{C}^2 = \chi_1(\chi_0 + \chi_2)(\chi_0 + \chi_3)\bar{u} = (\chi_0 + \chi_1 + \chi_2 + \chi_3)\bar{u} = \tfrac{1}{2}\bar{u},$$

since $\chi_0 + \chi_1 + \chi_2 + \chi_3$ is to be interpreted as $\frac{1}{2}\lambda + \frac{1}{2}\mu + \frac{1}{2}\nu + \frac{1}{2}\rho = \frac{1}{2}$. Similarly:

$$I\overline{vw} = \tfrac{1}{2}(\lambda + \mu)\overline{vw}, \qquad I\overline{uvw} = \tfrac{1}{2}\lambda\overline{uvw},$$
$$\bar{u}\bar{v} = \tfrac{1}{2}(\nu + \mu)\overline{uv}, \qquad \bar{u}\overline{vw} = \tfrac{1}{2}\mu\overline{uvw}, \qquad \bar{u}^2 = \bar{u}\overline{uv} = \bar{u}\overline{uvw} = 0.$$

These results are typical, all other products in the transformed multiplication table being obtainable from them by cyclic permutation of $u$, $v$, $w$ and $\mu$, $\nu$, $\rho$ and 1, 2, 3.

It is now readily verifiable that the algebra has the structure of a special train algebra as defined in § 4, with

$$\mathsf{U} = (\bar{u}, \bar{v}, \bar{w}, \overline{vw}, \overline{wu}, \overline{uv}, \overline{uvw}), \qquad \mathsf{U}^{(1)} = (\overline{vw}, \overline{wu}, \overline{uv}, \overline{uvw}),$$
$$\mathsf{U}^{(2)} = (\overline{uvw}), \qquad \mathsf{U}^{(3)} = 0.$$

Many of its properties can be most easily deduced from this transformed form. It can be shown that its principal and plenary train roots, other than unity, are the results of

$$\chi_0, \qquad \chi_0 + \chi_1, \qquad \chi_0 + \chi_2, \qquad \chi_0 + \chi_3,$$

operating respectively on $\frac{1}{2}\lambda$ and $\lambda$. Further details are postponed until the properties of special train algebras have been studied elsewhere.

## § 15. POLYPLOIDY.

A single example—the simplest possible—will illustrate the occurrence of special train algebras in this connection. The gametic algebra with multiplication table

$$\left.\begin{array}{lll} A^2 = A, & B^2 = AC = \frac{1}{6}A + \frac{2}{3}B + \frac{1}{6}C, & \\ C^2 = C, & BC = \frac{1}{2}B + \frac{1}{2}C, & AB = \frac{1}{2}A + \frac{1}{2}B, \end{array}\right\} \qquad (15.1)$$

refers to the inheritance of a single autosomal gene difference in autotetraploids. (*Cf.* Haldane, 1930, the case $m = 2$, with A, B, C written for $A^2$, $Aa$, $a^2$.)

This is a special train algebra, as may be seen by performing the transformation

$$A = A, \qquad A - B = u, \qquad A - 2B + C = p. \qquad (15.2)$$

It has the principal and plenary train equations

$$P(P - 1)(P - \tfrac{1}{6}) = 0, \qquad P[P - 1][P - \tfrac{1}{3}] = 0. \qquad (15.3)$$

## SUMMARY.

A population can be classified genetically at various levels, according to the frequencies of the gametic types which it produces, of the zygotic types of individuals which it contains, of types of mating pairs in the preceding generation, and so on. It is represented accordingly by means of hypercomplex numbers in one or other of a series of linear algebras (gametic, zygotic, copular, . . .), each algebra being isomorphic with the quadratic forms of the preceding algebra. Such a series of *genetic algebras* exists for any mode of genetic inheritance which is symmetrical in the sexes. (Genetic algebras for unsymmetrical inheritance also exist, but are not considered here.) Many calculations which occur in theoretical genetics can be expressed as manipulations within these algebras.

The algebras which arise in this way are all commutative non-associative linear algebras of a special kind. Firstly, they are *baric algebras*, *i.e.* they possess a scalar representation; secondly, they are *train algebras*, *i.e.* the rank equation of a suitably normalised hypercomplex number has constant coefficients. Some theorems concerning such algebras are enunciated.


---

## REFERENCES TO LITERATURE.


AITKEN, A. C., 1935. "The normal form of compound and induced matrices," *Proc. Lond. Math. Soc.* (2), vol. xxxviii, pp. 354–376.

DICKSON, L. E., 1914 (reprinted 1930). *Linear algebras*, Cambridge Tract, No. 16.

ETHERINGTON, I. M. H., 1939. "On non-associative combinations," *Proc. Roy. Soc. Edin.*, vol. lix, pp. 153–162.

GEPPERT, H., and KOLLER, S., 1938. *Erbmathematik*, Leipzig.

GLIVENKO, V., 1936. "Algèbre mendelienne," *Moscow Acad. Sci. C.R.*, vol. iv, pp. 385–386.

HALDANE, J. B. S., 1918. "The combination of linkage values," *Journ. Gen.*, vol. viii, pp. 299–309.

——, 1930. "Theoretical genetics of autopolyploids," *Journ. Gen.*, vol. xxii, pp. 359–372.

JENNINGS, H. S., 1935. *Genetics*, London.

WEDDERBURN, J. H. M., 1908 (*a*). "On hypercomplex numbers," *Proc. Lond. Math. Soc.* (2), vol. vi, pp. 77–118.

——, 1908 (*b*). "On certain theorems in determinants," *Proc. Edin. Math. Soc.* (1), vol. xxvii, pp. 67–69.


(*Issued separately November* 9, 1939.)

# 8



Reprinted from *Proc. Edinburgh Math. Soc.*, Ser. 2, **6**, Pt. 4, 222–230 (1941)

## Duplication of linear algebras

By I. M. H. Etherington.

(*Received and read* 3rd *May*, 1940.)

The process of duplication of a linear algebra was defined in an earlier paper[1], where its occurrence in the symbolism of genetics was pointed out. The definition will now be repeated with an amplification. Although for purpose of illustration it is applied to the algebra of complex numbers, duplication will seem of no special significance if attention is fixed on algebras with associative multiplication and unique division; for duplication generally destroys these properties. The results to be proved, however, show that it is significant in connection with various other conceptions which appeared in the discussion of genetic algebras; namely *baric algebras* and *train algebras* (defined in G.A.), also nilpotent algebras, linear transformation and direct multiplication of algebras.

### § 1. *Meaning of duplication.*

Let $X$ be a linear algebra of order $n$ over the field $F$, with basis $a^1, a^2, \ldots . a^n$, having the multiplication table

$$a^\mu a^\nu = \sum_{\sigma=1}^{n} \gamma_\sigma^{\mu\nu} a^\sigma, \qquad (\mu, \nu = 1, \ldots . n), \qquad (\gamma_\sigma^{\mu\nu} < F). \tag{1.1}$$

The commutative and associative laws of multiplication are not assumed. We shall write

$$X = (a^1, a^2, \ldots . a^n). \tag{1.2}$$

Except for the positive integer $n$, italic letters will be used consistently for hypercomplex numbers, or as they will be called *elements*; and except in the enumerating indices (which always run from 1 to $n$) greek letters (other than $\Sigma$) will be used consistently for elements of $F$, which will be called *numbers*. Also $\Sigma$ will always denote summation with respect to repeated indices. Thus we may without ambiguity

[1] Etherington, "Genetic algebras," *Proc. Roy. Soc., Edin.*, **59** (1939), 242-258. Reference will also be made to "On non-associative combinations," *ibid.*, 153-162. These papers will be referred to as G.A. and N.C. *Cf.* also *ibid.* (B), **61** (1941), 24-42.

define $X$ by writing instead of (1.2) and (1.1)

$$X = (a^{\mu}), \tag{1.3}$$

where

$$a^{\mu}\, a^{\nu} = \Sigma\, \gamma_{\sigma}^{\mu\nu}\, a^{\sigma}. \tag{1.4}$$

We have then

$$a^{\mu}\, a^{\nu}\, .\, a^{\theta}\, a^{\phi} = \Sigma\, \gamma_{\sigma}^{\mu\nu}\, a^{\sigma}\, .\, \Sigma\, \gamma_{\tau}^{\theta\phi}\, a^{\tau} = \Sigma\, \gamma_{\sigma}^{\mu\nu}\, \gamma_{\tau}^{\theta\phi}\, a^{\sigma}\, a^{\tau}. \tag{1.5}$$

Writing

$$a^{\mu}\, a^{\nu} = a^{\mu\nu}, \tag{1.6}$$

this becomes

$$a^{\mu\nu}\, a^{\theta\phi} = \Sigma\, \gamma_{\sigma}^{\mu\nu}\, \gamma_{\tau}^{\theta\phi}\, a^{\sigma\tau}, \tag{1.7}$$

which may be regarded as the multiplication table of another linear algebra over the same field $F$, denoted

$$X' = (a^{\mu\nu}) \tag{1.8}$$

and called the *duplicate* of $X$.

It was assumed in G. A. that $X$ was commutative; accordingly no distinction was drawn between $a^{\mu\nu}$ and $a^{\nu\mu}$, and $X'$ was a commutative algebra of order $\frac{1}{2}n\,(n+1)$. It was also pointed out that when $X$ is non-commutative, the non-commutative algebra $X'$ is of order $n^2$.

In the case when $X$ is commutative, however, it is still possible in carrying out the process (1.5, 6, 7) to draw a formal distinction between $a^{\mu\nu}$ and $a^{\nu\mu}$, and thus to obtain a *non-commutative duplicate algebra* of order $n^2$ instead of $\frac{1}{2}n\,(n+1)$. Its multiplication table will still be (1.7), but these equations will now number $n^4$ instead of

$$\tfrac{1}{2}\,.\,\tfrac{1}{2}n\,(n+1)\,.\,\{\tfrac{1}{2}n\,(n+1)+1\} = \tfrac{1}{8}n\,(n+1)\,(n^2+n+2). \tag{1.9}$$

(Provided that the order of the subalgebra $X^2$ is not less than 2, multiplication will be non-commutative in the non-commutative duplicate algebra.)

Consider, for example, the algebra of complex numbers, $Z = (1, i)$ where $1^2 = 1$, $1i = i1 = i$, $i^2 = -1$. Its commutative duplicate is $Z' = (1^2, 1i = i1, i^2)$, and its non-commutative duplicate is $Z' = (1^2, 1i, i1, i^2)$, with multiplication tables respectively

| | $a$ | $b$ | $c$ |
|---|---|---|---|
| $1^2 = a$ | $a$ | $b$ | $-a$ |
| $1i = i1 = b$ | | $c$ | $-b$ |
| $i^2 = c$ | | | $a$ |

(1.10)

| | $a$ | $b_1$ | $b_2$ | $c$ |
|---|---|---|---|---|
| $1^2 = a$ | $a$ | $b_1$ | $b_1$ | $-a$ |
| $1i = b_1$ | $b_2$ | $c$ | $c$ | $-b_2$ |
| $i1 = b_2$ | $b_2$ | $c$ | $c$ | $-b_2$ |
| $i^2 = c$ | $-a$ | $-b_1$ | $-b_1$ | $a$ |

(1.11)

These may be contrasted with the "direct square," or direct product of two algebras isomorphic with $Z$, say

$$ZZ_1 = (1, i) \times (I, j) = (1I, iI, 1j, ij),$$

having the commutative multiplication table

$$\begin{array}{r|cccc} & a & b_1 & b_2 & c \\ \hline 1I = a & a & b_1 & b_2 & c \\ iI = b_1 & & -a & c & -b_2 \\ 1j = b_2 & & & -a & -b_1 \\ ij = c & & & & a \end{array} \; . \tag{1.12}$$

Like $Z$, this is commutative and associative, and possesses a 1-element (having the properties of 1 and $I$ in the factor algebras), namely $a$; and it has the property of unique division. On the other hand both duplicate algebras are non-associative; and it follows Theorem II (i), (ii), *infra*, that except in the trivial case $n = 1$ a duplicate algebra cannot be a division algebra or possess a 1-element.

Returning to the general commutative algebra $X$, and supposing its order $> 2$, we can if desired draw distinctions between $a^{\mu\nu}$ and $a^{\nu\mu}$ in some cases but not all (*e.g.* regard $a^{12} = a^{21}$, but other $a^{\mu\nu} \neq a^{\nu\mu}$), and thus obtain intermediate *part-commutative duplicate algebras*, of any order between $n^2$ and $\frac{1}{2}n(n+1)$.

In the rest of this paper, except where otherwise indicated, it is optional whether we assume that $X$ is non-commutative, in which case $X'$ is unique; or that $X$ is commutative and that one of its duplicates is selected as *the* duplicate and denoted $X'$. The meaning of the phrase *quadratic form* is fixed accordingly: a quadratic form in $X$ means a linear combination (coefficients in $F$) of those products of base elements which are distinguished as corresponding to the base elements of $X'$.

§ 2. *General properties of a duplicate algebra.*

To any element

$$x' = \Sigma\, \alpha_{\mu\nu}\, a^{\mu\nu} \tag{2.1}$$

of $X'$, there corresponds the quadratic form $\Sigma\, \alpha_{\mu\nu}\, a^{\mu}\, a^{\nu}$ in $X$. The element $x'$ and the quadratic form will be called *isomorphs* of each other. The correspondence is unique both ways, and under it the operations of addition and multiplication both hypercomplex and scalar are conserved.

Also, in virtue of (1.6) and the multiplication table (1.4), to any element (2.1) of $X'$, there corresponds a unique element of $X$, called the *homomorph* of $x'$. Again, under this correspondence addition and both kinds of multiplication are conserved; but the correspondence is not unique in the opposite direction. It is nevertheless sometimes convenient (especially in the genetical symbolism) to use " = " for both correspondences, and thus to write:

$$\Sigma\, \alpha_{\mu\nu}\, a^{\mu\nu} = \Sigma\, \alpha_{\mu\nu}\, a^{\mu}\, a^{\nu} \text{ (its isomorph)} \tag{2.2}$$

$$= \Sigma\, \alpha_{\mu\nu}\, \gamma^{\mu\nu}_{\sigma}\, a^{\sigma} \text{ (its homomorph).} \tag{2.3}$$

Not all elements of $X$ are homomorphs: in order that $x$ should be a homomorph, it is necessary and sufficient that it should be a linear combination of the elements $\Sigma\, \gamma^{\mu\nu}_{\sigma}\, a^{\sigma}$, *i.e.* of $a^{\mu}\, a^{\nu}$; in other words it must belong to the invariant subalgebra $X^2$. Thus the homomorphism is a mapping of $X'$ on $X^2$.

When forming a product in $X'$, we may replace the elements to be multiplied by their homomorphs in $X$, and then multiply, leaving the product in quadratic form and taking its isomorph in $X'$. In symbols, if

$$x' = \Sigma\, \alpha_{\mu\nu}\, a^{\mu\nu}, \quad y' = \Sigma\, \beta_{\theta\phi}\, a^{\theta\phi},$$

then

$$x'\, y' = \Sigma\, \alpha_{\mu\nu}\, \gamma^{\mu\nu}_{\sigma}\, a^{\sigma} \,.\, \Sigma\, \beta_{\theta\phi}\, \gamma^{\theta\phi}_{\tau}\, a^{\tau} = \Sigma\, \alpha_{\mu\nu}\, \beta_{\theta\phi}\, \gamma^{\mu\nu}_{\sigma}\, \gamma^{\theta\phi}_{\tau}\, a^{\sigma\tau},$$

which is evidently the correct result. We deduce immediately[1]

THEOREM I. *In forming any product, power or continued product in $X'$, we can perform all the operations on the homomorphs in $X$, only in the final multiplication leaving the product in quadratic form: its isomorph in $X'$ will then be the result required.* (The operations have to be performed in a definite order since multiplication is non-associative.)

Elements of $X'$ whose homomorphs are zero will be called *o*-elements; they form a linear set which will be denoted $O$.

THEOREM II. (i) *Assuming $n > 1$, $X'$ necessarily contains o-elements other than zero, so that $O \neq 0$.* (ii) *In $X'$, any product which contains an o-element as one factor is zero.* (iii) *$O$ is an invariant subalgebra of $X'$.* (iv) *The difference algebra $(X' - O)$ is isomorphic with $X^2$.*

---

[1] In the genetical symbolism, this theorem corresponds to the fact that in order to obtain the distribution of zygotic types of an $r^{th}$ filial generation, provided that no selection acts on the zygotes, it is sufficient to trace only the gametic distribution through the $r - 1$ intervening generations.

For if $n > 1$, the elements $\Sigma \gamma_\sigma^{\mu\nu} a^\sigma$ (*i.e.* $a^\mu a^\nu$) are more than $n$ in number, and therefore cannot be linearly independent; this is equivalent to the statement (i). (ii) follows from Theorem I, and (iii) is an immediate consequence. Or (iii) and (iv) together follow from the general properties of homomorphisms[1].

In the algebras $Z'$ of § 1, write

$$a + c = o, \qquad b_1 - b_2 = o': \tag{2.4}$$

these are $o$-elements. The multiplication tables (1.10, 11) become

| | $a$ | $b$ | $o$ |
|---|---|---|---|
| $a$ | $a$ | $b$ | 0 |
| $b$ | | $o-a$ | 0 |
| $o$ | | | 0 |

(2.5)

| | $a$ | $b_1$ | $o$ | $o'$ |
|---|---|---|---|---|
| $a$ | $a$ | $b_1$ | 0 | 0 |
| $b_1$ | $b_1 - o'$ | $o - a$ | 0 | 0 |
| $o$ | 0 | 0 | 0 | 0 |
| $o'$ | 0 | 0 | 0 | 0 |

(2.6)

The zeros in the tables illustrate Theorem II (ii), (iii); while the results of suppressing all the $o$'s illustrate the isomorphism of $(Z' - O)$ with $Z^2$, *i.e.* with $Z$.

By a *polynomial* in an element $x$, we shall mean a finite linear combination of powers of $x$, with coefficients in $F$. Since $X$ does not in general contain a 1-element to serve as an interpretation of $x^0$ (and even if $X$ does, $X'$ does not), we shall exclude from consideration polynomials with a constant term. Thus when multiplication is (*a*) associative, (*b*) commutative and non-associative, (*c*) non-commutative and non-associative, a polynomial means a finite expression (for the index notation see N.C., § 2)

$$\left.\begin{array}{ll} (a) & \alpha x + \beta x^2 + \gamma x^3 + \delta x^4 + \epsilon x^5 + \dots, \\ (b) & \alpha x + \beta x^2 + \gamma x^3 + \delta x^4 + \epsilon x^{2\cdot 2} + \zeta x^5 + \dots, \\ (c) & \alpha x + \beta x^2 + \gamma x^{2+1} + \delta x^{1+2} + \epsilon x^{(2+1)+1} + \zeta x^{(1+2)+1} + \dots \end{array}\right\} \tag{2.7}$$

If $x$ is the homomorph of $x'$, then a polynomial in $x'$ has as homomorph the same polynomial in $x$ (perhaps compressed, if multiplication is associative or commutative in $X$ and not in $X'$).

---

[1] van der Waerden, *Moderne Algebra* (Berlin, 1930), I, pp. 56-57, where, since the postulate of associative multiplication in rings is not used, the results apply to non-associative algebras. "Invariant subalgebra" is here called *Ideal*, and "difference algebra" *Restklassenring*.

Suppose that every element $x = \Sigma a_\sigma a^\sigma$ of $X$ satisfies the identity

$$f(x, a_\sigma) = 0, \tag{2.8}$$

where $f(x, a_\sigma)$ is a polynomial in $x$ whose coefficients are functions of the coordinates $a_\sigma$ of $x$. Then the function $f(x', \Sigma a_{\mu\nu} \gamma^{\mu\nu}_\sigma)$, formed from any element $x'$ of $X'$ and the coordinates of its homomorph in the same way as $f(x, a_\sigma)$ is formed from $x$ and *its* coordinates, is an element of $X'$ whose homomorph is zero. Hence by Theorem II (ii),

$$x' . f(x', \Sigma a_{\mu\nu} \gamma^{\mu\nu}_\sigma) = 0, \qquad f(x', \Sigma a_{\mu\nu} \gamma^{\mu\nu}_\sigma) . x' = 0. \tag{2.9}$$

Thus we have

THEOREM III. *If every element* $x = \Sigma a_\sigma a^\sigma$ *of* $X$ *satisfies an identity* (2.8), *then every element* $x' = \Sigma a_{\mu\nu} a^{\mu\nu}$ *of* $X'$ *satisfies the identities* (2.9). *If multiplication is associative or commutative in* $X$ *and not in* $X'$, *the function* $f$ *in* (2.9) *may be interpretable in various ways.*

For example, every element $z = \alpha 1 + \beta i$ of $Z$ satisfies the rank equation

$$z^2 - 2\alpha z + \alpha^2 + \beta^2 = 0.$$

Theorem III can be applied not to this identity but to

$$z^3 - 2\alpha z^2 + (\alpha^2 + \beta^2) z = 0.$$

Using the notation (2.4), any element $z' = \alpha a + \beta b + \gamma o$ of the commutative duplicate algebra has the homomorph $\alpha 1 + \beta i$, and therefore satisfies

$$z'^4 - 2\alpha z'^3 + (\alpha^2 + \beta^2) z'^2 = 0,$$

which is in fact the rank equation of $Z'$. Similarly the element $z' = \alpha a + \beta b + \gamma o + \delta o'$ of the non-commutative duplicate satisfies the left and right rank equations

$$z'^{\,1+(1+2)} - 2\alpha z'^{\,1+2} + (\alpha^2 + \beta^2) z'^{\,2} = 0,$$
$$z'^{\,(2+1)+1} - 2\alpha z'^{\,2+1} + (\alpha^2 + \beta^2) z'^{\,2} = 0;$$

and also satisfies

$$z'^{\,1+(2+1)} - 2\alpha z'^{\,1+2} + (\alpha^2 + \beta^2) z'^{\,2} = 0,$$
$$z'^{\,(1+2)+1} - 2\alpha z'^{\,2+1} + (\alpha^2 + \beta^2) z'^{\,2} = 0.$$

§ 3. *Related algebras duplicated.*

THEOREM IV. *If part-commutative duplicate algebras are excluded, the duplicate of a linear transform of* $X$ *is a linear transform of* $X'$.

For if the equations of transformation of $X$ are

$$b^\alpha = \Sigma \lambda^\alpha_\mu a^\mu, \qquad a^\mu = \Sigma \Lambda^\mu_\alpha b^\alpha, \tag{3.1}$$

the multiplication table (1.4) becomes

$$b^\alpha b^\beta = \Sigma \lambda^\alpha_\mu \lambda^\beta_\nu \gamma^{\mu\nu}_\sigma \Lambda^\sigma_\epsilon b^\epsilon. \tag{3.2}$$

Duplicating (commutatively or non-commutatively), we obtain

$$b^{\alpha\beta} b^{\gamma\delta} = \Sigma \, (\lambda^{\alpha}_{\mu} \lambda^{\beta}_{\nu} \gamma^{\mu\nu}_{\sigma} \Lambda^{\sigma}_{\epsilon}) \, (\lambda^{\gamma}_{\theta} \lambda^{\delta}_{\phi} \gamma^{\theta\phi}_{\tau} \Lambda^{\tau}_{\xi}) \, b^{\epsilon\xi}$$

$$= \Sigma \, (\lambda^{\alpha}_{\mu} \lambda^{\beta}_{\nu}) \, (\lambda^{\gamma}_{\theta} \lambda^{\delta}_{\phi}) \, \gamma^{\mu\nu}_{\sigma} \gamma^{\theta\phi}_{\tau} \, (\Lambda^{\sigma}_{\epsilon} \Lambda^{\tau}_{\xi} b^{\epsilon\xi}); \qquad (3.3)$$

and this is precisely the result which would be obtained by applying to the duplicate multiplication table (1.7) (commutative or non-commutative correspondingly) the transformation

$$b^{\alpha\beta} = \Sigma \, \lambda^{\alpha}_{\mu} \lambda^{\beta}_{\nu} a^{\mu\nu}, \qquad a^{\mu\nu} = \Sigma \, \Lambda^{\mu}_{\alpha} \Lambda^{\nu}_{\beta} b^{\alpha\beta}. \qquad (3.4)$$

If $X'$ is (*a*) a commutative duplicate algebra, or (*b*) a non-commutative duplicate algebra, it will be seen[1] that the matrix of the induced transformation (3.4) in either direction is (*a*) the Schläflian (or second induced matrix), or (*b*) the direct square (or second Burnside matrix), of the matrix of the original transformation (3.1).

Similarly it is easy to prove

THEOREM V. *The commutative or non-commutative duplicate of the direct product of two algebras coincides with the direct product of their commutative or non-commutative duplicates.* Conversely, *the direct product of any two duplicate algebras coincides with a duplicate of their direct product.*

## § 4. *Algebras of special type duplicated.*

(*a*) *Nilpotent algebras*[2].

Suppose that $X$ is nilpotent of degree[3] $2\delta$; *i.e.* in $X$ all products of $2\delta$ factors vanish. It will be shown that $X'$ is nilpotent of degree $2\delta - 1$.

---

[1] See, *e.g.*, Aitken, *Proc. London Math. Soc.* (2), **38** (1935), 354-376.

[2] In this section, as in N.C. § 3, $\delta$, $\alpha$ denote positive integers.

[3] *Index* is the usual word in this context: *cf.* Wedderburn, *Proc. London Math. Soc.* (2), **6** (1908), 77-118; p. 111. But having drawn a distinction in N.C. between *index* and *degree*, I find the latter word more appropriate here. It is perhaps not irrelevant to point out an error in Wedderburn's paper, concerning nilpotent non-associative algebras. It is stated (*loc. cit.*, p. 111) that the sum of all the $r^{\text{th}}$ powers of such an algebra is less than (*i.e.* is contained in but is not equal to) the sum of the $(r-1)^{\text{th}}$ powers. This is not true of the commutative algebra $X = (a, b, c)$ where $a^2 = b$, $ab = b^2 = c$, $ac = bc = c^2 = 0$; for which $X^2 = (b, c)$, $X^3 = (c)$, $X^4 = 0$, $X^{2 \cdot 2} = (c)$, $X^5 = X^{2 \cdot 2 + 1} = X^{3+2} = 0$. For $X$ is nilpotent of degree 5, whereas $X^4 + X^{2 \cdot 2} = X^3$. *Cf.* Etherington, "Special train algebras," *Quart. Journ. Math.* (in press).

Consider first a product $a'$ in $X'$ containing $\delta$ factors. It is a linear combination of products of $\delta$ base elements. Each product of $\delta$ base elements is isomorphic with a product of $2\delta$ base elements in $X$, and is therefore an $o$-element. Now consider a product $x'$ of $2\delta-1$ factors in $X'$: it is expressible as $a'b'$, where either $a'$ or $b'$ contains at least $\delta$ factors; *i.e.* either $a'$ or $b'$ is an $o$-element, and hence (Theorem II) $x' = 0$; as was to be proved. The same argument applies *a fortiori* if $X$ is nilpotent of degree $2\delta - 1$.

Or suppose (*cf.* N.C., p. 156) that $X$ is nilpotent of altitude $\alpha$; *i.e.* in $X$ all products of altitude $\alpha$ vanish. A product $a'$ in $X'$ of altitude $\alpha - 1$ is a linear combination of products of base elements having the same altitude, each isomorphic with a product of altitude $\alpha$ in the base elements of $X$; and is therefore an $o$-element. A product $x'$ of altitude $\alpha$ is expressible as $a'b'$, where either $a'$ or $b'$ is of altitude $\alpha - 1$ and is thus an $o$-element; so that $x' = 0$.

We have thus proved:

THEOREM VI. *If $X$ is nilpotent* (i) *of degree $2\delta - 1$ or $2\delta$, or* (ii) *of altitude $\alpha$, then $X'$ is nilpotent* (i) *of degree $2\delta - 1$, or* (ii) *of altitude $\alpha$, accordingly.*

(*b*) *Baric and train algebras.*

If $X$ is a baric algebra (G.A., § 3), there exists for any element $x$ a number $\xi(x)$, the weight of $x$, such that

$$\xi(x+y) = \xi(x) + \xi(y), \quad \xi(xy) = \xi(x)\,\xi(y), \quad \xi(\alpha x) = \alpha\xi(x).$$

If $x'$ is any element of $X'$, with homomorph $x$, and we define $\xi(x')$ as being equal to $\xi(x)$, then it follows that

$$\xi(x'+y') = \xi(x') + \xi(y'), \quad \xi(x'y') = \xi(x')\,\xi(y'), \quad \xi(\alpha x') = \alpha\xi(x');$$

so that $\xi(x')$ is a weight function of $X'$. Moreover, if

$$\xi(a^{\mu}) = \xi^{\mu},$$

then

$$\xi(a^{\mu\nu}) = \xi(a^{\mu}a^{\nu}) = \xi^{\mu}\xi^{\nu}.$$

Thus we have

THEOREM VII. *If $X$ is a baric algebra with weight vector $\xi^{\mu}$, then $X'$ is a baric algebra with weight vector $\xi^{\mu}\xi^{\nu}$, and the weight of any element in $X'$ is equal to the weight of its homomorph in $X$.*

Combining this with Theorem III, we obtain

Theorem VIII. *If X is a train algebra with (left or right) principal train roots* 1, $\lambda$, $\mu$, ...., *then X' is a train algebra with (left or right) principal train roots included in* 1, 0, $\lambda$, $\mu$, .... Instances of this theorem were observed in G.A.

It may be stated that the duplicate of a special train algebra (G.A., p. 246), although a train algebra, is not always a special train algebra[1]. The question, which was left open in G.A., whether a train algebra is necessarily a special train algebra, is thus to be answered in the negative.

---

[1] The statement (G.A., p. 247) that "all the fundamental genetic algebras are special train algebras" refers to gametic algebras, not to the zygotic algebras which are derived from them by duplication.

Mathematical Institute,
16 Chambers Street,
Edinrurgh.

# 9



Reprinted from *Proc. Roy. Soc. Edinburgh,* **61**, 24–42 (1941)

## II.—Non-Associative Algebra and the Symbolism of Genetics.

By **I. M. H. Etherington,** B.A., Ph.D., Mathematical Institute, University of Edinburgh.

(MS. received August 14, 1940. Read December 2, 1940.)

CONTENTS.



### § 1. INTRODUCTION.

THE statistical material of genetics usually consists of frequency distributions—of genes, zygotes and mating couples—from which new distributions referring to their progeny arise. Combination of distributions by random mating is usually symbolised by the mathematical sign for multiplication; but this sign is not taken literally for the simple reason that the genetical laws connecting the distributions of progenitors and progeny are inconsistent with the laws governing multiplication in ordinary algebra. This is explained more fully in § 2.

However, there is no insuperable reason why the genetical sign of multiplication should not be taken literally; for it is possible with any particular type of inheritance to construct an "algebra"—distinct from ordinary algebra but of a type well known to mathematicians—such that the laws governing multiplication shall represent exactly the underlying genetical situation. These "genetic algebras" are of a kind known as "linear algebras," of which a simple description is given in § 4.

It is not suggested that the use of ordinary algebraic methods in conjunction with the specific principles of genetics will not lead to correct results. It seems, however, that the systematic use of genetic algebras would simplify and shorten the way to their attainment, and perhaps enable much more difficult problems to be tackled with equal ease.

The construction of genetic algebras has been described in a somewhat abstract way in a previous paper (Etherington, 1939 *b*), to which I shall refer as G.A. Here I propose to consider the symbolism more from the geneticist's point of view, applying it to some simple population problems, without going into the details of the mathematical background. It will be recognised that the current treatment of such problems does in reality make use of genetic algebras without noticing them explicitly. By elaborating the symbolism and adapting it to more complicated genetical premises (*e.g.* in the manner indicated in G.A. § 14), it should be possible to avoid the laborious complexity which other methods in such cases would involve.

Only elementary mathematical knowledge is assumed, and it is hoped that this paper will be found understandable by geneticists whose mathematical knowledge is quite limited.

## § 2. Genetical Multiplication.

Capital letters will be used to represent frequency or probability distributions, referring to either a population, a single individual, or a single gamete; such as (in the case of autosomal allelomorphs)

$P = DD$ = homozygous dominant individual, or population consisting of such;

$P = \alpha DD + \beta DR + \gamma RR$ = population with assigned frequencies $\alpha : \beta : \gamma$ of genotypes, or individual with assigned probabilities $\alpha$, $\beta$, $\gamma$ of belonging to one or other genotype;

$P = \delta D + \rho R$ = population which produces D and R gametes in given numerical ratio, or gamete which has probability $\delta$ of containing D, $\rho$ of containing R.

The *multiplication* of populations—individuals—gametes—means the calculation of progeny distribution resulting from their random mating—mating—fusion. Defining a *population* as a probability distribution of genetic types, we may say in all cases that we are multiplying populations.

Now multiplication in ordinary algebra obeys three laws: (1) the commutative law $PQ = QP$, (2) the associative law $P(QR) = (PQ)R$, (3) the distributive law $P(Q + R) = PQ + PR$.

The validity of the distributive law in the genetic symbolism is sufficiently obvious; it forms the basis of the method of "chess-board diagrams" often used as visual aids in the calculation of progeny distributions.

The associative law is not obeyed in genetical multiplication. This is seen by comparing the progeny of a mating between the offspring from two individuals or populations, denoted as PQ, and a third individual or population R (*i.e.* the product (PQ)R), with the progeny from P and the hybrid population QR (*i.e.* the product P(QR)). There is clearly no reason why they should be the same, and in fact unless P = R they are found to be different. Thus genetical multiplication is non-associative.

Regarding the commutative law, (i) if we are considering autosomal characters it will be obvious that this law applies, since the results of reciprocal matings are generally speaking identical, although we shall see below that in certain cases non-commutative multiplication can occur.

(ii) One might be tempted to say that with sex-linked characters multiplication is non-commutative, since the results of reciprocal matings are different. But it must be remembered that with sex-linked characters we can only speak of reciprocal matings in connection with the phenotype classification of a population; whereas the calculation of progeny distribution is only possible on the basis of the genotype classification. A given genotype (either involving the Y-chromosome or not) is either female or male, so that a reciprocal mating between genotypes is impossible. Suppose that we are multiplying a male genotype M and a female genotype F: then MF and FM both mean the same thing—the genotype distribution of their offspring; and so multiplication is commutative.

(iii) On the other hand, returning to autosomal inheritance, it is possible for this to be unsymmetrical in the sexes, through either crossing-over values or gametic selection being different in male and female. In such cases it is really optional whether we treat corresponding male and female genotypes as the same type (since their relevant gene content is the same) or as distinct types (since they produce different series of gametes). In the former case, PQ and QP have distinct meanings, referring to reciprocal crosses which do not produce similar distributions of offspring; and multiplication is non-commutative. In the latter case, the situation is as with sex-linkage.

To sum up, genetical multiplication is non-associative, but obeys the commutative and distributive laws; except that in certain cases we have the option of using a varied form of the symbolism in which the multiplication is non-commutative as well as non-associative.

## § 3. Non-Associative Products and Powers.

Non-commutative algebra of a special kind (matrix algebra) is widely familiar by reason of its many applications in geometry and physics. (Also in genetics: *cf.* Hogben, 1933; Geppert and Koller, 1938, Chap. 4.)

Hence there is no reason to fear that an algebra which does not obey all the usual laws will necessarily prove unmanageable.

But with non-associative algebra some precautions are required to avoid confusion, especially when dealing with products or powers involving many factors. With such an expression, brackets inserted in different ways would indicate different orders of association of the factors; and the corresponding interpretations of the whole product would refer to the various pedigrees which could be constructed with given ancestors. For example, the product $\{P^2(QR)\}S$ represents the pedigree below. The

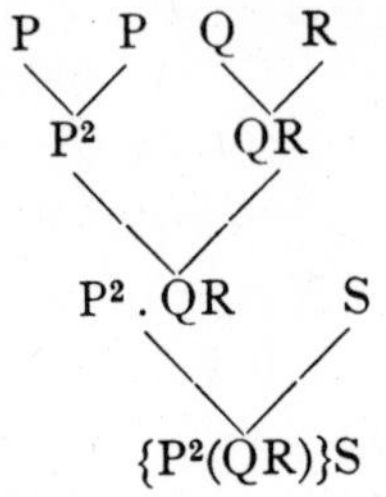


separate factors or ancestors P, P, Q, R, S may be thought of as given genotypes or distributions of genotypes. The factor $P^2$ may arise through self-fertilisation of an individual P, or from the mating of two individuals of the same genetic type, or from random mating within one population P or between two similar populations. The partial products $P^2$, QR, $P^2(QR)$, and the final result $\{P^2(QR)\}S$ are probability distributions which, for any particular type of inheritance, can be calculated when P, Q, R, S are known.

To avoid clumsiness of notation, it is convenient to use groups of dots in place of brackets, fewness of dots between factors conferring precedence in multiplication. Thus the above product would be written $P^2 \,.\, QR : S$. On putting $P=Q=R=S$, it becomes a power of P. (I have discussed elsewhere a notation and nomenclature for non-associative powers (1939 *a*, § 2); *e.g.* the power in question is denoted $P^{2.2+1}$. We shall be concerned, however, with only two simple types of non-associative powers, namely, the "principal" and "plenary" powers described below.)

Similarly, the product appearing at (10.1) below denotes

$$[\{(ab)(cd)\}(ef)][\{(ab)(cd)\}(gh)].$$

The pedigree for this is easily constructed; but it should be noted that in the context *a*, *b*, *c*, . . . denote gametes, so that *ab*, *cd*, . . . are the ancestral zygotes.

We shall find it important to distinguish between, *e.g.*, $P^{2.2}=(P^2)^2$ and $P^4=P\{P(P^2)\}=P : P \,.\, PP$. If mating takes place at random in a population, initially P, the successive generations, supposed discrete, are represented by the sequence of *plenary powers*

$$P, \quad P^2, \quad P^{2.2}, \quad P^{2^3}, \ldots P^{2^{n-1}}, \ldots, \qquad (3.1)$$

each the square of the preceding; while the sequence of *principal powers*

$$P, \quad P^2, \quad P^3, \ldots P^n, \ldots, \qquad (3.2)$$

each obtained from the preceding by multiplication with P, refers similarly to a mating system in which each generation is mated back to one original ancestor or ancestral population.

## § 4. Linear Algebras.

Linear algebras have been studied for some ninety years, and there is an extensive literature of the subject. The following brief description will be sufficient for the present purpose. Attention is confined to algebras "over the field of real numbers"; that is to say, the Greek letters below denote ordinary real numbers, and this convention will be observed throughout the paper.

Beginning with a simple case, a *commutative* * *linear algebra of order* 2 is determined when two given symbols or *units* A, B are subject to a multiplication table consisting of product rules of the form

$$A^2 = \alpha A + \beta B, \qquad AB = \gamma A + \delta B, \qquad B^2 = \epsilon A + \zeta B, \qquad (4.1)$$

the coefficients being given numerical constants. The algebra then consists of all possible expressions of the form

$$P = \lambda A + \mu B, \qquad (4.2)$$

which are called *hypercomplex numbers*.† Addition and multiplication of hypercomplex numbers are carried out as in ordinary algebra, the multiplication table (4.1) being used to reduce a product to the "linear" form (4.2). Thus if

$$P = \lambda A + \mu B, \qquad Q = \nu A + \rho B,$$

then

$$P \pm Q = (\lambda \pm \nu)A + (\mu \pm \rho)B, \qquad (4.3)$$

$$\left.\begin{aligned} PQ &= \lambda\nu A^2 + (\lambda\rho + \mu\nu)AB + \mu\rho B^2 \\ &= \lambda\nu(\alpha A + \beta B) + (\lambda\rho + \mu\nu)(\gamma A + \delta B) + \mu\rho(\epsilon A + \zeta B) \\ &= (\lambda\nu\alpha + \lambda\rho\gamma + \mu\nu\gamma + \mu\rho\epsilon)A + (\lambda\nu\beta + \lambda\rho\delta + \mu\nu\delta + \rho\mu\zeta)B. \end{aligned}\right\} \qquad (4.4)$$

* *Commutative* refers to the nature of multiplication; the *order* is the number of units on which the algebra is based.

† So called because they are a generalisation of the more familiar *complex numbers*. The algebra of complex numbers possesses a *real unit* 1 and an *imaginary unit* i, which are subject to the multiplication table $1^2 = 1$, $1i = i$, $i^2 = -1$.

It was implied in (4.1) that BA = AB. The linear algebra would be *non-commutative*, however, if different formulæ were prescribed for AB and BA; and then we should have $PQ \neq QP$. Unless special conditions are satisfied by the coefficients in (4.1), multiplication is non-associative.

A linear algebra of order $n$ is defined in an analogous way. It will be based on $n$ units, and will consist of hypercomplex numbers: a hypercomplex number is an expression which is linear (*i.e.* of the first degree throughout) in the $n$ units; and the algebra will have a multiplication table giving a linear formula for the square of each unit and for the product of each pair of units. (See, for example, the multiplication tables (5.3) and (11.10), which determine linear algebras of orders 3 and 5 respectively.)

The commutative and associative laws of addition,

$$P+Q=Q+P, \qquad (P+Q)+R=P+(Q+R),$$

always hold; so do the distributive laws

$$P(Q+R)=PQ+PR, \qquad (Q+R)P=QP+RP;$$

but multiplication may be non-commutative, non-associative, or both.

It will be seen that a linear algebra is completely determined when its multiplication table is known.

Given any two linear algebras of orders $m$ and $n$ (*i.e.* given their multiplication tables), it is possible by combining their multiplication tables in a certain way to deduce another linear algebra, of order $mn$, which is known as their *direct product*. This is of fundamental importance in the general theory of linear algebras, and we shall find (§ 11; *cf.* G.A. § 9) that it is also fundamental in the symbolism of genetics. If the units on which the first algebra is based are A, B, . . ., and those of the second A′, B′, . . ., then the units of the direct product may be interpreted as AA′, AB′, BA′, BB′, . . ..

Also (of less importance in the mathematical theory, but equally fundamental in genetics), from any linear algebra of order $n$ a closely related linear algebra called its *duplicate* can be derived, of order $\frac{1}{2}n(n+1)$ if the original algebra is commutative. If the original units are A, B, . . ., those of the duplicate algebra may be interpreted as $A^2$, $B^2$, AB, . . .. (The process of duplication was described in G.A. § 5; *cf.* also Etherington, 1941; it occurs here in §§ 5–7.)

## § 5. The Mendelian Gametic and Zygotic Algebras.

Consider a pair of autosomal allelomorphs D, R and the corresponding genotypes

$$A=DD, \qquad B=DR, \qquad C=RR. \quad . \quad . \quad . \quad (5.1)$$

We shall write optionally DD or $D^2$, RR or $R^2$. In accordance with mendelian principles and with the notation described at the beginning of § 2, we have the two sets of formulæ:

$$D^2 = D, \qquad DR = \tfrac{1}{2}D + \tfrac{1}{2}R, \qquad R^2 = R; \qquad (5.2)$$

$$\left.\begin{array}{lll} A^2 = A, & B^2 = \tfrac{1}{4}A + \tfrac{1}{2}B + \tfrac{1}{4}C, & C^2 = C, \\ BC = \tfrac{1}{2}B + \tfrac{1}{2}C, & CA = B, & AB = \tfrac{1}{2}A + \tfrac{1}{2}B. \end{array}\right\} \qquad (5.3)$$

These give the series of gametes produced by each type of zygote, and the series of zygotes produced by each type of mating couple, with coefficients denoting relative frequencies. *E.g.*, the second of equations (5.2) mean that a heterozygote produces D and R gametes in equal numbers; the second of equations (5.3) means that the offspring of a mating DR × DR are 25 per cent. DD, 50 per cent. DR, 25 per cent. RR.

A population P can be described by the frequencies either of the gametes which it produces, or of the zygotes which it contains, and accordingly we write:

(*Gametic representation*) $P = \delta D + \rho R$, . . . . (5.4)

(*Zygotic representation*) $P = \alpha A + \beta B + \gamma C$ . . . . . (5.5)

$= \alpha DD + \beta DR + \gamma RR$, . . . . (5.6)

in which we may assume

(*Normalising conditions*) $\delta + \rho = 1, \qquad \alpha + \beta + \gamma = 1.$ . . . . (5.7, 8)

The two representations are connected by (5.2); *i.e.* (5.6) implies (5.4) with

$$\delta = \alpha + \tfrac{1}{2}\beta, \qquad \rho = \tfrac{1}{2}\beta + \gamma. \qquad (5.9)$$

An examination of the above formulæ in the light of § 4 will show that by using this symbolism we are really dealing with two distinct linear algebras, both having commutative and non-associative multiplication, namely:

(1) The algebra of the symbols D, R with multiplication table (5.2). This will be called the *gametic algebra for simple mendelian inheritance*, and referred to as **G**. A hypercomplex number in this algebra has the form (5.4).

(2) The algebra of the symbols A, B, C with multiplication table (5.3). Call this the *zygotic algebra for simple mendelian inheritance*, and denote it **Z**. A hypercomplex number in **Z** has the form (5.5). Hypercomplex numbers in **G** and **Z** are interpreted as populations only if their coefficients are all positive; and it is generally convenient to require that the coefficients shall satisfy the normalising conditions (5.7, 8).

The relation between the two algebras is given by (5.1), which means that a hypercomplex number (or *linear* form) (5.5) in **Z** is equivalent to a *quadratic* form (5.6) in **G**. The quadratic form is reduced to a hypercomplex number in **G** by using the multiplication table (5.2). That is to say (*cf.* 5.9), the zygotic representation determines the gametic; but not *vice versa*, owing to the extra degree of freedom in the zygotic algebra.

Starting from the gametic multiplication table (5.2), the equations (5.3) are built up by the following process: we take the symbols A, B, C defined by (5.1) as units of a new algebra, and then

$$\left.\begin{aligned} A^2 &= DD\,.\,DD = D\,.\,D = A, \\ B^2 &= (\tfrac{1}{2}D + \tfrac{1}{2}R)^2 = \tfrac{1}{4}DD + \tfrac{1}{2}DR + \tfrac{1}{4}RR = \tfrac{1}{4}A + \tfrac{1}{2}B + \tfrac{1}{4}C, \\ AB &= DD\,.\,DR = D(\tfrac{1}{2}D + \tfrac{1}{2}R) = \tfrac{1}{2}A + \tfrac{1}{2}B, \end{aligned}\right\} \quad . \quad (5.10)$$

and so on. Thus the zygotic multiplication table is constructed from the gametic. This is the process of duplication referred to in § 4, and **Z** is thus the duplicate of **G**.

Suppose that we wish to find the progeny distribution of two mating populations P, Q, whose representations, either gametic or zygotic, are given. We have merely to form the product of the two hypercomplex numbers; that is to say (*cf.* 4.4), we multiply two corresponding representations together as in ordinary algebra, substitute (5.2) or (5.3), and simplify. The validity of the process follows from the fact that it is simply a translation into symbols of the more self-explanatory procedure of chess-board diagrams: in other words, it follows from the fact that genetical multiplication obeys the distributive law.

## § 6. "Shortcircuited" Multiplication.

By a fundamental property of duplicate algebras (Etherington, 1941, Theorem I), multiplication in **Z** can be "shortcircuited" by working in **G**: that is to say, to find PQ when P and Q are given zygotically, we first apply (5.2) to obtain the gametic representations, and then multiply *without* applying (5.2). Similarly, to evaluate in **Z** a complicated non-associative product involving any number of factors, all the operations can be performed in the simpler algebra **G**, only the final product being left in quadratic form and interpreted as a hypercomplex number in **Z**.

This corresponds to a well-known fact in genetics (*cf.* Jennings, 1917, pp. 101–102): in order to obtain the zygotic frequencies of an $n$th generation, provided that no selection acts on the zygotes, and in the absence of inbreeding, it is sufficient to trace only the gametic frequencies through the $n-1$ intervening generations.

To consider, for example, random mating of a population P *inter se*, suppose

$$P = \delta D + \rho R, \qquad (6.1)$$

where (5.9) holds if the zygotic representation is given in the first place. Then the next generation is

$$\left. \begin{aligned} F_1 = P^2 &= \delta^2 DD + 2\delta\rho DR + \rho^2 RR \\ &= \alpha_1 A + \beta_1 B + \gamma_1 C, \end{aligned} \right\} \qquad (6.2)$$

where

$$\alpha_1 = \delta^2, \qquad \beta_1 = 2\delta\rho, \qquad \gamma_1 = \rho^2. \qquad (6.3)$$

This is evidently simpler than evaluating $P^2$ in **Z** directly. The conclusion

$$\beta_1^2 = 4\alpha_1\gamma_1 \qquad (6.4)$$

is the well-known Pearson-Hardy law.

Evaluating (6.2) in **G** by use of (5.2),

$$\begin{aligned} P^2 &= \delta(\delta + \rho)D + \rho(\delta + \rho)R \\ &= \delta D + \rho R \end{aligned}$$

if (6.1) is normalised. Thus in **G** any normalised hypercomplex number satisfies

$$P^2 = P; \qquad (6.5)$$

hence all powers of P are equal, showing that the gene frequencies are undisturbed by random mating, or by random mating followed by any system of intermating of the generations. The zygotic distribution, however, in such cases, comes into equilibrium after one generation of random mating, since in **Z** we find

$$P^3 = P^2, \qquad P^{2.2} = P^2, \qquad (6.6,\ 7)$$

and all higher powers of P are equal to $P^2$. These equations follow immediately from (6.5) if $P^3$, $P^{2.2}$ are found by short-circuited multiplication.

## § 7. The Mendelian Copular Algebra.

The procedure of duplication (5.10), by which **Z** was derived from **G**, can be applied to an algebra repeatedly. Let us form **K**, the duplicate of **Z**, and then consider its genetical significance. By analogy with (5.1) we begin by taking

$$AA, \quad BB, \quad CC, \quad BC, \quad CA, \quad AB \qquad (7.1)$$

as the units of a new algebra. There is no need to introduce fresh symbols. The multiplication table will consist of 21 equations derived by manipulation of the equations (5.3), for example:

$$(AA)^2 = AA, \quad (BB)^2 = \tfrac{1}{16}AA + \tfrac{1}{4}BB + \tfrac{1}{16}CC + \tfrac{1}{4}BC + \tfrac{1}{8}CA + \tfrac{1}{4}AB, \text{ etc.} \qquad (7.2)$$

The interpretation is as follows: the coupled symbols (7.1) stand for the types of family into which the population can be sorted, classified according to the parental genotypes: or, we may say, they are the types of *couple* mated in the preceding generation. Hence (7.2) means that if a population of offspring of matings A × A is mated at random with itself or with a similar population, all the matings are of this type A × A; but if the parental matings were all B × B, then the six couple types occur in numerical proportions $\frac{1}{16}:\frac{1}{4}:\ldots$; and so on.

A population for which the relative frequencies of the couple types are

$$\lambda:\mu:\nu:\theta:\phi:\psi \quad . \quad . \quad . \quad . \quad . \quad (7.3)$$

is represented by a hypercomplex number

(*Copular representation*) $\mathrm{P}=\lambda\mathrm{AA}+\mu\mathrm{BB}+\nu\mathrm{CC}+\theta\mathrm{BC}+\phi\mathrm{CA}+\psi\mathrm{AB}$, . (7.4)

wherein

(*Normalising condition*) $\lambda+\mu+\nu+\theta+\phi+\psi=1$. . . . . (7.5)

From this we can pass by (5.3) and (5.1, 2) to the zygotic and gametic representations.

As in **G** and **Z**, the product of two hypercomplex numbers in **K** denoting populations gives in the same representation their offspring by random mating. This statement assumes that the couple types are not selected, *i.e.* they are of equal average surviving fertility; just as in **Z** and **G** we supposed no selection on zygotes or gametes. As before, multiplication in **K** can be short-circuited by working in **Z** or **G**.

Corresponding to the Pearson-Hardy law in the zygotic algebra, we have the following facts: a population, as a distribution of copular types, comes into equilibrium after *two* generations of amphimixis; after *one* generation, the equations

$$\theta^2=4\mu\nu, \qquad \phi^2=4\nu\lambda, \qquad \psi^2=4\lambda\mu \quad . \quad . \quad . \quad (7.6)$$

are satisfied; after two generations the further equation

$$\mu^2=16\nu\lambda \quad . \quad . \quad . \quad . \quad . \quad (7.7)$$

is satisfied; these four are the necessary and sufficient conditions for equilibrium in amphimixis, and imply also other relations such as

$$4\phi^2=\theta\psi=\mu^2. \quad . \quad . \quad . \quad (7.8)$$

These results are obtained very simply by using short-circuited multiplication, observing that $\mathrm{P}^2$ is necessarily of the form $(\alpha\mathrm{A}+\beta\mathrm{B}+\gamma\mathrm{C})^2$, and the next generation $\mathrm{P}^{2.2}$ of the form $((\delta\mathrm{D}+\rho\mathrm{R})^2)^2$.

## § 8. Systems of Mating.

Four systems of mating will be considered. The object in each case is to obtain the distribution of types in a filial generation from the distribution in the preceding generation; also, when it can be done simply, to find the distribution in the *n*th filial generation, and the equilibrium distribution which this approaches as *n* increases. For other treatment of these and similar problems, *cf.* Jennings (1916, 1917), Wentworth and Remick (1916), Robbins (1917, 1918), Hogben (1931, Chap. 6; 1933), Geppert and Koller (1938, § 20).

### (a) *Self-fertilisation, or Assortative Mating in Absence of Dominance.*

Starting from the zygotic distribution

$$P=\alpha A+\beta B+\gamma C \qquad . \quad . \quad . \quad . \quad (8a.1)$$

(where A = DD, B = DR, C = RR), if mating proceeds in successive generations by self-fertilisation, or by each individual mating with another of the same type, the first filial generation $F_1$ will consist of the offspring of A × A, B × B, C × C, occurring in proportions $\alpha : \beta : \gamma$; so that

$$F_1=\alpha A^2+\beta B^2+\gamma C^2 . \qquad (8a.2)$$

$$=\alpha A+\beta(\tfrac{1}{4}A+\tfrac{1}{2}B+\tfrac{1}{4}C)+\gamma C \qquad (8a.3)$$

$$=(\alpha+\tfrac{1}{4}\beta)A+\tfrac{1}{2}\beta B+(\tfrac{1}{4}\beta+\gamma)C. \qquad (8a.4)$$

It will be seen that the frequency of heterozygotes is halved; so if the *n*th filial generation is denoted

$$F_n=\alpha_n A+\beta_n B+\gamma_n C, \qquad (8a.5)$$

we shall have

$$\beta_1=\tfrac{1}{2}\beta, \qquad \beta_2=\tfrac{1}{4}\beta, \qquad \beta_3=\tfrac{1}{8}\beta, \ . \ . \ ., \qquad \beta_n=\frac{1}{2^n}\beta. \qquad (8a.6)$$

Also

$$\alpha_1=\alpha+\tfrac{1}{4}\beta, \qquad \gamma_1=\tfrac{1}{4}\beta+\gamma. \qquad (8a.7)$$

Let us find the quantities $u_1$, $u_2$, $u_3$, . . . by which the hypercomplex number representing the population increases in the successive generations. We have from (8*a*.1) and (8*a*.4):

$$u_1=F_1-P=\tfrac{1}{2}\beta(\tfrac{1}{2}A-B+\tfrac{1}{2}C); \qquad (8a.8)$$

and similarly we shall have

$$u_2=\tfrac{1}{2}\beta_1(\tfrac{1}{2}A-B+\tfrac{1}{2}C)=\tfrac{1}{4}\beta(\tfrac{1}{2}A-B+\tfrac{1}{2}C),$$

$$u_3=\tfrac{1}{8}\beta(\tfrac{1}{2}A-B+\tfrac{1}{2}C), \ . \ . \ ., \qquad u_n=\frac{1}{2^n}\beta(\tfrac{1}{2}A-B+\tfrac{1}{2}C). \qquad (8a.9)$$

The total increase in $n$ generations is therefore

$$u_1+u_2+u_3+\ .\ .\ .\ +u_n=\left(\tfrac{1}{2}+\tfrac{1}{4}+\tfrac{1}{8}+\ .\ .\ .\ +\frac{1}{2^n}\right)\beta(\tfrac{1}{2}\mathrm{A}-\mathrm{B}+\tfrac{1}{2}\mathrm{C}). \quad (8a.10)$$

The sum of the geometrical progression in brackets is $1-(\frac{1}{2})^n$. Hence the $n$th filial generation is

$$\mathrm{F}_n=\alpha\mathrm{A}+\beta\mathrm{B}+\gamma\mathrm{C}+\left(1-\frac{1}{2^n}\right)\beta(\tfrac{1}{2}\mathrm{A}-\mathrm{B}+\tfrac{1}{2}\mathrm{C}) \quad .\ .\ .\ (8a.11)$$

$$=\left(\alpha+\tfrac{1}{2}\beta-\frac{1}{2^{n+1}}\beta\right)\mathrm{A}+\frac{1}{2^n}\beta\mathrm{B}+\left(\tfrac{1}{2}\beta+\gamma-\frac{1}{2^{n+1}}\beta\right)\mathrm{C}. \quad .\ (8a.12)$$

As the number of generations increases, this quickly approaches the limiting stable distribution

$$(\alpha+\tfrac{1}{2}\beta)\mathrm{A}+(\gamma+\tfrac{1}{2}\beta)\mathrm{C}. \quad .\ .\ .\ .\ (8a.13)$$

(b) *Assortative Mating (Dominants × Dominants, Recessives × Recessives).*

The initial zygotic distribution

$$\mathrm{P}=\alpha\mathrm{A}+\beta\mathrm{B}+\gamma\mathrm{C} \quad .\ .\ .\ .\ .\ .\ (8b.1)$$

may be written phenotypically

$$\mathrm{P}=(\alpha+\beta)\mathfrak{D}+\gamma\mathrm{C}, \quad .\ .\ .\ .\ .\ (8b.2)$$

Here

$$\mathfrak{D}=\frac{\alpha}{\alpha+\beta}\mathrm{A}+\frac{\beta}{\alpha+\beta}\mathrm{B}=\frac{\alpha\mathrm{A}+\beta\mathrm{B}}{\alpha+\beta}, \quad .\ .\ .\ .\ (8b.3)$$

representing the genotype distribution of the dominants in P.

With the system of mating under consideration, the first filial generation is

$$\mathrm{F}_1=(\alpha+\beta)\mathfrak{D}^2+\gamma\mathrm{C}^2 \quad .\ .\ .\ .\ .\ .\ (8b.4)$$

$$=\frac{(\alpha\mathrm{A}+\beta\mathrm{B})^2}{\alpha+\beta}+\gamma\mathrm{C}^2. \quad .\ .\ .\ .\ .\ (8b.5)$$

Therefore

$$(\alpha+\beta)\mathrm{F}_1=(\alpha^2\mathrm{A}^2+2\alpha\beta\mathrm{AB}+\beta^2\mathrm{B}^2)+(\alpha+\beta)\gamma\mathrm{C}^2$$

$$=\alpha^2\mathrm{A}+2\alpha\beta(\tfrac{1}{2}\mathrm{A}+\tfrac{1}{2}\mathrm{B})+\beta^2(\tfrac{1}{4}\mathrm{A}+\tfrac{1}{2}\mathrm{B}+\tfrac{1}{4}\mathrm{C})+(\alpha+\beta)\gamma\mathrm{C}$$

$$=(\alpha^2+\alpha\beta+\tfrac{1}{4}\beta^2)\mathrm{A}+(\alpha\beta+\tfrac{1}{2}\beta^2)\mathrm{B}+(\tfrac{1}{4}\beta^2+\alpha\gamma+\beta\gamma)\mathrm{C}. \quad .\ (8b.6)$$

It will be found that $\mathrm{F}_1-\mathrm{P}$ is a multiple of $\frac{1}{2}\mathrm{A}-\mathrm{B}+\frac{1}{2}\mathrm{C}$, and hence that $\mathrm{F}_n$ can be found by summation of a series, just as in Case (*a*). The series in this case is not a geometrical progression, but it is of a type whose sum is easily obtained. Following the procedure of Case (*a*), it will be found that the total increase in $n$ generations can be expressed as

$$\left\{\frac{\frac{1}{2}\beta}{(\alpha+\frac{1}{2}\beta)(\alpha+\beta)}+\frac{\frac{1}{2}\beta}{(\alpha+\beta)(\alpha+\frac{3}{2}\beta)}+\ldots+\frac{\frac{1}{2}\beta}{(\alpha+\frac{1}{2}n\beta)[\alpha+\frac{1}{2}(n+1)\beta]}\right\}\beta(\alpha+\tfrac{1}{2}\beta)(\tfrac{1}{2}A-B+\tfrac{1}{2}C)$$

$$=\left\{\left(\frac{1}{\alpha+\frac{1}{2}\beta}-\frac{1}{\alpha+\beta}\right)+\left(\frac{1}{\alpha+\beta}-\frac{1}{\alpha+\frac{3}{2}\beta}\right)+\ldots+\left(\frac{1}{\alpha+\frac{1}{2}n\beta}-\frac{1}{\alpha+\frac{1}{2}(n+1)\beta}\right)\right\}\beta(\alpha+\tfrac{1}{2}\beta)(\tfrac{1}{2}A-B+\tfrac{1}{2}C)$$

$$=\left\{\frac{1}{\alpha+\frac{1}{2}\beta}-\frac{1}{\alpha+\frac{1}{2}(n+1)\beta}\right\}\beta(\alpha+\tfrac{1}{2}\beta)(\tfrac{1}{2}A-B+\tfrac{1}{2}C)$$

$$=\left\{\beta-\frac{\beta(\alpha+\frac{1}{2}\beta)}{\alpha+\frac{1}{2}(n+1)\beta}\right\}(\tfrac{1}{2}A-B+\tfrac{1}{2}C). \qquad (8b.7)$$

The $n$th filial generation is obtained by adding (8*b*.7) to (8*b*.1). We obtain

$$F_n=\alpha A+\beta B+\gamma C+\beta(\tfrac{1}{2}A-B+\tfrac{1}{2}C)-\frac{\beta(\alpha+\frac{1}{2}\beta)}{\alpha+\frac{1}{2}(n+1)\beta}(\tfrac{1}{2}A-B+\tfrac{1}{2}C)$$

$$=(\alpha+\tfrac{1}{2}\beta)A+(\tfrac{1}{2}\beta+\gamma)C-\frac{\beta(\alpha+\frac{1}{2}\beta)}{\alpha+\frac{1}{2}(n+1)\beta}(\tfrac{1}{2}A-B+\tfrac{1}{2}C). \qquad (8b.8)$$

As $n$ increases, the fraction with $\alpha+\frac{1}{2}(n+1)\beta$ in the denominator approaches zero. Hence $F_n$ approaches a stable distribution, namely, $(\alpha+\frac{1}{2}\beta)A+(\frac{1}{2}\beta+\gamma)C$, the same as in Case (*a*). (*Cf.* 8*a*.13.)

### (c) *Fraternal Mating.*

In this and the following case it is necessary to use the copular representation (7.4), from which of course the zygotic representation can be deduced. The determination of $F_n$ is much more difficult than in Cases (*a*) and (*b*). It is best obtained with the aid of matrix algebra; and as this is beyond the scope of this paper, I content myself with showing only in each case how the copular representation of any generation is deduced from the preceding.

Suppose that initially

$$P=\lambda AA+\mu BB+\nu CC+\theta BC+\phi CA+\psi AB, \qquad (8c.1)$$

and that brothers and sisters are mated at random. Then the filial generation is

$$F_1=\lambda(AA)^2+\mu(BB)^2+\nu(CC)^2+\theta(BC)^2+\phi(CA)^2+\psi(AB)^2. \qquad (8c.2)$$

Using short-circuited multiplication (*i.e.* (5.3) instead of (7.2)),

$$F_1=\lambda(A)^2+\mu(\tfrac{1}{4}A+\tfrac{1}{2}B+\tfrac{1}{4}C)^2+\nu(C)^2+\theta(\tfrac{1}{2}B+\tfrac{1}{2}C)^2+\phi(B)^2+\psi(\tfrac{1}{2}A+\tfrac{1}{2}B)^2$$

$$=\lambda AA+\mu(\tfrac{1}{16}AA+\tfrac{1}{4}BB+\tfrac{1}{16}CC+\tfrac{1}{4}BC+\tfrac{1}{8}CA+\tfrac{1}{4}AB)+\nu CC$$
$$+\theta(\tfrac{1}{4}BB+\tfrac{1}{2}BC+\tfrac{1}{4}CC)+\phi BB+\psi(\tfrac{1}{4}AA+\tfrac{1}{2}AB+\tfrac{1}{4}BB)$$

$$=(\lambda+\tfrac{1}{16}\mu+\tfrac{1}{4}\psi)AA+(\tfrac{1}{4}\mu+\tfrac{1}{4}\theta+\phi+\tfrac{1}{4}\psi)BB+(\tfrac{1}{16}\mu+\nu+\tfrac{1}{4}\theta)CC$$
$$+(\tfrac{1}{4}\mu+\tfrac{1}{2}\theta)BC+\tfrac{1}{8}\mu CA+(\tfrac{1}{4}\mu+\tfrac{1}{2}\psi)AB. \qquad (8c.3)$$

(d) *Filial Mating.*

Starting from an arbitrary copular distribution as in (*c*), suppose that each individual (or each individual of one sex) is mated with the parent of opposite sex. Then

$$F_1 = \lambda AA.A + \mu BB.B + \nu CC.C + \theta BC.(\tfrac{1}{2}B + \tfrac{1}{2}C) + \phi CA.(\tfrac{1}{2}C + \tfrac{1}{2}A) + \psi AB.(\tfrac{1}{2}A + \tfrac{1}{2}B) \quad (8d.1)$$

$$= \lambda A.A + \mu(\tfrac{1}{4}A + \tfrac{1}{2}B + \tfrac{1}{4}C)B + \nu C.C + \theta(\tfrac{1}{2}B + \tfrac{1}{2}C)^2 + \phi B(\tfrac{1}{2}C + \tfrac{1}{2}A) + \psi(\tfrac{1}{2}A + \tfrac{1}{2}B)^2$$

$$= \lambda AA + \mu(\tfrac{1}{4}AB + \tfrac{1}{2}BB + \tfrac{1}{4}BC) + \nu CC + \theta(\tfrac{1}{4}BB + \tfrac{1}{2}BC + \tfrac{1}{4}CC) + \phi(\tfrac{1}{2}BC + \tfrac{1}{2}AB) + \psi(\tfrac{1}{4}AA + \tfrac{1}{2}AB + \tfrac{1}{4}BB)$$

$$= (\lambda + \tfrac{1}{4}\psi)AA + (\tfrac{1}{2}\mu + \tfrac{1}{4}\theta + \tfrac{1}{4}\psi)BB + (\nu + \tfrac{1}{4}\theta)CC + (\tfrac{1}{4}\mu + \tfrac{1}{2}\theta + \tfrac{1}{2}\phi)BC + 0.CA + (\tfrac{1}{4}\mu + \tfrac{1}{2}\phi + \tfrac{1}{2}\psi)AB. \quad (8d.2)$$

## § 9. Compact Multiplication Tables.

If P and Q are any two normalised hypercomplex numbers in **G** (say $\delta D + \rho R$, $\delta' D + \rho' R$, where $\delta + \rho = \delta' + \rho' = 1$), then

$$PQ = \tfrac{1}{2}P + \tfrac{1}{2}Q. \quad (9.1)$$

This may be shown directly by multiplying and applying (5.2); or more briefly by observing that $\tfrac{1}{2}P + \tfrac{1}{2}Q$ is also normalised, so that by (6.5)

$$P^2 = P, \qquad Q^2 = Q,$$

$$\tfrac{1}{2}P + \tfrac{1}{2}Q = (\tfrac{1}{2}P + \tfrac{1}{2}Q)^2 = \tfrac{1}{4}P^2 + \tfrac{1}{2}PQ + \tfrac{1}{4}Q^2 = \tfrac{1}{4}P + \tfrac{1}{2}PQ + \tfrac{1}{4}Q,$$

from which (9.1) follows.

The result (9.1) may be regarded as a compact form of the gametic multiplication table, since it includes the three equations (5.2) as special cases. (It must be noted that (9.1) only applies if P, Q are normalised. The more general result is: $PQ = \tfrac{1}{2}(\delta' + \rho')P + \tfrac{1}{2}(\delta + \rho)Q$.)

It will be convenient to use the letters *a*, *b*, *c*, . . . to denote each either D or R, and then the compact multiplication table may be written

$$ab = \tfrac{1}{2}a + \tfrac{1}{2}b. \quad (9.2)$$

Applying to this the process of duplication, we obtain

$$ab.cd = (\tfrac{1}{2}a + \tfrac{1}{2}b)(\tfrac{1}{2}c + \tfrac{1}{2}d),$$

*i.e.*

$$ab.cd = \tfrac{1}{4}ac + \tfrac{1}{4}ad + \tfrac{1}{4}bc + \tfrac{1}{4}bd, \quad (9.3)$$

which gives a compact form of the zygotic multiplication table: for it includes all the six equations (5.3) as special cases. *E.g.* on putting $a = c = D$, $b = d = R$, we get from (9.3) the formula for $B^2$.

Similarly, the compact copular multiplication table is

$$ab.cd : ef.gh = \tfrac{1}{16}(a+b)(c+d).(e+f)(g+h),$$

*i.e.*

$$ab.cd : ef.gh = \tfrac{1}{16}ac.eg + \tfrac{1}{16}ac.eh + \ldots \quad \text{(16 terms),.} \qquad (9.4)$$

which includes the 21 equations (7.2).

## § 10. Offspring of Consanguineous Marriages.

Formulæ for the probability of RR offspring of various kinds of consanguineous marriages were given by Dahlberg (1929), and his verbal arguments may be translated into non-associative algebra. As an example, consider the distribution of genotypes DD, DR, RR in the offspring of a marriage between first cousins. This distribution will be found from a non-associative product of the form

$$ab.cd : ef :. ab.cd : gh. \quad . \quad . \quad . \quad . \quad (10.1)$$

Here each of the letters stands for either D or R; *ab* and *cd* denote the genetic constitutions of the common grandparents of the cousins; the two sibs, parents of cousins, are both represented by *ab . cd*; and the cousins themselves by *ab . cd : ef* and *ab . cd : gh* respectively.

Simplifying (10.1) by repeated use of (9.1), we have

$$ab.cd : ef = \left\{\frac{1}{2}\left(\frac{a+b}{2} + \frac{c+d}{2}\right)\right\}\frac{e+f}{2} = \left(\frac{a+b+c+d}{4}\right)\left(\frac{e+f}{2}\right)$$

$$= \frac{1}{2}\left(\frac{a+b+c+d}{4} + \frac{e+f}{2}\right) = \tfrac{1}{8}(a+b+c+d+2e+2f).$$

Similarly,

$$ab.cd : gh = \tfrac{1}{8}(a+b+c+d+2g+2h).$$

Therefore

$$ab.cd : ef :. ab.cd : gh = \tfrac{1}{64}(a^2+b^2+c^2+d^2+60 \text{ product terms}). \quad . \quad (10.2)$$

We must now take into account whatever information is given about the genetic constitution of the four grandparents. We might, for example, be given the genotype of one of them. Assuming, however, that they are merely random members of a stable population,

$$P = \delta^2 DD + 2\delta\rho DR + \rho^2 RR = \delta D + \rho R, \qquad (\delta + \rho = 1)$$

then the probability of $a$ being D or R is $\delta$ or $\rho$, and so for each of the ancestral gametes. Hence (10.2) yields for the offspring of first cousins the probability distribution

$$\tfrac{1}{64}\{(4\delta + 60\delta^2)DD + 60\delta\rho DR + (4\rho + 60\rho^2)RR\}$$
$$= (\tfrac{1}{16}\delta + \tfrac{15}{16}\delta^2)DD + \tfrac{15}{8}\delta\rho DR + (\tfrac{1}{16}\rho + \tfrac{15}{16}\rho^2)RR, \quad (10.3)$$

agreeing with Dahlberg's result.

## § 11. Further Genetic Algebras.

Consider inheritance depending on two pairs of autosomal allelomorphs, say D, R and D′, R′. The corresponding gametic algebras **G**, **G′** have multiplication tables:

$$\left.\begin{array}{lll} D^2 = D, & DR = \tfrac{1}{2}D + \tfrac{1}{2}R, & R^2 = R; \\ D'^2 = D', & D'R' = \tfrac{1}{2}D' + \tfrac{1}{2}R', & R'^2 = R'. \end{array}\right\} \quad . \quad . \quad (11.1)$$

Taking both pairs into account, there are four gametic types:

$$DD', \quad DR', \quad RD', \quad RR', \quad . \quad . \quad . \quad (11.2)$$

whose multiplication table is constructed as follows:—

$$\left.\begin{array}{l} DD'.DD' = D^2.D'^2 = DD', \\ DD'.DR' = D^2.D'R' = D(\tfrac{1}{2}D' + \tfrac{1}{2}R') = \tfrac{1}{2}DD' + \tfrac{1}{2}DR', \\ DR'.RD' = DR.D'R' = (\tfrac{1}{2}D + \tfrac{1}{2}R)(\tfrac{1}{2}D' + \tfrac{1}{2}R') = \tfrac{1}{4}DD' + \tfrac{1}{4}DR' + \tfrac{1}{4}RD' + \tfrac{1}{4}RR', \end{array}\right\} \quad (11.3)$$

and so on. (10 equations.)

(It will be seen that although multiplication is non-associative we assume, *e.g.*, $DD' . DR' = D^2 . D'R'$. This is justified because the combination of dashed and undashed symbols is mere juxtaposition, not genetical multiplication.) This is precisely the process referred to in § 4 of forming the direct product of the two algebras **G**, **G′**, which is well known in the theory of linear algebras.

Alternatively, let us use $a$, $b$ to denote each either D or R, and $a'$, $b'$ similarly for D′ or R′, so that, for example, $aa'$ can denote any of the four gametic types. Then we can write the joint multiplication table in the compact form:

$$aa'.bb' = ab.a'b' = (\tfrac{1}{2}a + \tfrac{1}{2}b)(\tfrac{1}{2}a' + \tfrac{1}{2}b'), \quad . \quad . \quad . \quad (11.4)$$

*i.e.*

$$aa'.bb' = \tfrac{1}{4}aa' + \tfrac{1}{4}ab' + \tfrac{1}{4}ba' + \tfrac{1}{4}bb'. \quad . \quad . \quad . \quad (11.5)$$

The zygotic algebra is obtained by duplicating the gametic, and is the direct product **ZZ′**. That is to say, it is immaterial whether the process of duplication is carried out before or after that of forming the direct product (Etherington, 1941, Theorem V). There is one point, in this connection, which requires elucidation. It has been pointed out that by pairing the four gametic types (11.2) we obtain the ten types of zygote, namely:

DD′.DD′, DD′.DR′, DD′.RD′, DD′.RR′, DR′.DR′, DR′.RD′, DR′.RR′, RD′.RD′, RD′.RR′, RR′.RR′, (11.6)

which figure in (11.3). There are, however, only nine genotypes, namely:

DDD′D′, DDD′R′, DDR′R′, DRD′D′, DRD′R′, DRR′R′, RRD′D′, RRD′R′, RRR′R′, (11.7) there being no distinction between the double heterozygotes DD′ . RR′ and DR′ . RD′, which both give rise to the genotype DRD′R′. The fact is that for calculating progeny distributions it is really optional whether we use the *zygotic algebra* based on the ten zygotic types (11.6), or the *genotypic algebra* based on the nine genotypes (11.7). The latter is obtained from the former simply by suppressing the distinction between DD′ . RR′ and DR′ . RD′. To geneticists the genotypic algebra would seem to be the obvious one to use; but the zygotic algebra is mathematically much simpler—firstly, because it is a direct product of simpler algebras; secondly, because it is a duplicate algebra; thirdly, because its multiplication table can be written in the compact form (11.5). In the final interpretation of any results obtained by use of the zygotic algebra, the distinction between the equivalent double heterozygotic types can be suppressed; just as with any zygotic or genotypic algebra, in the final interpretation of any calculation, the distinction between genotypes which are the same phenotype may be dropped in order to obtain a result true for phenotypes.

Some genetic algebras representing more complicated types of symmetrical inheritance were considered in G.A., including (§ 14) a group of three linked series of multiple allelomorphs, and (§ 15) inheritance in tetraploids. These algebras can all be manipulated on the lines illustrated for simple mendelian inheritance, the extra complication being to some extent offset by the consistent use of compact multiplication tables. As long as only symmetrical inheritance is considered, and zygotic types (differing in their gametic formation) are used rather than genotypes (differing in their relevant gene content), the corresponding gametic, zygotic and copular algebras are related by duplication; and two or more independent genetic algebras of the same kind (**G**, **Z** or **K**) can be combined by forming their direct product.

Let us finally consider briefly an *unsymmetrical* genetic algebra, *i.e.* representing inheritance which is not symmetrical in the sexes.

Consider a single gene difference D, R on the X-chromosome in a species where the male is heterogametic. The gametic types are

$$\textit{Ova},\ \mathrm{D},\ \mathrm{R};\quad \textit{Sperm},\ \mathrm{D},\ \mathrm{R},\ \mathrm{Y};\ . \quad . \quad . \qquad (11.8)$$

and the zygotic types with the gametes which they produce give the multiplication rules:

$$\left.\begin{array}{llll} \textit{Female}, & a \equiv \mathrm{DD}=\mathrm{D}, & b \equiv \mathrm{DR}=\tfrac{1}{2}\mathrm{D}+\tfrac{1}{2}\mathrm{R}, & c \equiv \mathrm{RR}=\mathrm{R}; \\ \textit{Male}, & d \equiv \mathrm{DY}=\tfrac{1}{2}\mathrm{D}+\tfrac{1}{2}\mathrm{Y}, & e \equiv \mathrm{RY}=\tfrac{1}{2}\mathrm{R}+\tfrac{1}{2}\mathrm{Y}. & \end{array}\right\} \qquad (11.9)$$

Hence $ad=\mathrm{D}(\tfrac{1}{2}\mathrm{D}+\tfrac{1}{2}\mathrm{Y})$, $bd=(\tfrac{1}{2}\mathrm{D}+\tfrac{1}{2}\mathrm{R})(\tfrac{1}{2}\mathrm{D}+\tfrac{1}{2}\mathrm{Y})$, and so on; whence we have the zygotic multiplication table:

$$\left.\begin{array}{lll} ad=\tfrac{1}{2}a+\tfrac{1}{2}d, & bd=\tfrac{1}{4}a+\tfrac{1}{4}b+\tfrac{1}{4}d+\tfrac{1}{4}e, & cd=\tfrac{1}{2}b+\tfrac{1}{2}e, \\ ae=\tfrac{1}{2}b+\tfrac{1}{2}d, & be=\tfrac{1}{4}b+\tfrac{1}{4}c+\tfrac{1}{4}d+\tfrac{1}{4}e, & ce=\tfrac{1}{2}c+\tfrac{1}{2}e. \end{array}\right\} \quad (11.10)$$

Since two males or two females produce no offspring, we must write also

$$a^2=b^2=c^2=d^2=e^2=ab=bc=ca=de=0.$$

It will be seen that the zygotic algebra is not obtained entirely by the process of duplication, since this would give, *e.g.*, $a^2=a$. If a population is denoted

$$P=\alpha a+\beta b+\gamma c+\delta d+\epsilon e, \quad (11.11)$$

and the male and female components are normalised separately:

$$\alpha+\beta+\gamma=\delta+\epsilon=1, \quad (11.12)$$

and if Q is another population represented in the same manner, then the product PQ describes the population of offspring, and will be automatically normalised.

An equally satisfactory scheme is to write instead of (11.9)

$$\left.\begin{array}{lll} a\equiv DD=D, & b\equiv DR=\tfrac{1}{2}D+\tfrac{1}{2}R, & c\equiv RR=R, \\ d\equiv DY=D+Y, & e\equiv RY=R+Y, & \end{array}\right\} \quad (11.13)$$

giving

$$ad=a+d, \quad bd=\tfrac{1}{2}a+\tfrac{1}{2}b+\tfrac{1}{2}d+\tfrac{1}{2}e, \quad \text{etc.} \quad (11.14)$$

We deal in this case with the female and male components of a population separately:

$$F=\alpha A+\beta b+\gamma c, \quad M=\delta d+\epsilon e. \quad (\alpha+\beta+\gamma=\delta+\epsilon=1.) \quad (11.15)$$

The offspring by random mating is given by their product, separated similarly into two components.

The numerical coefficients which appear on the right sides of (11.14) correspond to the asymmetrical extensors of Hogben's matrix notation (1933), just as the coefficients in (5.3) correspond to the symmetrical extensors.

## Summary.

The sign × is used by geneticists to indicate crossing of types. Literal interpretation of this as a symbol of multiplication leads to a type of algebra in which the associative law $P\times(Q\times R)=(P\times Q)\times R$ is not obeyed. The different "algebras" which in this way correspond to the various possible modes of inheritance known in genetics are therefore necessarily different from the algebra of ordinary numbers. They are of a kind known as "linear algebras," and it is shown that various genetical problems can be conveniently treated by means of a symbolism based on this fact.

I am indebted to Dr J. Ffoulkes Edwards for a lengthy correspondence in which this paper germinated; and to Dr Charlotte Auerbach, of the Institute of Animal Genetics, University of Edinburgh, for much constructive criticism.

---

## REFERENCES TO LITERATURE.

DAHLBERG, G., 1929. "Inbreeding in man," *Genetics*, vol. xiv, pp. 421–454.

ETHERINGTON, I. M. H., 1939 *a*. "On non-associative combinations," *Proc. Roy. Soc. Edin.*, vol. lix, pp. 153–162.

——, 1939 *b*. "Genetic algebras," *ibid.*, pp. 242–258. (Referred to as "G.A.")

——, 1941. "Duplication of linear algebras," *Proc. Edin. Math. Soc.* (*In press.*)

GEPPERT, H., and KOLLER, S., 1938. *Erbmathematik*, Leipzig.

HOGBEN, L., 1931. *Genetic Principles in Medicine and Social Science*, London.

——, 1933. "A matrix notation for mendelian populations," *Proc. Roy. Soc. Edin.*, vol. liii, pp. 7–25.

JENNINGS, H. S., 1916, 1917. "The numerical results of diverse systems of breeding," *Genetics*, vol. i, pp. 53–89; vol. ii, pp. 97–154.

ROBBINS, R. B., 1917, 1918. "Some applications of mathematics to breeding problems," *Genetics*, vol. ii, pp. 489–504; vol. iii, pp. 73–92, 375–389.

WENTWORTH, E. N., and REMICK, B. L., 1916. "Some breeding properties of the generalised mendelian population," *Genetics*, vol. i, pp. 608–616.

(*Issued separately March* 18, 1941.)

# 10

Reprinted from *Math. Scand.*, **10**, 25–44 (1962) by permission of the Mathematical Societies of Denmark, Iceland, Norway, Finland and Sweden

# GENETIC ALGEBRAS STUDIED RECURSIVELY AND BY MEANS OF DIFFERENTIAL OPERATORS

OLAV REIERSÖL

## 1. Introduction.

An algebra may be defined as a vector space together with a multiplication rule for the vectors such that

(1.1) The vector space is closed with respect to multiplication.
(1.2) Multiplication is distributive with respect to vector addition.
(1.3) A scalar factor may be moved freely within a product of vectors.

We shall consider a particular kind of non-associative algebras which have been called genetic algebras because of their application in population genetics. The study of these algebras and their genetical interpretations was initiated by I. M. H. Etherington [2]-[7]. Papers on genetic algebras have also been published by Gonshor [12], Raffin [17] and Schafer [18].

In this paper I shall present a new method of studying genetic algebras. One aspect of the method is that the multiplication rules are expressed by means of differential operators and that these operators are used in the study of the algebras. This makes it possible to avoid an explicit consideration of the components of the elements. Another aspect of the method is that the algebras are studied recursively. In the study of a genetic algebra corresponding to $k$ linked loci we make use of certain homomorphisms of this algebra onto algebras corresponding to smaller numbers of loci. In this way we may use results already found for the latter algebras in the derivation of results for the algebra corresponding to $k$ linked loci.

In the present paper we shall consider genetic algebras in the case of haploid gametes only, and further restrict the study to the case when the linkage distribution is the same for both sexes. It is evident, however, that the method is applicable to other genetic algebras.

## 2. Preliminaries on differential operators.

Let $f$ and $g$ be functions of a set of variables $x_1, \ldots, x_m$ such that the

Received June 18, 1960.

derivatives considered in the following exist. In the present paper the differential operators will be applied to polynomials only, so that all derivatives exist. We note that derivatives of polynomials may be defined algebraically, and in this paper all operators may be regarded as purely algebraic operators.

Let $D_{x_i}$ be defined by

$$D_{x_i}f = \frac{\partial f}{\partial x_i} \tag{2.1}$$

and let a linear form in the $D_{x_i}$ be defined by

$$\left(\sum_{i=1}^{m} c_i D_{x_i}\right) f = \sum_{i=1}^{m} c_i (D_{x_i}f) , \tag{2.2}$$

where $c_1, \ldots, c_m$ are constants. The operator

$$D = \sum_{i=1}^{m} c_i D_{x_i} \tag{2.3}$$

has the same properties as a differential operator with respect to a single variable,

$$D(f+g) = Df+Dg , \tag{2.4}$$
$$D(cf) = c(Df) , \tag{2.5}$$
$$D(fg) = f(Dg)+g(Df) . \tag{2.6}$$

We may similarly define polynomials in the operators $D_{x_i}$. If $D$ is such a polynomial, (2.4) and (2.5) are still valid. If $D_1$ and $D_2$ are two polynomials in the $D_{x_i}$ with constant coefficients we have in addition

$$(D_1+D_2)f = D_1f+D_2f , \tag{2.7}$$
$$(D_1D_2)f = D_1(D_2f) , \tag{2.8}$$
$$(cD)f = c(Df) . \tag{2.9}$$

Let us next consider $k$ differential operators $D_1, \ldots, D_k$ each of which is a linear form in the operators $D_{x_i}$ with constant coefficients. If each $D_i$ operates on a product $fg$ we may instead of $D_i$ consider two operators $D_{i1}$ and $D_{i2}$ defined by

$$D_{i1}(fg) = g(D_if) , \tag{2.10}$$
$$D_{i2}(fg) = f(D_ig) . \tag{2.11}$$

(Compare Stephens [19, p. 33].) Then

$$D_i = D_{i1}+D_{i2} . \tag{2.12}$$

If we have an expression of the form

(2.13) $$D_1^{\alpha_1}\dots D_k^{\alpha_k}(D_1^{\beta_1}\dots D_k^{\beta_k}f)(D_1^{\gamma_1}\dots D_k^{\gamma_k}g)\,,$$

where each $D_i$ within a parenthesis operates only within the parenthesis, while the $D_i^{\alpha_i}$ operate on the product of the two parentheses, then we may replace each $D_i$ which operates on both factors by $D_{i1}+D_{i2}$, each $D_i$ which operates on $f$ only by $D_{i1}$, and each $D_i$ which operates on $g$ only by $D_{i2}$. Formula (2.13) may thus be rewritten in the form

(2.14) $$\prod_{i=1}^{k}(D_{i1}+D_{i2})^{\alpha_i}D_{i1}^{\beta_i}D_{i2}^{\gamma_i}fg\,.$$

If we have a sum of terms of the form (2.13), each multiplied by a constant, then we may write each term in the form (2.14) and get an expression of the form

(2.15) $$P(D_{11}, D_{12}, \dots, D_{k1}, D_{k2})fg\,,$$

where $P$ denotes a polynomial with constant coefficients.

If $\Omega,\Omega_1,\Omega_2$ are polynomials in $D_{11},D_{12},\dots,D_{k1},D_{k2}$ we have

(2.16) $$\Omega(f_1g_1+f_2g_2) = \Omega(f_1g_1)+\Omega(f_2g_2)\,,$$

(2.17) $$\Omega(cfg) = c\Omega(fg)\,,$$

(2.18) $$(\Omega_1+\Omega_2)fg = \Omega_1fg+\Omega_2fg\,,$$

(2.19) $$(\Omega_1\Omega_2)fg = \Omega_1(\Omega_2fg)\,,$$

(2.20) $$(\Omega_1\Omega_2)fg = (\Omega_2\Omega_1)fg\,.$$

## 3. Preliminaries on linear difference equations with constant coefficients.

We shall give a summary of those results which are needed for the purpose of the present paper. For a more detailed treatment the reader is referred to textbooks on finite differences or difference equations, for instance the textbook by Jordan [13].

We consider a linear difference equation

(3.1) $$\sum_{i=0}^{k} b_i f(n+i) = g(n)\,,$$

where $g(n)$ is a known function and $b_0,b_1,\dots,b_k$ are constants. The variable $n$ will be supposed to take integer values only. Using the operator $E$, defined by $Ef(n)=f(n+1)$, we may write (3.1) in the form

(3.2) $$\psi(E)f(n) = g(n)\,,$$

where

(3.3) $$\psi(E) = \sum_{i=0}^{k} b_iE^i\,.$$

Let us first consider the homogeneous equation

$$(3.4) \qquad \psi(E)f(n) = 0 .$$

If the equation $\psi(x)=0$ has roots $r_1, r_2, \ldots, r_k$ which are all distinct, the general solution of (3.4) is

$$(3.5) \qquad f(n) = \sum_{i=1}^{k} c_i r_i^n ,$$

where the $c_i$ are arbitrary constants (real or complex).

If $r_1 = r_2 = \ldots = r_m$ the $m$ first terms on the right-hand side of (3.5) will be replaced by

$$(3.6) \qquad (c_1 + c_2 n + \ldots + c_m n^{m-1}) r_1^n .$$

A similar rule holds for any multiple root.

Let us next consider equation (3.2) when $g(n)$ has the form

$$(3.7) \qquad g(n) = \sum_{i=1}^{h} q_i a_i^n .$$

If $\psi(a_i) \neq 0$ for every $i$, then (3.2) has the solution

$$(3.8) \qquad f(n) = \sum_{i=1}^{h} \frac{q_i}{\psi(a_i)} a_i^n ,$$

and the general solution of (3.2) is the sum of (3.8) and the general solution of (3.4). The case when $g(n)$ contains a constant term is a special case of (3.7) which we obtain by setting one of the $a_i$ equal to 1.

If $g(n)$ is of the form

$$(3.9) \qquad g(n) = P(n) a^n ,$$

where $P(n)$ is a polynomial in $n$ and $a$ is a root of $\psi(x)$ of multiplicity $m$, then (3.2) has a solution of the form

$$(3.10) \qquad f(n) = Q(n) a^n ,$$

where $Q(n)$ is a polynomial whose degree is greater than the degree of $P(n)$ by $m$. The $m$ coefficients of the terms of degree lower than $m$ in $Q(n)$ will be arbitrary. The other coefficients of $Q(n)$ may be determined by insertion of (3.10) with undetermined coefficients in the difference equation and comparing coefficients on both sides.

If we have a solution $f_i(n)$ of the equation

$$\psi(E) f(n) = g_i(n)$$

for $i = 1, 2, \ldots, q$, then the equation

$$\psi(E)\,f(n) = \Sigma_{i=1}^{q} g_i(n)$$

has the solution $f(n) = \Sigma_{i=1}^{q} f_i(n)$.

## 4. Recombination operators.

Let us consider a set $S$ of integers and the partitions of this set into two disjoint subsets. When we include the empty set and the set $S$ itself as subsets of $S$, the number of such partitions is $2^{k-1}$, where $k$ is the number of elements of $S$. Let $U'$ and $U''$ be two complementary disjoint subsets of $S$ and let

$$U = (U', U'') = (U'', U')$$

denote the partition of $S$ defined by the subsets $U'$ and $U''$. Let $A = a_1 a_2 \ldots a_k$ and $B = b_1 b_2 \ldots b_k$. Let $U$ be a partition of the set $S = (1, 2, \ldots, k)$. The *recombination operator* $R(U)$ will be defined by

$$R(U)(A,B) = \tfrac{1}{2}\left((\textstyle\prod_{i\in U'} a_i)(\prod_{i\in U''} b_i) + (\prod_{i\in U''} a_i)(\prod_{i\in U'} b_i)\right). \tag{4.1}$$

Let $D_{a_i}$ and $D_{b_i}$ denote the differential operators $\partial/\partial a_i$ and $\partial/\partial b_i$. Since

$$\textstyle\prod_{i\in U'} a_i = (\prod_{i\in U''} D_{a_i})A \tag{4.2}$$

and

$$\textstyle\prod_{i\in U''} a_i = (\prod_{i\in U'} D_{a_i})A\,, \tag{4.3}$$

we get from (4.1)

$$R(U)(A,B) = \tfrac{1}{2}\left((\textstyle\prod_{i\in U'} D_{a_i})(\prod_{i\in U''} D_{b_i}) + (\prod_{i\in U''} D_{a_i})(\prod_{i\in U'} D_{b_i})\right)AB$$

or

$$R(U)\,(A,B) = R(U)(\textstyle\prod_{i\in S} D_{a_i}, \prod_{i\in S} D_{b_i})AB\,, \tag{4.4}$$

where $R(U)$ on the right hand side operates on the differential operators only, not on $AB$.

## 5. Presentation of the algebras.

We shall consider an algebra $\mathscr{A}_k$ whose elements $G$ are multilinear forms

$$G = \sum_{i_1=1}^{m_1} \cdots \sum_{i_k=1}^{m_k} g_{i_1\ldots i_k}\, a_{1i_1} \cdots a_{ki_k}\,, \tag{5.1}$$

where the $g_{i_1\ldots i_k}$ are real numbers and the $a_{ji_j}$ are variables. Each element of the form (5.1) with real coefficients will be supposed to belong to the algebra.

Addition of two elements of the algebra is taken to mean addition of polynomials in the usual sense, i.e.

$$(5.2)\quad \begin{cases} \sum_{i_1=1}^{m_1}\cdots\sum_{i_n=1}^{m_k} g_{i_1\ldots i_k} a_{1i_1}\cdots a_{ki_k} + \sum_{i_1=1}^{m_1}\cdots\sum_{i_k=1}^{m_k} g'_{i_1\ldots i_k} a_{1i_1}\cdots a_{ki_k} \\ = \sum_{i_1=1}^{m_1}\cdots\sum_{i_k=1}^{m_k} (g_{i_1\ldots i_k}+g'_{i_1\ldots i_k}) a_{1i_1}\cdots a_{ki_k}\,. \end{cases}$$

Multiplication of $G$ by a constant (a real number) is taken to mean multiplication of the polynomial by the real number in the usual sense, i.e.

$$(5.3)\qquad cG = \sum_{i_1=1}^{m_1}\cdots\sum_{i_k=1}^{m_k} c g_{i_1\ldots i_k} a_{1i_1}\cdots a_{ki_k}\,.$$

We shall next define a multiplication rule for the algebra. We shall use $\times$ as a symbol for this multiplication and we shall call it cross multiplication. We shall let $G_1G_2$ denote the product of the two polynomials $G_1$ and $G_2$ in the usual sense. If $G_1$ and $G_2$ belong to $\mathscr{A}_k$, then $G_1G_2$ does not belong to $\mathscr{A}_k$ because $\mathscr{A}_k$ contains polynomials of a fixed degree $k$, while the degree of $G_1G_2$ is $2k$.

The multiplication $G_1 \times G_2$ will be defined in such a way that:

(5.4) If $G_1$ and $G_2$ belong to $\mathscr{A}_k$, then $G_1 \times G_2$ belongs to $\mathscr{A}_k$.

(5.5) $G_1 \times G_2 = G_2 \times G_1$.

(5.6) $(cG_1) \times G_2 = c(G_1 \times G_2)$.

(5.7) $G_1 \times (G_2+G_3) = (G_1 \times G_2)+(G_1 \times G_3)$.

The $m_1m_2\ldots m_k$ elements of the form $A = a_{1i_1}a_{2i_2}\cdots a_{ki_k}$ form a basis of the algebra $\mathscr{A}_k$. Because of (5.6) and (5.7) the multiplication rule of the algebra will be determined if it is given for any pair of basis elements. The product of two basis elements $A_1$ and $A_2$ is defined by

$$(5.8)\qquad A_1 \times A_2 = \sum_{U \in W(S)} \lambda(U)\, R(U)\, A_1 A_2\,,$$

where $W(S)$ is the set of all partitions of the set $S = (1, 2, \ldots, k)$ into two disjoint subsets, where $\lambda(U)$ is a real number which is a function of $U$, and where $\sum_{U \in W(S)} \lambda(U) = 1$.

Let

$$(5.9)\qquad D_j = \sum_{i=1}^{m_j} D_{a_{ji}}$$

and let us use the notation

$$D_{j1}(A_1A_2) = A_2(D_jA_1)\,, \qquad D_{j2}(A_1A_2) = A_1(D_jA_2)\,,$$

which we used previously in Section 2. The multiplication rule may be rewritten in the form

$$(5.10) \quad A_1 \times A_2 = \sum_{U \in W(S)} \lambda(U) \left( R(U) \left( \prod_{j \in S} D_{j1}, \prod_{j \in S} D_{j2} \right) \right) A_1 A_2 ,$$

where $R(U)$ operates on the differential operators, not on $A_1 A_2$. The proof is as follows: Since $D_{j1}$ operates on $A_1$ only, and since $A_1$ contains only one $a_{ji}$, say $a_{ji_j}$, for each $j$, the operator $D_{j1}$ is in this case equivalent to the operator $D_{a_{ji_j}}$ and the equivalence between (5.8) and (5.10) follows from (4.4).

Since the operator $R(U)(\prod_{j \in S} D_{j1}, \prod_{j \in S} D_{j2})$ is independent of $A_1 A_2$, we get by means of (2.16)–(2.18)

$$(5.11) \quad G_1 \times G_2 = \sum_{U \in W(S)} \lambda(U) \, R(U) (\prod_{j \in S} D_{j1}, \prod_{j \in S} D_{j2}) G_1 G_2 ,$$

where $G_1$ and $G_2$ are any two elements of the algebra $\mathscr{A}_k$. It is easy to see that the multiplication defined by (5.11) actually satisfies (5.4)–(5.7). We may thus state

THEOREM 1. *The set of all multilinear forms of the form* (5.1) *with real coefficients together with an addition defined by* (5.2), *a multiplication by a scalar defined by* (5.3) *and a cross multiplication defined by* (5.11) *forms a commutative algebra. The multiplication rule of the basis elements is given by* (5.8).

We have considered a set $S$ consisting of the integers $1, 2, \ldots, k$. It is a formal generalization only to consider a set $S$ consisting of any given set of $k$ positive integers. With obvious modifications of (5.1)–(5.3), the results of this section are still valid.

We have considered an algebra with parameters which are real numbers. If we vary the parameters we get a family of algebras. Let us denote a particular algebra by $\mathscr{A}(\boldsymbol{m}_S, \boldsymbol{\lambda}_S)$, where $\boldsymbol{m}_S$ denotes the set of $m_j$ for which $j$ belongs to $S$, and where $\boldsymbol{\lambda}_S$ denotes the set of all $\lambda(U)$ for which $U$ belongs to $W(S)$. Let us denote by $(\mathscr{A}_k)$ the class of all algebras where the set $S$ consists of $k$ integers.

We shall write down more explicitly the multiplication rule for the first values of $k$. Instead of $\lambda(U) = \lambda((U', U''))$ we shall write $\lambda(U', U'')$, and we shall write $\lambda(S)$ when $U$ is the partition consisting of $S$ and the empty set.

In the algebra $\mathscr{A}_1(m_1)$

$$(5.12) \quad G_1 \times G_2 = \tfrac{1}{2}(G_1(D_1 G_2) + G_2(D_1 G_1)) .$$

In an algebra of the class $(\mathscr{A}_2)$

$$(5.13) \quad \begin{aligned} G_1 \times G_2 = {} & \tfrac{1}{2}\lambda(12)\left(G_1(D_1 D_2 G_2) + G_2(D_1 D_2 G_1)\right) + \\ & + \tfrac{1}{2}\lambda(1,2)\left((D_1 G_1)(D_2 G_2) + (D_2 G_1)(D_1 G_2)\right) . \end{aligned}$$

In an algebra of the class $(\mathscr{A}_3)$

$$(5.14)\qquad G_1\times G_2 = \tfrac{1}{2}\lambda(123)\big(G_1(D_1D_2D_3G_2)+G_2(D_1D_2D_3G_1)\big) + \\ + \tfrac{1}{2}\sum_{i=1}^{3}\lambda(i,jk)\,\big((D_iG_1)(D_jD_kG_2)+(D_iG_2)(D_jD_kG_1)\big),$$

where $(ijk)$ is a permutation of (123).

In an algebra of the class $(\mathscr{A}_4)$

$$(5.15)\qquad G_1\times G_2 = \tfrac{1}{2}\lambda(1234)\big(G_1(D_1D_2D_3D_4G_2)+G_2(D_1D_2D_3D_4G_1)\big) + \\ +\tfrac{1}{2}\sum_{i=1}^{4}\lambda(i,jkh)\,\big((D_iG_1)(D_jD_kD_hG_2)+(D_iG_2)(D_jD_kD_hG_1)\big)+ \\ +\tfrac{1}{2}\sum_{i<j}\lambda(ij,kh)\,\big((D_iD_jG_1)(D_kD_hG_2)+(D_iD_jG_2)(D_kD_hG_1)\big),$$

where $(ijkh)$ is a permutation of (1234).

In the formulae (5.13)–(5.15) each $D_i$ operates only within the parenthesis in which it is situated.

We note finally that the effect of the operator $D_j$ on an element $G$ of the form (5.1) is to remove the factor $a_{ji_j}$ from each term. We have for instance

$$(5.16)\qquad D_kG = \sum_{i_1=1}^{m_1}\cdots\sum_{i_{k-1}=1}^{m_{k-1}} g'_{i_1\ldots i_{k-1}}a_{1i_1}\cdots a_{k-1,\,i_{k-1}},$$

where

$$(5.17)\qquad g'_{i_1\ldots i_{k-1}} = \sum_{i_k=1}^{m_k} g_{i_1\ldots i_k}.$$

The operator $D_j$ is therefore essentially a summation operator. From the point of view of probability distributions the operator $D_j$ may be said to perform a marginalization of a distribution.

## 6. Genetic interpretations of Section 5.

A basis element $a_{1i_1}a_{2i_2}\ldots a_{ki_k}$ is interpreted as representing a gamete having the allele $a_{1i_1}$ at locus 1, the allele $a_{2i_2}$ at locus 2, and so on. An element (5.1) of the algebra whose coefficients are non-negative with sum one is interpreted as a probability distribution of the gametic types. The product (5.8) represents the probability distribution of gametes resulting from an individual of genotype $A_1A_2$. The set of numbers $\lambda(U)$ represents what H. Geiringer [8] has called the *linkage distribution.* In the case $k=2$, $\lambda(1,2)$ represents the recombination probability. In the genetic interpretation the $\lambda(U)$ must be non-negative. Moreover they are restricted by other inequalities.

The recombination operator $R(U)$ applied to two given gametes $A_1$ and $A_2$ represents the effect of crossing over during meiosis.

The product $G_1 \times G_2$ represents the probability distribution of gametes of the offspring when one population with gametic probability distribution $G_1$ mates at random with a population with gametic probability distribution $G_2$.

The linear combination $cG_1+(1-c)G_2$, where $0 \leqq c \leqq 1$, represents the gametic probability distribution of a mixture of two populations with gametic probability distributions $G_1$ and $G_2$.

If two or more loci are completely linked, they may be treated as one locus. In the case $k=2$ complete linkage means that $\lambda(1,2)=0$, and we get the multiplication rule

$$G_1 \times G_2 = \tfrac{1}{2}\big(G_1(D_1D_2G_2)+G_2(D_1D_2G_1)\big) , \tag{6.1}$$

and if $a_{1i_1}a_{2i_2}$ is replaced by a single symbol $b_{1v}$ then the multiplication rule will be the same as in the case of one locus.

If in the case $k=3$ we have complete linkage between loci 2 and 3, then $\lambda(2,13)=\lambda(3,12)=0$, and we get the multiplication rule

$$\begin{aligned} G_1 \times G_2 = {} & \tfrac{1}{2}\lambda(123)\big(G_1(D_1D_2D_3G_2)+G_2(D_1D_2D_3G_1)\big)+ \\ & +\tfrac{1}{2}\lambda(1,23)\big((D_1G_1)(D_2D_3G_2)+(D_1G_2)(D_2D_3G_1)\big) . \end{aligned} \tag{6.2}$$

Replacing $a_{2i_2}a_{3i_3}$ by a single variable $b_{2v}$ we get a multiplication rule of the form (5.13).

## 7. Sequences of powers and products connected with the different generations under panmixia.

Since multiplication is non-associative in the algebras defined in Section 5, there are different powers of the same degree having different *shapes* (Etherington [2]). We shall consider powers and products of two particular shapes. We shall consider the sequence of *plenary powers* defined by

$$G(n+1) = G(n) \times G(n), \qquad n = 0,\ 1,\ 2,\ \ldots\,. \tag{7.1}$$

Secondly, we shall consider the sequence of products defined by

$$H(n+2) = H(n) \times H(n+1), \qquad n = 0,\ 1,\ 2,\ \ldots\,. \tag{7.2}$$

If $H(0)=H(1)$ the sequence $\{H(n)\}$ will also be a sequence of powers. These powers are, however, different from the plenary powers.

The plenary power $G(n)$ represents the probability distribution of the gametic types in the $n$'th generation when the following conditions hold:

(7.3) The population is infinite.
(7.4) There is panmixia in each generation.
(7.5) All loci considered are in autosomes (i.e. chromosomes which are not sex chromosomes).
(7.6) The gametic probability distribution of generation 0 is $G(0)$.

The sequence $\{H(n)\}$ gives the probability distributions of the $X$ chromosomes if (7.3) and (7.4) hold. The $X$ chromosome of a male in the $n$'th generation comes from a female of the $(n-1)$'th generation. Of the two $X$ chromosomes of a female in the $n$'th generation one comes from a male and the other from a female of the $(n-1)$'th generation. When (7.3) and (7.4) hold, the female $X$ chromosome in the $n$'th generation coming from a female in the preceding generation must have the same probability distribution as the $X$ chromosome of a male in the $n$'th generation. Assuming that generation 0 has also been generated by panmixia, we may set

(7.7) $H(0) =$ the probability distribution of the female $X$ chromosome in generation 0 which comes from a male in the preceding generation.
(7.8) $H(1) =$ the probability distribution of the other female $X$ chromosome and the male $X$ chromosome in generation 0.

Then in the $n$'th generation the male $X$ chromosome will have the probability distribution $H(n+1)$ and the female $X$ chromosomes will have the probability distributions $H(n)$ and $H(n+1)$.

Other sequences of powers and products which have a genetic interpretation are the sequence of *principal powers* and the sequence of *primary products* (Etherington [3]). These sequences will not be considered in the present paper.

**8. Homomorphisms between the algebras.**

A mapping $G \to \eta(G)$ of an algebra $\mathscr{A}$ into an algebra $\mathscr{B}$ is called a *homomorphism* if

$$\eta(G_1+G_2) = \eta(G_1)+\eta(G_2)\,,$$
$$\eta(cG) = c\eta(G)\,,$$
$$\eta(G_1\times G_2) = \eta(G_1)\times\eta(G_2)\,,$$

for any elements $G_1$, $G_2$, $G$ which belong to $\mathscr{A}$ and any real number $c$.

We shall show that each of the differential operators $D_j$ defined by (5.9) generates a homomorphism of an algebra of the class $(\mathscr{A}_k)$ onto an algebra of the class $(\mathscr{A}_{k-1})$.

We have noted at the end of Section 5 that the effect of $D_j$ on $G$ is to remove the factor $a_{ji_j}$ from each term such that we get a multilinear form of a degree which is one less than the degree of $G$. If $G_1$ and $G_2$ are two elements of the algebra $\mathscr{A}(\boldsymbol{m}_S, \boldsymbol{\lambda}_S)$ and if $j$ is an integer belonging to the set $S$, then the elements $D_j G_1$ and $D_j G_2$ will belong to an algebra $\mathscr{A}(\boldsymbol{m}_{S-j}, \boldsymbol{\lambda}_{S-j})$ where $S-j$ denotes the set of integers which we obtain when $j$ is removed from $S$. From (2.4) and (2.5) follows that

$$(8.1) \qquad D_j(G_1+G_2) = D_j G_1 + D_j G_2 \,,$$

$$(8.2) \qquad D_j(cG) = cD_j G \,,$$

when $c$ is a constant. We shall next consider

$$(8.3) \quad D_j(G_1 \times G_2) = \sum_U \lambda(U)\, D_j\big(R(U)(\textstyle\prod_{h\in S} D_{h1}, \prod_{h\in S} D_{h2})\big)\, G_1 G_2 \,.$$

Using (2.20) and (4.1) we get

$$(8.4) \quad D_j R(U)(\textstyle\prod_{h\in S} D_{h1}, \prod_{h\in S} D_{h2})$$
$$= \tfrac{1}{2}(\textstyle\prod_{h\in U'} D_{h1} \prod_{h\in U''} D_{h2} + \prod_{h\in U''} D_{h1} \prod_{h\in U'} D_{h2})\, D_j \,.$$

Each of the two terms in the parenthesis on the right-hand side of (8.4) contains one and only one of the factors $D_{j1}$ and $D_{j2}$. Furthermore

$$D_{j1} D_j G_1 G_2 = D_{j2} D_j G_1 G_2 = D_{j1} D_{j2} G_1 G_2 \,.$$

Hence (8.4) multiplied by $G_1\, G_2$ is equal to

$$(8.5) \quad \big(R(U'-j,\, U''-j)(\textstyle\prod_{h\in S-j} D_{h1}, \prod_{h\in S-j} D_{h2})\big)\, D_{j1} D_{j2} G_1 G_2 \,,$$

where, of course, only one of the sets $U'$ and $U''$ contains $j$, such that if $j$ belongs to $U'$ then $U''-j = U''$. From (8.3)–(8.5) we get

$$(8.6) \quad D_j(G_1 \times G_2) = \sum_{U\in W(S-j)} \lambda(U) R(U)(\textstyle\prod_{h\in S-j} D_{h1}, \prod_{h\in S-j} D_{h2})\, D_{j1} D_{j2} G_1 G_2 \,,$$

where

$$(8.7) \qquad \lambda(U', U'') = \lambda(U_j', U'') + \lambda(U', U_j'')$$

if we write $U'_j$ for the union of $U'$ and the set which contains the element $j$ only. Evidently (8.6) means that

$$(8.8) \qquad D_j(G_1 \times G_2) = (D_j G_1) \times (D_j G_2) \,,$$

where $\boldsymbol{\lambda}_{S-j}$ is expressed in terms of $\boldsymbol{\lambda}_S$ by (8.7). We have thus shown that $D_j$ generates a homomorphism. Evidently any basis element of $\mathscr{A}(\boldsymbol{m}_{S-j}, \boldsymbol{\lambda}_{S-j})$ is the image of a basis element of $\mathscr{A}(\boldsymbol{m}_S, \boldsymbol{\lambda}_S)$. Hence any element of the former algebra is the image of an element of the latter algebra. This means that the mapping generated by $D_j$ is a mapping *onto* the algebra $\mathscr{A}(\boldsymbol{m}_{S-j}, \boldsymbol{\lambda}_{S-j})$. We shall formulate our results in the following theorem:

THEOREM 2. *If $G$ belongs to the algebra $\mathscr{A}(\boldsymbol{m}_S, \boldsymbol{\lambda}_S)$, where $S$ contains the integer $j$ and at least one other integer, then the mapping $G \to D_j G$ is a homomorphism of the algebra $\mathscr{A}(\boldsymbol{m}_S, \boldsymbol{\lambda}_S)$ onto the algebra $\mathscr{A}(\boldsymbol{m}_{S-j}, \boldsymbol{\lambda}_{S-j})$, where $\boldsymbol{\lambda}_{S-j}$ is expressed in terms of $\boldsymbol{\lambda}_S$ by* (8.7).

Let $T$ be a proper subset of $S$ and let $D_T$ denote the product

$$D_T = \prod_{j \in T} D_j \,. \tag{8.9}$$

Since the product of two homomorphisms is a homomorphism, we get

THEOREM 3. *If $G$ belongs to the algebra $\mathscr{A}(\boldsymbol{m}_S, \boldsymbol{\lambda}_S)$, then the mapping $G \to D_T G$ is a homomorphism of $\mathscr{A}(\boldsymbol{m}_S, \boldsymbol{\lambda}_S)$ onto $\mathscr{A}(\boldsymbol{m}_{S-T}, \boldsymbol{\lambda}_{S-T})$, where $\boldsymbol{\lambda}_{S-T}$ may be expressed in terms of $\boldsymbol{\lambda}_S$ by repeated application of* (8.7).

If $G$ belongs to $\mathscr{A}(m_1)$ the mapping $G \to D_1 G$ evidently is a homomorphism of $\mathscr{A}(m_1)$ onto the set of real numbers. Combining this with Theorem 3 we see that if $G$ belongs to the algebra $\mathscr{A}(\boldsymbol{m}_S, \boldsymbol{\lambda}_S)$, then the mapping $G \to D_S G$ is a homomorphism of $\mathscr{A}(\boldsymbol{m}_S, \boldsymbol{\lambda}_S)$ onto the set of real numbers. An algebra for which there exists a non-trivial homomorphism into the set of real numbers is called a *baric algebra* (Etherington [3]), and the real number which is the image of the element $G$ is called the *weight* of $G$. Using this terminology we may state

THEOREM 4. *The algebra $\mathscr{A}(\boldsymbol{m}_S, \boldsymbol{\lambda}_S)$ is a baric algebra with weight function $\xi(G) = D_S G$.*

The weight $\xi(G)$ is the sum of the coefficients in the expression (5.1). The weight of an element $G$ which represents a probability distribution is thus equal to 1.

Since $D_{S-T} D_T G = D_S G$, we have

$$\xi(D_T G) = \xi(G) \,. \tag{8.10}$$

The homomorphism $G \to D_T G$ thus preserves the weights of the elements.

It is evident that a sequence of products or powers is mapped onto a sequence of products or powers of the same shape. For instance a sequence of plenary powers is mapped onto a sequence of plenary powers, and a sequence $\{H(n)\}$ defined by (7.2) is mapped onto a sequence of the same type.

## 9. Explicit expressions and recurrence formulae for $G(n)$.

**9.1.** The sequence $\{G(n)\}$ of plenary powers was defined by (7.1). We noted that a sequence of plenary powers gives the gametic probability distribution in successive generations when there is panmixia, when the population is infinite, and when the loci are located in autosomes.

In the rest of this paper we shall consider algebras which have a genetic interpretation. This means that we consider the case when $\boldsymbol{\lambda}_S$ is a probability distribution, in other words the case when all $\lambda(U)$ are non-negative. Furthermore we shall suppose that $G(0)=G$ is a probability distribution. Then $G(n)$ is a probability distribution for every $n$ and $\xi(G(n))=1$ for every $n$.

In the case of one single locus we get from (5.12)

$$G \times G = G . \tag{9.1}$$

In this case any power of $G$ of any degree and of any shape will be equal to $G$. In particular

$$G(n) = G \tag{9.2}$$

for every $n$.

**9.2. The case of two loci.** In the following we shall always assume that no two loci are completely linked, for if they were completely linked we could regard them as one single locus. (See Section 6).

Setting $G_1=G_2=G(n)$ in (5.13) we get

$$G(n+1) = \lambda(12)G(n) + \lambda(1,2)(D_1G(n))(D_2G(n)) . \tag{9.3}$$

According to Theorem 2, $D_1G(n)$ and $D_2G(n)$ are plenary powers in the algebras $\mathscr{A}(m_2)$ and $\mathscr{A}(m_1)$, respectively. According to (9.2) we thus have $D_1G(n)=D_1G$ and $D_2G(n)=D_2G$ for every $n$. The difference equation (9.3) may thus be rewritten in the form

$$(E-\lambda(12))G(n) = \lambda(1,2)(D_1G)(D_2G) . \tag{9.4}$$

In this difference equation the values of the function $G(n)$ are not numbers but elements of an algebra. It is, however, easy to see that the results summarized in Section 3 apply to this case with a change of interpretation of the symbols. In (3.5) for instance the $c_i$ will now mean elements of the algebra, while the $r_i$ are still real or complex numbers. If some of the $r_i$ are complex numbers we shall also have to consider elements of the algebra where the coefficients of (5.1) are complex numbers, i.e. we must consider an algebra having as elements all multilinear forms (5.1) with complex coefficients. The properties of the algebras which we have found in the real case will hold also in the complex case. In the examples given in the present paper no complex roots occur.

The general solution of (9.4) is

$$G(n) = (D_1G)(D_2G) + C(\lambda(12))^n\,, \tag{9.5}$$

where $C$ is an element of the algebra which does not depend on $n$. Setting $n=0$ in (9.5) we get $C=G-(D_1G)(D_2G)$ which we insert in (9.5) to get

$$G(n) = (D_1G)(D_2G) + (\lambda(12))^n\big(G-(D_1G)(D_2G)\big)\,. \tag{9.6}$$

If a particular $G$ is given and if we wish to calculate $G(n)$ for some separate values of $n$, we should use (9.6). If we wish to calculate all $G(n)$ for a sequence of successive generations it is easier to use the difference equation (9.4) for recurrent computation of successive $G(n)$. Alternately we may use the homogeneous difference equation of the second order

$$(E-1)\big(E-\lambda(12)\big)G(n) = 0$$

which for the purpose of recurrent computation of $G(n)$ may be written in the form

$$G(n) = G(n-1) + \lambda(12)\big(G(n-1)-G(n-2)\big)\,.$$

**9.3. Three loci.** Setting $G_1=G_2=G(n)$ in (5.14) we get

$$G(n+1) = \lambda(123)G(n) + \sum_{i=1}^{3}\lambda(i,jk)\big(D_iG(n)\big)\big(D_jD_kG(n)\big)\,. \tag{9.7}$$

According to Theorems 2 and 3, $D_jD_kG(n)$ is a plenary power in the algebra $\mathscr{A}(m_i)$, and $D_iG(n)$ is a plenary power in the algebra

$$\mathscr{A}\big(m_j,\,m_k,\,\lambda(jk),\,\lambda(j,\,k)\big)\,.$$

Using (9.2) and (9.6) we thus get

$$D_jD_kG(n) = D_jD_kG\,, \tag{9.8}$$

$$D_iG(n) = (D_iD_jG)(D_iD_kG) + (\lambda(jk))^n\big((D_iG)-(D_iD_jG)(D_iD_kG)\big). \tag{9.9}$$

Inserting (9.8) and (9.9) in (9.7) we get the difference equation

$$\begin{aligned}\big(E-\lambda(123)\big)G(n) &= \big(1-\lambda(123)\big)(D_2D_3G)(D_1D_3G)(D_1D_2G) + \\ &+ \sum_{i=1}^{3}\lambda(i,jk)(D_jD_kG)\big(D_iG-(D_iD_jG)(D_iD_kG)\big)\big(\lambda(jk)\big)^n.\end{aligned} \tag{9.10}$$

If $\lambda(123)\neq\lambda(jk)$ for every pair of values $j,k$, then the general solution of (9.10) is

$$\begin{aligned}G(n) &= (D_2D_3G)(D_1D_3G)(D_1D_2G) + C\big(\lambda(123)\big)^n + \\ &+ \sum_{i=1}^{3}(D_jD_kG)\,\big(D_iG-(D_iD_jG)(D_iD_kG)\big)\,\big(\lambda\,(jk)\big)^n,\end{aligned} \tag{9.11}$$

where $C$ is an arbitrary element of the algebra. Setting $n=0$ we find

$$(9.12)\quad C = G + 2(D_2D_3G)(D_1D_3G)(D_1D_2G) - \sum_{i=1}^{3}(D_iG)(D_jD_kG)\,.$$

If for instance $\lambda(123)=\lambda(23)$, then $\lambda(1,23)=0$. The coefficient of $(\lambda(23))^n$ in (9.10) thus is equal to zero. The solution will then be (9.11) after deletion of the term containing $(\lambda(23))^n$.

As in the case $k=2$ it is clear that we should use the explicit expression for computation if we wish to compute $G(n)$ for some separate values of $n$. If we wish to compute all $G(n)$ for a sequence of successive generations we may compute the $D_iG(n)$ recurrently by means of (9.4) and use (9.7) for recurrent computation of $G(n)$. An alternative method of recurrent computation is to use the difference equation

$$(9.13)\quad (E-\lambda(123))(E-\lambda(23))(E-\lambda(13))(E-\lambda(12))\,G(n)$$
$$= (1-\lambda(123))(1-\lambda(23))(1-\lambda(13))(1-\lambda(12))(D_2D_3G)(D_1D_3G)(D_1D_2G)\,.$$

That $G(n)$ satisfies this difference equation is seen by application of the operator $(E-\lambda(23))(E-\lambda(13))(E-\lambda(12))$ to both sides of (9.10).

Since (9.13) is a difference equation of order four, it cannot be used for computation of $G(1)$, $G(2)$ and $G(3)$. These values may, however, be computed in the manner previously described, and after that we may use (9.13) for recurrent computation of $G(4)$, $G(5)$, and so on.

There is no great difference between the amount of numerical work required by the two methods. If we compute each $G(n)$ by both methods we get a good checking of each individual value.

If we apply the operator $E-1$ to (9.13) we see that $G(n)$ satisfies the homogeneous difference equation

$$(9.14)\quad (E-1)(E-\lambda(123))(E-\lambda(23))(E-\lambda(13))(E-\lambda(12))G(n) = 0$$

which is called the *train equation* of the plenary powers.

**9.4. Remarks on the general case.** The method we have used may in principle be used for calculation of explicit expressions and linear difference equations for $G(n)$ for any number of loci. The explicit expression for $G(n)$ in the algebra $\mathscr{A}(\boldsymbol{m}_S, \boldsymbol{\lambda}_S)$ will be of the form $\sum_i C_i r_i^n$, where the $C_i$ are independent of $n$ or polynomials in $n$. In accordance with the terminology of Etherington the $r_i$ will be called the *train roots* of $G(n)$ in the algebra $\mathscr{A}(\boldsymbol{m}_S, \boldsymbol{\lambda}_S)$. They are roots of the characteristic equation of the homogeneous linear difference equation (*train equation*) of $G(n)$ in the algebra $\mathscr{A}(\boldsymbol{m}_S, \boldsymbol{\lambda}_S)$.

In the general case we get

$$(9.15)\qquad G(n+1) = \sum_{U\in W(S)} \lambda(U)\left(\prod\nolimits_{j\in U'} D_j G(n)\right)\left(\prod\nolimits_{j\in U''} D_j G(n)\right).$$

One of the terms on the right-hand side of (9.15) is $\lambda(S)G(n)$. The other terms are products of plenary powers corresponding to smaller numbers of loci. If we have already found explicit formulae for these powers, (9.15) gives an explicit formula for $G(n)$ in the algebra $\mathscr{A}(\boldsymbol{m}_S, \boldsymbol{\lambda}_S)$, except for a single coefficient which is independent of $n$ and which may be determined by setting $n=0$.

THEOREM 5. *The train roots of* $G(n)$ *in the algebra* $\mathscr{A}(\boldsymbol{m}_S, \boldsymbol{\lambda}_S)$ *are* $\lambda(S)$, *the train roots of the* $G(n)$ *in the algebras* $\mathscr{A}(\boldsymbol{m}_{S-j}, \boldsymbol{\lambda}_{S-j})$ *for every* $j\in S$ *and the products of the train roots of* $G(n)$ *in* $\mathscr{A}(\boldsymbol{m}_T, \boldsymbol{\lambda}_T)$ *with the train roots of* $G(n)$ *in* $\mathscr{A}(\boldsymbol{m}_{S-T}, \boldsymbol{\lambda}_{S-T})$ *for every subset* $T$ *of* $S$ *which contains at least two elements and for which the number of elements of* $T$ *does not exceed the number of elements of* $S-T$. *The train roots are all real and are situated in the interval* $]0,1]$.

The proof of the first part of this theorem is obvious from a consideration of (9.15). The last part of the theorem is a consequence of the fact that $0<\lambda(S)\leqq 1$ for every algebra which has a genetical interpretation.

## 10. Explicit expressions and recurrence formulae for $H(n)$.

**10.1.** The sequence of products $\{H(n)\}$ defined by (7.2) gives the probability distributions of an $X$ chromosome in successive generations when there is panmixia and the population is infinite.

Let us set $H(0)=G_0$ and $H(1)=G_1$ and let us suppose that $G_0$ and $G_1$ are probability distributions. Then $\xi(H(n))=1$ for every $n$.

In the case of a single locus we get

$$H(n+2) = H(n+1)\times H(n) = \tfrac{1}{2}H(n+1)+\tfrac{1}{2}H(n).$$

We thus already have a difference equation for $H(n)$ which may be rewritten in the form

$$(10.1)\qquad (E-1)(E+\tfrac{1}{2})H(n) = 0.$$

The general solution of this equation is

$$(10.2)\qquad H(n) = C_1+C_2(-\tfrac{1}{2})^n.$$

Setting $n=0$ and $n=1$ in (10.2) we get

$$(10.3)\qquad C_1 = \tfrac{1}{3}(G_0+2G_1),$$

$$(10.4)\qquad C_2 = \tfrac{2}{3}(G_0-G_1).$$

**10.2. Two loci.** For the sake of brevity we shall write $\theta$ instead of $\lambda(1, 2)$. Then

$$(10.5)\quad H(n+2) = \tfrac{1}{2}(1-\theta)\big(H(n)+H(n+1)\big) + \\ + \tfrac{1}{2}\theta\big((D_1H(n)(D_2H(n+1))+(D_2H(n))(D_1H(n+1))\big),$$

Applying (10.2) to $D_1H(n)$ and $D_2H(n)$ we get

$$(10.6)\quad \big(E^2-\tfrac{1}{2}(1-\theta)E-\tfrac{1}{2}(1-\theta)\big)H(n) = \theta\big(C_3+C_4(-\tfrac{1}{2})^n+C_5(\tfrac{1}{4})^n\big),$$

where

$$(10.7)\quad C_3 = (D_1C_1)(D_2C_1),$$

$$(10.8)\quad C_4 = \tfrac{1}{4}((D_1C_1)(D_2C_2)+(D_2C_1)(D_1C_2)),$$

$$(10.9)\quad C_5 = -\tfrac{1}{2}(D_1C_2)(D_2C_2).$$

Let $r_1$ and $r_2$ be the two roots of the equation

$$(10.10)\quad x^2-\tfrac{1}{2}(1-\theta)x-\tfrac{1}{2}(1-\theta) = 0.$$

In the genetic interpretation $\theta$ is situated in the interval $[0, 1]$ and cannot be much greater than $\frac{1}{2}$. Then $r_1$ and $r_2$ must be real and have opposite signs since their product is $-\frac{1}{2}(1-\theta)$. Equation (10.10) has the roots 1 and $-\frac{1}{2}$ if and only if $\theta=0$. This means that the two loci are completely linked and can be analyzed as one single locus. Equation (10.10) has the root $\frac{1}{4}$ if and only if $\theta=0{,}9$. This is an impossible value in the genetic interpretation. In all cases of genetic interest we can therefore assume that the numbers $r_1$, $r_2$, $-\frac{1}{2}$, $\frac{1}{4}$ and 1 are all different. Then the solution of the difference equation (10.6) is given by

$$(10.11)\quad H(n) = C_3+C_4(-\tfrac{1}{2})^{n-2}+\frac{\theta C_5}{10\theta-9}\left(\frac{1}{4}\right)^{n-2}+C_6r_1{}^n+C_7r_2{}^n,$$

where $C_6$ and $C_7$ are determined by setting $n=0$ and $n=1$.

Applying the operator $(E+\frac{1}{2})(E-\frac{1}{4})$ to (10.6) we get

$$(10.12)\quad (E+\tfrac{1}{2})(E-\tfrac{1}{4})\big(E^2-\tfrac{1}{2}(1-\theta)E-\tfrac{1}{2}(1-\theta)\big)H(n) = \tfrac{9}{8}\theta C_3.$$

Applying the operator $E-1$ to this equation we get the homogeneous difference equation

$$(10.13)\quad (E-1)(E+\tfrac{1}{2})(E-\tfrac{1}{4})\big(E^2-\tfrac{1}{2}(1-\theta)E-\tfrac{1}{2}(1-\theta)\big)H(n) = 0.$$

**10.3. Train roots and asymptotic distribution in the general case.** It is clear from the method we have used that the probability distribution $H(n)$ in the general case will be a sum of exponential terms $r_i{}^n$ with coefficients which are independent of $n$ or polynomials in $n$. We shall call the $r_i$ the train roots of $H(n)$ in the algebra considered. Two of the train roots of $H(n)$ in $\mathscr{A}(\boldsymbol{m}_S, \boldsymbol{\lambda}_S)$ will be the roots of

$$x^2 - \tfrac{1}{2}\lambda(S)x - \tfrac{1}{2}\lambda(S) = 0 . \tag{10.14}$$

These roots are real and have opposite signs since their product is $-\frac{1}{2}\lambda(S)$ where $\lambda(S) > 0$. Let $r_1$ be the positive root and $r_2$ the negative root. We see that $r_1 \leqq 1$ and that $r_1 + r_2 = \frac{1}{2}\lambda(S)$ is positive. Hence $r_2 > -1$. We get

THEOREM 6. *The train roots of $H(n)$ in $\mathscr{A}(\boldsymbol{m}_S, \boldsymbol{\lambda}_S)$ are the roots of* (10.14), *the train roots of $H(n)$ in $\mathscr{A}(\boldsymbol{m}_{S-j}, \boldsymbol{\lambda}_{S-j})$ for every $j \in S$, and the products of the train roots of $H(n)$ in the algebra $\mathscr{A}(\boldsymbol{m}_T, \boldsymbol{\lambda}_T)$ with the train roots of $H(n)$ in $\mathscr{A}(\boldsymbol{m}_{S-T}, \boldsymbol{\lambda}_{S-T})$ for every subset $T$ of $S$ which has at least two elements and for which the number of elements of $T$ does not exceed the number of elements of $S-T$. The train roots are all real and are situated in the interval* $]-1, 1]$.

The positive root of (10.14) is equal to 1 if and only if $\lambda(S) = 1$. Then $\lambda(U) = 0$ for any other element $U$ of $W(S)$. Using (5.11) we again get (10.1) corresponding to the fact that all loci are completely linked and can be analyzed as one single locus. The term in the expansion of $H(n)$ corresponding to a train root 1 will thus always be a constant, not a polynomial in $n$.

We conclude from these results that $H(n)$ converges to a limit when $n$ tends to infinity. This limit $H(\infty)$ is found by letting $n \to \infty$ in (7.2). We then get $H(\infty) = H(\infty) \times H(\infty)$. Thus $H(\infty)$ is an idempotent element of the algebra. An idempotent element must, however, have the form $\prod_{j \in S} \Gamma_j$, where $\Gamma_j$ is a linear form in the variables $a_{ji_j}$. This is easily proved by induction using (9.15).

## 11. Concluding remarks.

In the introduction I noted that one new aspect of the present paper is the use of recursion in the study of genetic algebras. Geiringer [8]–[11] has, however, applied a recursive approach to the individual probabilities without considering algebras. She obtained the scalar version of formula (9.15) [8, formula (31)]. She notes that this formula gives a system of difference equations. She does not, however, seem to be aware of the possibility of solving this system recursively. She indicates a method of solution which is unnecessarily complicated.

Bennet [1] indicated another method of solving the system. His method is also more complicated than the method given in the present paper.

Geiringer [9] carried through the solution in the case of three loci and got an explicit expression which is the scalar version of (9.11) and (9.12) of the present paper. It may be noted, however, that this explicit formula follows easily from the train equation (9.14) which had previously been published by Etherington [5].

The results given for one locus and two loci in the case of autosomes and one locus in the case of $X$-chromosomes have been known for a long time and can be found in textbooks (Li [15, Chapters 4, 5, 8], Kempthorne [14, Chapter 2]).

The sequence $H(n)$ defined by (7.2) and its genetic interpretation do not seem to have been considered before, and the results of Sections 10.2 and 10.3 seem to be new.

The use of the sequence $H(n)$ in the study of sex-linked traits presupposes that the traits are represented by loci in the $X$ chromosomes only, not in the $Y$ chromosomes. This seems to be the usual case. (See for instance [16, p. 64]).

## REFERENCES

1. J. H. Bennet, *On the theory of random mating*, Ann. Eugenics 18 (1954), 311–317.
2. I. M. H. Etherington, *On non-associative combinations*, Proc. Roy. Soc. Edinburgh 59 (1939), 153–162.
3. I. M. H. Etherington, *Genetic algebras*, Proc. Roy. Soc. Edinburgh 59 (1939), 242–258.
4. I. M. H. Etherington, *Commutative train algebras of ranks* 2 *and* 3, J. London Math. Soc. 15 (1940), 136–148.
5. I. M. H. Etherington, *Special train algebras*, Quart. J. Math. Oxford Ser. (2) 12 (1941), 1–8.
6. I. M. H. Etherington, *Duplication of linear algebras*, Proc. Edinburgh Math. Soc. (2) 6 (1941), 222–230.
7. I. M. H. Etherington, *Non-associative algebra and the symbolism of genetics*, Proc. Roy. Soc. Edinburgh Sect. B 61 (1941), 24–42.
8. H. Geiringer, *On the probability theory of linkage in Mendelian heredity*, Ann. Math. Statist. 15 (1944), 25–57.
9. H. Geiringer, *Further remarks on linkage theory in Mendelian heredity*, Ann. Math. Statist. 16 (1945), 390–393.
10. H. Geiringer, *On the mathematics of random mating in case of different recombination values for males and females*, Genetics 33 (1948), 548–564.
11. H. Geiringer, *On some mathematical problems arising in the development of Mendelian genetics*, J. Amer. Statist. Assoc. 44 (1949), 526–547.
12. H. Gonshor, *Special train algebras arising in genetics*, Proc. Edinburgh Math. Soc. (2) 12 (1960), 41–53.
13. C. Jordan, *Calculus of finite differences*, Budapest, 1939.
14. O. Kempthorne, *An introduction to genetic statistics*, New York, 1957.
15. C. C. Li, *Population genetics*, Chicago, 1955.

16. N. E. Morton, *Further scoring types in sequential linkage tests, with a critical review of autosomal and partial sex linkage in man*, Amer. J. Human Genetics 9 (1957), 55–75.
17. R. Raffin, *Axiomatisation des algèbres génétiques*, Acad. Roy. Belg. Bull. Cl. Sci. (5) 37 (1951), 359–366.
18. R. D. Schafer, *Structure of genetic algebras*, Amer. J. Math. 71 (1949), 121–135.
19. E. Stephens, *The elementary theory of operational mathematics*, New York, 1937.

UNIVERSITY OF OSLO, NORWAY

# IV

# Calculus for Statistico-genetics

# Editor's Comments on Paper 11

**11 Cotterman:** A Calculus for Statistico-genetics
Unpublished dissertation

The following material was originally presented by C. W. Cotterman as a Ph.D. dissertation in 1940 at Ohio State University. Had it been properly published at that time, it would have advanced the progress of genetics by a decade or more on several fronts: it predates the widespread publication of Malécot's probability model for descent by eight years; it elaborates on operator models for descent structure previously used only by Haldane and Moshinsky (1939), simultaneously in modified form by Etherington (Papers 6 through 9), and later in modified form as well by R. A. Fisher (1949); it develops apparently from first principles a method of bayesian statistics and applies it to conditional probabilities in descent, thereby predating modern works like Karlin (1968); by enumerating types of identity by descent it predates in certain ways the work of authors mentioned in the introduction to Jacquard (Paper 20); and it most certainly justifies Cotterman's claim in his introduction that the *method* of axiomatic derivation has a place in genetic theory.

The justification, if one is needed, for including this work is that it uses so many of the techniques that have since proved invaluable in both genetic and social structural work. As an example of a practical application of this material, the reader should see Cotterman (1951, 1954). For a demonstration that the genius of Cotterman is not limited to a single topic, see Cotterman (1969), a study that is, incidentally, well within the province of modern structural mathematics and thus is itself unique in a world of stochastic (if not diffuse) thinkers. Besides this material, Cotterman apparently has unpublished materials that treat allelic systems as systems of algebras in order to study combinatorial problems in the relation of genotype to phenotype.

It is easiest to see the importance of Cotterman's work if it is compared to the papers of Etherington (Papers 6 through 9) or to a number of more recent papers. In particular, the paper of Hilden (1970) provides both a sensible interpretation of Etherington's material (together with algorithms for computation of various products) and an exposition in a form very comparable to the present paper. Likewise, Lillestøl (1971) is a treatment of a closely related problem, with very similar techniques.

# 11

This article, which is appearing in print here for the first time, was written by Mr. Cotterman in 1940.

# A CALCULUS FOR STATISTICO-GENETICS

DISSERTATION

Presented in Partial Fulfillment of the Requirements for the Degree of Doctor of Philosophy in the Graduate School of The Ohio State University

By

CHARLES WILLIAM COTTERMAN, B.A., M.A.

The Ohio State University

1 9 4 0

Approved by:

Lawrence H. Snyder

Adviser

# TABLE OF CONTENTS

# A CALCULUS FOR STATISTICO-GENETICS

## INTRODUCTION

This study represents an attempt to systematize the application of mathematical probability to genetics through the introduction of a terminology and symbolism especially designed for that purpose. The usefulness of the system is tested by applying it to a wide variety of statistico-genetic problems, embracing some 460 mathematical relationships, but it is the method, rather than the results, which must be regarded as the prime objective of the study. It has indeed been necessary to limit its extension in the present writing to certain selected topics, and particular attention has been paid in this regard to those mathematical consequences of Mendelism which lead directly to statistical methods for the study of human heredity.

The treatment has also been restricted to the case of monofactorial autosomal inheritance - the simplest mechanism of hereditary transmission - which, however, serves as a ready pattern for extending the method to more complicated cases (multiple alleles, sex-linkage, multiple factors, etc.). Two further topics which have been omitted entirely are selection and mutation, the two agencies capable of altering gene frequencies. The present study is therefore confined to those conditions which exist in a population which is stable with respect to its supply of genes and, especially, to the consequences of genetic equilibrium in a random-breeding population.

Thanks to the fact that the study of heredity has attracted the attention of several eminent mathematicians, much has been written in the field of statistico-genetics. In fact, mathematical theory has already outstepped its practical application, both in the field of human genetics and in the biometrical study of evolution. If this be so, then genetics is perhaps the exception to Woodger's claim that "Biology exemplifies the case of a science in which theory has never yet caught up with practice". This must be attributed largely, perhaps, to the fact that genetics, of all biological sciences, is the most susceptible of mathematical treatment. But there is at least one other good reason; genetic experiment is also the most costly in time and materials, so that the guidance of careful experimental planning and refined statistical analysis is all the more appreciated. In human genetics this is especially true. Scarcity of data has contributed much to a desire for greater variety and efficiency in statistical methods designed for that study.

Whatever novelty may exist in the methods described herein will probably be found in the distinction between several grades of gene relationship which is drawn in the next section and employed consistently throughout the paper. This, together with the chosen method of symbolization and system of derivation, lends to the analysis a superficial resemblance to the formulations of "symbolic logic". The resemblance is, however, sufficiently real to warrant a few remarks about this general subject.

The <u>axiomatic</u> or <u>logistic</u> method or, as it is commonly called, the method of "symbolic logic", is a quasi-mathematical science which makes possible the exact symbolic representation of laws governing phenomena which are generally thought incapable of mathematical treatment. Logistic is therefore a generalized kind of mathematics which deals with quality and form in the same way that mathematics proper deals with quantity and number. It concerns itself primarily with this question: "If I have a number of definite statements in an axiom system, what other statements must it also contain and what statements are excluded?" The word "axiom", as used in logistic, is taken to mean any preliminary proposition, which may or may not be capable of experimental verification. It does not therefore imply a self-evident truth. An axiom system may thus be either interpreted or uninterpreted, that is to say, it may concern itself with theorems about conditions which are known to exist in Nature or, on the other hand, it may merely constitute a system of non-numerical mathematics applicable to purely hypothetical situations. In the latter case it will demonstrate not truth but merely form. The main braches of logistic deal with rules governing (1) elementary statements, (2) statements with operators, (3) identity, (4) classes, and (5) relations.

Though at present the logistic method is not generally

familiar to biologists, it is becoming recognized as an indispensable tool for the construction of any exact science and should therefore find special recommendation for biology. Thus far its application to biology has been attempted by only one author, J. H. Woodger, who, in a book entitled "The Axiomatic Method in Biology", outlines a system patterned after the classical work, "Principia Mathematica", of Whitehead and Russell (1925). Woodger's system deals with general biological growth, cell division, reproduction, Mendelian theory, embryology and taxonomy. Being a foundational work, most of the book is devoted to the very difficult task of finding the most general phenomena of life and fashioning these into the form of exact symbolic statements. Consequently, the section on Mendelian theory is not much developed and is limited to certain relationships which seem to be of little more than academic interest. Woodger deals with the number of different sorts of gametes, genotypes and matings possible with sets of varying numbers of alleles and with a general system of classifying Mendelian ratios for the purpose of cataloging "cases" of heredity.

As has already been mentioned, the resemblance which the present analysis may bear to symbolic logic is entirely superficial. No attempt has been made to set down a series of statements in symbols such that their logic and truth is completely revealed in the symbols themselves. In other words, the mathematical formulations of the following pages could not go unaccompanied by verbal explanation, genetic and mathematical, of the rules whereby these statements are reached; nor would the system have any meaning for one unfamiliar with the basic principles of reproduction and genetics. It cannot therefore be regarded either as an interpreted or an uninterpretated axiom system. What perhaps can be claimed, however, is that something of an approach to the method of logistic has been achieved. In particular, the symbolism employed has been adopted with three views in mind, that of greater precision of thought, of greater generality of statement, and of greater revelation of genetic logic. These are, of course, the same objectives sought in the use of the axiomatic method and which prompted Bertrand Russell to say: "A good notation has a subtlety and suggestiveness which at times makes it seem almost like a live teacher ... and a perfect notation would be a substitute for thought".

It is therefore hoped that the distinction between different kinds of "likeness" or "identity" in genes, which is one of the fundamental features of the following analysis, will prove to be a departure in the direction of a more precise language for dealing with genetic problems and an approach to the more exact method of logistic, which, it will be recalled, deals with identity as one of its chief problems. These two features, the notation and the more explicit definition of gene similarities, are therefore aimed to accomplish much of what is sought in the use of symbolic logic, namely, a shorter and more precise derivation of theorems from the fundamental axioms. Indeed, many of the theorems and formulae which have required pages for their derivation in the original articles are often obtainable in one, two, or three steps by the methods of the present calculus.

In conclusion, it is hoped that the methods put forth in this paper will prove helpful as an aid in teaching the elementary principles of statistico-genetics to students as well as a tool for the further discovery of new laws and regularities inherent in the Mendelian theory. Finally, it is felt that whatever success might have been achieved by the comparatively crude methods used here will only serve to convince others of a very real need in genetics for the more rigorous methods of logistic.

Following two preliminary chapters, the main body of the paper is divided into six sections. In each of these a general discussion of the contents of the section, including a definition of all new symbols, precedes the mathematical part. This consists of a series of mathematical statements which have been numbered according to the decimal system commonly employed in writings on logistic. After each theorem are placed in parentheses the numbers of those axioms and theorems which are directly involved in the derivation. It will be understood, of course, that if the axiomatic or logistic method had actually been used, these catalogs of axioms and theorems would present completely intelligible, though perhaps uninterpretable, stories in themselves. As it stands, however, it will be seen that little verbal explanation is required, so that a considerable measure of conciseness has been achieved. This becomes increasingly apparent as the sections unfold.

For their constant encouragement and criticism and for the inspiration furnished by their own studies in human genetical methodology the author wishes to express his sincere gratitude to Professors Laurence H. Snyder and D. Cecil Rife.

## THREE TYPES OF GENE RELATIONSHIP

The word "identity" is one which has acquired in biology a variety of meanings. First, it may be used to imply strict or "logico-mathematical" identity, which is usually symbolized by '=' or 'I' and means that the two things are actually one and the same. This is the sense involved when the biologist says that poliomyelitis and infantile paralysis are identical diseases, that levulose is identical with fructose, or that $i = \sqrt{-1}$. More often, however, "identity" is used in speaking of things which are merely "exactly alike" or "of a similar sort". Woodger (1937) discusses this particular meaning at considerable length and points out that there is no need, as he once thought, for a special relation of "intrinsic identity" to handle the concepts of "identital twins", "identical organs", "identical genes", etc., but that these are only special cases of logico-mathematical identity in which the two things, while not the same thing, have strictly identical properties of one or more kind. One can therefore express this symbolically by stating that '$xI_sy$', which means that properties of the sort '$s$' are identical in '$x$' and '$y$'.

The geneticist who speaks of genes as being "identical" may mean any of at least three different things. He may say that "two genes producing chlorophyll deficiency in maize were formerly believed to occupy different chromosome loci but are now known to be identical". This is obviously a strict identity. Or he may say that "two albinos possess identical pairs of genes. Here, of course, the inference is not that they are the *same* genes but merely that they are equivalent in the sense that, if they could be exchanged, there is reason to believe that no change would be wrought in the phenotypes or breeding behaviors of their possessors. To this class of properties, that is, physiological action and breeding behavior, Woodger has assigned the symbol '**genet**', and identity with respect to **genet** is indicated by the Gothic letter '$\mathfrak{J}$'.

$$\mathfrak{J} =_{Df} \hat{x}\hat{y}\{(\alpha):\alpha\,\varepsilon\,\mathbf{genet}.\supset.x\,\varepsilon\,\alpha \equiv y\,\varepsilon\,\alpha.\}.$$

The above definition from Woodger is reproduced here merely to acquaint the reader who is not familiar with symbolic logic with the appearance of that language. It may be translated: $\mathfrak{J}$ is a relation such that, for every $\alpha$ which is a member of **genet**, if x is $\alpha$, then y is $\alpha$, and if y is $\alpha$, then x is $\alpha$. This definition of identity in genes will be subscribed to throughout this paper.

Further, the geneticist may say that two genes are "identical" if they result one from the other or both from a common gene through the process of gene reproduction, as, for example, two genes, one in the fertilized egg and one in a somatic cell of the adult, or two genes in brother and sister, or in a single cell of one individual as a result of inbreeding. This type of "identity" might seem to include the last type since genes which are derived one from the other or both from one are, of course, ordinarily perfect "duplicates" and therefore identical in the sense of $\mathfrak{J}$. However, this need not be the case and, accordingly, we propose the term "derivative" to replace this last kind of "identity". Mutation may cause two derivative genes to be non-identical, and, in fact, the mutation rate might be defined in terms of the probability that two derivative genes should be non-identical.

A further apparent complication, however, arises from the fact that all genes are, of course, ultimately derivative (provided one ascribes to a monophyletic theory of evolution), which, then, would seem to obviate the necessity of the term "derivative". Although this is not actually the case, it nevertheless is readily seen that a more complete definition of "derivative" is needed and that this will necessarily be a relative one. It should also be, if possible, a mathematically exact one, but so far the author has been unable to fulfill this requirement. We may say that derivative genes shall be taken to mean two or more genes derived recently, in terms of generations of adults, from some common gene or one from the other. But precisely how recently? Again, in the absence of a definite criterion we may say 5 or 6 generations for the human population. Though this is quite arbitrary, it is nevertheless serviceable for several reasons.

First, the chance that mutation should have occurred during this time is in most cases quite negligible, whereas it would not be so for some longer period. Hence, in the solution of many statistico-genetic problems we may choose to assume that mutation is absent and that all derivative genes must be identical with but little loss of accuracy. Secondly, inbreeding which comes about through the occurrence of a common ancestor more distantly removed than 5 or 6 generations will have entirely negligible genetic effects. (Reference is again made to the human species.) Thirdly, if a more precise definition is to be found it will probably involve the average genetic relationship between pairs of individuals living in the population at any given time, and this must certainly be more remote than fourth or fifth cousins, so that the necessary correction will probably be even smaller than imagined above. Moveover, it would be extremely difficult to obtain genealogical information on a larger number of generations even though this were desirable. Finally, other writers in the field of statistico-genetics, while not employing an explicit distinction between derivative and identical genes, have nevertheless employed the same approximate reasoning in deriving their formulae and theorems. This will be evidenced later by the fact that their formulae can also be obtained by means of the methods of the present calculus, which employs that distinction.

Finally we come to the concept of "alleles" (or allelomorphs), which has purposely been avoided until this stage. Alleles are, of course, genes which occupy corresponding (strictly identical) loci in homologous chromosomes and therefore regularly segregate at meiosis. However, there again arises a difference in meaning depending on whether one excludes the class of identical alleles from this definition. In the customary symbolism, B is an allele of b, but b may or may not be regarded as an allele of another b, say b′. Now, it was perhaps the original intention to exclude identical genes from the class of alleles, but there appears to be a decided disadvantage to this, for it inclines one to think of "allelic" as the disjunction of "identical". But "allelic" could not mean "non-identical", for non-identical genes do not have to be alleles, nor does "identical" mean "non-allelic". Non-allelic genes, however, might be identical. This could come about as a result of unequal crossing-over producing "duplicate genes" in the same chromosome or as a result of other types of chromosomal aberration, provided these were unaccompanied by "position effects". But "identical genes will usually occupy corresponding chromosome loci.

At this point it may be remarked that this vagueness in definition of "allele" has very likely contributed to a sense of finality in the meaning of the term "identical" as applied to genes, so that the distinction between derivative and non-derivative identical alleles and its usefulness in the derivation of many genetic principles has frequently been overlooked. Certainly the reader will agree, in view of the above discussion and others which could possibly arise over many other genetic concepts, that there is much room for application of the perfect precision of symbolic logic in genetics.

To summarize, the following definitions will be adopted

in this paper: <u>Alleles</u> are genes occupying corresponding (strictly identical) chromosome loci, and therefore segregating at meiosis; <u>identical</u> genes are genes which are physiologically equivalent or substitutable (possess strictly identical physiological action); and <u>derivative</u> genes are genes which are relatively recently descended one from the other or both from some common gene. That this classification is entirely logical or that it would be adaptable to a logistic axiom system cannot be claimed, but it at least serves the purposes of the present study. It will be noticed, incidentally, that the classification is based upon the three essential properties whereby the existence of the gene can be inferred, namely, its (1) behavior in meiosis, (2) physiological action, and (3) reproduction.

It will now be seen that these three criteria of classification are mutually independent in the sense that a gene may be related to another gene in any of the 8 relations formed by combination of the three classes and their negations. Thus two genes may be:

(1) allelic, derivative, identical

(2) allelic, derivative, non-identical

(3) allelic, non-derivative, identical

(4) allelic, non-derivative, non-identical

(5) non-allelic, derivative, identical

(6) non-allelic, derivative, non-identical

(7) non-allelic, non-derivative, identical

(8) non-allelic, non-derivative, non-identical

Let a given gene be denoted by $A_1$. Then any of its derivative identical alleles (case 1) will be denoted by the same symbol. A derivative but non-identical allele of $A_1$ (2) would, of course, represent a recent mutation and would require a different symbol, say $a_1$, the subscript being retained to indicate its origin with respect to $A_1$. Other alleles identical with $A_1$ but not derivative (3) may be given the symbols $A_2$, $A_3$, $A_4$, etc. "Old" mutant or non-derivative and non-identical alleles of $A_1$ (4) will similarly be denoted by $a_2$, $a_3$, $a_4$, etc. Case 5 requires the recent occurrence of chromosomal aberration and case 6 requires, in addition to this, a mutation or position effect. These are ordinarily excessively rare events and their occurrence will not be postulated in the following sections. In fact, since the present study is restricted to the case of a single pair of alleles, we shall not require any symbolism for dealing with cases 5-8.

We shall, however, require symbols to denote genes the relationship of which to $A_1$ is not completely specified. For example, an identical allele which is either derivative or non-derivative with respect to $A_1$ may be denoted by $A_{12}$; a mutant allele of unspecified origin, by $a_{12}$. Further, an allele which is either identical or non-identical with $A_1$ will be labelled $\xi_1$ if it is a derivative of $A_1$ but $\xi_2$, $\xi_3$, $\xi_4$, etc., if it is not a derivative of $A_1$. Finally, any allele of $A_1$ may be designated by $\xi_{12}$. Thus, $\xi$ is the logical sum of A and a (and all other mutant alleles, if such exist) and $\xi_{12}$ is the logical sum of $\xi_1$ and $\xi_2$. The following table will help make these definitions clear.

Symbols for the Alleles of a Gene $A_1$.

| | Derivative | Non-derivative | Derivative or Non-derivative |
|---|---|---|---|
| Identical | $A_1$ | $A_2$ $A_3$ ... | $A_{12}$ $A_{13}$ ... |
| Non-identical | $a_1$ | $a_2$ $a_3$ ... | $a_{12}$ $a_{13}$ ... |
| Identical or Non-identical | $\xi_1$ | $\xi_2$ $\xi_3$ ... | $\xi_{12}$ $\xi_{13}$ ... |

In this paper, the symbol '$\xi$', standing for any one of a set of non-identical alleles, will serve two general purposes. In the first place, it will be used whenever it is desired to express a genetic law which holds irrespective of the type (A, a, etc.) of the genes involved. The symbol $[\xi_1\xi_2]$ thus merely indicates a pair of non-identical alleles and cannot properly be called a "genotype", for the "types" or physiological properties of the genes are completely and often purposefully unspecified. $[\xi_1\xi_2]$ will therefore be referred to as a "gene-pair", while only such symbols as [Aa] will be termed "genotypes". The second function, which is closely allied to the first, is that of indicating incompletely specified genotypes, such as [A$\xi$] and [$\xi$a]. The '$\xi$' here replaces the "dash" in the commonly used equivalents, [A_] and [_a]. In the event that complete penetrance of the genes is postulated (cf. Section 3), then [A$\xi$] and [aa] will also serve to designate dominant and recessive "phenotypes". The dash, on the other hand, will be used in this paper to indicate a "missing" gene; thus, [a_] would represent an abnormal haploid condition, while [_] would signify a gamete lacking a $\xi$-allele.

## FURTHER NOTES ON SYMBOLISM

The mathematical principles involved in this study are of the very simplest nature; we require only the two Fundamental Theorems of Probability. The first is the well-known and almost self-evident Addition or "Either-Or" Theorem, which can be stated: If an event can occur in several different or mutually exclusive ways, then the chance that it should occur in any way whatsoever is the sum of the probabilities of the several alternatives. In the calculus of logistic the symbol '$\varepsilon$' replaces the word "is" and the sign '$\cup$' is the so-called "exclusive-or" function. Thus

$$E \; \varepsilon \; E_1 \cup E_2$$

says that E is either $E_1$ or $E_2$, but not both at once. The logician would also say that E is the "logical sum" of $E_1$ and $E_2$. The point of interest here is that the expression was doubtlessly prompted by the Addition Theorem of probability, which states that

$$\text{if } E \; \varepsilon \; E_1 \cup E_2, \text{ then } P[E] = P[E_1] + P[E_2],$$

where 'P' stands for "probability of".

The second is the Multiplication or "As-Well-As" Theorem, which can be stated: The chance that several different independent events should all occur simultaneously is the continuous product of their separate probabilities. In symbolic logic '$\cap$' is the sign for "and" or "as-well-as". Hence

$$\text{if } E \; \varepsilon \; E_1 \cap E_2, \text{ then } P[E] = P[E_1] \cdot P[E_2],$$

and E is said to be the "logical product" of $E_1$ and $E_2$.

In many genetic problems, however, we shall have to deal with combined events which are not independent, such that the Multiplication Theorem will not apply. Moreover, we must clearly distinguish between three types of problem which may be illustrated by the following questions: (1) What is the probability that a man will be of genotype [aa] if his aunt is known to be of genotype [Aa]? (2) What is the probability that a particular aunt-nephew pair will have the aunt of genotype [Aa] and the nephew of genotype [aa]? (3) What is the probability that a particular aunt-nephew pair will have the one individual of genotype [Aa] and the other of genotype [aa]?

The distinction between questions (1) and (2) is one which is typical of many problems in probability, biological and otherwise. In general, if we have two events X and Y which are not independent, then the probability that if X has occurred Y should also occur is not equal to the probability that X and Y together should occur. The author is not familiar with any phrases in mathematical probability for differentiating these two types of problem and therefore suggests the term "implicative probability" for the first case, which may be symbolized by

$$(X)P[Y] = \phi$$

and may be read: "the probability of Y implied by X is $\phi$" or "given X, then the probability of Y is $\phi$". We shall call the second probability a "combinatory probability", denoted by

$$P\underline{[X][Y]} = \psi$$

and read: "the probability that X should occur, and then Y, is $\psi$". The relationship between the two is then expressed by

$$P\underline{[X][Y]} = P[X]\cdot(X)P[Y],$$

where P[X] is the initial probability of X.

The distinction between questions (2) and (3) is merely one of order, and the purpose of the line under the above symbols is to emphasize that a particular order is being specified. Thus $P\underline{[X][Y]}$ is the probability of first X then Y and $P\underline{[Y][X]}$ is the probability of first Y then X. Question (3) may therefore be answered by

$$P[X][Y] = P\underline{[X][Y]} + P\underline{[Y][X]}$$

where P[X][Y] therefore stands for the probability of X and Y in either order. Incidentally, it will be seen that in

order to have $$P\underline{[X][Y]} = P\underline{[Y][X]}$$

and therefore $$P[X][Y] = 2\cdot P\underline{[X][Y]}$$

we must have $P[X] : P[Y] = (X)P[Y] : (Y)P[X]$.

If, however, the two events are identical, then the probability of the combined occurrence will be denoted simply by

$$P[X][X] = P[X]\cdot(X)P[X],$$

and an expression such as $P\underline{[X][X]}$ would be meaningless since the combined event can occur in only one way.

One further convention will now be introduced. Rather than take 'P' to stand for "probability of", we shall select some other capital letter which, in addition to standing for "probability of", will also describe the genetic relationship of the thing Y to the preceeding thing X.

Thus (X)H[Y] = ø will be read: "given the thing X, then the probability that the thing Y, related to X in the manner described by H, should occur is ø". For example, the symbol (Aa)G[A] will later be used to denote the probability that an individual of genotype [Aa] should produce a (G)amete containing the gene A. The sign (AA)O[Aa] denotes the probability that, if a person is of genotype [AA], his (O)ffspring will be of genotype [Aa]; this then is the probability of a genotype [Aa] when information about one parent is completely lacking but where the genotype of the other parent is completely specified as [AA].

The general utility for genetics of such a symbol, showing explicitly in the parentheses just what information is provided for the reckoning of the event in brackets, will be readily foreseen. In human genetics particularly, where one is forced to make use of fragmentary data of many sorts, we shall wish to find probability expressions for certain events under a wide range and variety of known facts. For instance, biological relationship may be wholly or only partially specified (e. g., twins); phenotypes may be known in neither, one or both parents (or in other relatives); genotypes may be completely, partially, or not at all specified in any combination of relatives; sex may or may not be known; etc. Furthermore, the symbol may be used repeatedly to indicate a chain of genetic events, as in

$$(X)F[W] = (X)H[Y] \cdot (Y)J[Z] \cdot (Z)K[U] \cdot (U)I[W],$$

which shows not only how the probability of the event W is

inferred from the event X, but also that the relation F is the "relative product" of the relations H, J, K and L, as defined in "Principia Mathematica". We shall also require statements of the sort

$$H[X] = \theta,$$

which, lacking a pair of parentheses before it, simply indicates that no information of any sort is given relevant to the thing to which X stands in the relation H. Thus, G[A] would be the probability of a gamete possessing the gene A derived from a completely unspecified genotype or phenotype.

Use of the above notation makes it hardly necessary to define any of the compound symbols employed throughout the paper provided one has first made himself familiar with the meanings of the various "relation signs". For convenience these will be listed here together with a brief description of their meanings. In some cases a more complete definition will be found in the section where that relation is first used. The numbers following the descriptions in the following list refer to these sections.

G gamete. (1)

H mating. (1)

J an individual who is not inbred. (1)

I an individual who is inbred. (1)

O offspring. (1)

P parent. (1)

C ancestor of sort J common to two Z-relatives. (1)

Z relatives both of sort J. (1)

Q subclass of Z wherein no two paths of descent are completely independent. (1)

PO subclass of Q which is the logical sum of P and O. (1)

$P^n$ a lineal ancestor of the n-th generation. (5)

$O^n$ a lineal descendent of the n-th generation. (1)

$PO^n$ logical sum of $P^n$ and $O^n$. (5)

R subclass of Z wherein two paths of descent are completely independent. (1)

S full sibs, a subclass of R. (1)

F fraternal twins, a subclass of S. (6)

E monozygotic twins, a subclass of R. (6)

Y the relation in which an I individual stands to a relative of sort J. (4)

Q', Q" "half-relations" formed by the two lines of descent connecting an I individual to its ancestor. (4)

L the relation in which the terminal I member of an "inbreeding loop" stands to any other member of the loop; a subclass of Y. (4)

K subclass of L in which the second member is a parent of the terminal member. (4)

M relation of an offspring to its self-fertilized parent. (4)

B the relation in which the terminal I member stands to the starting member of an inbreeding loop when there is a second loop starting with the mate of the starting member of the first loop, a subclass of Y. (4)

X relatives both of sort I. (4)

Qs sibs whose parents are Q-relatives, a subclass of X. (4)

T twins, the logical sum of E and F. (6)

U the universal relation, the logical sum of Z, Y and X. (4).

## SECTION 1. GENIC RESEMBLANCE

This section deals with what might be termed the pure mechanics of Mendelism. Its axioms and theorems will be applicable to any "genotype" in the ordinary sense of that word. That is to say, whether a gene is of one physiological action or another does not matter in this section, so that the symbol '$\xi$', standing for any one of a set of non-identical alleles, can be used throughout. Because of this generality the theorems of this section are of great utility in the derivation of statistico-genetic formulae. The fact that they are theorems (derivable from the "Law of Segregation") and not axioms suggests that they could be dispensed with in the sections which follow. This is true, and, in fact, many of the genetic principles of the subsequent sections have been previously formulated by other writers (including the author) without recourse to them. But this has generally entailed a much more involved derivation, usually a painful enumeration of "possibilities and probabilities" which can easily be avoided by direct methods which make use of the theorems of this section.

First let us enumerate all possible ways in which a pair of alleles in one individual might be related to those of a second individual with respect solely to their "derivativeness". We shall adopt the expression "genic resemblance" in speaking of such relationship. We must include the possibility of inbreeding whereby an individual may receive a given gene in two ways through a common ancestor of its parents. This would make the two genes comprising its gene-pair derivatives, which would be symbolized by $[\xi_1^*]$. (The alternative symbol $[\xi_1\xi_1]$ unfortunately has certain disadvantages.) Thus, we are dealing with four genes in two genotypes, and, if the two individuals are related and both inbreds, then any one of the four genes may be a derivative of any or all of the remaining three.

There are exactly 12 possibilities as shown in the following diagram:

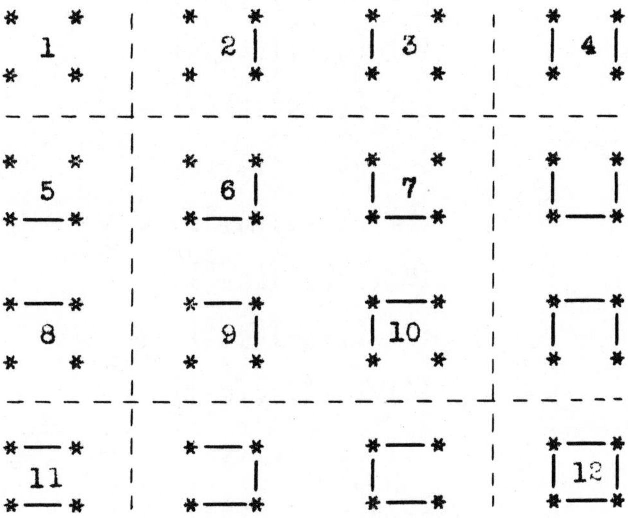


In this diagram the four genes are represented by asterisks arranged in the form of a square. The two in the top row belong to one individual; the two in the bottom row, to the second individual. Derivativeness is indicated by a line connecting the two genes thus related; genes not connected are non-derivatives. The 16 figures actually represent only 12 possibilities since four of them (these being unnumbered in the diagram) are necessarily identical with case 12. It will be noticed that the four columns illustrate the cases wherein, of the two genes possessed by one individual, the second individual possesses neither, the one, the other, or both of their derivatives. The four rows illustrate those types of genic resemblance possible in pairs of relatives wherein neither, the one, the other, or both are inbred.

Representing the gene-pairs of the two relatives by a pair of brackets, the 12 sorts of genic resemblance may be written as follows:

(1) $[\xi_1\xi_2][\xi_3\xi_4]$

(2) $[\xi_1\xi_2][\xi_1\xi_3]$

(3) $[\xi_1\xi_2][\xi_2\xi_3]$

(4) $[\xi_1\xi_2][\xi_1\xi_2]$

(5) $[\xi_1\xi_2][\xi_1^2]$

(6) $[\xi_1\xi_2][\xi_2^2]$

(7) $[\xi_1\xi_2][\xi_3^2]$

(8) $[\xi_1^2][\xi_1\xi_2]$

(9) $[\xi_2^2][\xi_1\xi_2]$

(10) $[\xi_3^2][\xi_1\xi_2]$

(11) $[\xi_1^2][\xi_1^2]$

(12) $[\xi_1^2][\xi_2^2]$

Although it is not appropriate to do so for some purposes, the above list will now be shortened by combining the types as indicated by the dotted partitions in the diagram. We thereby ignore, in cases where but one gene and its derivative are shared, whether it is the one or the other gene, and, in cases where but one individual has a pair of derivative genes, whether it is the one or the other individual. This gives 7 major types of genic resemblance:

$$\text{(i)}\ [\xi_1\xi_2][\xi_3\xi_4]$$
$$\text{(ii)}\ [\xi_1\xi_2][\xi_{12}\xi_3]$$
$$\text{(iii)}\ [\xi_1\xi_2][\xi_1\xi_2]$$
$$\text{(iv)}\ [\xi_1\xi_2][\xi_{12}^2]$$
$$\text{(v)}\ [\xi_1\xi_2][\xi_3^2]$$
$$\text{(vi)}\ [\xi_1^2][\xi_1^2]$$
$$\text{(vii)}\ [\xi_1^2][\xi_2^2]$$

It might be mentioned that the number seven (7), which is a rather strange number in genetics, has previously been discovered by Woodger (1937) as representing the number of functionally different types of matings which can be made when one has a set of at least four multiple alleles. The analogy of matings with relatives and of alleles with derivatives will readily be seen. Thus, if $\xi_1$, $\xi_2$, $\xi_3$, $\xi_4$ are taken to represent non-identical alleles and if $[\xi\xi][\xi\xi]$ means a mating, then the same 7 types would exist. The significance of the classification in its present, more general, form, is, however, somewhat more interesting.

One application which immediately obtrudes itself is the possibility of employing the classification as a systematic and objective basis for naming and describing different "kinds" of biological relationship in bisexually reproducing diploids. Thus, the genic resemblance between full sibs can only be of types (i), (ii) or (iii); half-sibs may have either (i) or (ii); parent-child relationship allows only (ii); monozygotic twins, only (iii), or (iii) and (vi) if inbred; two selflings from a plant can be (iii), (iv), (vi) and (vii); a "father" and a child by his own daughter may be of types (ii), (iii) and (iv); and so on. If it is possible under bisexual diploidy to have two relatives whose genic resemblance might include any one, any two, any three, etc. types, then there must be exactly 128 (= $2^7$) different "kinds" of relatives. Whether Nature has undertaken to "invent" all of these cannot be vouchsafed, but at least a good number of them have been enumerated by the author.

Further indication that the suggested classification may be a logical one is furnished by the fact that each one of many kinds of relatives which can be enumerated on the basis of possible types of genic resemblance is also found to have one, and only one, set of "laws" specifying the probabilities of the various possibilities which characterize that kind of relationship. At least, such has been found to be the case for all types of relatives yet examined. A large part of this section and of Section 4 is concerned with these probability functions for various types of relatives. Only a few of the many types of relatives are examined in these two sections; they include, however, almost all types of relationship which occur in human families.

Woodger, in his biological axiom system (1937), defines an individual's pedigree as "regular up to the $\kappa$-th generation" if, for every cardinal number up to $\kappa$, the individual possesses $2^k$ amcestors in that generation. A regular pedigree (symbolized 'pedreg') is therefore one exhibiting no inbreeding. In our terminology an individual is a 'J' individual if his pedigree is regular up to a number of generations equal to that adopted in the definition of "derivative"; otherwise he is an inbred or 'I' individual. In this section we shall consider only those sorts of relationship in which inbreeding is not involved. This limits the number of possible major types of genic resemblance to three, or to 4 in the extended list of 12, namely, $[\xi_1\xi_2][\xi_3\xi_4]$, $[\xi_1\xi_2][\xi_1\xi_3]$, $[\xi_1\xi_2][\xi_2\xi_3]$, and $[\xi_1\xi_2][\xi_1\xi_2]$. The letter 'Z' will now be used to designate any sort of biological relationship between individuals neither of whom are inbred, i. e., J individuals. Before considering this class of relatives, however, it will be necessary to state the basic axioms of Mendelism.

Axiom 1.1.1 states that all matings in the population are assumed to be between "unrelated" individuals, if we mean by this individuals who do not share derivative genes. Axiom 1.2.1 states the "Law of Segregation", while its corollaries, 1.2.2 and 1.2.3, state explicitly that non-disjunction and other irregularities of segregation are assumed non-existant. Although this principle, due to Mendel, has received ample experimental verification in a multitude of cases of inheritance, it may be well to reiterate that the use of the word "axiom" does not imply that the Law of Segregation is a universal truth. It is merely a proposition, which, if true, will give rise to all genetic consequences described in the theorems of this and later sections which are traceable to 1.2.1.

Axiom 1.3.1, involving the symbol $(\xi_1\xi_2\text{-}\xi_3\xi_4)O[\xi_1\xi_3]$, states that, given a mating between individuals of the sorts $[\xi_1\xi_2]$ and $[\xi_3\xi_4]$, the probability of an offspring having the gene-pair $[\xi_1\xi_3]$ is one-quarter. This axiom therefore contains the assumptions of equal viability of the two sorts of gametes formed by any individual, of random fertilization, and of equal viability of the resulting zygotes. It gives rise immediately to the basic Mendelian ratios of 1.3.2-4 and to theorem 1.4.1, which shows that under the system of mating postulated by 1.1.1, all individuals must possess two non-derivative alleles. Axioms 1.2.1 and 1.3.1 in combination also enable one to state what might be considered the converse of the Law of Segregation: The probability that any particular gene in an individual's gene-pair was derived from a particular parent is one-half (1.5.1, 1.5.2).

We may now write the very important theorem

1.7.1 $(\xi_1\xi_2)Z[\xi_1\xi_{23}]r = \Sigma\{(\xi_1\xi_2)C[\xi_1\xi_3]\cdot(\xi_1\xi_3)O^{n_2}[\xi_1\xi_{24}]\}$

$= \Sigma\{(\xi_1\xi_2)P[\xi_1\xi_4]\cdot(\xi_1\xi_4)P[\xi_1\xi_{24}]\cdots(n_1 \text{ times})$

$\cdot(\xi_1\xi_{24})O[\xi_1\xi_5]\cdot(\xi_1\xi_5)O[\xi_1\xi_{23}]\cdots(n_2 \text{ times})\}$

$= \Sigma(\tfrac{1}{2})^{n_1}(\tfrac{1}{2})^{n_2} = \Sigma(\tfrac{1}{2})^{n} =_{Df} r.$

In this expression '$Z[\xi_1\xi_{23}]r$' denotes a relative of sort Z and of "degree r", 'C' denotes a common ancestor of the two Z-relatives, the ancestor also being a J individual, and '$O^{n_2}$' denotes a descendent of a J who is also a J. In the terminology of logistic, the relation C would be defined as the $n_1$-th power of the relation P, i. e., the relative product of $n_1$ P-relations, and the relation $O^{n_2}$ would be defined as the $n_2$-th power of the relation O. We may also define $n_1$ and $n_2$ as the numbers of "links" or meioses in a "path of descent" connecting the common ancestor with the two Z-relatives, n being their total $(n_1 + n_2)$. '$\Sigma$' indicates that the summation is to be extended over all possible paths of descent.

This principle is fundamental for the derivation of all succeeding ones in this section. The terms "path" and "link" are suggested because of the customary use of arrow diagrams for portraying pedigrees. This method has been used extensively by Wright (1921a, 1921b, 1934) in connection with a variety of genetic problems and with his general method of "path coefficients". The quantity r in 1.7.1 has been termed by Wright the "coefficient of relationship" of the two relatives of type Z. As is shown by the left-hand member of 1.7.1, $(\zeta_1\xi_2)Z[\xi_1\xi_{23}]r$, it is the probability that two Z-relatives will share derivatives of a single gene, irrespective of whether the second gene of the pair is shared. The coefficient of relationship can also be defined as the most likely fraction of all genes to be shared by two relatives, but this meaning will not concern us here.

Two important subclasses of the relation Z must now be distinguished. The following series of diagrams illustrate several simple degrees of relationship. The relatives in question, whose coefficients of relationship are indicated, are those represented by the solid circles in the diagrams.

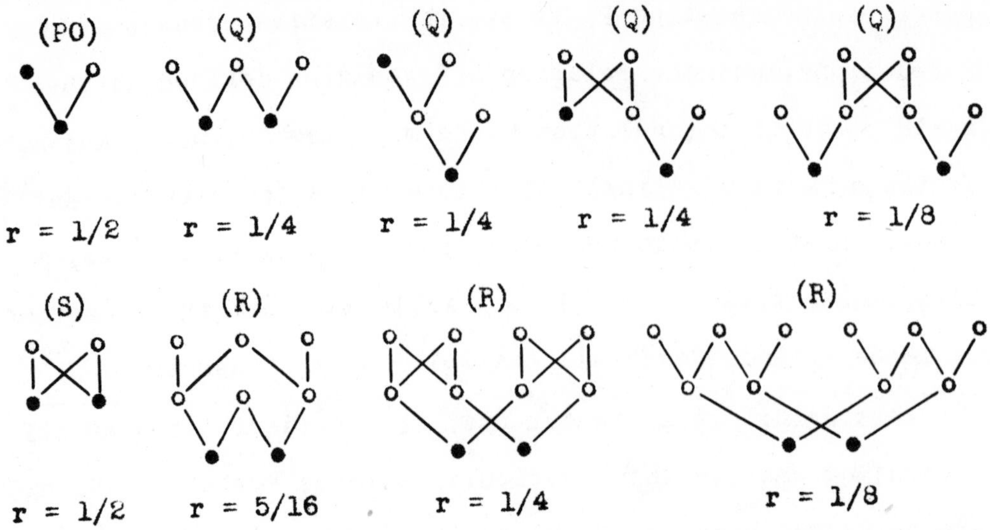


The last four of these examples, namely, full sibs, half-sib half first cousins, double first cousins, and double half first cousins, are seen to have one feature in common which is not possessed by the other relatives pictured - they are connected by two completely separate and independent paths of descent. Uncle-nephew and single first cousins also have two connecting paths, but these are identical in their terminal links. The author is not familiar with any terms for differentiating these two sorts of relatives and therefore suggests the terms "bilineal" and "unilineal" to apply to relatives of the sort Z having two or only one independent path of descent, respectively. They will be symbolized by 'R' and 'Q', respectively, these being subclasses of 'Z'.

The importance of the distinction is readily seen. Unilineal relatives cannot possibly share both genes of a pair, or, more precisely, be of the genic resemblance $[\xi_1\xi_2][\xi_1\xi_2]$, since, having but one independent path of descent, this would require the occurrence of non-disjunction in at least one instance along that path (1.8.1). For relatives of the sort Q, the remaining types of genic resemblance must therefore take probabilities as indicated in 1.8.2-4. Bilineal relatives, possessing two completely different paths of descent, can, on the other hand, be of all four types of genic resemblance, with probabilities as indicated in 1.9.1-4.

Interesting special cases of 'Q' and 'R' are provided by Z-relatives of degree r = 1/2, these being, respectively, parent-offspring and full sibs. Because of their special interest in human genetics they are given the special designations 'PO' and 'S', according to the definitions of 1.10.1 and 1.11.1. In the case of parent-offspring relationship, the quantity (1 - 2r) becomes zero, so that we have the well recognized fact (1.10.2-5) that parent and offspring must share one or the other of their genes. Full sibs do not suffer any such restriction upon the number of possible types of genic resemblance, but their special interest in connection with twinning and other genetic problems merits their special designation. Mathematically, 'S' is not a special case of 'R' in the sense that certain types of genic resemblance disappear but is of interest rather on account of the equality in the ratio of the four types (1.11.2-5).

The following table summarizes the probability relationships for the classes of relatives described above. For the purpose of simplifying notation in subsequent sections, the formal symbols for the probabilities of the 4 types of genic resemblance in Z-relatives of any class are replaced by the symbols $c_2$, $c_1$, $c_1$, $c_0$, as defined in subsection 1.12 and shown in the following table. In this table it will be noticed that $(c_2 + c_1)$ equals r and $(c_1 + c_0)$ equals $(1 - r)$ in every case, as is, of course, demanded by 1.7.1. It will also be seen that 'Q' might be regarded as a special case of 'R' in which the term, $r^2$, vanishes because of the single independent path of descent. Furthermore, identical twins, which will later be denoted by 'E', might also be regarded as a special case of 'R' in which r = 1, while "unrelated" individuals are, of course, Z-relatives of degree r = 0.

Probability Functions for Types of Genic Resemblance in Subclasses of Z-Relatives

| Type of Resemblance | Z | R | $r^2=0$ Q | $r^2 = 0$ $r = \frac{1}{2}$ PO |
|---|---|---|---|---|
| $[\xi_1\xi_2][\xi_1\xi_2]$ | $c_2$ | $r^2 = r^2$ | 0 | 0 |
| $[\xi_1\xi_2][\xi_1\xi_3]$ | $c_1$ | $r(1-r) = r - r^2$ | r | $\frac{1}{2}$ |
| $[\xi_1\xi_2][\xi_2\xi_3]$ | $c_1$ | $r(1-r) = r - r^2$ | r | $\frac{1}{2}$ |
| $[\xi_1\xi_2][\xi_3\xi_4]$ | $c_0$ | $(1-r)^2 = 1 - 2r + r^2$ | $1 - 2r$ | 0 |

## SECTION 1

1.1.1 $H[\xi_1\xi_2\text{-}\xi_3\xi_4] = 1.$ **Axiom**

1.2.1 $(\xi_\alpha\xi_\beta)G[\xi_\alpha] = (\xi_\alpha\xi_\beta)G[\xi_\beta] = \frac{1}{2}.$ **Axiom**

1.2.2 $(\xi_\alpha\xi_\beta)G[\xi_{\alpha\beta}] =_{\text{Df}} (\xi_\alpha\xi_\beta)G[\xi_\alpha] + (\xi_\alpha\xi_\beta)G[\xi_\beta] = 1.$

(1.2.1)

1.2.3 $(\xi_\alpha\xi_\beta)G[\xi_\alpha\xi_\beta] = (\xi_\alpha\xi_\beta)G[\_] = 0.$ (1.2.2)

1.3.1 $(\xi_1\xi_2\text{-}\xi_3\xi_4)O[\xi_1\xi_3] = (\xi_1\xi_2)G[\xi_1]\cdot(\xi_3\xi_4)G[\xi_3].$ Axiom

1.3.2 $(\xi_1\xi_2\text{-}\xi_3\xi_4)O[\xi_1\xi_3] = (\xi_1\xi_2\text{-}\xi_3\xi_4)O[\xi_1\xi_4]$

$= (\xi_1\xi_2\text{-}\xi_3\xi_4)O[\xi_2\xi_3] = (\xi_1\xi_2\text{-}\xi_3\xi_4)O[\xi_2\xi_4] = \frac{1}{4}.$

(1.3.1,1.2.1)

1.3.3 $(\xi_1\xi_2\text{-}\xi_3\xi_4)O[\xi_1\xi_{34}] =_{\text{Df}} (\xi_1\xi_2\text{-}\xi_3\xi_4)O[\xi_1\xi_3]$

$+ (\xi_1\xi_2\text{-}\xi_3\xi_4)O[\xi_1\xi_4] = \frac{1}{2}.$ (1.3.2)

1.3.4 $(\xi_1\xi_2\text{-}\xi_3\xi_4)O[\xi_{12}\xi_{34}] =_{\text{Df}} (\xi_1\xi_2\text{-}\xi_3\xi_4)O[\xi_1\xi_{34}]$

$+ (\xi_1\xi_2\text{-}\xi_3\xi_4)O[\xi_2\xi_{34}] = 1.$ (1.3.3)

1.3.5 $(\xi_1\xi_2\text{-}\xi_3\xi_4)O[\xi_\alpha\xi_\beta] =_{\text{Df}} (\xi_1\xi_2\text{-}\xi_3\xi_4)O[\xi_{12}\xi_{34}] = 1.$

(1.3.4)

1.4.1 $J[\xi_\alpha\xi_\beta] = H[\xi_1\xi_2\text{-}\xi_3\xi_4]\cdot(\xi_1\xi_2\text{-}\xi_3\xi_4)O[\xi_\alpha\xi_\beta] = 1.$

(1.1.1,1.3.5)

1.4.2 $J[\xi_\alpha{}^2] = J[\xi_\alpha\xi_{\alpha\beta}] - J[\xi_\alpha\xi_\beta] = 1 - 1 = 0.$ (1.4.1)

1.5.1 $(\xi_1\xi_2)P[\xi_1\xi_3] = (\xi_1\xi_3\text{-}\xi_2\xi_4)O[\xi_1\xi_2]:$

$\{(\xi_1\xi_3\text{-}\xi_2\xi_4)O[\xi_1\xi_2] + (\xi_1\xi_3\text{-}\xi_2\xi_4)O[\xi_1\xi_2]\}$

$= \frac{1}{4}/(\frac{1}{4} + \frac{1}{4}) = \frac{1}{2}.$ (1.3.2)

1.5.2 $(\xi_1\xi_2)P[\xi_2\xi_3] = (\xi_1\xi_2)P[\xi_1\xi_3] = \frac{1}{2}.$ (1.5.1)

1.5.3 $(\xi_1\xi_2)P[\xi_{12}\xi_3] = (\xi_1\xi_2)P[\xi_1\xi_3] + (\xi_1\xi_2)P[\xi_2\xi_3] = 1.$

(1.5.2)

1.6.1 $(\xi_1\xi_2)O[\xi_1\xi_3] = (\xi_1\xi_2)G[\xi_1] = \frac{1}{2}.$ (1.2.1)

1.6.2 $(\xi_1\xi_2)O[\xi_2\xi_3] = (\xi_1\xi_2)G[\xi_2] = \frac{1}{2}.$ (1.2.1)

1.6.3 $(\xi_1\xi_2)O[\xi_{12}\xi_3] = (\xi_1\xi_2)O[\xi_1\xi_3] + (\xi_1\xi_2)O[\xi_2\xi_3] = 1.$

(1.6.1-2)

1.7.1 $(\xi_1\xi_2)Z[\xi_1\xi_{23}]r = \Sigma\{(\xi_1\xi_2)C[\xi_1\xi_4]\cdot(\xi_1\xi_4)O^{n_2}[\xi_1\xi_{23}]\}$

$= \Sigma\{(\xi_1\xi_2)P[\xi_1\xi_3]\cdot(\xi_1\xi_3)P[\xi_1\xi_{24}]\cdots\cdot(n_1 \text{ times})$

$\cdot(\xi_1\xi_{24})O[\xi_1\xi_5]\cdot(\xi_1\xi_5)O[\xi_1\xi_{23}]\cdots\cdot(n_2 \text{ times})\}$

$= \Sigma(\tfrac{1}{2})^{n_1}(\tfrac{1}{2})^{n_2} = \Sigma(\tfrac{1}{2})^{n} =_{Df} r.$ (1.5.1-3,1.6.1-3)

1.7.2 $(\xi_1\xi_2)Z[\xi_{14}\xi_{23}]r = (\xi_1\xi_2)Z[\xi_\alpha\xi_\beta]r = J[\xi_\alpha\xi_\beta] = 1.$

(1.4.1)

1.7.3 $(\xi_1\xi_2)Z[\xi_4\xi_{23}]r = (\xi_1\xi_2)Z[\xi_{14}\xi_{23}]r - (\xi_1\xi_2)Z[\xi_1\xi_{23}]r$

$= 1 - r.$ (1.7.1-2)

1.8.1 $(\xi_1\xi_2)Q[\xi_1\xi_2]r = (\xi_1\xi_2)G[\xi_1\xi_2] = 0.$ (1.2.3)

1.8.2 $(\xi_1\xi_2)Q[\xi_1\xi_3]r = (\xi_1\xi_2)Z[\xi_1\xi_{23}]r - (\xi_1\xi_2)Q[\xi_1\xi_2]r$

$= r - 0 = r.$ (1.7.1,1.8.1)

1.8.3 $(\xi_1\xi_2)Q[\xi_2\xi_3]r = (\xi_1\xi_2)Z[\xi_{13}\xi_2]r - (\xi_1\xi_2)Q[\xi_1\xi_2]r$

$= r - 0 = r.$ (1.7.1,1.8.1)

1.8.4 $(\xi_1\xi_2)Q[\xi_3\xi_4]r = (\xi_1\xi_2)Z[\xi_{24}\xi_3]r - (\xi_1\xi_2)Q[\xi_2\xi_3]r$

$= 1 - r - r = 1 - 2r.$ (1.7.3,1.8.3)

1.8.5 $(\xi_1\xi_2)Q[\xi_{12}\xi_3]r = (\xi_1\xi_2)Q[\xi_1\xi_3]r + (\xi_1\xi_2)Q[\xi_2\xi_3]r$

$= r + r = 2r.$ (1.8.2,1.8.3)

1.9.1 $(\xi_1\xi_2)R[\xi_1\xi_2]r = (\xi_1\xi_2)Z[\xi_1\xi_{23}]r\cdot(\xi_1\xi_2)Z[\xi_{14}\xi_2]r$

$= r^2.$ (1.7.1)

1.9.2 $(\xi_1\xi_2)R[\xi_1\xi_3]r = (\xi_1\xi_2)Z[\xi_1\xi_{23}]r\cdot(\xi_1\xi_2)Z[\xi_{14}\xi_3]r$

$= r(1 - r).$ (1.7.1,1.7.3)

1.9.3 $(\xi_1\xi_2)R[\xi_2\xi_3]r = (\xi_1\xi_2)Z[\xi_{24}\xi_3]r\cdot(\xi_1\xi_2)Z[\xi_2\xi_{13}]r$

$= r(1 - r).$ (1.7.1,1.7.3)

1.9.4 $(\xi_1\xi_2)R[\xi_3\xi_4]r = (\xi_1\xi_2)Z[\xi_{23}\xi_4]r\cdot(\xi_1\xi_2)Z[\xi_{14}\xi_3]r$

$= (1 - r)^2.$ (1.7.3)

1.9.5 $(\xi_1\xi_2)R[\xi_{12}\xi_3]r = (\xi_1\xi_2)R[\xi_1\xi_3]r + (\xi_1\xi_2)R[\xi_2\xi_3]r$

$= 2r(1 - r).$ (1.9.2,1.9.3)

1.10.1 $(\xi_1\xi_2)PO[\xi_\alpha\xi_\beta] =_{Df} (\xi_1\xi_2)Q[\xi_\alpha\xi_\beta]\frac{1}{2}.$

1.10.2 $(\xi_1\xi_2)PO[\xi_1\xi_2] = 0.$ (1.8.1)

1.10.3 $(\xi_1\xi_2)PO[\xi_1\xi_3] = \frac{1}{2}.$ (1.8.2)

1.10.4 $(\xi_1\xi_2)PO[\xi_2\xi_3] = \frac{1}{2}.$ (1.8.3)

1.10.5 $(\xi_1\xi_2)PO[\xi_3\xi_4] = 0.$ (1.8.4)

1.10.6 $(\xi_1\xi_2)PO[\xi_{12}\xi_3] = 1.$ (1.8.5)

1.11.1 $(\xi_1\xi_2)S[\xi_\alpha\xi_\beta] =_{Df} (\xi_1\xi_2)R[\xi_\alpha\xi_\beta]\frac{1}{2}.$

1.11.2 $(\xi_1\xi_2)S[\xi_1\xi_2] = \frac{1}{4}.$ (1.9.1)

1.11.3 $(\xi_1\xi_2)S[\xi_1\xi_3] = \frac{1}{4}.$ (1.9.2)

1.11.4 $(\xi_1\xi_2)S[\xi_2\xi_3] = \frac{1}{4}.$ (1.9.3)

1.11.5 $(\xi_1\xi_2)S[\xi_3\xi_4] = \frac{1}{4}.$ (1.9.4)

1.11.6 $(\xi_1\xi_2)S[\xi_{12}\xi_3] = \frac{1}{2}.$ (1.9.5)

1.12.1 $(\xi_1\xi_2)Z[\xi_1\xi_2]r =_{Df} c_2.$

1.12.2 $(\xi_1\xi_2)Z[\xi_1\xi_3]r = (\xi_1\xi_2)Z[\xi_2\xi_3]r =_{Df} c_1.$

1.12.3 $(\xi_1\xi_2)Z[\xi_{12}\xi_3]r = 2c_1.$ (1.12.2)

1.12.4 $(\xi_1\xi_2)Z[\xi_3\xi_4]r =_{Df} c_0.$

1.12.5 $c_2 + c_1 = r.$ (1.12.1,1.12.2,1.7.1)

1.12.6 $c_1 + c_0 = 1 - r.$ (1.12.2,1.12.4,1.7.3)

## SECTION 2. GENOTYPE RESEMBLANCE

This section is devoted to genotype resemblance in Z-relatives and is limited to the case of a pair of autosomal alleles, labelled 'A' and 'a', as is stated in Axiom 2.1.1. The gametic ratio of A : a is defined as p : q in 2.1.2 and 2.1.3. Axiom 2.2.1 then asserts that mutation in either direction is non-existant.

It is quite possible to derive many of the theorems relevant to genotype resemblances while ignoring the distinction between derivative and non-derivative identical alleles. Recourse to the theorems of Section 1, however, makes even the most complicated problem in genotype resemblance of the greatest simplicity, while at the same time usually providing a much more general solution. The standard procedure in this section, therefore, is to first write down the genotype or genotypes to be considered, for example, Aa, then assign subscripts to the genes to indicate their origin, such as $A_1a_2$, and, without ever changing this designation, to consider $A_1a_2$ merely a special case of $\xi_1\xi_2$ and therefore subject to all of the theorems of Section 1. It will be remembered that the use of the general allele '$\xi$' in Section 1 implied that any specific allele, such as A or a, might be substituted for $\xi$ at any time. However, it did not specify whether different alleles might be substituted at different times, as would be possible through mutation. Consequently, we require Axiom 2.2.1 to insure that any derivative $\xi_1$ of $A_1$ will also be $A_1$ and not $a_1$, and similarly for $a_2$. Theorems 2.3.1-4, which give the gametic ratios of the three genotypes, offer the simplest illustration of this general method.

Theorems 2.4.1-2 express the gene frequencies as functions of the genotype frequencies, these relationships holding irrespective of the system of mating. The ordinary monohybrid Mendelian ratios are given in subsection 2.5. Use of the superscript 0 in such symbols as $(AA \overset{0}{-} AA)$ merely denotes that the mates are unrelated (r = 0). Although a relationship would not affect the ratios of 2.5, it would affect those of later subsections. Genotype expectations for consanguine matings will be dealt with in Section 4, where the superscript 0 of the present section will be replaced by r. No attention need be paid in this section to phenotypes, but it will be noticed that the case of dominance of A over a or of a over A can easily be handled with the aid of the general allele '$\xi$' as in $[A\xi]$ and $[\xi a]$. Thus, 2.5.6 gives the classical 3:1 monohybrid ratio appropriate to the case of dominance.

We now introduce the axiom of random mating. The symbols $\xi^1$, $\xi^2$, $\xi^3$, $\xi^4$ are taken to represent any four alleles which may or may not be identical. Then

2.6.1 $$H[\xi^1\xi^2 - \xi^3\xi^4] = J[\xi^1\xi^2]\cdot J[\xi^3\xi^4]$$

states that the probability of a mating between any two genotypes in specified order, say $[\xi^1\xi^2]$ in the female and $[\xi^3\xi^4]$ in the male, is the product of the relative frequencies of these same genotypes in the population. It is then easy to demonstrate the fact (2.7.1-4), first noted by Hardy (1908), that random mating produces an equilibrium in the distribution of the three genotypes, AA, Aa, aa, the frequencies of which are then proportional to the terms of the squared binomial of the gametic proportions of the two genes. The derivation also reveals the additional fact, later noted by Wentworth and Remick (1916), that such an equilibrium is reached in a single generation of random mating. In the theorems of subsection 2.7 and following, such symbols as J[<u>Aa</u>] are used to designate the probability of a heterozygous genotype of specified order, say A of maternal and a of paternal origin, while [Aa] is the logical sum of [<u>Aa</u>] and [<u>aA</u>].

The gamete proportions of incompletely specified genotypes are given by 2.9.1-3. The proportion, $q/(1+q)$, of recessive-carrying gametes derived from genotypes showing the dominant character has previously been assigned the symbol '$\zeta$' (Cotterman, 1937a). That it merits this special designation is shown by the numerous instances in which $\zeta$ appears in this and later sections. In fact, in Section 5 it will be found that $\zeta$ is merely a special case of a still more general function of the recessive gene frequency. It is thought unnecessary to add to this subsection such expressions as $(\xi a)G[A]$ or $(\xi a)G[a]$, which would be of interest in a genetic analysis wherein a was assumed dominant to A. However, it is easily seen that such functions would be like those for (A$\xi$)G[a] and (A$\xi$)G[A] but with q replaced by p. In fact, in all theorems of this and later sections which involve genotype frequencies, it will be seen that the series of functions are symmetrical with respect to p and q. Thus, if we have the probability of (AA)Z[Aa]r expressed as a function of p, we can write immediately that (aa)Z[Aa]r will be an identical function of q. By redefining our gene symbols and interchanging p and q we could therefore shorten many series of formulae in this and other sections. This has not generally been done, however, since it has seemed desirable to record all useful formulae for the ready reference of those who might wish to employ them in a genetic analysis.

We have already expressed in 2.5 the genotype expectations of matings wherein (1) genotypes are completely specified in both parents. With the aid of 2.8 it is then easy to supply similar expectations for the following other cases: (2) genotype complete in one parent, partial in the other; (3) genotypes partial in both parents; (4) genotype complete in one parent, unknown in the other; and (5) genotype partial in one parent, unknown in the other. The sixth case: genotypes completely unspecified in both parents, has, of course, been supplied by the general genotype proportions of 2.7. Cases (2) and (3) are worked out in subsections 2.10 and 2.11, but cases (4) and (5), although they could be treated more simply here, are actually included in the more general statements of 2.16, which give the probabilities of specific genotypes when the genotype of either one parent or one offspring is known. Formulae 2.10.8-9 and 2.11.3-4 are those appropriate to the study of dominant traits in randomly collected human families, as was first shown by Snyder (1934).

The remainder of the section is devoted to the specification of genotype probabilities when information is available concerning a single relative and to the combinatory probabilities of various genotype combinations in pairs of relatives. The general case of implicative probabilities for all classes of Z-relatives of any degree are given in subsection 2.12 and the corresponding combinatory probabilities in 2.15. For the special class of Q-relatives, implicative probabilities are supplied in 2.14 and combinatory probabilities in 2.15. The subclass 'PO' of 'Q' is worked out in subsections 2.16 and 2.17. Similarly, R-relatives are treated in subsections 2.18 and 2.19, but the special case of full sibs ('S'), which has previously been worked out by Cotterman (1937) and Rife (1938), is deferred to Section 6, since the relationships are identical with those for "fraternal twins", which are treated there under the relation 'F'.

## SECTION 2

2.1.1 $G[A] + G[a] = G[\xi] = 1.$ **Axiom**

2.1.2 $G[A] =_{Df} p.$

2.1.3 $G[a] =_{Df} q = 1 - p.$ (2.1.1,2.1.2)

2.2.1 $(A_\alpha \xi_\beta)G[a_\alpha] = (a_\alpha \xi_\beta)G[A_\alpha] = 0.$ **Axiom**

2.3.1 $(AA)G[A] = (A_1A_2)G[A_{12}] = 1.$ (1.2.2,2.2.1)

2.3.2 $(Aa)G[A] = (A_1a_2)G[A_1] = \frac{1}{2}.$ (1.2.1,2.2.1)

2.3.3 $(Aa)G[a] = (A_1a_2)G[a_2] = \frac{1}{2}.$ (1.2.1,2.2.1)

2.3.4 $(aa)G[a] = (a_1a_2)G[a_{12}] = 1.$ (1.2.2,2.2.1)

2.4.1 $G[A] = J[AA]\cdot(AA)G[A] + J[Aa]\cdot(Aa)G[A]$

$= J[AA] + \frac{1}{2}\cdot J[Aa].$ (2.3.1,2.3.2)

2.4.2 $G[a] = J[aa]\cdot(aa)G[a] + J[Aa]\cdot(Aa)G[a]$

$= J[aa] + \frac{1}{2}\cdot J[Aa].$ (2.3.3,2.3.4)

2.5.1 $(AA \overset{o}{-} AA)O[AA] = (A_1A_2 \overset{o}{-} A_3A_4)O[A_{12}A_{34}] = 1.$ (1.3.4)

2.5.2 $(AA \overset{o}{-} Aa)O[AA] = (AA \overset{o}{-} Aa)O[Aa] = (A_1A_2 \overset{o}{-} A_3a_4)O[A_{12}a_4]$

$= (A_1A_2 \overset{o}{-} A_3a_4)O[A_{12}A_3] = \frac{1}{2}.$ (1.3.3)

2.5.3 $(AA \overset{o}{-} aa)O[Aa] = (A_1A_2 \overset{o}{-} a_3a_4)O[A_{12}a_{34}] = 1.$ (1.3.4)

2.5.4 $(Aa \overset{o}{-} Aa)O[AA] = (Aa \overset{o}{-} Aa)O[aa] = (A_1a_2 \overset{o}{-} A_3a_4)O[A_1A_3]$

$= (A_1a_2 \overset{o}{-} A_3a_4)O[a_2a_4] = \frac{1}{4}.$ (1.3.2)

2.5.5 $(Aa \overset{o}{-} Aa)O[Aa] = (A_1a_2 \overset{o}{-} A_3a_4)O[A_1a_4] + (A_1a_2 \overset{o}{-} A_3a_4)O[A_3a_2]$

$= \frac{1}{2}.$ (1.3.2)

2.5.6 $(Aa \overset{o}{-} Aa)O[A\xi] = (Aa \overset{o}{-} Aa)O[AA] + (Aa \overset{o}{-} Aa)O[Aa] = \frac{3}{4}.$

(2.5.4,2.5.5)

2.5.7 $(Aa \overset{o}{-} aa)O[Aa] = (Aa \overset{o}{-} aa)O[aa] = (A_1a_2 \overset{o}{-} a_3a_4)O[A_1a_{34}]$

$= (A_1a_2 \overset{o}{-} a_3a_4)O[a_2a_{34}] = \frac{1}{2}.$ (1.3.3)

2.5.8 $(aa \overset{o}{-} aa)O[aa] = (a_1a_2 \overset{o}{-} a_3a_4)O[a_{12}a_{34}] = 1.$ (1.3.4)

2.6.1 $H[\xi^1\xi^2 \overset{o}{-} \xi^3\xi^4] = J[\xi^1\xi^2]\cdot J[\xi^3\xi^4].$ **Axiom**

$$2.7.1\ J[AA] = H[AA \overset{\circ}{-} AA]\cdot(AA \overset{\circ}{-} AA)O[AA] + H[\underline{AA \overset{\circ}{-} Aa}]\cdot(AA \overset{\circ}{-} Aa)O[AA]$$
$$+ H[\underline{Aa \overset{\circ}{-} AA}]\cdot(Aa \overset{\circ}{-} AA)O[AA] + H[Aa \overset{\circ}{-} Aa]\cdot(Aa \overset{\circ}{-} Aa)O[AA]$$
$$= H[AA \overset{\circ}{-} AA] + H[\underline{AA \overset{\circ}{-} Aa}] + \tfrac{1}{4} H[Aa \overset{\circ}{-} Aa]$$
$$= (J[AA])^2 + J[AA]\cdot J[Aa] + \tfrac{1}{4}\cdot(J[Aa])^2$$
$$= \{J[AA] + \tfrac{1}{4}\cdot J[Aa]\}^2 = (G[A])^2 = p^2.$$ [Hardy, 1908] (2.5.1,2.5.2,2.5.4,2.6.1,2.4.1,2.1.2)

$$2.7.2\ J[\underline{Aa}] = J[\underline{aA}] = G[A]\ G[a] = pq.$$ (2.1.2,2.1.3,2.7.1)

$$2.7.3\ J[Aa] = J[\underline{Aa}] + J[\underline{aA}] = 2pq.$$ [Hardy, 1908] (2.7.2)

$$2.7.4\ J[aa] = (G[a])^2 = q^2.$$ [Hardy, 1908] (2.1.3,2.7.1)

$$2.7.5\ J[\underline{A\xi}] = J[\underline{\xi A}] = J[AA] + J[\underline{Aa}] = p.$$ (2.7.1-2)

$$2.7.6\ J[A\xi] = J[AA] + J[Aa] = 1 - q^2.$$ (2.7.1,2.7.3)

$$2.7.7\ J[\underline{a\xi}] = J[\underline{\xi a}] = J[aa] + J[\underline{aA}] = q.$$ (2.7.2,2.7.4)

$$2.7.8\ J[a\xi] = J[aa] + J[Aa] = 1 - p^2.$$ (2.7.3-4)

$$2.8.1\ (\xi\xi)J[AA] = J[AA] = p^2.$$ (2.7.1)

$$2.8.2\ (\xi\xi)J[\underline{Aa}] = (\xi\xi)J[\underline{aA}] = J[\underline{Aa}] = J[\underline{aA}] = pq.$$ (2.7.2)

$$2.8.3\ (\xi\xi)J[Aa] = J[Aa] = 2pq.$$ (2.7.3)

$$2.8.4\ (\xi\xi)J[aa] = J[aa] = q^2.$$ (2.7.4)

$$2.8.5\ (\underline{A\xi})J[AA] = J[AA] : J[\underline{A\xi}] = p.$$ (2.7.1,2.7.5)

$$2.8.6\ (\underline{A\xi})J[\underline{Aa}] = J[\underline{Aa}] : J[\underline{A\xi}] = q.$$ (2.7.2,2.7.5)

$$2.8.7\ (\underline{\xi A})J[AA] = J[AA] : J[\underline{\xi A}] = p.$$ (2.7.1,2.7.5)

$$2.8.8\ (\underline{\xi A})J[\underline{aA}] = J[\underline{aA}] : J[\underline{\xi A}] = q.$$ (2.7.2,2.7.5)

$$2.8.9\ (\underline{a\xi})J[\underline{aA}] = J[\underline{aA}] : J[\underline{a\xi}] = p.$$ (2.7.2,2.7.7)

$$2.8.10\ (\underline{a\xi})J[aa] = J[aa] : J[\underline{a\xi}] = q.$$ (2.7.4,2.7.7)

$$2.8.11\ (\underline{\xi a})J[\underline{Aa}] = J[\underline{Aa}] : J[\underline{\xi a}] = p.$$ (2.7.4,2.7.7)

$$2.8.12\ (\underline{\xi a})J[aa] = J[aa] : J[\underline{\xi a}] = q.$$ (2.7.4,2.7.7)

$$2.8.13\ (A\xi)J[AA] = J[AA] : J[A\xi] = p/(1+q).$$ (2.7.1,2.7.6)

$$2.8.14\ (A\xi)J[Aa] = J[Aa] : J[A\xi] = 2q/(1+q) = \mathrm{Df}\ 2\zeta.$$

(2.7.3,2.7.6)

2.8.15 $(a\xi)J[Aa] = J[Aa]:J[a\xi] = 2p/(1+p).$ (2.7.3,2.7.8)

2.8.16 $(a\xi)J[aa] = J[aa]:J[a\xi] = q/(1+p).$ (2.7.4,2.7.8)

2.9.1 $(\underline{A\xi})G[A] = (\underline{A\xi})J[AA]\cdot(AA)G[A] + (\underline{A\xi})J[\underline{Aa}]\cdot(Aa)G[A]$

$= p + \frac{1}{2}q.$ (2.8.5,2.3.1,2.8.6,2.3.2)

2.9.2 $(\underline{A\xi})G[a] = (\underline{A\xi})J[\underline{Aa}]\cdot(Aa)G[a] = \frac{1}{2}q.$ (2.8.6,2.3.3)

2.9.3 $(A\xi)G[A] = (A\xi)J[AA]\cdot(AA)G[A] + (A\xi)J[Aa]\cdot(Aa)G[A]$

$= 1/(1+q). = 1-\zeta.$ (2.8.9,2.3.1,2.8.10,2.3.2)

2.9.4 $(A\xi)G[a] = (A\xi)J[Aa]\cdot(Aa)G[a] = \zeta.$ (2.8.10,2.3.3)

2.10.1 $(A\xi \overset{o}{-} AA)O[AA] = (A\xi)G[A]\cdot(AA)G[A] = 1-\zeta.$ (2.9.3,2.3.1)

2.10.2 $(A\xi \overset{o}{-} AA)O[Aa] = (A\xi)G[a]\cdot(AA)G[A] = \zeta.$ (2.9.4,2.3.1)

2.10.3 $(A\xi \overset{o}{-} AA)O[A\xi] = (A\xi \overset{o}{-} AA)O[AA] + (A\xi \overset{o}{-} AA)O[Aa] = 1.$

(2.10.1,2.10.2)

2.10.4 $(A\xi \overset{o}{-} Aa)O[AA] = (A\xi)G[A]\cdot(Aa)G[A] = \frac{1}{2}(1-\zeta).$

(2.9.3,2.3.2)

2.10.5 $(A\xi \overset{o}{-} Aa)O[Aa] = (A\xi)G[A]\cdot(Aa)G[a] + (A\xi)G[a]\cdot(Aa)G[A]$

$= \frac{1}{2}.$ (2.9.3-4,2.3.2-3)

2.10.6 $(A\xi \overset{o}{-} Aa)O[A\xi] = (A\xi \overset{o}{-} Aa)O[AA] + (A\xi \overset{o}{-} Aa)O[Aa] = 1 - \frac{1}{2}\zeta.$

(2.10.4,2.10.5)

2.10.7 $(A\xi \overset{o}{-} Aa)O[aa] = (A\xi)G[a]\cdot(Aa)G[a] = \frac{1}{2}\zeta.$ (2.9.4,2.3.3)

2.10.8 $(A\xi \overset{o}{-} aa)O[A\xi] = (A\xi \overset{o}{-} aa)O[Aa] = (A\xi)G[A]\cdot(aa)G[a]$

$= 1-\zeta.$ [Snyder, 1934] (2.9.3,2.3.4)

2.10.9 $(A\xi \overset{o}{-} aa)O[aa] = (A\xi)G[a]\cdot(aa)G[a] = \zeta.$ [Snyder, 1934]

(2.9.4,2.3.4)

2.11.1 $(A\xi \overset{o}{-} A\xi)O[AA] = (A\xi)G[A]\cdot(A\xi)G[A] = (1-\zeta)^2.$ (2.9.3)

2.11.2 $(A\xi \overset{o}{-} A\xi)O[Aa] = (A\xi)G[A]\cdot(A\xi)G[a] + (A\xi)G[a]\cdot(A\xi)G[A]$

$= 2\zeta(1-\zeta).$ (2.9.3,2.9.4)

2.11.3 $(A\xi \overset{o}{-} A\xi)O[A\xi] = (A\xi \overset{o}{-} A\xi)O[AA] + (A\xi \overset{o}{-} A\xi)O[Aa] = 1 - \zeta^2$.

[Snyder, 1934] (2.11.1,2.11.2)

2.11.4 $(A\xi \overset{o}{-} A\xi)O[aa] = (A\xi)G[a] \cdot (A\xi)G[a] = \zeta^2$. (2.9.4)

[Snyder, 1934]

2.12.1 $(A_1A_2)Z[A_1A_2]r = c_2$. (1.12.1)

2.12.2 $(A_1A_2)Z[A_{12}A_3]r = (A_1A_2)Z[A_{12}\xi_3]r \cdot (\underline{A\xi})J[\underline{AA}] = 2c_1p$.

(1.12.3,2.7.5)

2.12.3 $(A_1A_2)Z[A_3A_4]r = (A_2A_2)Z[\xi_3\xi_4]r \cdot (\xi\xi)J[AA] = c_0p^2$.

(1.12.4,2.8.1)

2.12.4 $(AA)Z[AA]r = (A_1A_2)Z[A_{13}A_{24}]r = c_2 + 2c_1p + c_0p^2$.

(2.12.1-3)

2.12.5 $(A_1A_2)Z[A_{12}a_3]r = (A_1A_2)Z[A_{12}\xi_3]r \cdot (\underline{A\xi})J[\underline{Aa}] = 2c_1q$.

(1.12.3,2.8.6)

2.12.6 $(A_1A_2)Z[A_3a_4]r = (A_1A_2)Z[\xi_3\xi_4]r \cdot (\xi\xi)J[Aa] = 2c_0pq$.

(1.12.4,2.8.2)

2.12.7 $(AA)Z[Aa]r = (A_1A_2)Z[A_{123}a_4]r = 2q(c_1 + c_0p)$.

(2.12.5-6)

2.12.8 $(AA)Z[aa]r = (A_1A_2)Z[\xi_3\xi_4]r \cdot (\xi\xi)J[aa] = c_0q^2$. (1.12.1)

2.12.9 $(A_1a_2)Z[A_1A_3]r = (A_1a_2)Z[A_1\xi_3]r \cdot (\underline{A\xi})J[\underline{AA}] = c_1p$.

(1.12.2,2.8.5)

2.12.10 $(A_1a_2)Z[A_3A_4]r = (A_1a_2)Z[\xi_3\xi_4]r \cdot (\xi\xi)J[AA] = c_0p^2$.

(1.12.4,2.8.1)

2.12.11 $(Aa)Z[AA]r = (A_1a_2)Z[A_{14}A_3]r = p(c_1 + c_0p)$.

(2.12.9,2.12.10)

2.12.12 $(A_1a_2)Z[A_1a_2]r = c_2$. (1.12.1)

2.12.13 $(A_1a_2)Z[A_1a_3]r = (A_1a_2)Z[A_1\xi_3]r \cdot (\underline{A\xi})J[\underline{Aa}] = c_1q$.

(1.12.2,2.8.6)

2.12.14 $(A_1a_2)Z[A_3a_2]r = (A_1a_2)Z[\xi_3a_2]r\cdot(\underline{\xi a})J[\underline{Aa}] = c_1p.$

(1.12.2,2.8.11)

2.12.15 $(A_1a_2)Z[A_3a_4]r = (A_1a_2)Z[\xi_3\xi_4]r\cdot(\xi\xi)J[Aa] = 2c_0pq.$

(1.12.4,2.8.3)

2.12.16 $(Aa)Z[Aa]r = (A_1a_2)Z[A_{13}a_{24}]r = c_2 + c_1 + 2c_0pq.$

(2.12.12-15)

2.12.17 $(A_1a_2)Z[a_2a_3]r = (A_1a_2)Z[a_2\xi_3]r\cdot(\underline{a\xi})J[aa] = c_1q.$

(1.12.2,2.8.10)

2.12.18 $(A_1a_2)Z[a_3a_4]r = (A_1a_2)Z[\xi_3\xi_4]r\cdot(\xi\xi)J[aa] = c_0q^2.$

(1.12.4,2.8.4)

2.12.19 $(Aa)Z[aa]r = (A_1a_2)Z[\xi_{24}\xi_3]r = q(c_1 + c_0q).$

(2.12.17-18)

2.12.20 $(aa)Z[AA]r = (a_1a_2)Z[\xi_3\xi_4]r\cdot(\xi\xi)J[AA] = c_0p^2.$

(1.12.4,2.8.1)

2.12.21 $(a_1a_2)Z[A_3a_{12}]r = (a_1a_2)Z[\xi_3a_{12}]r\cdot(\underline{\xi a})J[\underline{Aa}] = 2c_1p.$

(1.12.3,2.8.11)

2.12.22 $(a_1a_2)Z[A_3a_4]r = (a_1a_2)Z[\xi_3\xi_4]r\cdot(\xi\xi)J[Aa] = 2c_0pq.$

(1.12.4,2.8.2)

2.12.23 $(aa)Z[Aa]r = (a_1a_2)Z[A_3a_{124}]r = 2p(c_1 + c_0q).$

(2.12.21-22)

2.12.24 $(a_1a_2)Z[a_1a_2]r = c_2.$ (1.12.1)

2.12.25 $(a_1a_2)Z[a_{12}a_3]r = (a_1a_2)Z[a_{12}\xi_3]r\cdot(\underline{a\xi})J[aa] = 2c_1q.$

(1.12.3,2.8.10)

2.12.26 $(a_1a_2)Z[a_3a_4]r = (a_1a_2)Z[\xi_3\xi_4]r\cdot(\xi\xi)J[aa] = c_0q^2.$

(1.12.4,2.8.4)

2.12.27 $(aa)Z[aa]r = (a_1a_2)Z[a_{14}a_{23}]r = c_2 + 2c_1q + c_0q^2.$

(2.12.24-26)

2.13.1 $Z[AA]r[AA] = J[AA]\cdot(AA)Z[AA]r = p^2(c_2 + 2c_1p + c_0p^2).$

(2.7.1,2.12.4)

2.13.2 $Z[\underline{AA]r[Aa}] = Z[\underline{Aa]r[AA}] = J[AA]\cdot(AA)Z[Aa]r$

$= J[Aa]\cdot(Aa)Z[AA]r = 2p^2q(c_1 + c_0p).$

(2.7.1,2.12.7,2.7.3,2.12.11)

2.13.3 $Z[AA]r[Aa] = Z[\underline{AA]r[Aa}] + Z[\underline{Aa]r[AA}] = 4p^2q(c_1 + c_0p).$

(2.13.2)

2.13.4 $Z[\underline{AA]r[aa}] = Z[\underline{aa]r[AA}] = J[AA]\cdot(AA)Z[aa]r$

$= J[aa]\cdot(aa)Z[AA]r = c_0p^2q^2.$

(2.7.1,2.12.8,2.7.4,2.12.20)

2.13.5 $Z[AA]r[aa] = Z[\underline{AA]r[aa}] + Z[\underline{aa]r[AA}] = 2c_0p^2q^2.$

(2.13.4)

2.13.6 $Z[Aa]r[Aa] = J[Aa]\cdot(Aa)Z[Aa]r = 2pq(c_2 + c_1 + 2c_0pq).$

(2.7.3,2.12.16)

2.13.7 $Z[\underline{Aa]r[aa}] = Z[\underline{aa]r[Aa}] = J[Aa]\cdot(Aa)Z[aa]r$

$= J[aa]\cdot(aa)Z[Aa]r = 2pq^2(c_1 + c_0q).$

(2.7.3,2.12.19,2.7.4,2.12.23)

2.13.8 $Z[Aa]r[aa] = Z[\underline{Aa]r[aa}] + Z[\underline{aa]r[Aa}] = 4pq^2(c_1 + c_0q).$

(2.13.7)

2.13.9 $Z[aa]r[aa] = J[aa]\cdot(aa)Z[aa]r = q^2(c_2 + 2c_1q + c_0q^2).$

(2.7.4,2.12.27)

2.14.1 $(AA)Q[AA]r = p(p + 2r - 2rp).$ (2.12.4,1.8.1-5)

2.14.2 $(AA)Q[Aa]r = 2q(p + r - 2rp).$ (2.12.7,1.8.1-5)

2.14.3 $(AA)Q[aa]r = q^2(1 - 2r).$ (2.12.8,1.8.1-5)

2.14.4 $(Aa)Q[AA]r = p(p + r - 2rp).$ (2.12.11,1.8.1-5)

2.14.5 $(Aa)Q[Aa]r = r + 2(1 - 2r)pq.$ (2.12.16,1.8.1-5)

2.14.6 $(Aa)Q[aa]r = q(q + r - 2rq).$ (2.12.19,1.8.1-5)

2.14.7 $(aa)Q[AA]r = p^2(1-2r)$. (2.12.20,1.8.1-5)

2.14.8 $(aa)Q[Aa]r = 2p(q+r-2rq)$. (2.12.23,1.8.1-5)

2.14.9 $(aa)Q[aa]r = q(q+2r-2rq)$. (2.12.27,1.8.1-5)

2.14.10 $(AA)Q[A\xi]r = (AA)Q[AA]r + (AA)Q[Aa]r = 1-q^2(1-2r)$.

(2.14.1-2)

2.14.11 $(Aa)Q[A\xi]r = (Aa)Q[AA]r + (Aa)Q[Aa]r$

$= 1-q(q+r-2rq)$. (2.14.4-5)

2.14.12 $(aa)Q[A\xi]r = (aa)Q[AA]r + (aa)Q[Aa]r$

$= 1-q(q+2r-2rq)$. (2.14.7-8)

2.14.13 $(A\xi)Q[AA]r = (A\xi)J[AA]\cdot(AA)Q[AA]r$

$+ (A\xi)J[Aa]\cdot(Aa)Q[AA]r = p(p+r-2rp) + \frac{p^2r}{1+q}$.

(2.8.13,2.14.1,2.8.14,2.14.4)

2.14.14 $(A\xi)Q[Aa]r = (A\xi)J[AA]\cdot(AA)Q[Aa]r$

$+ (A\xi)J[Aa]\cdot(Aa)Q[Aa]r = \{r+rp+p(1+q)(1-2r)\}\frac{2q}{1+q}$.

(2.8.13,2.14.2,2.18.14,2.14.5)

2.14.15 $(A\xi)Q[A\xi]r = (A\xi)Q[AA]r + (A\xi)Q[Aa]r$

$= \frac{(1+q)[1-q^2(1-2r)]-2q^2r}{(1+q)}$. (2.14.13-14)

2.15.1 $Q[AA]r[AA] = p^3(p+2r-2rp)$. (2.13.1,1.8.1-5)

2.15.2 $Q[\underline{AA}]r[\underline{Aa}] = Q[\underline{Aa}]r[\underline{AA}] = 2p^2q(p+r-2rp)$.

(2.13.2,1.8.1-5)

2.15.3 $Q[AA]r[Aa] = 4p^2q(p+r-2rp)$. (2.13.3,1.8.1-5)

2.15.4 $Q[\underline{AA}]r[\underline{aa}] = Q[\underline{aa}]r[\underline{AA}] = p^2q^2(1-2r)$.(2.13.4,1.8.1-5)

2.15.5 $Q[AA]r[aa] = 2p^2q^2(1-2r)$. (2.13.5,1.8.1-5)

2.15.6 $Q[Aa]r[Aa] = 2pq\{r+2(1-2r)pq\}$. (2.13.6,1.8.1-5)

2.15.7 $Q[\underline{Aa}]r[\underline{aa}] = Q[\underline{aa}]r[\underline{Aa}] = 2pq^2(q+r-2rq)$.

(2.13.7,1.8.1-5)

2.15.8 $Q[Aa]r[aa] = 4pq^2(q + r - 2rq)$. (2.13.8,1.8.1-5)

2.15.9 $Q[aa]r[aa] = q^3(q + 2r - 2rq)$. (2.13.9,1.8.1-5)

2.15.10 $Q[A\xi]r[aa] = Q[AA]r[aa] + Q[Aa]r[aa]$

$= 2pq^2(1 + q - 2qr)$. (2.15.5,2.15.8)

2.15.11 $Q[A\xi]r[A\xi] = Q[AA]r[AA] + Q[AA]r[Aa] + Q[Aa]r[Aa]$

$= (1 - q^2)^2(1 - 2r) + 2pr(1 + pq)$. (2.15.1,2.15.3,2.15.6)

2.16.1 $(AA)PO[AA] = p$. (2.14.1,1.10.1)

2.16.2 $(AA)PO[Aa] = q$. (2.14.2,1.10.1)

2.16.3 $(AA)PO[aa] = 0$. (2.14.3,1.10.1)

2.16.4 $(Aa)PO[AA] = \frac{1}{2}p$. (2.14.4,1.10.1)

2.16.5 $(Aa)PO[Aa] = \frac{1}{2}$. (2.14.5,1.10.1)

2.16.6 $(Aa)PO[aa] = \frac{1}{2}q$. (2.14.6,1.10.1)

2.16.7 $(aa)PO[AA] = 0$. (2.14.7,1.10.1)

2.16.8 $(aa)PO[Aa] = p$. (2.14.8,1.10.1)

2.16.9 $(aa)PO[aa] = q$. (2.14.9,1.10.1)

2.16.10 $(AA)PO[A\xi] = (AA)PO[AA] + (AA)PO[Aa] = 1$. (2.16.1-2)

2.16.11 $(Aa)PO[A\xi] = (Aa)PO[AA] + (Aa)PO[Aa] = \frac{1}{2}(1 + p)$.

(2.16.4-5)

2.16.12 $(aa)PO[A\xi] = (aa)PO[AA] + (aa)PO[Aa] = p$. (2.16.7-8)

2.16.13 $(A\xi)PO[AA] = (A\xi)J[AA]\cdot(AA)PO[AA] + (A\xi)J[Aa]$

$\cdot(Aa)PO[AA] = P/(1 + q) = 1 - 2\zeta$. (2.8.13-14;2.16.1-4)4

2.16.14 $(A\xi)PO[Aa] = (A\xi)J[AA]\cdot(AA)PO[Aa] + (A\xi)J[Aa]$

$\cdot(Aa)PO[Aa] = q(1 + p)/(1 + q) = \zeta(2 - 3\zeta)/(1 - \zeta)$

(2.8.13-14;2.16.2-5)

2.16.15 $(A\xi)PO[aa] = (A\xi)J[AA]\cdot(AA)PO[aa] + (A\xi)J[Aa]$

$\cdot(Aa)PO[aa] = q^2/(1 + q) = \zeta^2/(1 - \zeta)$.

(2.8.13-14;2.16.3-6)

2.16.16 $(A\xi)PO[A\xi] = (A\xi)PO[AA] + (A\xi)PO[Aa]$

$= (1+pq)/(1+q) = (1-\zeta-\zeta^2)/(1-\zeta)$. (2.16.13-14)

2.17.1 $PO[AA][AA] = p^3$. (2.15.1,1.10.1)

2.17.2 $PO\underline{[AA][Aa]} = PO\underline{[Aa][AA]} = p^2q$. (2.15.2,1.10.1)

2.17.3 $PO[AA][AA] = 2p^2q$. (2.15.3,1.10.1)

2.17.4 $PO\underline{[AA][aa]} = PO\underline{[aa][AA]} = 0$. (2.15.4,1.10.1)

2.17.5 $PO[AA][aa] = 0$. (2.15.5,1.10.1)

2.17.6 $PO[Aa][Aa] = pq$. (2.15.6,1.10.1)

2.17.7 $PO\underline{[Aa][aa]} = PO\underline{[aa][Aa]} = pq^2$. (2.15.7,1.10.1)

2.17.8 $PO[Aa][aa] = 2pq^2$. (2.15.8,1.10.1)

2.17.9 $PO[aa][aa] = q^3$. (2.15.9,1.10.1)

2.17.10 $PO[A\xi][aa] = PO[AA][aa] + PO[Aa][aa] = 2pq^2$.

(2.17.5,2.17.8)

2.17.11 $PO[A\xi][A\xi] = PO[AA][AA] + PO[AA][Aa] + PO[Aa][Aa]$

$= p(1+pq)$. (2.17.1,2.17.3,2.17.6)

2.18.1 $(AA)R[AA]r = (r+p-rp)^2$. (2.12.4,1.9.1-5)

2.18.2 $(AA)R[Aa]r = 2q(1-r)(r+p-rp)$. (2.12.7,1.9.1-5)

2.18.3 $(AA)R[aa]r = (1-r)^2q^2$. (2.12.8,1.9.1-5)

2.18.4 $(Aa)R[AA]r = p(1-r)(r+p-rp)$. (2.12.11,1.9.1-5)

2.18.5 $(Aa)R[Aa]r = r+2(1-r)^2pq$. (2.12.16,1.9.1-5)

2.18.6 $(Aa)R[aa]r = q(1-r)(r+q-rq)$. (2.12.19,1.9.1-5)

2.18.7 $(aa)R[AA]r = (1-r)^2p^2$. (2.12.20,1.9.1-5)

2.18.8 $(aa)R[Aa]r = 2p(1-r)(r+q-rq)$. (2.12.23,1.9.1-5)

2.18.9 $(aa)R[aa]r = (r+q-rq)^2$. (2.12.27,1.9.1-5)

2.19.1 $R[AA]r[AA] = p^2(r+p-rp)^2$. (2.13.1,1.9.1-5)

2.19.2 $R\underline{[AA]r[Aa]} = R\underline{[Aa]r[AA]} = 2p^2q(1-r)(r+p-rp)$.

(2.13.2,1.9.1-5)

2.19.3 $R[AA]r[Aa] = 4p^2q(1-r)(r+p-rp).$ (2.13.3,1.9.1-5)

2.19.4 $R[\underline{AA]r[aa}] = R[\underline{aa]r[AA}] = (1-r)^2p^2q^2.$ (2.13.4,1.9.1-5)

2.19.5 $R[AA]r[aa] = 2(1-r)^2p^2q^2.$ (2.13.5,1.9.1-5)

2.19.6 $R[Aa]r[Aa] = 2pq\{r+2(1-r)^2pq\}.$ (2.13.6,1.9.1-5)

2.19.7 $R[\underline{Aa]r[aa}] = R[\underline{aa]r[Aa}] = 2pq^2(1-r)(r+q-rq).$

(2.13.7,1.9.1-5)

2.19.8 $R[Aa]r[aa] = 4pq^2(1-r)(r+q-rq).$ (2.13.8,1.9.1-5)

2.19.9 $R[aa]r[aa] = q^2(r+q-rq)^2.$ (2.13.9,1.9.1-5)

2.19.10 $R[A\xi]r[aa] = R[AA]r[aa] + R[Aa]r[aa]$

$= 2pq^2(1-r)(1+r+q-rq).$ (2.19.4,2.19.8)

2.19.11 $R[A\xi]r[A\xi] = R[AA]r[AA] + R[AA]r[Aa] + R[Aa]r[Aa]$

$= p^2(1+q)(1-r)(1+r+q-rq) + pr(pr+2q).$

(2.19.1,2.19.3,2.19.6)

## SECTION 3. PHENOTYPIC RESEMBLANCE

The history of the gene concept has paralleled that of the molecule, the atom, and the electron in several important respects. First perceived by Mendel as a purely hypothetical unit to explain certain perceptual phenomena, it has long since ceased to play a subservient role in genetics. Although its intrinsic properties remain largely unexplained, the reality of the gene as a cytological entity has been firmly established, and while like other "atoms" it is not perfectly immutable, its extrinsic properties and behavior have been found to be much more stable than the phenomena for which it was invented. Thus, while it was natural to postulate an invariable association between gene and phenotype in the beginning, it is now not only unnecessary to do so, but, in view of the many cases of "irregular inheritance", is actually a dangerous form of generalization. It is therefore recognized that genes or genotypes need not be perfectly correlated with traits and that such variation in expression may be measured both quantitatively and qualitatively[2]. Timofeeff-Ressowski (1931) has introduced the terms "expressivity" and "penetrance" to describe the extent to which these two forms of variation in gene expression are minimized. Many gene substitutions produce phenotypic changes of varying degree or intensity in different individuals; constancy in quantitative expression being described in terms of high expressivity. Other genes may fail in some individuals to produce a detectable change of any sort, the percentage of changes being called the penetrance of the gene or genotype involved.

Incomplete expressivity of a gene does not introduce any statistical difficulties into the study of heredity apart from those resulting from misclassification and the remedy here is perhaps best sought in improved diagnosis. Incomplete penetrance, however, presents a definite statistical problem and the present section therefore contains a brief outline for the extension of the methods of the two previous sections to such problems of phenogenetics. There are, of course, a good number of hereditary characters in man which are for all practical purposes free from variation in penetrance and it is hoped to discover many more for these are of special usefulness. For such cases, the genotype formulae of Section 2 will serve equally well as phenotype expectations and it is not deemed necessary here to define new symbols for phenotypes with axioms postulating complete correlation with the genotypes followed by a repetition of the genotype formulae. With this purpose in view, incomplete genotypes such as [A¿] have been provided in every part of Section 2 in order to accomodate the hypothesis of dominance with complete penetrance.

The most general case of incomplete penetrance which might be considered in connection with a single pair of autosomal alleles would be one in which three phenotypes, which might be labelled 'U', 'V' and 'W', were possible in each of the three genotypes, [AA], [Aa] and [aa], with probabilities, $u_{AA}$, $v_{AA}$, $w_{AA}$, $u_{Aa}$, $v_{Aa}$, etc. These could be dealt with in the manner described below or a correlation coefficient might somehow be employed as a simplification. Numerous modifications of the general formulae could then be developed for situations wherein only two pehnotypes were involved and where various phenotypes were assumed invariably associated with certain genotypes. This general treatment will not be undertaken here, but rather the one case of greatest immediate interest in human genetics will alone be considered. This is the case of the "irregular dominant" wherein there are but two phenotypes which we shall designate 'N' and 'D' (suggested by the expressions "normal" and "defective") with the abnormality D appearing invariably in homozygotes of the sort [AA], occasionally in heterozygotes [Aa], and not at all in homozygotes of the sort [aa]. These conditions are stated in 3.1.1 and 3.2.1-6; 'e' is the penetrance of the gene in the heterozygous or simplex state. The abnormality is therefore assumed to be "completely hereditary" in the sense that it never appears unless the gene A is present. Or, at least, we may assume that any similar abnormality appearing also in [aa] homozygotes but due to other causes can be detected as of different etiology and thereby excluded from the analysis.

The fact that irregular dominance is a common phenomenon in human heredity is one to be expected in accordance with Fisher's theory of the evolution of dominance. This theory simply holds that a mutant gene will usually exhibit much more variability in its effect in the simplex state in which it originally exists through mutation than in the duplex state. This variability of the heterozygote is further assumed to be partly genetic and due to modifying genes which therefore furnish a basis for selective evolution toward dominance or recessiveness depending on whether the gene is beneficial or harmful. Variability in penetrance or expressivity will therefore be most pronounced in a population which is highly heterogeneous with respect to modifying genes, a condition to be expected in a large freely-interbreeding population like the human. As Roberts (1940) further suggests, a relationship should be anticipated between the severity of a gene and its penetrance. A dominant or semi-dominant gene of high penetrance which produces a severe abnormality in heterozygotes would be eliminated very rapidly after each mutation. A less drastic pathogenic gene, however, would persist through many generations and would thus have ample time to acquire lower penetrance and therefore approach recessiveness under the selective processes postulated by Fisher. Many rare abnormalities in man have been found to be irregular in their dominance, e. g., epiloia, Laurence-Moon-Biedl syndrome, brachydactyly, etc. In the opinion of Roberts, many commoner diseases, the inheritance of which is at present known to be somehow complicated, will probably be found complicated merely by low penetrance rather than by the interaction of two or more "main gene" effects.

Dominant genes which manifest themselves only occasionally will present rather puzzling pedigrees. The trait may appear in any combination of relatives and "skipping of a generation" will be a common observation. Moreover, the lower the penetrance, the smaller will be the correlation between relatives, the more irregular the transmission, and the more difficult the detection of the gene. It is therefore of considerable interest to search for simple and direct statistical relationships for demonstrating the existence of such genes. In cases of low penetrance, the problem of greatest interest is, of course, that of determining the nature of the other influences, genetic and environmental, which are capable of tipping the balance in the heterozygote. In this section, however, attention will be confined to the primary problem of the detection of the main gene under all degrees of penetrance.

In subsection 3.3, the symbol J[AA,D] denotes the probability that an individual will be a homozygote for the abnormal gene and also a defective. 3.4.1 and 3.4.2 then state the proportions of normal and defective persons in the general population as functions of the dominant gene frequency p and its penetrance e in the simplex condition. Such expressions as these, involving only the symbols D and N, are of special interest because they are directly observable ratios and have been marked by an asterisk (*). With two parameters, p and e, equations 3.4.1 and 3.4.2 will, of course, not suffice for their simultaneous estimation. In combination with other observable ratios, however, such equations of estimation might be constructed and the resulting estimates might then be used in still other equations of this section for testing the fit of the genetic hypothesis.

With the aid of the statements of 3.5, 3.6 and 3.7 we may easily write the expected ratios (3.8) of D and N children in each of the three mating aggregates, (D°D), (D°N) and (N°N). Unfortunately, these are all rather complex functions of p and e. In suitable combination, however, they might be found to provide rather direct equations of estimation. For irregularly dominant genes of moderate penetrance, a ratio of considerable interest because of its easy observation is the proportion of cases in which the abnormality "skips a generation" or fails to appear in either parent of a defective. This is symbolized by (O[D])H[N°N] and is again a complex function (3.9.1) of the parameters p and e. As is shown in 3.9.2, however, this approximates very closely to (1 - e) when $p^2$ approaches zero. That such a result is to be anticipated follows, of course, from the fact that the effect of putting $p^2$ equal to zero is equivalent to assuming that homozygous defectives do not exist and that matings between heterozygotes do not occur. In the case of very rare irregular dominants, the vast majority of matings producing defective children will have only one parent a heterozygote, and the proportion of cases in which this parent is also defective may therefore be taken as an estimate of the penetrance of the gene in simplex condition.

Theorem 3.10.1 is a very general one, giving, for all classes and degrees of Z-relatives, the proportion of defectives to be expected amongst the relatives of a given defective person. The case of the rare abnormality ($p \longrightarrow 0$) is again of much interest since we see that the expectation is then simply proportional to the penetrance, e, and the coefficient of relationship, r, (3.10.2). Thus, if the trait generally appears in but 20 per cent. of the brothers and sisters of defective propositi, then the penetrance of the gene is e = .40, and it should also be found that the trait appears in 20 per cent. of parents and offspring, in 10 per cent. of grandparents, grandchildren, aunts, uncles, nieces and nephews, in 5 per cent. of first cousins, and so on. Expressing this another way, we can say that the ratio of the incidences of a rare irregular dominant trait in Z-relatives of two different degrees, $r_1$ and $r_2$, is simply the ratio of the two coefficients of relationship:

$$\frac{r_1 e}{r_2 e} = \frac{r_1}{r_2}.$$

Equations of this sort in which the quantity e disappears are of much value and it would be extremely helpful if such could be found for common irregular dominants in which the quantity p or $p^2$ is not negligible. One equation which does not satisfy this requirement, but which is of some interest because of its comparartive simplicity is 3.12.1, giving the proportion of cases in which a given parent, say the mother, or an offspring of a D individual is expected to be of the same phenotype.

SECTION 3

3.1.1 $(\xi^1\xi^2)J[D] + (\xi^1\xi^2)J[N] = 1.$ **Axiom**

3.2.1 (AA)J[D] = 1. **Axiom**

3.2.2 (AA)J[N] = 0. (3.1.1,3.2.1)

3.2.3 (Aa)J[D] $=_{Df}$ e.

3.2.4 (Aa)J[N] = 1 - e. (3.1.1,3.2.3)

3.2.5 (aa)J[D] = 0. **Axiom**

3.2.6 (aa)J[N] = 1. (3.1.1,3.2.5)

3.3.1 J[AA,D] = J[AA]·(AA)J[D] = $p^2$. (2.7.1,3.2.1)

3.3.2 J[AA,N] = J[AA]·(AA)J[N] = 0. (2.7.1,3.2.2)

3.3.3 J[Aa,D] = J[Aa]·(Aa)J[D] = 2pqe. (2.7.4,3.2.3)

3.3.4 J[Aa,N] = J[Aa]·(Aa)J[N] = 2pq(1 - e). (2.7.4,3.2.4)

3.3.5 J[aa,D] = J[aa]·(aa)J[D] = 0. (2.7.2,3.2.5)

3.3.6 J[aa,N] = J[aa]·(aa)J[N] = $q^2$. (2.7.2,3.2.6)

3.4.1 *J[D] = J[AA,D] + J[Aa,D] + J[aa,D]

= p(p + 2qe). (3.3.1,3.3.3,3.3.5)

3.4.2 *J[N] = J[AA,N] + J[Aa,N] + J[aa,N]

= q(1 + p - 2pe). (3.3.2,3.3.4,3.3.6)

3.5.1 (D)J[AA] = J[AA,D]:J[D] = p/(p + 2qe). (3.3.1,3.4.1)

3.5.2 (D)J[Aa] = J[Aa,D]:J[D] = 2qe/(p + 2qe). (3.3.3,3.4.1)

3.5.3 (D)J[aa] = J[aa,D]:J[D] = 0. (3.3.5,3.4.1)

3.5.4 (N)J[AA] = J[AA,N]:J[N] = 0. (3.3.2,3.4.2)

3.5.5 (N)J[Aa] = J[Aa,N]:J[N] = 2p(1 - e)/(1 + p - 2pe).

(3.3.4,3.4.2)

3.5.6 (N)J[aa] = J[aa,N]:J[N] = q/(1 + p - 2pe). (3.3.6,3.4.2)

3.6.1 (D)G[A] = (D)J[AA]·(AA)G[A] + (D)J[Aa]·(Aa)G[A]

= (p + qe)/(p + 2qe). (3.5.1,2.3.1,3.5,2,2.3.2)

$$3.6.2\quad (D)G[a] = (D)J[Aa]\cdot(Aa)G[a] + (D)J[aa]\cdot(aa)G[a]$$
$$= qe/(p + 2qe). \qquad (3.5.2, 2.3.3, 3.5.3, 2.3.4)$$

$$3.6.3\quad (N)G[A] = (N)J[AA]\cdot(AA)G[A] + (N)J[Aa]\cdot(Aa)G[A]$$
$$= p(1 - e)/(1 + p - 2pe). \qquad (3.5.4, 2.3.1, 3.5.5, 2.3.2)$$

$$3.6.4\quad (N)G[a] = (N)J[Aa]\cdot(Aa)G[a] + (N)J[aa]\cdot(aa)G[a]$$
$$= (1 - pe)/(1 + p - 2pe). \qquad (3.5.5, 2.3.3, 3.5.6, 2.3.4)$$

$$3.7.1\quad H[D\overset{o}{-}D] = H[AA\overset{o}{-}AA]\cdot(AA)J[D]\cdot(AA)J[D]$$
$$+ 2\cdot H[AA\overset{o}{-}Aa]\cdot(AA)J[D]\cdot(Aa)J[D]$$
$$+ H[Aa\overset{o}{-}Aa]\cdot(Aa)J[D]\cdot(Aa)J[D]$$
$$= (J[AA,D])^2 + 2\cdot J[AA,D]\cdot J[Aa,D] + (J[Aa,D])^2$$
$$= (J[AA,D] + J[Aa,D])^2 = (J[D])^2.$$
$$(2.6.1, 3.3.1, 3.3.3)$$

$$3.7.2\quad H[N\overset{o}{-}N] = H[aa\overset{o}{-}aa]\cdot(aa)J[N]\cdot(aa)J[N]$$
$$+ 2\cdot H[Aa\overset{o}{-}aa]\cdot(Aa)J[N]\cdot(aa)J[N]$$
$$+ H[Aa\overset{o}{-}Aa]\cdot(Aa)J[N]\cdot(Aa)J[N]$$
$$= (J[aa,N])^2 + 2\cdot J[Aa,N]\cdot J[aa,N] + (J[Aa,N])^2$$
$$= (J[aa,N] + J[Aa,N])^2 = (J[N])^2.$$
$$(2.6.1, 3.3.4, 3.3.6)$$

$$3.7.3\quad H[D\overset{o}{-}N] = 1 - H[D\overset{o}{-}D] - H[N\overset{o}{-}N] = 1 - (J[D])^2 - (J[N])^2$$
$$= 2\cdot J[D]\cdot J[N]. \qquad (3.7.1, 3.7.2)$$

$$3.8.1\quad {*}(D\overset{o}{-}D)O[D] = (D)G[A]\cdot(D)G[A]\cdot(AA)J[D] + 2\cdot(D)G[A]$$
$$\cdot(D)G[a]\cdot(Aa)J[D] = \frac{(p + qe)(p + qe + 2qe^2)}{(p + 2qe)^2}.$$
$$(3.6.1, 3.6.2, 3.2.2, 3.2.3)$$

$$3.8.2\quad {*}(D\overset{o}{-}D)O[N] = 2\cdot(D)G[A]\cdot(D)G[a]\cdot(Aa)J[N] + (D)G[a]$$
$$\cdot(D)G[a]\cdot(aa)J[N] = \frac{2qe(1 - e)(p + qe) + q^2e^2}{(p + 2qe)^2}.$$

(3.6.1,3.6.2,3.2.4,3.2.6)

3.8.3 $$*(D\overset{0}{-}N)O[D] = (D)G[A]\cdot(N)G[A]\cdot(AA)J[D] + (D)G[A]\cdot(N)G[a]\cdot(Aa)J[D] + (D)G[a]\cdot(N)G[A]\cdot(Aa)J[D] = \frac{p(1-e)(p+2qe)+e(1-pe)(p+qe)}{(p+2qe)(1+p-2pe)}.$$

(3.6.1-4,3.2.2,3.2.3)

3.8.4 $$*(D\overset{0}{-}N)O[N] = (D)G[A]\cdot(N)G[a]\cdot(Aa)J[N] + (D)G[a]\cdot(N)G[A]\cdot(Aa)J[N] + (D)G[a]\cdot(N)G[a]\cdot(aa)J[N] = \frac{(p+qe)(1-pe)(1-e)+pqe(1-e)^2+qe(1-pe)}{(p+2qe)(1+p-2pe)}.$$

(3.6.1-4,3.2.4,3.2.6)

3.8.5 $$*(N\overset{0}{-}N)O[D] = 2\cdot(N)G[A]\cdot(N)G[a]\cdot(Aa)J[D] + (N)G[A]\cdot(N)G[A]\cdot(AA)J[D] = \frac{2pe(1-e)(1-pe)+p^2(1-e)^2}{(1+p-2pe)^2}.$$

(3.6.3-4,3.2.1,3.2.3)

3.8.6 $$*(N\overset{0}{-}N)O[N] = 2\cdot(N)G[A]\cdot(N)G[a]\cdot(Aa)J[N] + (N)G[a]\cdot(N)G[a]\cdot(aa)J[N] = \frac{2p(1-e)^2(1-pe)+(1-pe)^2}{(1+p-2pe)^2}.$$

(3.6.3-4,3.2.4,3.2.6)

3.9.1 $$*(O[D])H[N\overset{0}{-}N] = \{H[N\overset{0}{-}N]\cdot(N\overset{0}{-}N)O[D]\} : \{H[D\overset{0}{-}D]\cdot(D\overset{0}{-}D)O[D] + H[D\overset{0}{-}N]\cdot(D\overset{0}{-}N)O[D] + H[N\overset{0}{-}N]\cdot(N\overset{0}{-}N)O[D]\}$$
$$= q^2(1-e)(p-pe+2e-2pe^2) : \{q(1-e)[2qe(1-pe) + pq(1-e) + 2p(p+2qe)] + (p+qe)[2qe(1-pe) + p(p+qe+2qe^2)]\}.$$ (3.7.1-3,3.8.1,3.8.3,3.8.5)

3.9.2 $$*(O[D])H[N\overset{0}{-}N],(p^2\to 0) = 1-e.$$ (3.9.1)

3.10.1 $$*(D)Z[D]r = (D)J[AA]\cdot(AA)Z[AA]r\cdot(AA)J[D] + (D)J[AA]\cdot(AA)Z[Aa]r\cdot(Aa)J[D]$$

$$+ (D)J[Aa]\cdot(Aa)Z[AA]r\cdot(AA)J[D]$$

$$+ (D)J[Aa]\cdot(Aa)Z[Aa]r\cdot(Aa)J[D]$$

$$= \{p(c_2 + 2c_1p + c_0p^2) + 4pqe(c_1 + c_0p) + 2qe^2(c_2 + c_1 + 2c_0pq)\}/(p + 2qe).$$

(3.5.1,2;2.12.4,7,11,16;3.2.1,3)

3.10.2 $*(D)Z[D]r,(p \to 0) = 2qe^2(c_2 + c_1)/2qe = re.$ (3.10.1)

3.10.3 $*(D)Z[N]r = (D)J[AA]\cdot(AA)Z[Aa]r\cdot(Aa)J[N]$

$$+ (D)J[AA]\cdot(AA)Z[aa]r\cdot(aa)J[N]$$

$$+ (D)J[Aa]\cdot(Aa)Z[Aa]r\cdot(Aa)J[N]$$

$$+ (D)J[Aa]\cdot(Aa)Z[aa]r\cdot(aa)J[N]$$

$$= \{2pq(1 - e)(c_1 + c_0p) + pq^2c_0 + 2qe(1 - e)(c_2 + c_1 + 2c_0pq) + 2q^2e(c_1 + c_0q)\}/(p + 2qe).$$

(3.5.1,2;2.12.7,8,16,19;3.2.4,6)

3.10.4 $*(N)Z[D]r = (N)J[Aa]\cdot(Aa)Z[AA]r\cdot(AA)J[D]$

$$+ (N)J[Aa]\cdot(Aa)Z[Aa]r\cdot(Aa)J[D]$$

$$+ (N)J[aa]\cdot(aa)Z[AA]r\cdot(AA)J[D]$$

$$+ (N)J[aa]\cdot(aa)Z[Aa]r\cdot(Aa)J[D]$$

$$= \{2p^2(1 - e)(c_1 + c_0p) + 2pe(1 - e)(c_2 + c_1 + 2c_0pq) + p^2qc_0 + 2pqe(c_1 + c_0q)\}/(1 + p - 2pe).$$

(3.5.5,6;2.12.11,16,20,23;3.2.1,3)

3.10.5 $*(N)Z[N]r = (N)J[Aa]\cdot(Aa)Z[Aa]r\cdot(Aa)J[N]$

$$+ (N)J[Aa]\cdot(Aa)Z[aa]r\cdot(aa)J[N]$$

$$+ (N)J[aa]\cdot(aa)Z[Aa]r\cdot(Aa)J[N]$$

$$+ (N)J[aa]\cdot(aa)Z[aa]r\cdot(aa)J[N]$$

$$= \{2p(1 - e)^2(c_2 + c_1 + 2c_0pq) + 4pq(1 - e)(c_1 + c_0q) + q(c_2 + 2c_1q + c_0q^2)\}/(1 + p - 2pe).$$

(3.5.5,6;2.12.16,19,23,27;3.2.4,6)

3.11.1 $*Z[D]r[D] = J[D]\cdot(D)Z[D]r = p\{p(c_2 + 2c_1p + c_0p^2)$
$+ 4pqe(c_1 + c_0p) + 2qe^2(c_2 + c_1 + 2c_0pq)\}$. (3.4.1,3.10.1)

3.11.2 $*Z[\underline{D]r[N}] = J[D]\cdot(D)Z[N]r = pq\{2p(1 - e)(c_1 + c_0p)$
$+ pqc_0 + 2e(1 - e)(c_2 + c_1 + 2c_0pq) + 2qe(c_1 + c_0q)\}$.
(3.4.1,3.10.3)

3.11.3 $*Z[\underline{N]r[D}] = J[N]\cdot(N)Z[D]r = pq\{2p(1 - e)(c_1 + c_0p)$
$+ pqc_0 + 2e(1 - e)(c_2 + c_1 + 2c_0pq) + 2qe(c_1 + c_0q)\}$.
(3.4.2,3.10.4)

3.11.4 $*Z[D]r[N] = Z[\underline{D]r[N}] + Z[\underline{N]r[D}] = 2pq\{2p(1 - e)(c_1$
$+ c_0p) + pqc_0 + 2e(1 - e)(c_2 + c_1 + 2c_0pq) + 2qe(c_1 + c_0q)\}$.
(3.11.2,3.11.3)

3.11.5 $*Z[N]r[N] = J[N]\cdot(N)Z[N]r = q\{q(c_2 + 2c_1q + c_0q^2)$
$+ 4pq(1 - e)(c_1 + c_0q) + 2p(1 - e)^2(c_2 + c_1 + c_0pq)\}$.
(3.4.2,3.10.5)

3.12.1 $*(D)PO[D] = (p^2 + 2pqe + qe^2)/(p + 2qe)$.
(3.10.1,1.10.1-6,1.12.1-6)

3.13.1 $*PO[D][D] = p(p^2 + 2pqe + qe^2)$.
(3.11.1,1.10.1-6,1.12.1-6)

## SECTION 4. CONSANGUINE MATING

Inbreeding systems may become quite complex, but fortunately only the simplest types of consanguinity are found in human pedigrees and the present section will be confined to these. First we shall define the "coefficient of inbreeding" as devised by Wright (1921a). This ratio, symbolized here by '$i$', is the probability that a gene will be paired with its derivative, giving a gene-pair of the sort [$\xi_1{}^2$]. It may also be defined as the most likely fraction of all gene-pairs to consist to such paired derivatives. Customarily the expression "homozygous pairs" is used in place of "derivative pairs" in defining the inbreeding coefficient. But this, of course, does not have the same meaning, and requires further that the definition be made relative to the "average heterozygosis" of the population, which, for any given gene pair at equilibrium, will be $2pq$. But $p$ will, of course, vary for different gene pairs. Use of the distinction between "identical" and "derivative" genes therefore removes the relativeness of the inbreeding coefficient and makes its meaning exactly the same for all gene pairs. Actually, however, a certain relativeness still remains due to the arbitrary definition of "derivative" adopted in this paper. Wright, in his classical work (1921a) on inbreeding, devised the coefficient for dealing with problems of homozygosis and not for the purposes to which it is put in this section. Moreover, in connection with many practical genetic problems, it is useful to think in terms of homozygosis, as in the following illustrative definition from Lush (1937): "...if the average Shorthorn is heterozygous for a thousand pairs of genes, then a Shorthorn showing an inbreeding coefficient of 25 percent would probably still be heterozygous for about 750 pairs."

In the simplest case, i may be computed as equal to half the coefficient of relationship, $r_p$, of the parents:

$$i = \tfrac{1}{2}r_p = \tfrac{1}{2}\Sigma(\tfrac{1}{2})^{n_1+n_2} = \Sigma(\tfrac{1}{2})^{n_1+n_2+1},$$

where $n_1$ and $n_2$ are the numbers of links in paths of descent connecting a common ancestor with each parent. Thus, in the following somewhat complex pedigree of inbreeding, the calculation is carried out as indicated:

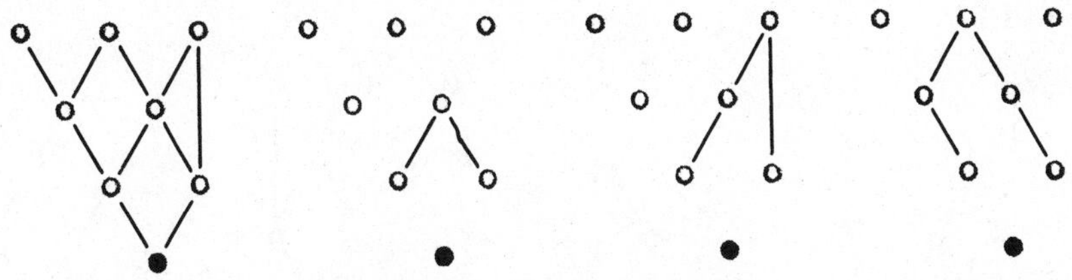

$$i = 7/32 = \tfrac{1}{2}(\tfrac{1}{2})^2 + \tfrac{1}{2}(\tfrac{1}{2})^3 + \tfrac{1}{2}(\tfrac{1}{2})^4$$

In cases where one or more common ancestor of the parents is inbred, the parental coefficient of relationship must be multiplied by the factor $(1 + i_c)$, $i_c$ being the inbreeding coefficient of the common ancestor, and the expression then summed for all common ancestors:

$$i = \Sigma\{(\tfrac{1}{2})^{n_1+n_2+1}(1 + i_c)\}.$$

Thus, in the pedigree below, the coefficient of relationship of the parents 'X' and 'Y' is $(\tfrac{1}{2})^2(1 + \tfrac{1}{8})$ since the coefficient of inbreeding of their common parent is $\tfrac{1}{2}(\tfrac{1}{2})^2$.

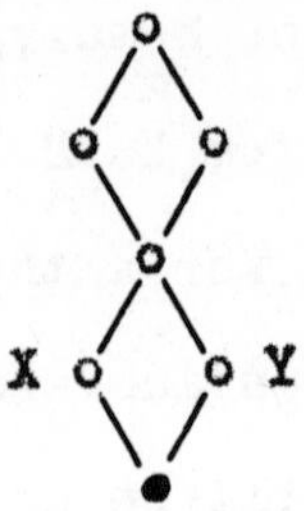


$$i = 9/64 = \tfrac{1}{2}(\tfrac{1}{2})^2(1 + \tfrac{1}{8})$$

The author now suggests an equivalent but somewhat more easily visualized procedure and one which introduces a terminology useful in other connections. To determine the coefficient of inbreeding, take ($\frac{1}{2}$) to the power of the number of links in a "loop" of descent closing upon the inbred individual in question, sum for all such loops, and multiply by two. The following illustration will clarify the meaning of "loop":

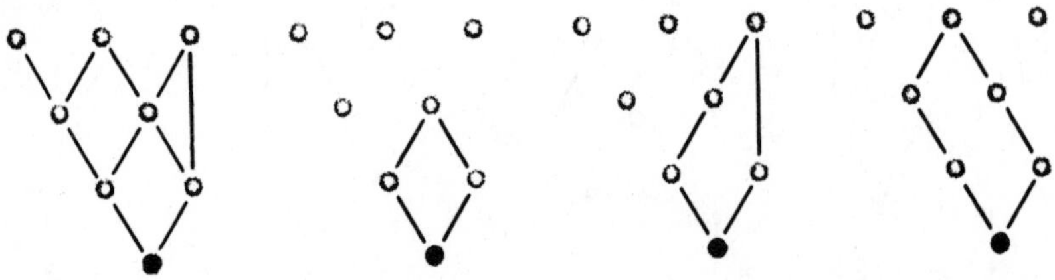

$$i = 7/32 = 2(\tfrac{1}{2})^4 + 2(\tfrac{1}{2})^5 + 2(\tfrac{1}{2})^6$$

The second illustration may be dealt with similarly if we permit the term "loop" to include a constricted loop, that is, one passing through an ancestor to open into a second loop:

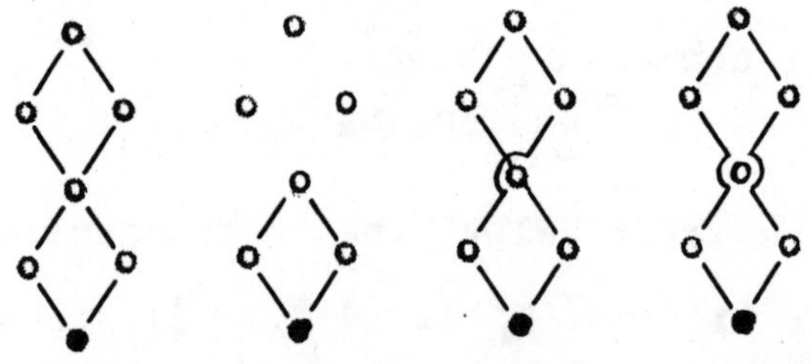

$$i = 9/64 = 2(\tfrac{1}{2})^4 + 2(\tfrac{1}{2})^8 + 2(\tfrac{1}{2})^8$$

Notice that the 8-linked loop must be counted twice since it may "twist" in either direction. Similarly, when ancestors are inbred, the coefficient of relationship of the parents must allow paths of descent to pass through the inbred ancestors to form loops. Thus:

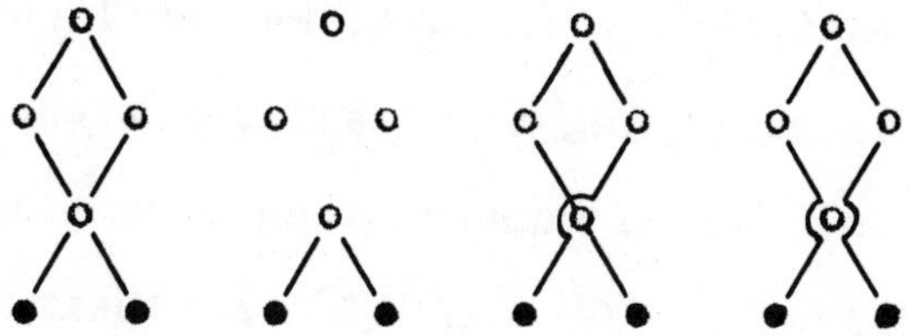

$$r = 9/32 = (\tfrac{1}{2})^2 + (\tfrac{1}{2})^6 + (\tfrac{1}{2})^6$$

The derivation of i in the case of the loop interpretation may now be reasoned as follows. Consider any member of a loop other than the terminal inbred member, having the gene-pair $[\xi_1\xi_2]$:

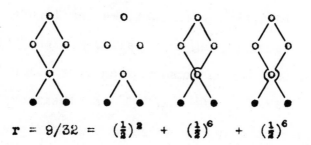

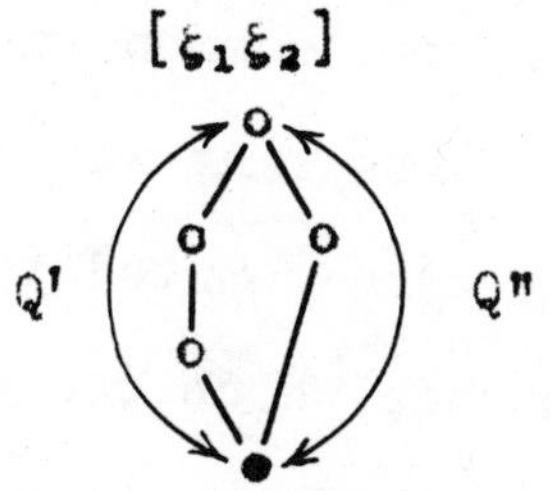


In this example, the gene $\xi_1$ might pass to the inbred individual around one "half" of the loop with the probability $(\frac{1}{2})^3$ and around the other "half-loop" with the probability $(\frac{1}{2})^2$, so that the probability of $[\xi_1{}^2]$ in the terminal inbred member is therefore $(\frac{1}{2})^5$. But the same might happen with equal probability in the case of gene $\xi_2$, hence the factor 2, and

$$i = 2\Sigma(\tfrac{1}{2})^{l_1}(\tfrac{1}{2})^{l_2} = 2\Sigma(\tfrac{1}{2})^{l},$$

where $l_1$ and $l_2$ are the numbers of links in the two "half-loops" and $l$ is their total. It is easily seen that the same reasoning would apply in connection with any member of the loop besides the starting member pictured in the above example. Hence, any other member of a loop is related to the terminal inbred member in what might be described as a "double-Q" relationship. In fact, we shall refer to the two "half-loops" as supplying two "half-relations" between the terminal and any second member of an inbreeding loop. They will be symbolized by 'Q'' and 'Q"' and their degrees by '$r_1$' and '$r_2$', respectively. As is shown in the above example, $l_1$ and $l_2$ need not be equal, hence $r_1$ need not equal $r_2$ and the expression "half-relation" must therefore not be taken too literally.

It was previously pointed out that, if the two parents were inbred or had inbred common ancestors, the inbreeding coefficient of their offspring was still $\frac{1}{2}r_p$ provided that all inbreeding loops were counted in computing $r_p$. Thus, for _any_ type of relationship which can exist between two relatives, their coefficient of relationship, $r_p$, is always twice the inbreeding coefficient of an offspring of their union. This fact permits us to make use of a new relation 'U' (suggested by the word "universal"), which is defined as any sort of relationship which can exist between bisexually reproducing diploids. Now, 'U' is therefore the logical sum of three major types of relatives: 'Z' relatives, both of which are J individuals; 'Y' relatives, one being a J and the other an I individual; and 'X' relatives, both of which are I individuals. The only general law yet discovered covering the universal class 'U' is the one just stated, viz.: for any mating between U-relatives of degree r, the i of their offspring is $\frac{1}{2}$r. It must be hastily added, however, that r does not have exactly the same meaning in connection with Y and X-relatives as it has for Z-relatives. For Z-relatives (and some classes of Y and X), r is the probability that one gene and its derivative of a given pair will be shared. That is, $r = (\xi_1\xi_2)Z[\xi_1\xi_{23}]r = (\xi_1\xi_2)Z[\xi_{13}\xi_2]r$. This interpretation of r must be modified in the case of many subclasses of 'Y' and 'X'. However, r is computed the same way in every case.

This leads to the problem of genic resemblance in Y and X-relatives. The first question which might be asked is: how many different subclasses of these classes exist? The author does not presume to know the answer here; the number must certainly be quite large, but possibly not as large as supposed in the hypothetical reasoning of Section 1. Furthermore, one would not hope to discover general sets of probability functions for possible types of genic resemblance (7 in the case of Y-relatives, 12 in X-relatives) applicable to all cases of 'Y' and 'X', for the relations 'Y' and 'X' are certainly more complicated that 'Z' and even here two sets of laws were required ('R' and 'Q'). The problem is evidently one of considerable complexity, for Woodger, when confronted with inbreeding in another type of problem, has only this to say: "How to extend this theorem to cases where this ('pedreg') is not the case is at present an unsolved problem. But every member of ped which is not regular up to the $\kappa$th generation can be depicted as a regular one in which certain terms are identical, and then the above calculation is possible. The problem is, therefore, what is the minimum information we require about a non-regular ped in order that such a transformation into a regular one may be carried out, and how can it be expressed in the calculus of peds? In other words, how can we characterize classes of members of ped for which these properties are invariant?" The present problem of genic resemblances in members of irregular pedigrees presents essentially the same difficulties. In some cases of 'Y' and 'X' the author has found that the minimum information for specifying the probabilities of types of genic resemblance consists solely of the two coefficients r and i. This is true of the two subclasses of 'Y' and one subclass of 'X' defined in the succeeding paragraphs. In other, more complicated cases, however, it is almost certain that further "parameters" are needed, or, what amounts to the same thing, much more complicated definitions of the subclasses.

Consanguine matings in man will for the most part be isolated cases, although the same mating may be repeated, that is, produce several sibs. Consequently, the pedigree of a single inbred individual will usually contain but one main inbreeding loop. A good many cases of consanguine mating in man will therefore give rise to relatives whose relationship is of the sort depicted in the following series of examples:

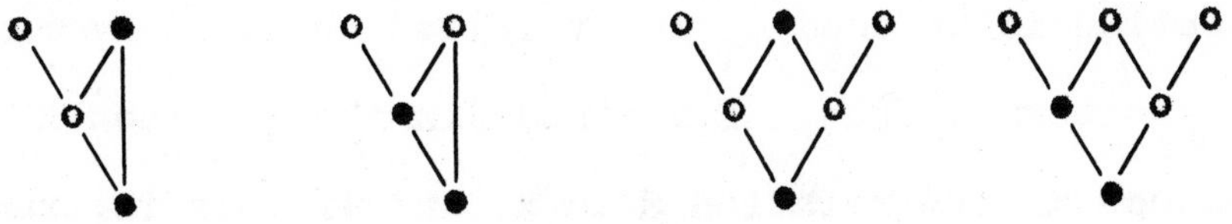

| | | | |
|---|---|---|---|
| r = 3/4 | r = 3/4 | r = 1/2 | r = 5/8 |
| i = 1/4 | i = 1/4 | i = 1/8 | i = 1/8 |

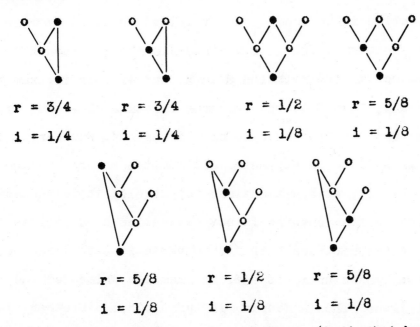

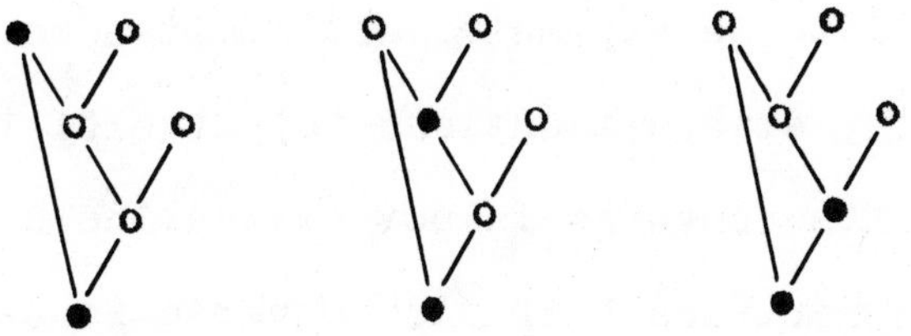

| | | |
|---|---|---|
| r = 5/8 | r = 1/2 | r = 5/8 |
| i = 1/8 | i = 1/8 | i = 1/8 |

In all of these examples, the relatives (indicated by solid circles) whose coefficients of relationship are specified are two members of the loop and one of these is the terminal inbred member. Such relatives, consisting of the terminal and any other member of a single inbreeding loop (there being no other loops in the pedigree) will now be designated as 'L' relatives. This is, of course, a subclass of 'Y' since one individual is a J and the other an I. Any other two members of a single inbreeding loop exclusive of the terminal member will, of course, be both J's and therefore simply Q-relatives. As is seen in the above diagrams, i must be equal for loops having the same number of links but r may vary in loops of the same link number due to variations in the link numbers of the half-loops.

At this point a word must be interpolated about relations in general. In the "Principia Mathematica", the "converse" of a relation is defined as the relation obtained by reversing the two terms. Thus, if x stands in the relation E to y and y stands in the relation Ǝ to x, and if this is true for any x and y, then Ǝ is the converse relation of E. For example, 'parent' is the converse of 'offspring'. Further, a relation is said to be "symmetrical" if its converse is included in (that is, always identical to) itself; it is "not-symmetrical" if its converse is not always included in itself; and it is "asymmetrical" if its converse is never identical to (mutually excludes) itself. As examples Whitehead and Russell give: "Thus, 'cousin' is symmetrical, 'brother' is not-symmetrical (because when x is the brother of y, y may be either the brother or the sister of x), and 'husband' is asymmetrical." It might be remarked here that these are but the first definitions concerning relations in a book which devotes several hundred pages to that subject. Moreover, biological relationships are frequently quoted as examples, which would seem to suggest plainly enough that the geneticist, whose whole subject involves relations, could profit much from a thorough knowledge of the laws governing this branch of logistic.

The point of interest here is that the reason for the alleged non-symmetry of 'brother' and asymmetry of 'husband' is obviously merely a matter of sex. Replacing them by 'sib' and 'spouse' we obtain symmetrical relations. Further, the asymmetry of 'parent' and 'descendent' must be claimed purely on temporal grounds. The important thing for us, however, is that, with respect to what must be considered their most fundamental biological property, viz., genic resemblance, all of the above-mentioned biological relations must be regarded as symmetrical. For it is inherent in the proofs of Section 1 that

if $\qquad (\zeta_1\zeta_2)_x Z[\xi_1\zeta_{23}]_y r = r,$

then also $\qquad (\zeta_1\xi_2)_y Z[\xi_1\xi_{23}]_x r = r,$ etc.

That is to say, if x is an aunt and y is a nephew, the implicative probabilities for Q-relatives (subsection 2.14) will serve in either direction. The same applies to R-relatives and, in fact, we may say that all Z-relations are symmetrical with respect to the probabilities of their autosomal genic resemblances. They may or may not be symmetrical as regards temporal or sexual considerations, but this is of no importance in our genetic problems. In connection with sex-linked inheritance, however, temporal and sexual asymmetry does cause an asymmetry in genic resemblance.

Returning to the subject of inbreeding, it is now clear that, unlike Z-relations, Y-relations are fundamentally asymmetrical, that is, asymmetrical with respect to genic resemblances, as will be seen immediately from the fact that the J individual can only be of the sort $[\xi_1\xi_2]$ while the I individual can possess a gene-pair of the sort $[\xi_1{}^2]$ or $[\xi_1\xi_2]$. Consequently, it is necessary to alter the above definition of 'L' and state that 'L' is the relation in which the terminal (inbred) member of a single inbreeding loop stands to any other member of the loop, there being only one such loop in the pedigree. Then, in a symbol such as $(\xi_1\xi_2)L[\xi_3\xi_4]ri$, it follows from the definition of 'L' that $(\xi_1\xi_2)$ refers to the J member and $[\xi_3\xi_4]$ to the I member. Moreover, since there is only one I individual involved, we need not take care to specify that i is the inbreeding coefficient of the individual represented in the brackets.

The maximum number of types of genic resemblance between relatives of the numerous subclasses of 'Y' is 7, but only 6 of these are possible in the subclass 'L'. Their probabilities are completely specifiable in terms of r and i and are derived in subsection 4.2. For convenience they are listed below in the same order used in Section 1 together with an example.

(1) $(\xi_1\xi_2)L[\xi_3\xi_4]ri = 1 - 2r + 2i$ (3/8)

(2) $(\xi_1\xi_2)L[\xi_1\xi_3]ri = r - 2i$ (1/4)

(3) $(\xi_1\xi_2)L[\xi_2\xi_3]ri = r - 2i$ (1/4)

(4) $(\xi_1\xi_2)L[\xi_1\xi_2]ri = i$ (1/16)

(5) $(\xi_1\xi_2)L[\xi_1^2]ri = \frac{1}{2}i$ (1/32)

(6) $(\xi_1\xi_2)L[\xi_2^2]ri = \frac{1}{2}i$ (1/32)

(7) $(\xi_1\xi_2)L[\xi_3^2]ri = 0$

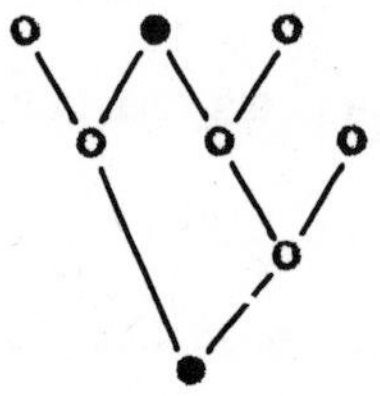

r = 3/8

i = 1/16

Having found such a set of functions covering all cases of a given relation, one can then set about discovering its special cases deductively by imposing all possible restrictions upon the values or r and i. If in the above set we put $i = 0$, this requires $l = \infty$, that is, a loop of infinite links or, in other words, no inbreeding at all. Then as might be expected, the types (4), (5) and (6) vanish while the first three take the values $(1 - 2r)$, r, r, characteristic of the Q-relation. This accords with the view taken earlier that 'L' is to be regarded as a sort of inbred Q-relation, or, as was suggested, a "double-Q" relation. If we put $r = 0$, then, of course, we have no relation at all. But another special case and in that sense a subclass of 'L' is found by setting $(1 - 2r + 2i) = 0$. This makes $r = i + \frac{1}{2}$. But $i = 2(\frac{1}{2})^{l} = (\frac{1}{2})^{l-1}$; therefore $r = (\frac{1}{2})^{l-1} + (\frac{1}{2})$, showing that $r_1 = (\frac{1}{2})^{l-1}$ and $r_2 = (\frac{1}{2})^1$, or that $l_1 = l - 1$ and $l_2 = 1$. In other words, type (1) genic resemblance will be impossible if the terminal member is connected to the second member by half-loops one of which has but a single link, thus making that member a parent (or perhaps we should say a "parent-plus") of the terminal member. This subclass, which we shall designate K', is therefore the relation of a terminal member of a loop to either of its parents, and is a special case of 'L' in the same way that the symmetrical relation 'PO' is a special case of 'Q'. Incidentally, it will be observed that 'K' is the only kind of L-relation possible in the case of the 3-linked loop of parent-child incest.

The final restriction which can be placed upon r and i is to cause (2) and (3) to vanish by setting $(r - 2i) = 0$.

Then $r = 2i$; but since we know (4.1.1,4.1.3) that $i = \frac{1}{2}r_p = 2r_1r_2$, we must also have $r = r_p = 4r_1r_2$. But it is always true that $r = r_1 + r_2$, hence we will have $r_1 + r_2 = 4r_1r_2$. This last equation is satisfied under two conditions, when $r_1 = r_2 = 0$ or when $r_1 = r_2 = \frac{1}{2}$. The first solution tells us the obvious fact that types (2) and (3) vanish when there isn't any relationship. The second solution is more interesting, for we must then have $r_1 = (\frac{1}{2})^{l_1} = \frac{1}{2}$ and $l_1 = 1$, and similarly $l_2 = 1$. That is to say, we must have a 2-linked loop. Further, the parents must be perfectly related ($r_p = r = 1$) and the coefficient of inbreeding must be $i = \frac{1}{2}r_p = \frac{1}{2}$. This is obviously the case of self-fertilization:

$r = 1$
$i = 1/2$

$$(\zeta_1\zeta_2)M[\zeta_1{}^2] = \tfrac{1}{2}i = (1/4)$$
$$(\zeta_1\zeta_2)M[\zeta_1\zeta_2] = i = (1/2)$$
$$(\zeta_1\zeta_2)M[\zeta_2{}^2] = \tfrac{1}{2}i = (1/4)$$

Again, putting $r = 1$ and $i = \frac{1}{2}$ in the general 'L' equations we find as expected that types (1), (2) and (3) vanish while (5), (4) and (6) take the ratio 1:2:1. These formulae would not, of course, be applicable to a case of self-pollination where that mode of reproduction was commonly practiced, but only to the isolated case of selfing in a plant population which was normally random-breeding like the human population. We shall call this special subclass of 'K' the relation 'M' (suggested by "momentarily monoecious").

Another subclass of 'Y' which is needed for describing relatives arising from many cases of consanguinity in man is depicted in the following examples:

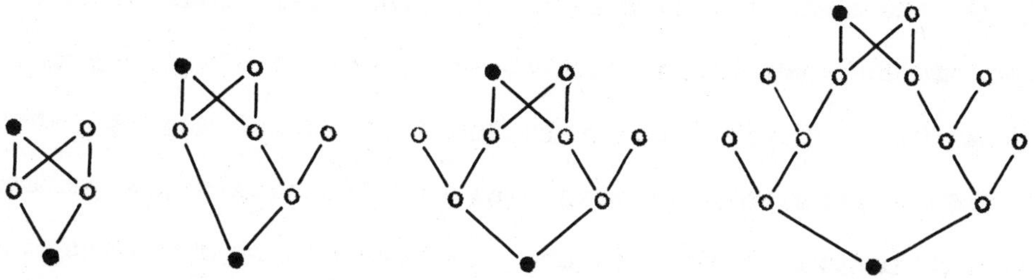

| | | | |
|---|---|---|---|
| r = 1/2 | r = 3/8 | r = 1/4 | r = 1/8 |
| i = 1/4 | i = 1/8 | i = 1/16 | i = 1/64 |

These cases differ from those used above in describing the 'L' relation in that the terminal member in this instance possesses two ancestors who are starting members of loops closing upon the terminal inbred member. These two are mates in every case, and consequently, the terminal inbred individual possesses two loops of equal link number, or, as it might also be regarded, a single branched loop, each of branches comprising two links. We therefore define 'B' as the relation in which a terminal inbred member stands to the starting member of a loop when the terminal member also possesses a second loop starting with the mate of the starting member of the first loop. It is easily seen that all other relationships in the above pedigrees are described by 'L', 'Q' and 'R'. Due to the presence of two loops the 'B' relation permits the terminal member to receive a pair of derivative genes $[\xi_3^2]$ which were not present in the one starting member. Consequently, all 7 types of genic resemblance limited to 'Y' are possible in B-relatives. Their probabilities are derived in subsection 4.5. There are no subclasses of 'B'.

Of the many subclasses of 'X', that is, kinds of relationship between inbred individuals, only one will be discussed here. This is the relation 'Qs', which may be defined as the relationship between sibs or offspring whose parents are Q-relatives. Examples are provided in the following diagrams:

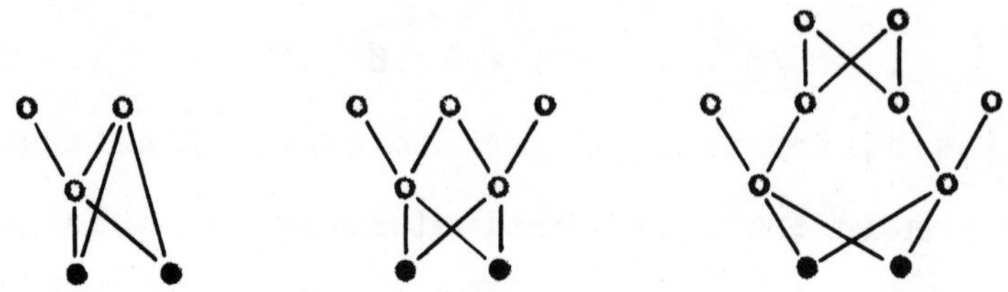

r = 3/4     r = 5/8     r = 9/16

i = 1/4     i = 1/8     i = 1/16

Inbred sibs must, of course, both possess the same coefficient, i, and the coefficient of relationship will always be $r = i + \frac{1}{2}$. Consequently, the relation 'Qs' is completely described in terms of i. Moreover, it is a symmetrical relation in the sense that implicative probabilities of genic resemblance will serve in either direction, a characteristic not exhibited by many subclasses of 'X'.

Probability functions governing the relation 'Qs' are derived in subsections 4.6 and 4.7. A new complication is encountered here in the case of X-relatives as a consequence of the fact that both relatives are I individuals and can therefore be of the gene-pair $[\xi_1{}^2]$ as well as $[\xi_1\xi_2]$. We therefore require combinatory probabilities of genic resemblance as well as implicative probabilities if genotype resemblances are to be worked out for Qs-relatives as was done for Z-relatives in Section 2. Actually, it would have been preferable in Section 2 to have used combinatory probabilities, such as $Z[\xi_1\xi_2][\xi_3\xi_4]$, in place of implicative probabilities. For example,

2.12.20 $$(aa)Z[AA]r = (a_1a_2)Z[\xi_3\xi_4]r\cdot(\xi_3\xi_4)J[AA]$$

would be replaced by

$$(aa)Z[AA]r = Z[a_1a_2]r[\xi_3\xi_4]\cdot(\xi_3\xi_4)J[AA],$$

wherein $$Z[a_1a_2]r[\xi_3\xi_4] = J[\xi_1\xi_2]\cdot(\xi_1\xi_2)Z[\xi_3\xi_4]r.$$

The equivalence of the two procedures, of course, follows from the fact that $J[\xi_1\xi_2]$ is, by definition, equal to unity (1.4.1).

Axioms 4.8.1-5 state that the frequencies of [AA] and [aa] homozygotes in inbred individuals possessing paired derivatives are equal to those for the gametic population, while the three genotypes, [AA], [Aa] and [aa], occur with the same frequencies in I individuals of the sort $[\xi_1\xi_2]$ as in J individuals. This involves the assumption that inbreeding is equally probable in all kindreds irrespective of the genotypes of their members. In reality this condition may not be precisely realized, since the presence of hereditary abnormalities in close relatives may sometimes discourage, or perhaps in some instances encourage, the performance of consanguinity. These axioms lead immediately to a general expression (4.10.1,4.10.3) for the probability of a recessive (or dominant) homozygote in the child produced through matings of relatives of any degree or kind. This naturally involves only the frequency of the gene concerned and the coefficient of inbreeding and constitutes a general expression covering the specific examples previously formulated by Hogben (1933).

The formulae of subsection 3.10 assume nothing to be known concerning the parents other than the degree of their relationship. On the other hand, if the genotypes of the parents are completely specified, then, of course, knowledge as to the relationship of these parents cannot alter the expectations of their children. When, however, as in the case of dominance, the genotypes are incompletely specified in one or both parents, the result is affected by both the kind and the degree of the parental relationship. Subsections 4.11 and 4.12 are therefore added for the purpose of extending the theorems of 2.10 and 2.11 to consanguine matings involving Q- and R-relatives. It will be noticed that when i is put equal to zero expressions 4.11.1 and 4.12.1 reduce to $\zeta$ while 4.11.2 and 4.12.2 reduce to $\zeta^2$ as is expected from the formulae of Snyder (2.10.9, 2.11.4).

## SECTION 4

4.1.1 $I[\xi_\alpha^2] = \Sigma\{[\xi_1^2] + [\xi_2^2]\}$

$= \Sigma\{(\xi_1\xi_2)Q'[\xi_1\xi_{123}]\cdot(\xi_1\xi_2)Q''[\xi_{123}\xi_1]\}$

$+ \Sigma\{(\xi_1\xi_2)Q'[\xi_2\xi_{123}]\cdot(\xi_1\xi_2)Q''[\xi_{123}\xi_2]\}$

$= \Sigma(r_1r_2) + \Sigma(r_1r_2) = 2\Sigma(r_1r_2)$

$= 2\Sigma\{(\frac{1}{2})^{l_1}(\frac{1}{2})^{l_2}\} = 2\Sigma(\frac{1}{2})^{l} = _{\text{Df}}\ i.$ (1.6.1)

4.1.2 $I[\xi_\alpha\xi_\beta] = I[\xi_\alpha\xi_{\alpha\beta}] - I[\xi_\alpha^2] = 1 - i.$ (4.1.1)

4.1.3 $i = 2\Sigma\{(\frac{1}{2})^{l_1}(\frac{1}{2})^{l_2}\} = \frac{1}{2}\Sigma\{(\frac{1}{2})^{l_1-1}(\frac{1}{2})^{l_2-1}\} = \frac{1}{2}r_p$ (4.1.1)

4.2.1 $(\xi_1\xi_2)L[\xi_1^2]ri = (\xi_1\xi_2)Q'[\xi_1\xi_{123}]\cdot(\xi_1\xi_2)Q''[\xi_{123}\xi_1]$

$= r_1r_2 = (\frac{1}{2})^{l_1}(\frac{1}{2})^{l_2} = (\frac{1}{2})^{l} = \frac{1}{2}i.$ (4.1.1)

4.2.2 $(\xi_1\xi_2)L[\xi_2^2]ri = (\xi_1\xi_2)Q'[\xi_2\xi_{123}]\cdot(\xi_1\xi_2)Q''[\xi_{123}\xi_2]$

$= r_1r_2 = (\frac{1}{2})^{l_1}(\frac{1}{2})^{l_2} = (\frac{1}{2})^{l} = \frac{1}{2}i.$ (4.1.1)

4.2.3 $(\xi_1\xi_2)L[\xi_1\xi_2]ri = (\xi_1\xi_2)Q'[\xi_1\xi_{123}]\cdot(\xi_1\xi_2)Q''[\xi_{123}\xi_2]$

$+ (\xi_1\xi_2)Q'[\xi_2\xi_{123}]\cdot(\xi_1\xi_2)Q''[\xi_{123}\xi_1]$

$= r_1r_2 + r_1r_2 = 2r_1r_2 = i.$ (4.1.1)

4.2.4 $(\xi_1\xi_2)L[\xi_1\xi_{123}]ri = (\xi_1\xi_2)Q'[\xi_1\xi_{123}]$

$+ (\xi_1\xi_2)Q''[\xi_{123}\xi_1] = r_1 + r_2 = r.$ (4.1.1)

4.2.5 $(\xi_1\xi_2)L[\xi_1\xi_{123}]ri = (\xi_1\xi_2)L[\xi_1\xi_3]ri + (\xi_1\xi_2)L[\xi_1\xi_2]ri$

$+ 2\cdot(\xi_1\xi_2)L[\xi_1^2]ri = r.$ (4.2.4)

4.2.6 $(\xi_1\xi_2)L[\xi_1\xi_3]ri = (\xi_1\xi_2)L[\xi_1\xi_{123}]ri - (\xi_1\xi_2)L[\xi_1\xi_2]ri$

$- 2\cdot(\xi_1\xi_2)L[\xi_1^2]ri = r - 2i.$ (4.2.5,4.2.3,4.2.1)

4.2.7 $(\xi_1\xi_2)L[\xi_2\xi_3]ri = (\xi_1\xi_2)L[\xi_2\xi_{123}]ri - (\xi_1\xi_2)L[\xi_1\xi_2]ri$

$- 2\cdot(\xi_1\xi_2)L[\xi_2^2]ri = r - 2i.$ (4.2.5,4.2.3,4.2.2)

4.2.8 $(\xi_1\xi_2)L[\xi_3^2]ri = (\xi_1\xi_2)L[\xi_{123}^2]ri - (\xi_1\xi_2)L[\xi_1^2]ri$

$- (\xi_1\xi_2)L[\xi_2^2]ri = i - \frac{1}{2}i - \frac{1}{2}i = 0.$ (4.2.1,4.2.2)

4.2.9 $(\xi_1\xi_2)L[\xi_3\xi_4]ri = (\xi_1\xi_2)L[\xi_{123}\xi_{1234}]ri - (\xi_1\xi_2)L[\xi_1^2]ri$

$- (\xi_1\xi_2)L[\xi_2^2]ri - (\xi_1\xi_2)L[\xi_1\xi_2]ri - (\xi_1\xi_2)L[\xi_3^2]ri$

$- (\xi_1\xi_2)L[\xi_1\xi_3]ri - (\xi_1\xi_2)L[\xi_2\xi_3]ri = 1 - 2r + 2i.$

(4.2.1-8)

4.3.1 $(\xi_\alpha\xi_\beta)K[\xi_\gamma\xi_\delta]i =_{Df} (\xi_\alpha\xi_\beta)L[\xi_\gamma\xi_\delta]ri, (r = i + \tfrac{1}{2}).$

4.3.2 $(\xi_1\xi_2)K[\xi_3\xi_4]i = 0.$ (4.2.9)

4.3.3 $(\xi_1\xi_2)K[\xi_1\xi_3]i = \tfrac{1}{2} - i.$ (4.2.6)

4.3.4 $(\xi_1\xi_2)K[\xi_2\xi_3]i = \tfrac{1}{2} - i.$ (4.2.7)

4.3.5 $(\xi_1\xi_2)K[\xi_1\xi_2]i = i.$ (4.2.3)

4.3.6 $(\xi_1\xi_2)K[\xi_1{}^2]i = \tfrac{1}{2}i.$ (4.2.1)

4.3.7 $(\xi_1\xi_2)K[\xi_2{}^2]i = \tfrac{1}{2}i.$ (4.2.2)

4.3.8 $(\xi_1\xi_2)K[\xi_3{}^2]i = 0.$ (4.2.8)

4.4.1 $(\xi_\alpha\xi_\beta)M[\xi_\gamma\xi_\delta] =_{Df} (\xi_\alpha\xi_\beta)K[\xi_\gamma\xi_\delta]i, (i = \tfrac{1}{2}).$

4.4.2 $(\xi_1\xi_2)M[\xi_3\xi_4] = 0.$ (4.3.2)

4.4.3 $(\xi_1\xi_2)M[\xi_1\xi_3] = 0.$ (4.3.3)

4.4.4 $(\xi_1\xi_2)M[\xi_2\xi_3] = 0.$ (4.3.4)

4.4.5 $(\xi_1\xi_2)M[\xi_1\xi_2] = \tfrac{1}{2}.$ (4.3.5)

4.4.6 $(\xi_1\xi_2)M[\xi_1{}^2] = \tfrac{1}{4}.$ (4.3.6)

4.4.7 $(\xi_1\xi_2)M[\xi_2{}^2] = \tfrac{1}{4}.$ (4.3.7)

4.4.8 $(\xi_1\xi_2)M[\xi_3{}^2] = 0.$ (4.3.8)

4.5.1 $(\xi_1\xi_2)B[\xi_1{}^2]ri = (\xi_1\xi_2)Q'[\xi_1\xi_{123}]\cdot(\xi_1\xi_2)Q''[\xi_{123}\xi_1]$

$= r_1r_2 =_{Df} \tfrac{1}{2}\Sigma(r_1r_2) = \tfrac{1}{4}[2\Sigma(r_1r_2)] = \tfrac{1}{4}i.$ (4.1.1)

4.5.2 $(\xi_1\xi_2)B[\xi_2{}^2]ri = (\xi_1\xi_2)Q'[\xi_2\xi_{123}]\cdot(\xi_1\xi_2)Q''[\xi_{123}\xi_2]$

$= r_1r_2 =_{Df} \tfrac{1}{2}\Sigma(r_1r_2) = \tfrac{1}{4}[2\Sigma(r_1r_2)] = \tfrac{1}{4}i.$ (4.1.1)

4.5.3 $(\xi_1\xi_2)B[\xi_3{}^2]ri = (\xi_1\xi_2)B[\xi_{123}{}^2]ri - (\xi_1\xi_2)B[\xi_1{}^2]ri$

$- (\xi_1\xi_2)B[\xi_2{}^2]ri = i - \tfrac{1}{4}i - \tfrac{1}{4}i = \tfrac{1}{2}i.$ (4.5.1,4.5.2)

4.5.4 $(\xi_1\xi_2)B[\xi_1\xi_2]ri = (\xi_1\xi_2)Q'[\xi_1\xi_{123}]\cdot(\xi_1\xi_2)Q''[\xi_{123}\xi_2]$

$+ (\xi_1\xi_2)Q'[\xi_2\xi_{123}]\cdot(\xi_1\xi_2)Q''[\xi_{123}\xi_1] = r_1r_2 + r_1r_2$

$= 2r_1r_2 = \Sigma(r_1r_2) = \frac{1}{2}i.$ (4.5.1)

4.5.5 $(\xi_1\xi_2)B[\xi_1\xi_{123}]ri = (\xi_1\xi_2)Q'[\xi_1\xi_{123}]$

$+ (\xi_1\xi_2)Q''[\xi_{123}\xi_1] = r_1 + r_2 = r.$ (1.6.1)

4.5.6 $(\xi_1\xi_2)B[\xi_1\xi_{123}]ri = (\xi_1\xi_2)B[\xi_1\xi_3]ri + (\xi_1\xi_2)B[\xi_1\xi_2]ri$

$+ 2\cdot(\xi_1\xi_2)B[\xi_1^2]ri = r.$ (4.5.5)

4.5.7 $(\xi_1\xi_2)B[\xi_1\xi_3]ri = (\xi_1\xi_2)B[\xi_1\xi_{123}]ri - (\xi_1\xi_2)B[\xi_1\xi_2]ri$

$- 2\cdot(\xi_1\xi_2)B[\xi_1^2]ri = r - i.$ (4.5.6,4.5.4,4.5.1)

4.5.8 $(\xi_1\xi_2)B[\xi_2\xi_3]ri = (\xi_1\xi_2)B[\xi_2\xi_{123}]ri - (\xi_1\xi_2)B[\xi_1\xi_2]ri$

$- 2\cdot(\xi_1\xi_2)B[\xi_2^2]ri = r - i.$ (4.5.6,4.5.4,4.5.2)

4.5.9 $(\xi_1\xi_2)B[\xi_3\xi_4]ri = (\xi_1\xi_2)B[\xi_{123}\xi_{1234}]ri$

$- (\xi_1\xi_2)B[\xi_1^2]ri - (\xi_1\xi_2)B[\xi_2^2]ri - (\xi_1\xi_2)B[\xi_3^2]ri$

$- (\xi_1\xi_2)B[\xi_1\xi_2]ri - (\xi_1\xi_2)B[\xi_1\xi_3]ri - (\xi_1\xi_2)B[\xi_2\xi_3]ri$

$= 1 - 2r + \frac{3}{2}i.$ (4.5.1-8)

4.6.1 $(\xi_1^2)Qs[\xi_1^2]i = (\xi_1\xi_2)S[\xi_1\xi_2] = \frac{1}{4}.$ (1.10.2)

4.6.2 $(\xi_1^2)Qs[\xi_1\xi_2]i = (\xi_1\xi_2)S[\xi_1\xi_3] = \frac{1}{4}.$ (1.10.3)

4.6.3 $(\xi_2^2)Qs[\xi_1\xi_2]i = (\xi_1\xi_2)S[\xi_2\xi_3] = \frac{1}{4}.$ (1.10.4)

4.6.4 $(\xi_1^2)Qs[\xi_2^2]i = (\xi_1\xi_2)S[\xi_2^2] = 0.$ (1.10.2-5)

4.6.5 $(\xi_1^2)Qs[\xi_2\xi_3]i = (\xi_1\xi_2)S[\xi_3\xi_4] = \frac{1}{4}.$ (1.10.5)

4.6.6 $(\xi_1\xi_2)Qs[\xi_1\xi_2]i = (\xi_1\xi_2)S[\xi_1\xi_2] = \frac{1}{4}.$ (1.10.2)

4.6.7 $(\xi_1\xi_2)Qs[\xi_1\xi_3]i = (\xi_1\xi_2)S[\xi_1\xi_3] = \frac{1}{4}.$ (1.10.3)

4.6.8 $(\xi_1\xi_2)Qs[\xi_2\xi_3]i = (\xi_1\xi_2)S[\xi_2\xi_3] = \frac{1}{4}.$ (1.10.4)

4.7.1 $Qs[\xi_1^2]i[\xi_1^2] = I[\xi_\alpha^2]\cdot(\xi_1^2)Qs[\xi_1^2] = \frac{1}{4}i.$

(4.1.1,4.6.1)

4.7.2 $Qs[\underline{\xi_1^2]i[\xi_1\xi_2}] = I[\xi_\alpha^2]\cdot(\xi_1^2)Qs[\xi_1\xi_2]i = \frac{1}{4}i.$

(4.1.1,4.6.2)

4.7.3 $Qs[\underline{\xi_2^2]i[\xi_1\xi_2}] = I[\xi_\alpha^2]\cdot(\xi_2^2)Qs[\xi_1\xi_2]i = \frac{1}{4}i.$

(4.1.1,4.6.3)

4.7.4 $Qs[\xi_1^2]i[\xi_2^2] = I[\xi_\alpha^2]\cdot(\xi_1^2)Qs[\xi_2^2]i = 0$ (4.1.1,4.6.4)

4.7.5 $\underline{Qs[\xi_1^2]i[\xi_2\xi_3]} = I[\xi_\alpha^2]\cdot(\xi_1^2)Qs[\xi_2\xi_3]i = \tfrac{1}{4}i.$

(4.1.1,4.6.5)

4.7.6 $\underline{Qs[\xi_1\xi_2]i[\xi_1^2]} = \underline{Qs[\xi_1^2]i[\xi_1\xi_2]} = \tfrac{1}{4}i.$ (4.7.2)

4.7.7 $\underline{Qs[\xi_1\xi_2]i[\xi_2^2]} = \underline{Qs[\xi_2^2]i[\xi_1\xi_2]} = \tfrac{1}{4}i.$ (4.7.3)

4.7.8 $\underline{Qs[\xi_1\xi_2]i[\xi_3^2]} = \underline{Qs[\xi_1^2]i[\xi_2\xi_3]} = \tfrac{1}{4}i.$ (4.7.5)

4.7.9 $Qs[\xi_1\xi_2]i[\xi_1\xi_2] = I[\xi_\alpha\xi_\beta]\cdot(\xi_1\xi_2)Qs[\xi_1\xi_2]i = \tfrac{1}{4}(1 - i).$

(4.1.2,4.6.6)

4.7.10 $Qs[\xi_1\xi_2]i[\xi_1\xi_3] = I[\xi_\alpha\xi_\beta]\cdot(\xi_1\xi_2)Qs[\xi_1\xi_3]i = \tfrac{1}{4}(1 - i).$

(4.1.2,4.6.7)

4.7.11 $Qs[\xi_1\xi_2]i[\xi_2\xi_3] = I[\xi_\alpha\xi_\beta]\cdot(\xi_1\xi_2)Qs[\xi_2\xi_3]i = \tfrac{1}{4}(1 - i).$

(4.1.2,4.6.8)

4.7.12 $Qs[\xi_1\xi_2]i[\xi_3\xi_4] = Qs[\xi_1\xi_{12}]i[\xi_{123}\xi_{1234}]$

$- Qs[\xi_1^2]i[\xi_1^2] - \underline{Qs[\xi_1^2]i[\xi_1\xi_2]} - \underline{Qs[\xi_2^2]i[\xi_1\xi_2]}$

$- Qs[\xi_1^2]i[\xi_2^2] - \underline{Qs[\xi_1^2]i[\xi_2\xi_3]} - \underline{Qs[\xi_1\xi_2]i[\xi_1^2]}$

$- \underline{Qs[\xi_1\xi_2]i[\xi_2^2]} - \underline{Qs[\xi_1\xi_2]i[\xi_3^2]} - Qs[\xi_1\xi_2]i[\xi_1\xi_2]$

$- Qs[\xi_1\xi_2]i[\xi_1\xi_3] - Qs[\xi_1\xi_2]i[\xi_2\xi_3] = \tfrac{1}{4} - i.$

(4.7.1-11)

4.8.1 $(\xi_\alpha^2)I[A_\alpha^2] = G[A] = p.$ Axiom

4.8.2 $(\xi_\alpha^2)I[a_\alpha^2] = G[a] = q.$ (4.8.1,2.1.3)

4.8.3 $(\xi_\alpha\xi_\beta)I[A_\alpha A_\beta] = (\xi_\alpha\xi_\beta)J[A_\alpha A_\beta] = J[AA] = p^2.$ Axiom

4.8.4 $(\xi_\alpha\xi_\beta)J[A_\alpha a_\beta] = (\xi_\alpha\xi_\beta)J[A_\alpha a_\beta] = J[Aa] = 2pq.$

(4.8.3,2.8.3)

4.8.5 $(\xi_\alpha\xi_\beta)J[a_\alpha a_\beta] = (\xi_\alpha\xi_\beta)J[a_\alpha a_\beta] = J[aa] = q^2.$

(4,8.3,2.8.4)

4.9.1 $I[A_\alpha^2] = I[\xi_\alpha^2]\cdot(\xi_\alpha^2)I[A_\alpha^2] = ip.$ (4.1.1,4.8.1)

4.9.2 $I[a_\alpha^2] = I[\xi_\alpha^2]\cdot(\xi_\alpha^2)I[a_\alpha^2] = iq.$ (4.1.1,4.8.2)

$$4.9.3\ I[A_\alpha A_\beta] = I[\xi_\alpha\xi_\beta]\cdot(\xi_\alpha\xi_\beta)I[A_\alpha A_\beta] = (1-i)p^2. \qquad (4.1.2, 4.8.3)$$

$$4.9.4\ I[A_\alpha a_\beta] = I[\xi_\alpha\xi_\beta]\cdot(\xi_\alpha\xi_\beta)I[A_\alpha a_\beta] = 2(1-i)pq. \qquad (4.1.2, 4.8.4)$$

$$4.9.5\ I[a_\alpha a_\beta] = I[\xi_\alpha\xi_\beta]\cdot(\xi_\alpha\xi_\beta)I[a_\alpha a_\beta] = (1-i)q^2. \qquad (4.1.2, 4.8.5)$$

$$4.10.1\ I[AA] = I[A_\alpha{}^2] + I[A_\alpha A_\beta] = ip + (1-i)p^2. \qquad (4.9.1, 4.9.3)$$

$$4.10.2\ I[Aa] = I[A_\alpha a_\beta] = 2(1-i)pq. \qquad (4.9.4)$$

$$4.10.3\ I[aa] = I[a_\alpha{}^2] + I[a_\alpha a_\beta] = iq + (1-i)q^2.$$

[Examples by Hogben, 1933] (4.9.2, 4.9.5)

$$4.11.1\ (A\xi\,\underline{^{Qr}}\,aa)O[aa] = \{Q[Aa]r[aa] : Q[A\xi]r[aa]\}\cdot(Aa\text{-}aa)O[aa]$$

$$= \frac{r+q-2rq}{1+q-2rq} = \frac{\zeta+2i-6i\zeta}{1-4i\zeta}. \qquad (2.15.8\text{-}10, 2.5.7, 4.1.3)$$

$$4.11.2\ (A\xi\,\underline{^{Qr}}\,A\xi)O[aa] = \{Q[Aa]r[Aa] : Q[A\xi]r[A\xi]\}\cdot(Aa\text{-}Aa)O[aa]$$

$$= \frac{q[r+2(1-2r)pq]}{2p(1+q)^2(1-2r)+4r(1+pq)}$$

$$= \frac{(1-4i)\zeta^2(1-2\zeta)+i\zeta(1-\zeta)^2}{(1-4i)(1-2\zeta)+4i(1-\zeta)(1-\zeta-\zeta^2)}.$$

(2.15.6-11, 2.5.4, 4.1.3)

$$4.12.1\ (A\xi\,\underline{^{Rr}}\,aa)O[aa] = \{R[Aa]r[aa] : R[A\xi]r[aa]\}\cdot(Aa\text{-}aa)O[aa]$$

$$= \frac{r+q-rq}{1+r+q-rq} = \frac{\zeta+2i-4i\zeta}{1+2i-4i\zeta}. \qquad (2.19.8\text{-}10, 2.5.7, 4.1.3)$$

$$4.12.2\ (A\xi\,\underline{^{Rr}}\,A\xi)O[aa] = \{R[Aa]r[Aa] : R[A\xi]r[A\xi]\}\cdot(Aa\text{-}Aa)O[aa]$$

$$= \frac{2(1-r)^2pq^2+qr}{2(1-r)(1-q^2)(1+r+q-rq)+2r(pr+2q)}$$

$$+ \frac{(1-2i)^2\zeta^2(1-2\zeta)+i\zeta(1-\zeta)^2}{(1-2i)(1-2\zeta)(1+2i-4i\zeta)+4i(1-\zeta)^2(1-2i\zeta+\zeta)}$$

(2.19.6-11, 2.5.4, 4.1.3)

## SECTION 5. GENETIC PROGNOSIS

This section contains a few suggestions which might prove helpful for the development of exact methods of genetic prognosis. By this is meant the specification of the probability of a particular genotype or phenotype in a child when information of varying degrees of completeness is available concerning many members of its pedigree. The methods necessary for this purpose will be analogous to but considerably more complex than those provided in Sections 2, 3 and 4. In those sections implicative probabilities were supplied for all degrees and types of relatives commonly encountered in human pedigrees, but they were limited in every case to a single relative or, at best, two relatives when these consisted of the two parents. We may presume that the genetic trait is one appearing late in life, in which case the prognosis is made for the individual at an earlier age, otherwise it will be made for a child yet unborn, that is, for the benefit of prospective parents. As in previous sections, the case of a single pair of autosomal alleles will alone be considered, and we shall assume that an estimate of the gene frequency, p or q, is available. Further, the treatment will be limited to pedigrees which are regular up to the earliest generation supplying information.

First we require an additional notion from the calculus of relations. In the "Principia Mathematica" the "relative product" of two relations E and H is defined as the relation which holds between x and z where there is an intermediate term y such that x has the relation E to y and y has the relation H to z and when this is true for any x, y and z. Thus, if V is the relation 'first cousin' and 'P', 'O' and 'S' are relations having the same meanings used previously in this calculus, then we can easily define a first cousin by the relative product

$$wVz =_{Df} wOx.xSy.yPz.$$

Further, the relative product of a relation with itself is called the "square" of the relation and the continuous product of n such relations is called the "n-th power" of the relation. In this section we shall have need for the n-th powers of 'P' and 'O', these being, of course, descriptive subclasses of 'Q' which include what are commonly termed the "lineal ancestors" and "lineal descendents". These are symbolized by '$P^n$' and '$O^n$' and are defined functionally (that is, with respect to implicative probability of genic resemblance) in subsection 5.1.

As was previously remarked, the 'PO' relationship is a unique one. Offspring must receive one gene of a pair from either parent, whereas r applies to other types and degrees of relatives only in an average sense. This can be expressed more precisely in terms of variance. If we let g represent the number (0, 1 or 2) of gene derivatives per pair shared by relatives and let $c_g$ stand for the logical sum of $c_0$, $2c_1$ and $c_2$, then the mean value of g is

$$m = \Sigma\{c_g(g)\} = 2c_1 + 2c_2 = 2r \text{ genes per pair,}$$

and its variance is given by

$$V(g) = \Sigma\{c_g(g^2)\} - m^2 = 2c_1 + 4c_2 - 4(c_1 + c_2)^2,$$

which, for Q-relatives, becomes

$$V(g)_Q = 2r(1 - 2r).$$

Thus for Q-relatives of degree $r = 1/2$ (that is, 'PO'), we see that the average number of genes per pair shared is $2r = 1$, but that this number has no variance ($V(g)_Q = 0$). The above formula shows one further point of interest. The maximum value of $V(g)_Q$ is reached when $\frac{d}{dr}\{2r(1-2r)\} = 2 - 8r = 0$ or $r = 1/4$. Hence, while parents are unique in that they contribute invariably and equally to the autosomal inheritance of an individual, grandparents enjoy the distinction of making the most variable contributions as far as absolute quantity of germ-plasm is concerned.

The above is, of course, an easily foreseen consequence of the mechanism of meiosis and gives rise to numerous "special cases" noted in the previous sections. It also makes possible a theorem of considerable generality and usefulness in the mathematics of genetic prognosis. This theorem was, to the knowledge of the writer, first formulated by Mr. W. F. Floyd in an appendix to Woodger (1937). Floyd refers to all of the $2^{n-1}$ parental ancestors of the n-th ancestral generation as comprising that individual's n-th paternal "half-generation", and then shows that if p is the proportion of [AA] homozygotes and q the proportion of [aa] homozygotes in the $\kappa$-th paternal half-generation and if p' and q' are the proportions of [AA] and [aa] homozygotes in the $\lambda$-th maternal half-generation and if all intermediate generations are unspecified, then the probability that the individual will be of the genotype [AA] is $\frac{1}{4}(1+p-q)(1+p'-q')$ and the probability that the individual will be of the genotype [aa] is $\frac{1}{4}(1-p+q)(1-p'+q')$.

We shall now change the above notation since it conflicts with that previously used in this calculus. Moreover, Floyd's theorems are but special cases of a more general theorem (5.2.2) which can be easily visualized in the following diagram:

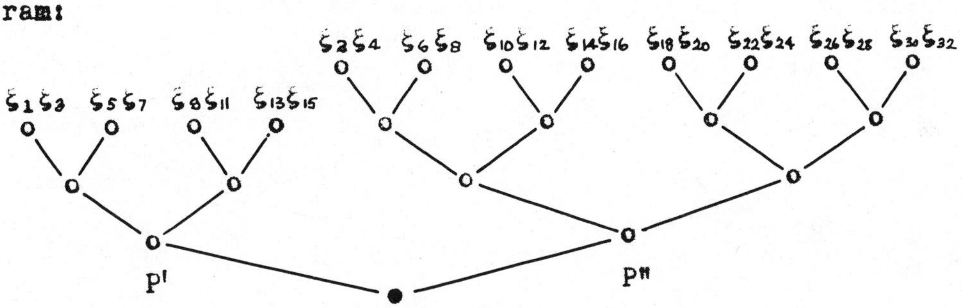


$$J^*[\xi_\mu \xi_\nu],(\mu = 1,3,5\ldots;\ \nu = 2,4,6\ldots) = (\tfrac{1}{2})^{3+4}.$$

This diagram shows the genes possessed by all members of the 3-rd paternal half-generation and all members of the 4-th maternal half-generation. Now, the terminal member of the pedigree must possess one paternal and one maternal gene (one even- and one odd-numbered $\xi$) but the parents, P' and P", will possess any one of the genes possessed by the terminal member's half-generation ancestors with equal probability, hence the terminal member will obtain any combination of odd- and even-numbered genes with equal probability. This isaactually a generalization of theorem 1.3.2.

If all of the $\xi$'s are specified in both half-generations and if we let $x'$ and $y'$ stand for the proportions of A and a genes in the n-th paternal half-generation and $x''$ and $y''$ for the corresponding proportions in the m-th maternal half-generation, then it follows directly from 5.2.2 that the probabilities of the three genotypes (denoted here by J*[AA], J*[Aa] and J*[aa]) will be given by $x'x''$, $x'y'' + x''y'$ and $y'y''$, respectively. It is esily seen that these three expectations correspond to the formulae of Floyd. The following pedigree provides an example of the calculation:

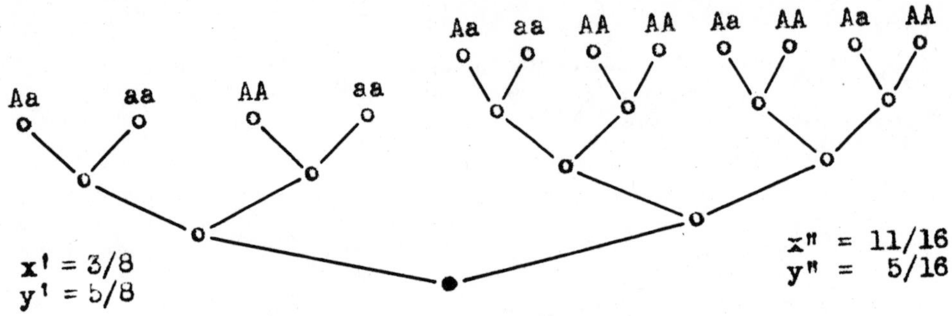


$$J^*[AA] = x'x'' = 33/128$$

$$J^*[Aa] = x'y'' + x''y' = 70/128$$

$$J^*[aa] = y'y'' = 25/128$$

The result is rather interesting for it shows that the probabilities of the three genotypes in the last member of the pedigree are independent of the distribution of the A and a genes within the two half-generations and also independent of the number of intervening generations. It is indeed doubtful whether the theorem as such will ever prove useful either in human or in experimental genetics; however, the writer is of the opinion that it might possibly serve as a skeleton procedure upon which exact probability methods for genetic prognosis could be elaborated. For this purpose the formulae would require modification in such a way as to provide for the following features which are to be expected in actual human pedigree analysis: (1) incompletely specified half-generations; (2) incompletely specified genotypes (dominance and consequent use of intermediate generations; (3) utilization of information provided by collateral relatives as well as lineal ancestors; and perhaps finally (4) incomplete penetrance.

The first of these modifications offers no difficulty. Consider the following pedigree:

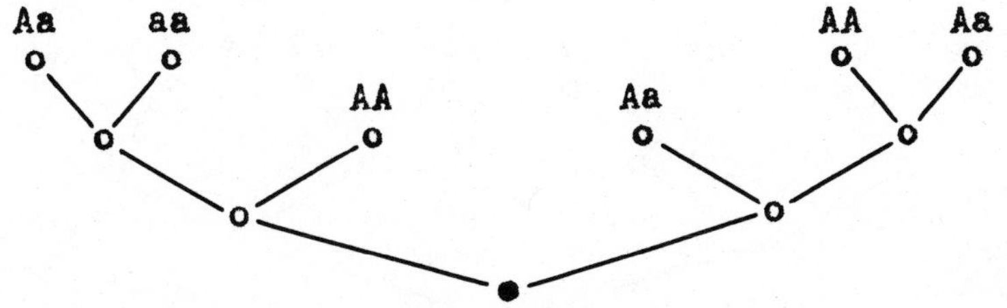


This can be expanded into a pedigree of two completely specified 3rd half-generations:

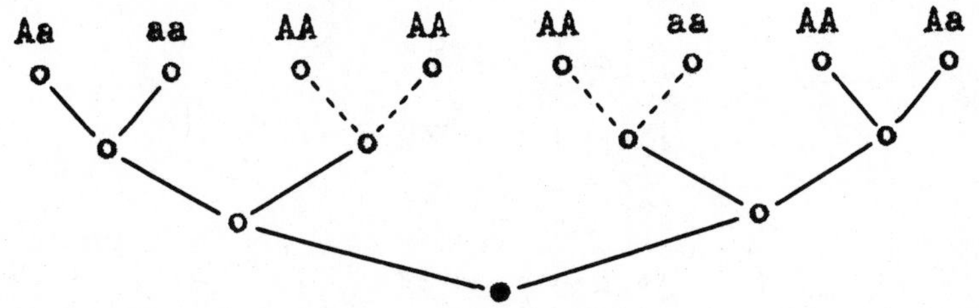


and the calculation will then proceed as in the previous example. In this example it will be noticed that every member whose genotype is specified also possesses a mate with a specified genotype. Consider next a pedigree in which this is not the case:

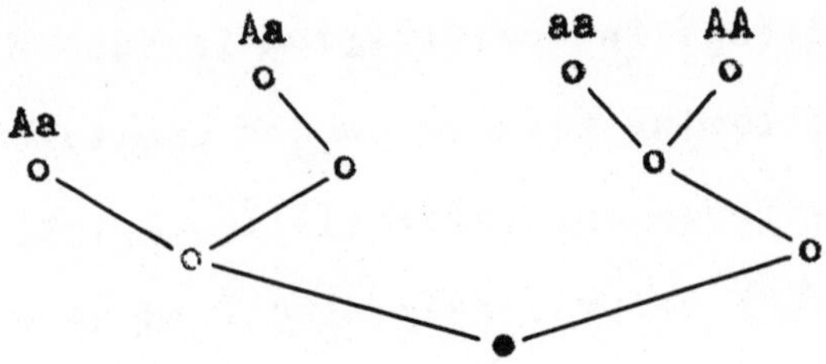


This can also be expanded into a pedigree of complete half-generations by including unspecified genotypes [ξξ]. If we then let z' and z" represent the proportions of such unspecified genes in the two half-generations, such that $x' + y' + z' = 1$ and $x'' + y'' + z'' = 1$, theorem 5.2.2 then permits the following easy modification of Floyd's formulae:

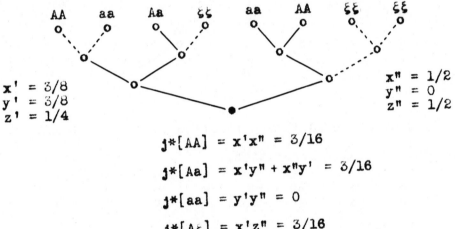

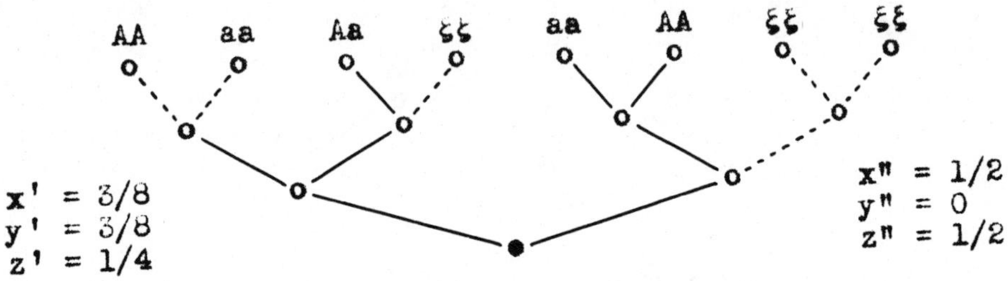


$$j^*[AA] = x'x'' = 3/16$$

$$j^*[Aa] = x'y'' + x''y' = 3/16$$

$$j^*[aa] = y'y'' = 0$$

$$j^*[\underline{A\xi}] = x'z'' = 3/16$$

$$j^*[\underline{\xi A}] = x''z' = 1/8$$

$$j^*[\underline{a\xi}] = y'z'' = 3/16$$

$$j^*[\underline{\xi a}] = y''z' = 0$$

$$j^*[\xi\xi] = z'z'' = 1/8$$

$$J^*[AA] = j^*[AA] + j^*[\underline{A\xi}]\cdot(\underline{A\xi})J[AA] + j^*[\underline{\xi A}]\cdot(\underline{\xi A})J[AA] + j^*[\xi\xi]\cdot(\xi\xi)J[AA] = \frac{3}{16} + \frac{5}{16}p + \frac{1}{8}p^2$$

$$J^*[Aa] = j^*[Aa] + j^*[\underline{A\xi}]\ (\underline{A\xi})J[\underline{Aa}] + j^*[\underline{\xi A}]\ (\underline{aA})J[\underline{Aa}]$$

$$+ j^*[\underline{a\xi}]\cdot(\underline{a\xi})J[\underline{aA}] + j^*[\underline{\xi a}]\cdot(\underline{\xi a})J[\underline{Aa}]$$

$$+ j^*[\xi\xi]\cdot(\xi\xi)J[Aa] = \frac{3}{16} + \frac{5}{16}q + \frac{3}{16}p + \frac{1}{8}pq$$

$$J^*[aa] = j^*[aa] + j^*[\underline{a\xi}]\cdot(\underline{a\xi})J[aa] + j^*[\underline{\xi a}]\cdot(\underline{\xi a})J[aa]$$

$$+ j^*[\xi\xi]\cdot(\xi\xi)J[aa] = \frac{5}{16}q + \frac{1}{8}q^2$$

Unfortunately the present section goes no further toward the achievement of its purpose except for the development in subsection 5.3 of certain theorems which are of interest in the prognosis of recessive abnormalities. Theorem 5.3.9 states that, if an individual's pedigree is regular up to the n-th generation and if all of its members are of the genotype [A$\xi$] including the individual himself (designated the 1st generation), then the probability that a gamete produced by this individual will carry the gene [a] is

$$\frac{q}{1 + nq} = \text{Df } \zeta_n.$$

Theorem 5.3.10 states that, if an individual's pedigree is regular up to the n-th generation and includes members all of whom are of the genotype [A$\xi$], then the probability that the individual (designated the 0-th generation) will be of the genotype [aa] is

$$\left(\frac{q}{1 + nq}\right)^2 = \text{Df } \zeta_n^2.$$

The proofs of subsection 5.3 follow the method of mathematical induction.

In the following diagram illustrating these theorems, open circles represent dominant phenotypes [A$\xi$] and the final probabilities given are those for the appearance of the recessive phenotype [aa] in the last member of the pedigree. The $\zeta$'s along the paths of descent indicate the probabilities of passage of the recessive gene through these links.

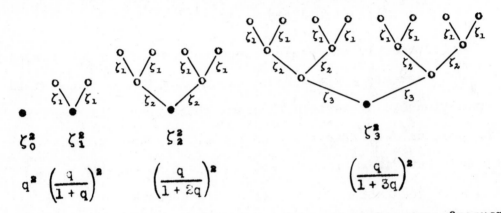

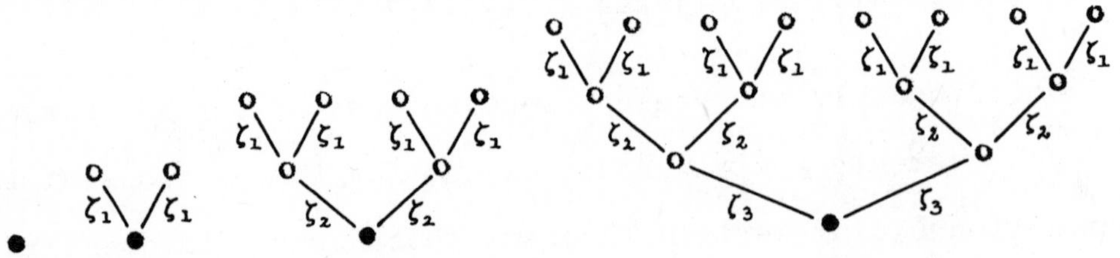


$\zeta_0^2$ $\qquad \zeta_1^2$ $\qquad \zeta_2^2$ $\qquad \zeta_3^2$

$q^2$ $\qquad \left(\frac{q}{1+q}\right)^2$ $\qquad \left(\frac{q}{1+2q}\right)^2$ $\qquad \left(\frac{q}{1+3q}\right)^2$

Thus $\zeta_n^2$ is a general function of the recessive gene frequency giving the probability of a recessive abnormality in a child when its pedigree is known to include normal ancestors up to the n-th generation. It reduces in $\zeta_1^2$ to the formula (2.11.4) of Snyder and in $\zeta_0^2$ to the incidence of the recessive trait in the general population.

## SECTION 5

5.1.1 $(\xi_1\xi_2)O^n[\xi_1\xi_3] = (\xi_1\xi_2)O^n[\xi_2\xi_3] = \text{Df } \{(\xi_1\xi_2)O[\xi_1\xi_3]$
$\cdot(\xi_1\xi_3)O[\xi_1\xi_4]\cdots\cdot(\text{n terms})\} = (\tfrac{1}{2})^n.$ (1.6.1,1.6.2)

5.1.2 $(\xi_1\xi_2)P^n[\xi_1\xi_3] = (\xi_1\xi_2)P^n[\xi_2\xi_3] = \text{Df } \{(\xi_1\xi_2)P[\xi_1\xi_3]$
$\cdot(\xi_1\xi_3)P[\xi_1\xi_4]\cdots\cdot(\text{n terms})\} = (\tfrac{1}{2})^n.$ (1.5.1,1.5.2)

5.1.3 $(\xi_1\xi_2)PO^n[\xi_{12}\xi_3] = (\xi_1\xi_2)P^n[\xi_{12}\xi_3] = (\xi_1\xi_2)O^n[\xi_{12}\xi_3]$
$= (\xi_1\xi_2)Q[\xi_{12}\xi_3]r, (r = (\tfrac{1}{2})^n).$ (1.8.5,5.1.1-2)

5.2.1 $([\xi_1\xi_3][\xi_5\xi_7],\ldots(2^n \text{ terms}))O^n[\xi_\mu\xi_2](\mu=1,3,5\ldots)$
$= (\xi_1\xi_3)O^n[\xi_1\xi_2] = (\xi_1\xi_3)O^n[\xi_3\xi_2] = \ldots = (\tfrac{1}{2})^n.$(5.1.3)

5.2.2 $([\xi_1\xi_3][\xi_5\xi_7],\ldots(2^n \text{ terms});[\xi_2\xi_4][\xi_6\xi_8],\ldots(2^m \text{ terms}))$
$O^{m,n}[\xi_\mu\xi_\nu] = ([\xi_1\xi_3][\xi_5\xi_7],\ldots(2^{n-1} \text{ terms}))O^{n-1}[\xi_\mu\xi_2]$
$\cdot([\xi_2\xi_4][\xi_6\xi_8],\ldots(2^{m-1} \text{ terms}))O^{m-1}[\xi_1\xi_\nu]$
$\cdot(\xi_\mu\xi_2 - \xi_1\xi_\nu)O[\xi_\mu\xi_\nu] = \tfrac{1}{4}(\tfrac{1}{2})^{n-1}(\tfrac{1}{2})^{m-1} = (\tfrac{1}{2})^{m+n}.$

(1.3.2,5.2.1)

5.3.1 $(P^u_{2u-1}[A\xi],P^{u-1}_{2u-2}[A\xi],\ldots P^2_2[A\xi],P^1_1[A\xi])G[a] = \text{Df } \dfrac{\chi}{1+u\chi}.$

5.3.2 $(P^u_{2u-1}[A\xi],P^{u-1}_{2u-2}[A\xi],\ldots P^2_2[A\xi],P^1_1[A\xi])G[A] = \dfrac{1+(u-1)\chi}{1+u\chi}.$

(5.3.1,2.1.1)

5.3.3 $(P^u_{2u}[A\xi],P^{u-1}_{2u-1}[A\xi],\ldots P^1_2[A\xi])O[aa] =$
$\{(P^u_{2u-1}[A\xi],P^{u-1}_{2u-2}[A\xi],\ldots P^2_2[A\xi],P^1_1[A\xi])G[a]\}^2 = \dfrac{\chi^2}{(1+u\chi)^2}$

(5.3.1)

5.3.4 $(P^u_{2u}[A\xi],P^{u-1}_{2u-1}[A\xi],\ldots P^1_2[A\xi])O[Aa] =$
$2\cdot(P^u_{2u-1}[A\xi],P^{u-1}_{2u-2}[A\xi],\ldots P^2_2[A\xi],P^1_1[A\xi])G[A]$
$\cdot(P^u_{2u-1}[A\xi],P^{u-1}_{2u-2}[A\xi],\ldots P^2_2[A\xi],P^1_1[A\xi])G[a]$
$= \dfrac{\chi\{1+(u-1)\chi\}}{(1+u\chi)^2}.$ (5.3.1-2)

5.3.5 $(P_{2^u}^u[A\xi], P_{2^{u-1}}^{u-1}[A\xi], \ldots P_{2}^1[A\xi])0[AA] =$

$\{(P_{2^{u-1}}^u[A\xi], P_{2^{u-2}}^{u-1}[A\xi], \ldots P_{2}^2[A\xi], P_{1}^1[A\xi])G[A]\}^2$

$$= \frac{\{1 + (u-1)\chi\}^2}{(1 + u\chi)^2}. \hspace{3cm} (5.3.2)$$

5.3.6 $(P_{2^u}^u[A\xi], P_{2^{u-1}}^{u-1}[A\xi], \ldots P_{2}^1[A\xi])0[A\xi] =$

$(P_{2^u}^u[A\xi], P_{2^{u-1}}^{u-1}[A\xi], \ldots P_{2}^1[A\xi])0[AA] +$

$(P_{2^u}^u[A\xi], P_{2^{u-1}}^{u-1}[A\xi], \ldots P_{2}^1[A\xi])0[Aa]$

$$= \frac{1 + 2u\chi + (u^2 - 1)\chi^2}{(1 + u\chi)^2}. \hspace{2cm} (5.3.4-5)$$

5.3.7 $(P_{2^u}^{u+1}[A\xi], P_{2^{u-1}}^u[A\xi], \ldots P_{2}^2[A\xi], P_{1}[A\xi])G[a] =$

$\{(P_{2^u}^u[A\xi], P_{2^{u-1}}^{u-1}[A\xi], \ldots P_{2}^1[A\xi])0[Aa]:$

$(P_{2^u}^u[A\xi], P_{2^{u-1}}^{u-1}[A\xi], \ldots P_{2}^1[A\xi])0[A\xi]\}(Aa)G[a]$

$$= \frac{1}{2} \cdot \frac{2\chi\{1 + (u-1)\chi\}}{(1 + u\chi)^2} \cdot \frac{(1 + u\chi)^2}{1 + 2u\chi + (u^2-1)\chi^2} = \frac{\chi}{1 + (u+1)\chi}.$$

$$(5.3.4, 5.3.6, 2.3.3)$$

5.3.8 $(P_{2^{u+1}}^{u+1}[A\xi], P_{2^u}^u[A\xi], \ldots P_{2}^1[A\xi])0[aa] =$

$(P_{2^u}^{u+1}[A\xi], P_{2^{u-1}}^u[A\xi], \ldots P_{2}^2[A\xi], P_{1}^1[A\xi])G[a]^2$

$$= \left(\frac{\chi}{1 + (u+1)\chi}\right)^2. \hspace{2.5cm} (5.3.7)$$

5.3.9 $G[a] = q/(1 + 0 \cdot q) = q = Df\ \zeta_0.$ $\hspace{2cm} (2.1.3)$

$$(P_1[A\xi])G[a] = (A\xi)G[a] = \frac{q}{1 + 1 \cdot q} = Df\ \zeta_1. \hspace{1cm} (2.9.4)$$

$.. (P_{2^{n-1}}^n[A\xi], P_{2^{n-2}}^{n-1}[A\xi], \ldots P_{2}^2[A\xi], P_{1}^1[A\xi])G[a]$

$$= \frac{q}{1 + nq} = Df\ \zeta_n. \hspace{2.5cm} (5.3.1, 5.3.7)$$

5.3.10 $(P^{0}[A\xi])O[aa] = \text{Df } J[aa] = q^{2} = \left(\frac{q}{1 + 0 \cdot q}\right)^{2} = \text{Df } \zeta_0^{2}.$

(2.7.4)

$(P_{2}^{1}[A\xi])O[aa] = \text{Df } (A\xi \overset{0}{-} A\xi)O[aa] = \left(\frac{q}{1 + 1 \cdot q}\right)^{2} = \text{Df } \zeta_1^{2}.$

(2.11.4)

$\therefore (P_{2^n}^{n}[A\xi], P_{2^{n-1}}^{n-1}[A\xi], \ldots P_{2}^{1}[A\xi])O[aa]$

$= \left(\frac{q}{1 + nq}\right)^{2} = \text{Df } \zeta_n^{2}.$ (5.3.3, 5.3.8)

# SECTION 6. TWINS

This section invloves no new procedures but is merely an extension of the methods of previous sections to the simple but rather special relation 'T' which is constituted of "twins". In the genetical study of this group one is usually confronted with a problem unlike that presented in the study of other kinds of relatives. In all previous sections we were able to assume that, whereas our knowledge of their genotypes or phenotypes might be partial or completely wanting, we were at least certain as to the kind and degree of the relatives involved in the study. Twins, however, are of two sorts, and the primary problem in connection with such relatives is usually that of deciding whether, in any given instance, we are dealing with one or the other kind of relation. Consequently, we require a new type of symbol, one indicating the probability that a particular instance of a relation is an example of one of its subclasses. For this purpose we shall use the sign '$\mathcal{R}$' and (T)$\mathcal{R}$[E] and (T)$\mathcal{R}$[F] are therefore the probabilities that, given the relation 'T', the relation is of the sort 'E' or 'F'. Axiom 6.1.1 therefore merely states that there are two subclasses comprising the undefined relation 'T', these being designated 'E' and 'F' and occurring in the population of twins in the proportions t and (1 - t), respectively (6.1.2,3).

'E' is the relationship between monozygotic or "identiccal" twins and 'F' is the relationship between dizygotic or "fraternal" twins. They can be defined genetically as R-relatives of degrees 1 and $\frac{1}{2}$, respectively (6.2.1,6.3.1). The implicative probabilities of subsection 6.5 then merely show that E-twins must be identical in their genotypes and the combinatory probabilities (6.6) show that E-twins of the three genotypes will occur in the same ratio to be expected for single individuals. Theorem 6.6.4 gives what is commonly referred to as the "percentage of concordancy", that is, the proportion of all pairs which are identical with respect to any genotype, this being, of course, 100 per cent. in the case of identical twins.

If two twins are found to belong to different genotypes or phenotypes, then provided we assume the conditions of Sections 1 and 2 invariable, viz., absence of mutation and complete penetrance, the pair may be definitely labelled as fraternal or dizygotic. If, however, they are identical with respect to all such rigidly controlled hereditary traits for which they have been inspected, then the "proof" of their monozygosity must be rested upon the smallness of their probability of being dizygotic and yet phenotypically identical in so many respects. That is to say, we adopt the null hypothesis that we are involved with the relation 'F' and then attempt to disprove it. If nothing is known concerning the phenotypes or genotypes of the parents, sibs, or other relatives, then the probabilities which are required in testing this hypothesis are those supplied by 6.7.1, 6.7.5 and 6.7.9 for the case of absence of dominance or 6.7.9 and 6.7.13

when dominance is complete. When, through knowledge of other relatives, it can be demonstrated that a pair of twins which are similar in some monofactorial trait could have been dissimilar, then the probability of their similarity can never be less than 1/2, if they are heterozygous, or less than 1/4, if they are homozygous. This is also shown by the formulae of subsection 6.7. As q varies from 0 to 1 the probabilities 6.7.1 and 6.7.9 vary from 1 to 1/4, while 6.7.5 has its maximum of 5/8 when p = q = ½ and its minimum of ½ when q = 0 or 1. Thus, absence of knowledge concerning the parents or other relatives may in many cases seriously impare the genetic diagnosis of type of twinning. The formulae of subsection 6.7 would, of course, serve equally well for sibs other than fraternal twins.

Combinatory probabilities of genotype and phenotype resemblance in sibs (6.8) were previously developed by Cotterman (1937b) and by Rife (1938). Theorem 6.8.12 gives the proportion of concordancy in fraternal twins when there are three distinguishable genotypes and 6.8.13 supplies the corresponding formula for the case of dominance. These probabilities are, of course, of interest not in the diagnosis of twins but in the testing of genetic hypotheses in randomly collected sib or twin data. If it were desired, complete sets of implicative and combinatory probabilities of genotype and phenotype resemblance could be developed for the general category of twins, as is illustrated by 6.9.1 and 6.10.1. These would then be applicable to mass data in which a classification as to type of twins had not been undertaken. Finally, subsections 6.11 to 6.14 furnish the implicative and combinatory probabilities of E- and F-twins appropriate to the case of incomplete penetrance. They are of particular interest because of their comparative simplicity, this being a result, of course, of the simple values of the coefficient of relationship ($r = 1$, $r = \frac{1}{2}$).

## SECTION 6

6.1.1 $(T)\mathcal{R}[E] + (T)\mathcal{R}[F] = 1.$ Axiom

6.1.2 $(T)\mathcal{R}[E] =_{Df} t.$

6.1.3 $(T)\mathcal{R}[F] = 1 - t.$ (6.1.1,6.1.2)

6.2.1 $(\xi_\alpha\xi_\beta)E[\xi_\gamma\xi_\delta] =_{Df} (\xi_\alpha\xi_\beta)R[\xi_\gamma\xi_\delta]r, (r = 1).$

6.3.1 $(\xi_\alpha\xi_\beta)F[\xi_\gamma\xi_\delta] =_{Df} (\xi_\alpha\xi_\beta)R[\xi_\gamma\xi_\delta]r, (r = \frac{1}{2})$

$= (\xi_\alpha\xi_\beta)S[\xi_\gamma\xi_\delta].$

6.4.1 $(\xi_\alpha\xi_\beta)T[\xi_\gamma\xi_\delta] = (T)\mathcal{R}[E]\cdot(\xi_\alpha\xi_\beta)E[\xi_\gamma\xi_\delta]$

$+ (T)\mathcal{R}[F]\cdot(\xi_\alpha\xi_\beta)F[\xi_\gamma\xi_\delta].$ (6.1.1)

6.5.1 $(AA)E[AA] = (A_1A_2)E[A_1A_2] = 1.$ (6.2.1,2.18.1)

6.5.2 $(Aa)E[Aa] = (A_1a_2)E[A_1a_2] = 1.$ (6.2.1,2.18.5)

6.5.3 $(aa)E[aa] = (a_1a_2)E[a_1a_2] = 1.$ (6.2.1,2.18.9)

6.6.1 $E[AA][AA] = J[AA]\cdot(AA)E[AA] = J[AA] = p^2.$ (2.7.1,6.5.1)

6.6.2 $E[Aa][Aa] = J[Aa]\cdot(Aa)E[Aa] = J[Aa] = 2pq.$

(2.7.3,6.5.2)

6.6.3 $E[aa][aa] = J[aa]\cdot(aa)E[aa] = J[aa] = q^2.$ (2.7.4,6.5.3)

6.6.4 $E[\xi^1\xi^2][\xi^1\xi^2] =_{Df} E[AA][AA] + E[Aa][Aa] + E[aa][aa]$

$= 1.$ (6.6.1-3)

6.7.1 $(AA)F[AA] = \frac{1}{4}(1 + p)^2.$ (6.3.1,2.18.1)

6.7.2 $(AA)F[Aa] = \frac{1}{2}(1 + p)(1 - p).$ (6.3.1,2.18.2)

6.7.3 $(AA)F[aa] = \frac{1}{4}(1 - p)^2.$ (6.3.1,2.18.3)

6.7.4 $(Aa)F[AA] = \frac{1}{4}p(1 + p).$ (6.3.1,2.18.4)

6.7.5 $(Aa)F[Aa] = \frac{1}{2}(1 + pq).$ (6.3.1,2.18.5)

6.7.6 $(Aa)F[aa] = \frac{1}{4}q(1 + q).$ (6.3.1,2.18.6)

6.7.7 $(aa)F[AA] = \frac{1}{4}(1 - q)^2.$ (6.3.1,2.18.7)

6.7.8 $(aa)F[Aa] = \frac{1}{2}(1 + q)(1 - q).$ (6.3.1,2.18.8)

6.7.9 $(aa)F[aa] = \frac{1}{4}(1 + q)^2.$ (6.3.1,2.18.9)

6.7.10 $(AA)F[A\xi] = (AA)F[AA] + (AA)F[Aa] = \frac{1}{4}(1+p)(3-p).$ (6.7.1-2)

6.7.11 $(Aa)F[A\xi] = (Aa)F[AA] + (Aa)F[Aa] = \frac{1}{4}(2+3p-p^2).$ (6.7.4-5)

6.7.12 $(aa)F[A\xi] = (aa)F[AA] + (aa)F[Aa] = \frac{1}{4}(1-q)(3+q).$ (6.7.7-8)

6.7.13 $(A\xi)F[A\xi] = (A\xi)J[AA]\cdot(AA)F[A\xi] + (A\xi)J[Aa]\cdot(Aa)F[A\xi]$

$$= \frac{4+4q-3q^2-q^3}{4(1+q)}.$$ (2.8.13-14,6.7.10-11)

6.7.14 $(A\xi)F[aa] = (A\xi)J[AA]\ (AA)F[aa] + (A\xi)J[Aa]\cdot(Aa)F[aa]$

$$= \frac{q^2(3+q)}{4(1+q)}.$$ (2.8.13-14,6.7.3,6.7.6)

6.8.1 $F[AA][AA] = \frac{1}{4}p^2(1+p)^2.$ [Rife, 1938] (6.3.1,2.19.1)

6.8.2 $F[\underline{AA}][\underline{Aa}] = F[\underline{Aa}][\underline{AA}] = \frac{1}{2}p^2q(1+p).$ (6.3.1,2.19.2)

6.8.3 $F[AA][Aa] = p^2q(1+p).$ [Rife, 1938] (6.3.1,2.19.3)

6.8.4 $F[\underline{AA}][\underline{aa}] = F[\underline{aa}][\underline{AA}] = \frac{1}{4}p^2q^2.$ (6.3.1,2.19.4)

6.8.5 $F[AA][aa] = \frac{1}{2}p^2q^2.$ [Rife, 1938] (6.3.1,2.19.5)

6.8.6 $F[Aa][Aa] = pq(1+pq).$ [Rife, 1938] (6.3.1,2.19.6)

6.8.7 $F[\underline{Aa}][\underline{aa}] = F[\underline{aa}][\underline{Aa}] = \frac{1}{2}pq^2(1+q).$ (6.3.1,2.19.7)

6.8.8 $F[Aa][aa] = pq^2(1+q).$ [Rife, 1938] (6.3.1,2.19.8)

6.8.9 $F[aa][aa] = \frac{1}{4}q^2(1+q)^2.$ [Cotterman, 1937b] (6.3.1,2.19.9)

6.8.10 $F[A\xi][aa] = \frac{1}{2}pq^2(4-p).$ [Cotterman, 1937b] (6.3.1,2.19.10)

6.8.11 $F[A\xi][A\xi] = 1 - \frac{1}{4}q^2(4-p)^2 + 2p^2q^2.$ [Cotterman, 1937b] (6.3.1,2.19.11)

6.8.12 $F[\xi^1\xi^2][\xi^1\xi^2] =_{Df} F[AA][AA] + F[Aa][Aa] + F[aa][aa]$

$= \frac{1}{4}(4 - 11p + 14p^2 - 6p^3 + 3p^4)$. (6.8.1,6,9)

6.8.13 $F[A\xi][A\xi] + F[aa][aa] = 1 - \frac{1}{2}pq^2(4 - p)$. (6.8.9,11)

6.9.1 $(AA)T[AA] = (T)\mathcal{R}[E]\cdot(AA)E[AA] + (T)\mathcal{R}[F]\cdot(AA)F[AA]$

$= t + (1 - t)\frac{1}{4}(1 + p)^2$. (6.1.2-3,6.5.1,6.7.1)

6.10.1 $T[AA][AA] = (T)\mathcal{R}[E]\cdot E[AA][AA] + (T)\mathcal{R}[F]\cdot F[AA][AA]$

$= tp^2 + (1 - t)\frac{1}{4}p^2(1 + p)^2$. (6.1.2-3,6.6.1,6.8.1)

6.11.1 $(D)E[D] = (p + 2qe^2)/(p + 2qe)$. (3.10.1,6.2.1)

6.11.2 $(D)E[N] = 2qe(1 - e)/(p + 2qe)$. (3.10.3,6.2.1)

6.11.3 $(N)E[D] = 2pe(1 - e)/(1 + p - 2pe)$. (3.10.4,6.2.1)

6.11.4 $(N)E[N] = 2p(1 - e)^2/(1 + p - 2pe)$. (3.10.5,6.2.1)

6.12.1 $E[D][D] = p^2 + 2pqe^2$. (3.11.1,6.2.1)

6.12.2 $E[\underline{D}][\underline{N}] = E[\underline{N}][\underline{D}] = 2pqe(1 - e)$. (3.11.2,6.2.1)

6.12.3 $E[D][N] = 4pqe(1 - e)$. (3.11.4,6.2.1)

6.12.4 $E[N][N] = q^2 + 2pq(1 - e)^2$. (3.11.5,6.2.1)

6.13.1 $(D)F[D] = \{\frac{1}{4}p(1 + p)^2 + pqe(1 + p) + qe^2(1 + pq)\}/(p + 2qe)$.

(3.10.1,6.3.1)

6.13.2 $(D)F[N] = \{\frac{1}{2}pq(1 - e)(1 + p) + \frac{1}{4}pq^2 + qe(1 - e)(1 + pq)$

$+ \frac{1}{2}q^2e(1 + q)\}/(p + 2qe)$. (3.10.3,6.3.1)

6.13.3 $(N)F[D] = \{\frac{1}{2}p^2(1 - e)(1 + p) + \frac{1}{4}p^2q + pe(1 - e)(1 + pq)$

$+ \frac{1}{2}pqe(1 + q)\}/(1 + p - 2pe)$. (3.10.4,6.3.1)

6.13.4 $(N)F[N] = \{p(1 - e)^2(1 + pq) + pq(1 - e)(1 + q)$

$+ \frac{1}{4}q(1 + q)^2\}/(1 + p - 2pe)$. (3.10.5,6.3.1)

6.14.1 $F[D][D] = \frac{1}{4}p^2(1 + p)^2 + p^2qe(1 + p) + pqe^2(1 + pq)$.

(3.11.1,6.3.1)

6.14.2 $F[\underline{D}][\underline{N}] = F[\underline{N}][\underline{D}] = \frac{1}{2}p^2q(1 - e)(1 + p) + \frac{1}{4}p^2q^2$

$+ pqe(1 - e)(1 + pq) + \frac{1}{2}pq^2e(1 + q)$. (3.11.2,6.3.1)

$$6.14.3\quad F[D][N] = p^2q(1-e)(1+p) + \tfrac{1}{2}p^2q^2 + 2pqe(1-e)(1+pq) + pq^2e(1+q). \qquad (3.11.4, 6.3.1)$$

$$6.14.4\quad F[N][N] = \tfrac{1}{4}q^2(1+q)^2 + pq(1-e)(1+q) + p(1-e)^2(1+\tfrac{1}{2}pq). \qquad (3.11.5, 6.3.1)$$

## CONCLUSION

Genetics is the study of similarity and variation in organisms related by descent. This definition, though easily comprehended and widely quoted, has nevertheless been inadequately appreciated from time to time in discussions on genetic methodology. Thus, the belief that human genetical study should necessarily involve the scrutiny of many individuals in several generations has but slowly given way to the realization that surprisingly large quantities of information are sometimes possessed by data comprising only two generations or even but a single generation. On the other hand, no one can perhaps claim to have undertaken a genetic investigation involving single or unrelated individuals. We are therefore brought back to our original definition: genetics is the study of relatives - but any kind of relatives will suffice.

What the author has attempted to supply in the preceeding pages is a classification of biological relations based upon what seems to be their most fundamental genetic characteristic, namely, the origin of the four genes comprising the two gene-pairs of the related individuals. Genes thus "related by descent" are termed "derivatives" in this paper and the 4-way system of similarity in respect to this criterion has been called the "genic resemblance" of the relatives in question. Under bisexual diploidy only 7 such types of genic resemblance can exist, but this number taken in all assortments gives rise to a theoretical total of 128 functionally different "kinds" of biological relationship, most of which may be exemplified by relatives of varying degrees (coefficients of relationship). In this paper 9 different kinds of relatives are defined and the probability functions for their permissable types of genic resemblance are set forth. This represents but a small portion of the many kinds of relationship which are known to be possible through existing mechanisms of reproduction; it includes, however, almost all relationships commonly encountered in human kindreds. Moreover, all of these relations are found to be completely described in terms of two parameters, the coefficients of inbreeding (i) and relationship (r) which were devised by Wright (1921b).

One of the objects of the present paper is to show that the solution of many statistico-genetic problems is greatly simplified when the analysis proceeds from a consideration of the probabilities of the various genic resemblances limited to the types of relatives involved. To the knowledge of the author, no systematic use of such procedure has heretofore been undertaken, with the result that earlier derivations of many genetic principles have often been made unnecessarily complex. Having set down the facts assumed to be known concerning the relatives in question, one first considers the permissable genic resemblances and their probabilities. The probabilities of specific genotypes are then easily obtained by superimposing on such terms the probabilities of the "types" of genes which are required. As a final step, the phenotypes may be inferred by applying a third set of probabilities to the genotype expressions. It is perhaps folly to ask which is the most important or fundamental property of a gene, its reproduction, chemical nature, or physiological manifestation. However, from the point of view of simplicity in the understanding of genetic principles it is seen that the greatest convenience is offered by considering these three properties separately and in that order. The criterion of "identity" in genes in its usual chemical-physiological sense would thus seem to lose much of its significance. In fact, there might appear to be some advantages in the teaching of genetics if all reference to "types" of genes were omitted in the early exposition of the subject. The concepts of mutant genes, multiple alleles, homozygotes, heterozygotes, and so on, would then be deferred until a thorough mastery of chromosome mechanics and genic resemblances in all types of relatives had first been acquired.

An attempt has been made to choose a notation which is suggestive and which also avoids the ambiguities which so easily arise in any subject involving the manipulation of probabilities. Success in this effort, however, has been far from complete. Moreover, it would be of considerable interest to extend the methods to cases involving multiple factors, multiple alleles, sex-linkage, mutation, selection, etc. Finally, most of the relations dealt with in this paper are dyadic, that is, involve but two members. The relation involving two parents and a single child is quite easily handled and some further suggestions are offered in Section 5 for the utilization of information concerning many relatives. A very extensive problem, however, awaits solution in connection with n-adic relations. In bringing about such improvements and extensions it is almost certain that increased use of the methods of symbolic logic will be required.

## REFERENCES

Cotterman, C. W. 1937a. The detection of sex-linkage in families collected at random. Ohio Jour. Sci. 37: 75-81.

Cotterman, C. W. 1937b. Indication of unit factor inheritance in data comprising but a single generation. Ohio Jour. Sci. 37: 127-140.

Hardy, G. H. 1908. Mendelian proportions in a mixed population. Sci. 28: 49-50.

Hogben, Lancelot. 1933. "Nature and Nurture". W. W. Norton. New York.

Lush, Jay L. 1937. "Animal Breeding Plans". Collegiate Press. Ames.

Rife, D. C. 1938. Simple modes of inheritance and the study of twins. Ohio Jour. Sci. 38: 281-293.

Roberts, J. A. F. 1940. "An Introduction to Medical Genetics". Oxford University Press. London.

Snyder, L. H. 1934. Studies in human inheritance X. A table to determine the proportion of recessives to be expected in various matings involving a unit character. Genetics 19: 1-17.

Timofeeff-Ressowsky, N. W. 1931. Gerichtetes Varieren in der phanotypischen Manifestierung einiger Genovariationen von D. funebris. Naturwiss. 19: 188-200.

Wentworth, E. N. and B. L. Remick. 1916. Some breeding properties of the generalized Mendelian population. Genetics 1: 608-616.

Whitehead, A. N. and Bertrand Russell. 1925. "Principia Mathematica". Second Edition. Cambridge, at the University Press.

Woodger, J. H. 1937. "The Axiomatic Method in Biology". Cambridge, at the University Press.

Wright, Sewall. 1921a. Systems of mating I-IV. Genetics 6: 111-178.

Wright, Sewall. 1921b. Coefficients of inbreeding and relationship. Amer. Nat. 56: 330-338.

Wright, Sewall, 1934. The method of path coefficients. Ann. Math. Stat. 5: 161.

# AUTOBIOGRAPHY

I, Charles William Cotterman, was born May 27, 1914 in Dayton, Ohio and received all of my secondary education in the public schools of that city. Entering the Ohio State University in 1932, I received the degree of Bachelor of Arts in 1935 and the degree of Master of Arts in 1937. During the school year of 1936-1937 I served as graduate assistant in the Department of Zoology and Entomology and in 1937 was appointed instructor in genetics and biometry in that Department. This position I held for three years while completing the requirements for the degree of Doctor of Philosophy. My final quarter of residence in the Graduate School was spent in off-campus research as Research Associate in the Laboratory of Vertebrate Genetics of the University of Michigan.

# V

# New Departures in Structures

# Editor's Comments on Papers 12 Through 16



As was mentioned in the introduction to Part I, the history of social theory and its primary concerns is complicated. To most geneticists and social theorists the relationship of this theory to genetic theory is tenuous. My argument here is that the structures which have been described by social theory are of immense value to genetic theory independently of their origin. Nonetheless, the origin is worth knowing, since if the same structures can be used to interpret *both* the biological and social reality of man, albeit with possible differences in their conditions of application, then we must surely consider their importance in a more general theory.

As in the introduction I proceed historically, and very selectively. The reader is now aware that much of this literature derives from attempts to interpret the cognitive structure of primitive or other non-Western groups. Two recently republished short studies should be read for historic perspective. Durkheim and Mauss (1963), together with its introduction by Needham, gives a view of the ethnographic origin of the study of social categories (totemism) and of their interpretation as behaviorally regulatory structures. Mauss (1967), with an introduction by Evans-Pritchard, is a still-recognized interpretation of exchange behavior, including exchange within the types of structures determined cognitively.

However, these early studies were not particularly mathematical. Chapple and Coon (1942, pp. 281–295) presented an early and crude attempt at axiomatization, but the appendix by André Weil in Lévi-Strauss (1969, but originally 1949) had much greater influence. A highly insightful paper by Ruheman (1945) was also ignored, apparently, but a related paper on the same topic by Elkin (1950) has received recognition. Both these papers, as well as that by Dumont (1966), are discussions of the marriage rules ascribed to certain Australian groups and should be read by readers who seek more background for the article by Livingstone that is reprinted here.

The entire problem was nicely summarized mathematically to its date by White (1963), of which the paper reprinted here is but a very brief summary. That text also included sections on interpretations of role structures and role trees which have yet to be adequately described mathematically [see articles by Lorrain, and by Witz and Earls, in Ballonoff (1974a)]. Exploration on the demographic correlates of these structures was carried out by Meggitt (1968), by Yengoyan (1968), and by Ballonoff (1973). It is useful to compare these demographic models to the cognitive work; the abstract structures are the same. This statement is most disputed by the cognitivists, whose recent concerns have, in any case, been more with language and semantics than with social structure. See Kay (1971) for a treatment of the semantic foundation of our structures, and Tyler (1969) for criticism of the general literature. See Liu (1973) for a current summary of the algebraic literature.

In the last two years, an interesting body of published, unpublished, and in-press papers have been completed. Following in the tradition of Macfarlane, Mukherjee (1972) provides an enumeration technique for family structures, while Greechie and Ottenheimer (unpublished) have studied kin-based relations over similar structures. The paper by Atkins (1974) brilliantly summarizes types of kin enumeration by numerical methods and shows that the more important kin measures can be derived from each other by simple transformations. This paper will find easy acceptance by geneticists and should be studied by them independently of other literature referenced here. Ballonoff (1973) shows, using relations similar to those in Atkins's paper, that minimal structures may be derived for quite general classes of marriage systems, which structures also have the group theoretical properties discussed in papers of this section. These structures are also those used by geneticists to study regular systems of mating.

The paper of Maruyama is included in this section because it depends on structural techniques in a manner unique to genetics but which is in fact closely related to the papers presented here. The relationship is principally through the fact that structural techniques may be used to generate incidence matrices (graphs) that show possible marriage patterns among groups. Although this paper interprets these graphs as geographic patterns, mathematically it is not important whether the basic units are geographically or otherwise defined. Revision needed in the technique for kin-based interaction is in the selection of coefficients of migration and of incidence matrix.

# 12



Reprinted from *Southwestern J. Anthropol.*, **15**, No. 4, 361–372 (1959)

## A FORMAL ANALYSIS OF PRESCRIPTIVE MARRIAGE SYSTEMS AMONG THE AUSTRALIAN ABORIGINES[1]

FRANK B. LIVINGSTONE

SINCE THE PUBLICATION of Lévi-Strauss' remarkable monograph *Les structures élémentaires de la parenté,* much attention has been directed toward prescriptive marriage systems and their structural implications. Such systems are quite widespread in Australia, and many different types of prescriptive marriage system seem to be represented there. Usually these systems are described in the ethnographic literature as prescribing marriage with a certain relative, e.g. mother's brother's daughter, father's mother's brother's son's daughter, etc., and in this way kinship is said to regulate marriage. Radcliffe-Brown has been the principal exponent of this point of view. For many tribes, however, other ethnologists have stated that marriage seems to be regulated by the section or sub-section system as the case may be. Lévi-Strauss, Leach, and others have contributed to the formal analysis of these systems, and, it seems to me, have shown that these systems become more comprehensible when marriage is considered as an exchange between various kinds of local groups. This paper will attempt a further analysis of the prescriptive marriage systems in Australia. It will not, however, attempt to answer the question how marriage is regulated there, nor will it attempt to demonstrate that any particular tribe has in fact this or that marriage system. It will instead be a completely formal analysis which will attempt to deduce the possible types of prescriptive marriage system which could exist there from certain assumptions of the nature of the local groups between which women are exchanged. Since in many cases the prescriptive marriage systems which are derived in this paper correspond rather closely to those of some of the tribes of Australia, they will be named after these tribes; but this should not be interpreted as an attempt to prove that these tribes actually have these particular types of prescriptive marriage systems.

The assumptions of the nature of the local group are as follows:

(1) The local group, which exchanges women with other similarly constituted groups, consists of co-resident, patrilineally-related males. Since women are exchanged between these groups, at times the female relatives are part of the group, and at times the spouses who were obtained in exchange are part of it.[2]

1 During the course of this study the author has been the recipient of a Postdoctoral Fellowship from the National Science Foundation. Drs David F. Aberle and Elman R. Service read drafts of this paper and I wish to acknowledge their helpful criticisms and suggestions.

2 Throughout this paper I will refer to groups of males as giving wives to or receiving wives from other groups of males. For those who feel that this reduces women to chattels I wish to

(2) Each local group is divided into sections, with the males of alternate generations belonging to the same section.

(3) For all the marriage systems which will be considered, the marriage rules apply between the sections, i.e. the males and females of any particular section *must* marry into one and only one other section, but this does not have to be the same section for the males as for the females.

(4) The total social system within which exchanges occur is divided into two patrilineal exogamous moieties, so that one-half of the local groups within the system exchange with the other half.

Although the above assumptions might seem quite restrictive, it should be noted that social systems which fulfill them are widespread in Australia. Thus, Radcliffe-Brown says "Everywhere in Australia the fundamental basis of social organization is a system of patrilineal local groups . . . ," and "The most widespread kind of kinship division in this sense, which is almost universal in Australian tribes, is into two alternating generation divisions. . . ."[3]

I will now consider as variables the number of different kinds of local group within the social system and the nature of the exchange relations between groups. Because of the moiety division of the society, the number of different kinds of local group must be a multiple of two; since the marriage rule is between sections, there must be as many sections in one moiety as in the other. The exchange relations can be either direct or indirect. In direct exchange the men and women of a particular section must marry into the same section, while in indirect exchange the men and women of a particular section marry into different sections.

The simplest marriage system resulting from these assumptions would consist of two kinds of local group, one in Moiety X and the other in Moiety Y, and direct exchange between them. The local group in Moiety X consists of the two sections, A and B, and the group in Moiety Y has the sections, C and D. In one generation there is direct exchange between the sections A and C, and in the next generation between B and D. The prescribed marriage for this system is with a woman who is structurally both mother's brother's daughter and father's sister's daughter. The basic working of the system is shown by the two diagrams on Figure 1. This has been called the Kariera system, since there is general agreement that this tribe's marriage system is the classic example of this type. Romney and Epling[4] have shown that the two different kinds of local group are arranged in checkerboard

point out that these marriage systems can be as easily derived by considering matrilineal lines and an exchange of husbands between them. I only use patrilineal lines because these remain in one place in Australia and the women move between them.

3 Radcliffe-Brown, 1951, pp. 38-39.

4 Romney and Epling, 1958, p. 61.

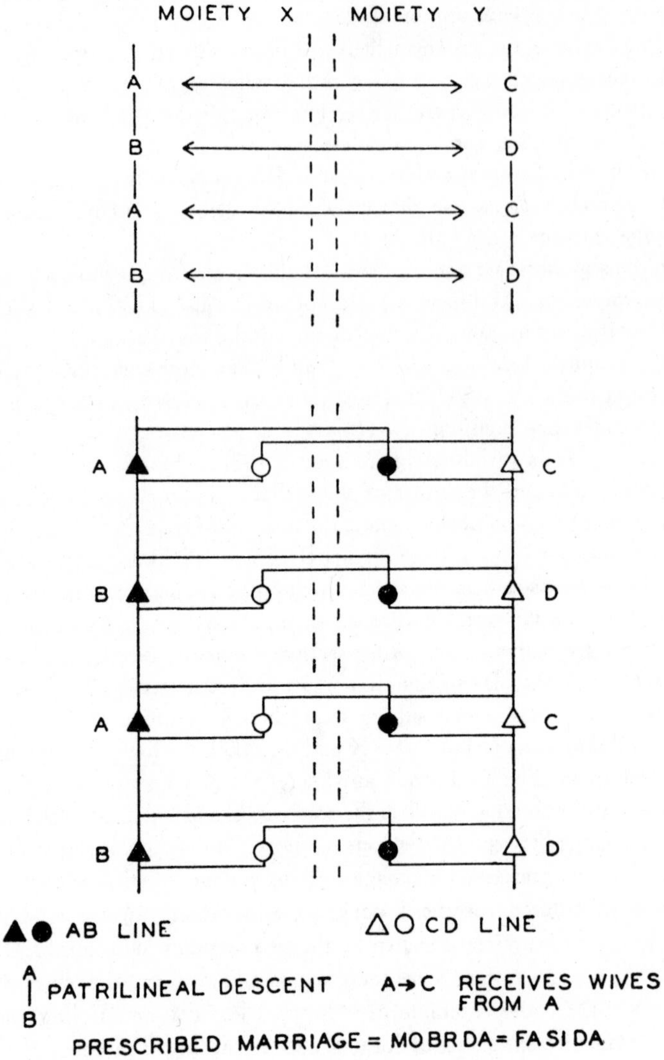


FIG. 1. The Kariera marriage system.

fashion throughout the Kariera territory. Thus, any group belonging to Moiety X is surrounded by groups belonging to Moiety Y, with which it exchanges women. Romney and Epling have made a detailed analysis of this system, to which the interested reader is referred.

The next possibility to be considered is a social system with four different kinds of local group and, as in the Kariera system, direct exchange between them. Two of these kinds of local group would belong to Moiety X and the other two to Moiety Y. In order to have a marriage system with four kinds of local group and direct exchange different from the preceding one, one kind of local group has to exchange in one generation with one of the two kinds in the opposite moiety and in the next generation with the other kind in the opposite moiety. If this were not the case, but instead one kind of local group in one moiety exchanged in every generation with one kind of local group in the opposite moiety, the result would be two superimposed Kariera systems. A system with four different kinds of local group and direct exchange is shown on Figure 2, and has been called Arunta. For the Arunta, Spencer and Gillen[5] enumerate the four kinds of local group, each of which consists of two sections. Their map of Arunta territory shows that the four kinds of local group are scattered throughout Arunta territory, although they do not seem to be in so exact a checkerboard arrangement as those of the Kariera.

For the marriage system shown on Figure 2, and also of course the Arunta, the prescribed marriage is with mother's mother's brother's daughter's daughter, who is not the same woman nor in the same section as father's sister's daughter or mother's brother's daughter. In this system the males of alternate generations, who belong to the same section, exchange women with the same kind of local group. For example for any ego on Figure 2, ego, his father's father, and his son's son obtain wives from—and their sisters go as wives to—the same kind of local group in the opposite moiety, while ego's father, his son, and his son's son's son have the same marriage alliance with the other kind of local group in the opposite moiety.

If the number of different kinds of local group is increased to more than four and direct exchange is still practiced, the resulting systems would not seem to have any basic differences from the two preceding systems, but would be variations of the Arunta or Kariera systems. There is, however, one other system with direct exchange, which does not fulfill the assumptions outlined at the beginning of this paper but which seems well worth mentioning. This is the six section system on Ambryn Island,[6] which has three different "patrisibs" with the males of alternate generations of any patriline belonging to the same section. Marriage has been

5 Spencer and Gillen, 1927, pp. 63-64.

6 Lane and Lane, 1958.

MOIETY  X     MOIETY  Y

▲● AB  LINE     △○ CD  LINE
△⊕ EF  LINE     △⊙ GH  LINE

PRESCRIBED  MARRIAGE  =  MO MO BR DA DA  =
FA FA SI SO DA = FA MO BR SO DA ≠ MO BR DA = FA SI DA

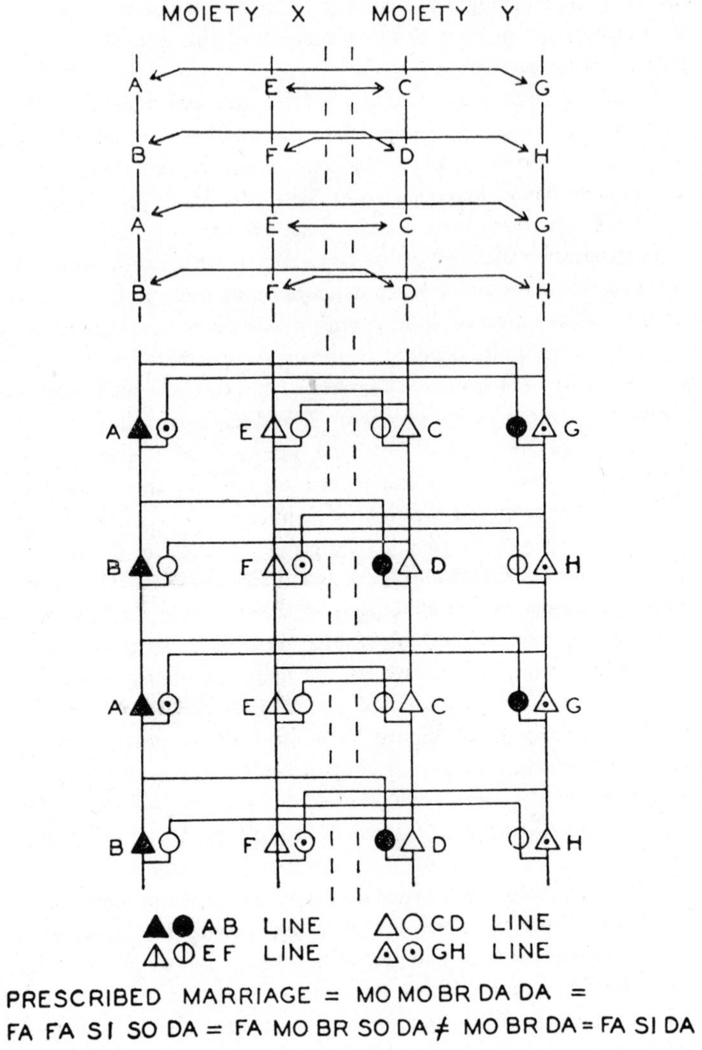

FIG. 2. The Arunta marriage system.

said to be prescribed with mother's brother's daughter's daughter, but it seems to me that the system is more understandable if the prescribed wife is described in another way, although, of course, she may well be this relative. Structurally the prescribed wife is also father's mother's brother's son's daughter, which means that she is a woman of ego's generation from the same patrisib from which ego's father's father obtained a wife. Since marriage is forbidden with mother's brother's daughter or with any woman from mother's patrisib, and prescribed with a woman of the same patrisib as father's mother, this system can be seen to be very similar to that of the Arunta. For any patriline, males of alternate generations obtain wives from —and their sisters go as wives to—the same patrisib. It should be noted in this case —and also that of the Arunta—that there is continuous exchange occurring between the local groups; it is just that any particular patriline must exchange alternately with the two other local groups.

The next systems to be considered are those having indirect exchange. Again the simplest system has two different kinds of local group, one in each of the moieties. But now the males of a particular section give wives to one section of the opposite moiety and receive wives from the other section of the opposite moiety. This system cannot be shown with just two patrilines, as was the Kariera system, but needs four. The marriage relations of a particular patriline would depend on how it is "synchronized" with the patrilines in the opposite moiety. This is shown on Figure 3; a patriline either gives women in every generation to a patriline of the opposite moiety or it receives women in every generation from a patriline in the opposite moiety. For example, the patriline on the left in Moiety X in Figure 3 which has the generations ABAB always gives women to the patriline on the left in Moiety Y and receives women from the patriline on the right in Moiety Y. Since the exchange is always in the same direction, this change in the marriage rule in the presence of only two different kinds of local group results in the development between the patrilines of matrilateral cross-cousin marriage. And in this case the mother's brother's daughter, who is the prescribed mate, is not the same woman as the father's sister's daughter. Marriage systems with matrilateral cross-cousin marriage are thought to exist in Australia among the Murngin and Yir Yiront, but these tribes appear to have four different kinds of local group. It would seem reasonable to suppose that with the adoption of this new marriage rule there would be a rapid change to four different kinds of local group because of the great differences in "social position" between them. With respect to any ego, the groups in the opposite moiety would be divided into "wife-givers" and "wife-takers" and as is usual in this type of situation there would be great differ-

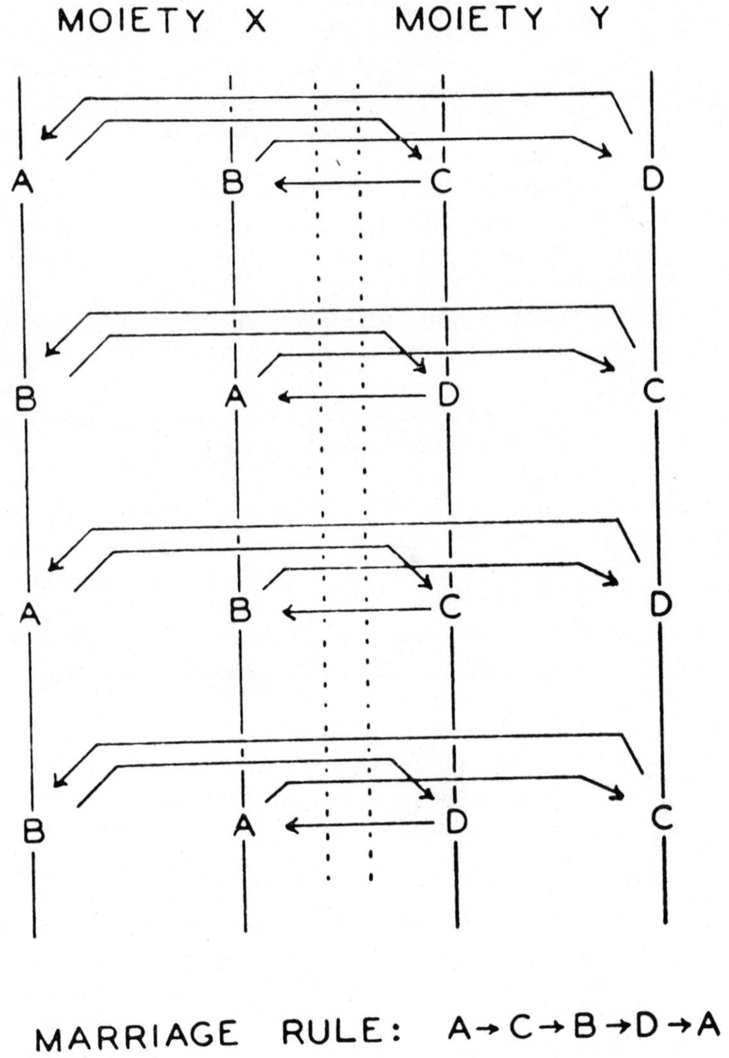


Fig. 3. The development of matrilateral cross-cousin marriage.

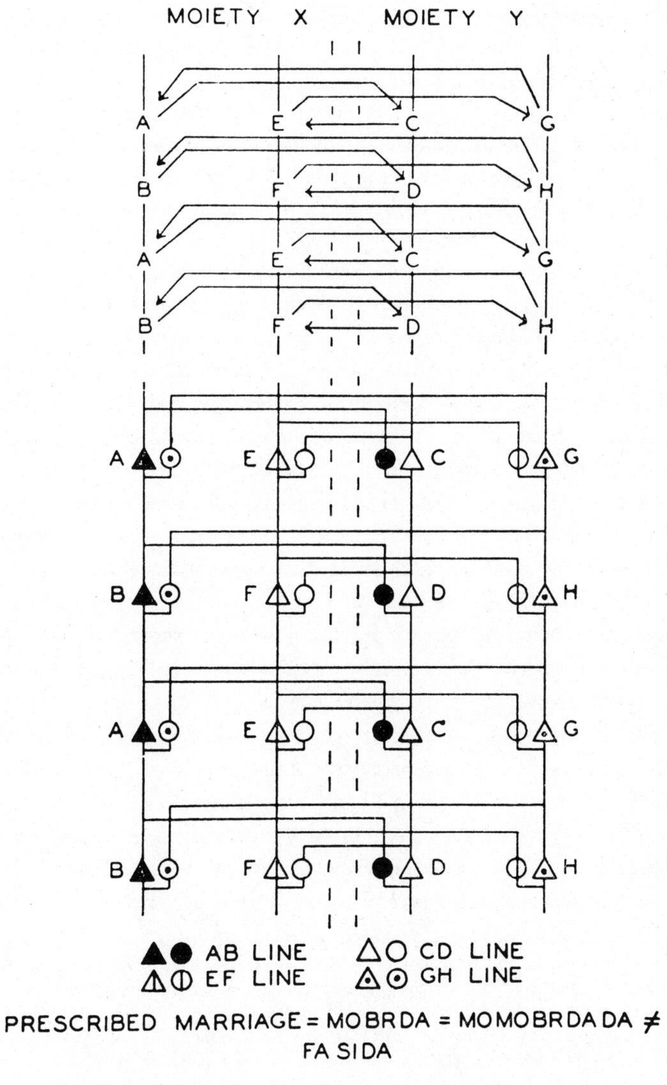


FIG. 4. A prescriptive matrilateral cross-cousin marriage system.

ences in how ego acted toward these groups. Figure 4 shows a matrilateral cross-cousin marriage system with four different kinds of local group. I have not named it after any tribe, although it does seem to be some anthropologists' conception of the Murngin marriage system. But I do not want to get involved in the famous "Murngin controversy" or even be accused of reopening this controversy; so others can believe it or not, as they see fit.[7] Whether or not any Australian tribe actually practices prescriptive matrilateral cross-cousin marriage, I conclude that it is a possibility and the only possibility when there are only two different kinds of local group and indirect exchange.

The final possible type of prescriptive marriage system, patrilateral cross-cousin marriage, can also be derived from a system with direct exchange, but this time it is that of the Arunta. For the previous system it was assumed that there were two kinds of local group which began to practice indirect exchange, and the result was a prescriptive matrilateral cross-cousin marriage system. Now we will consider the result when an Arunta type of system adopts indirect exchange. Indirect exchange would most probably occur in an Arunta system in the following way. Instead of sending their sisters in exchange for their wives, the men of a particular section now send their daughters to the local groups from which they obtained wives, and the son's of these men now obtain wives from the groups to which their father's sisters went as wives. The system resulting from this type of exchange is shown on Figure 5. In this system the prescribed marriage is still with mother's mother's brother's daughter's daughter as it is in the Arunta system, but this woman is now also father's sister's daughter. Thus, this is a system of prescriptive patrilateral cross-cousin marriage. In a recent issue of this journal Needham purported to prove that "a prescriptive marriage system of affinal alliances between lineal descent groups based on exclusive patrilateral cross-cousin marriage cannot exist,"[8] but the system shown on Figure 5 fulfills all the requirements of a prescriptive patrilateral cross-cousin marriage system and hence can very well exist. However, one of the assumptions of Needham's proof was that the "logically necessary

7 The diagram on Figure 4 which shows the individuals is identical with Leach's (1951, p. 33) conception of the Murngin system, except that I have gone a step farther. On Leach's diagram ego's patriline (B), which would be the EF patriline on Figure 4, is shown as having no female relatives, and the women married to the *natielker's* (mother's mother's brother) patriline (CCC), which would be the GH patriline on Figure 4, are shown as not related to any patriline. I have shown the women married to the GH patriline as related to the EF patriline. Since this is a closed system and these women must be related to some patriline, the EF patriline would appear to be the only one to which they could possibly be related.

8 Needham, 1958, p. 211.

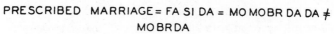

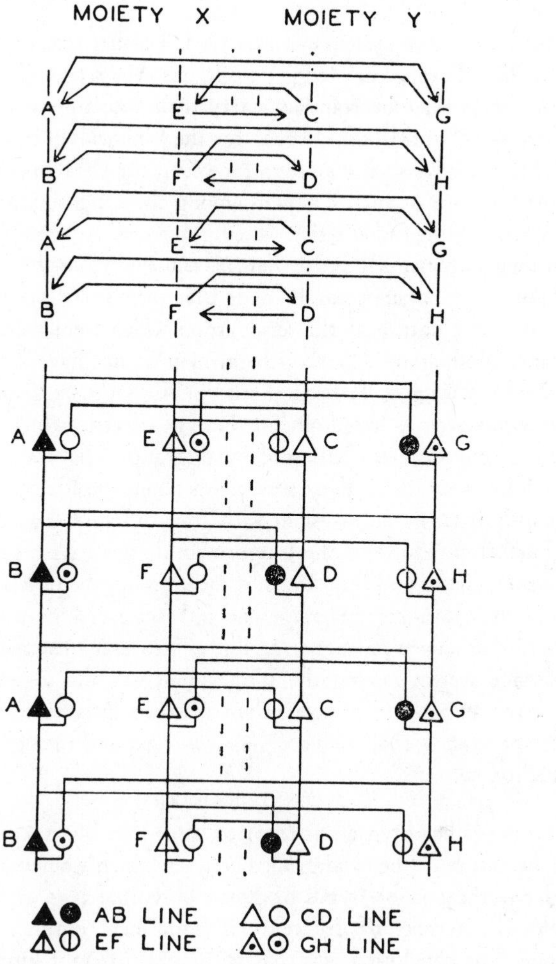


PRESCRIBED MARRIAGE = FA SI DA = MO MOBR DA DA ≠ MOBRDA

FIG. 5. A prescriptive patrilateral cross-cousin marriage system.

number of lines is three,"[9] and the system shown on Figure 5 needs four lines in order to function.

Although such a system can exist, it is another question as to whether or not one ever has. To my knowledge no one has ever reported one, although McConnel[10] indicates that the Kandyu marry their father's sister's daughter. But Radcliffe-Brown's[11] and Elkin's[12] data for the Ungarinyin tribe seem to suggest that this tribe may have such a system. However, their statements do not entirely support the existence of such a system among the Ungarinyin, and they even report that marriage with father's sister's daughter may be prohibited. Nevertheless several of their statements indicate that this is the only possible system the Ungarinyin could have *if* the assumption is made that their social system fulfills the requirements as to the nature of the local group which were made at the beginning of this paper. Both state: (1) the Ungarinyin do *not* have direct exchange; (2) the prescribed marriage is father's mother's brother's son's daughter; and (3) the Ungarinyin system is based on four lines of descent, which seems to indicate that they have four different kinds of local group. The marriage system shown on Figure 5 accords with all these statements about the Ungarinyin system, and more importantly it is the only system with four different kinds of local group which does. Furthermore, even if the Ungarinyin do not have a section system, but only patrilineal moieties and four kinds of local group, the practice of father's mother's brother's son's daughter marriage and indirect exchange would result in a system similar to that shown on Figure 5. The evidence against the existence of this type of marriage system among the Ungarinyin will not be presented, because this paper is not an attempt to prove that this tribe does have such a system but only an attempt to show that such a system can exist and to suggest the Ungarinyin as a possibility.

## SUMMARY

This paper has been an attempt to show that by making certain assumptions about the nature of the local groups between which women are exchanged, one can then derive the possible types of prescriptive marriage system by considering as variables the number of different kinds of local group and the nature of the exchange. The conclusion was that with four different kinds of local group prescriptive marriage systems with direct exchange and both types of indirect exchange, matrilateral and patrilateral cross-cousin marriage can exist among the tribes of Australia.

9 *Idem,* p. 207.
10 McConnel, 1939, p. 72.
11 Radcliffe-Brown, 1931, pp. 337-339.
12 Elkin, 1932, pp. 312-317; Elkin, 1954, pp. 74-75.

BIBLIOGRAPHY

ELKIN, A. P.
1932 *Social Organization in the Kimberly Division, Northwestern Australia* (Oceania, vol. 2, pp. 296-333).
1954 *The Australian Aborigines* (Angus and Robertson: London).

LANE, R., AND B. LANE
1958 *The Evolution of Ambryn Kinship* (Southwestern Journal of Anthropology, vol. 14, pp. 107-135).

LEACH, E. R.
1951 *The Structural Implications of Matrilateral Cross-Cousin Marriage* (Journal, Royal Anthropological Institute, vol. 81, pp. 23-55).

LÉVI-STRAUSS, C.
1949 *Les structures élémentaires de la parenté* (Presses Universitaires de France: Paris).

MCCONNEL, U. H.
1939 *The Social Organization of the Tribes of the Cape York Peninsula* (Oceania, vol. 10, pp. 54-72).

NEEDHAM, R.
1958 *The Formal Analysis of Prescriptive Patrilateral Cross-Cousin Marriage* (Southwestern Journal of Anthropology, vol. 14, pp. 199-219).

RADCLIFFE-BROWN, A. R.
1931 *The Social Organization of the Australian Tribes* (Oceania, vol. 1, pp. 34-63, 206-246, 322-341, 426-456).
1951 *Murngin Social Organization* (American Anthropologist, vol. 53, pp. 37-55).

ROMNEY, A. K., AND P. J. EPLING
1958 *A Simplified Model of Kariera Kinship* (American Anthropologist, vol. 60, pp. 59-74).

SPENCER, B., AND F. J. GILLEN
1927 *The Arunta* (Macmillan and Co.: London).

UNIVERSITY OF MICHIGAN MEDICAL SCHOOL
ANN ARBOR, MICHIGAN

# 13

*By permission of Mouton (Paris-La Haye) and Ecole Pratique des Hautes Etudes*

# A Mathematical Model of the Structure of Kinship

PHILIPPE COURRÈGE

*Translated expressly for this Benchmark volume by Dwight Read, University of California, Los Angeles, from "Un Modèle mathématique des structures élémentaires de parenté,"* L'Homme, **5**, *No. 3–4, 248–290 (1965)*

## INTRODUCTION

The purpose of the present work is to present, using the axiomatic method, a simple mathematical model permitting the study of the functioning of the kinship system in a society where the population is divided into disjoint matrimonial classes. By the kinship system is meant the rules of marriage and descent which are expressed in terms of the classes.

The mathematical model presented here was constructed in order to represent a certain number of the kinship structures studied in the book by C. Lévi-Strauss, Les structures élémentaires de la parenté[1]. In particular, the systems of the Kariera, the Aranda, and the Murngin will be presented and applications will primarily be made to these systems.

Consider the following two examples of application. The first is relevant to the well known ethnographic fact that the Kariera system seems to be a system composed of "two matrilineal moities" (op cit. p. 203). This fact is also expressed in the mathematical model (cf. 2.9): the elementary kinship structure (cf. 1.1) associated with the Kariera system is isomorphic (2.1) moities (1.4 and 2.9). Such a decomposition into a direct product is then established for the Aranda system (2.11) as a consequence of a more general criterion (2.10).

A second example concerns the functioning of the Murngin system. More precisely, it is shown, in the mathematical model, that the altering of marriages from one generation to the other according to the

---

[1] I wish to express here the great satisfaction I have received— and not just as a mathematician— first from the rigour and cultural richness of this book, and then from the interest of M. Lévi-Strauss in the present work. I also wish to thank Mme. Dejean who has read the manuscript and has suggested several revisions, as well as my friends Bernard and Robert Jaulin and Jacques Roubaud who directed me to this work and have aided me in its development.

normal and optional types (cf. op cit. p. 224) is a logical consequence of the fact that the system under consideration permits marriage with matrilateral cross-cousins and not with patrilateral cross-cousins (see § 3).

Another possible application would consist in classifying, by means of the corresponding mathematical models, all conceivable kinship systems with respect to a society with n matrimonial classes. This task has already been undertaken by H. White in his book: An anatomy of kinship. We will only indicate (cf. comments in § 1) in what way White's work can be fit into the form proposed here.

## METHODOLOGY

In conformation with the axiomatic method, the definition and derivation of the mathematical model are carefully and systematically distinguished from its ethnographic significance. Thus the present work is on two levels: the axiomatic development of the mathematical theory of Les structures élémentaire de la parenté is accompanied by a commentary that attempts to establish a correspondence, as much as possible bijective, between, on the one hand the mathematical objects and the equations that link them and on the other, the concepts proposed by the ethnographer which give the account of reality.

The mathematical text directs the development of the work. It is divided into paragraphs, themselves subdivided into numbers with two indices. The numbers for the ethnographic part of the text correspond to those for the mathematical portion and are introduced by the letter E and noted. The reader is asked, for each number, to first read the mathematical text in order to comprehend the abstract concepts, independent of any ethnographic interpretation, and then only to examine the ethnographic interpretation.

This method, which clearly separates the mathematical model from the ethnographic descriptions, is done in order to satisfy both the requirements of rigor and clarity for a mathematical theory (requirements which are its whole purpose, in fact), since the mathematical model is in no way intermingled with the confusion surrounding the ethnological concepts, and the requirements of flexibility and nuances for the ethnographic ideas. Since the mathematical model has its own existence and is not linked to the ethnographic ideas except by those that are locally significant, it cannot impose abusive rigidity or gross schematization.

Even though this work develops by the method described above, it is addressed essentially to a reader possessing the minimum mathematical knowledge necessary to understand the development of a mathematical text, at least in its mail points. The minimal level of mathematics needed is completely elementary and consists only of several simple concepts from set theory. These are given in § 0. Some of the passages require additional knowledge about permutation groups. These passages are bracketed by two asterisks * * and can be skipped without interrupting comprehension of the text as a whole.

The ethnographic terminology is that employed by C. Lévi-Strauss in his book Les structures élémentaires de la parenté.

## § 0. BASIC CONCEPTS AND MATHEMATICAL NOTATION

### 0.1 Applications

If E and F designate sets, and u is a mapping from E into F (we say that u is a function defined in E with values in F), we will denote by $u(x)$ the image in F of an element x in E by the mapping u.

If A is a subset of E, $A \subset E$, we will denote by $u[A]$ the image of A under the mapping u: $u[A]$ is the set $\{u(x) | x \in A\}$ of all elements in F of the form $u(x)$ where $x \in A$. We will denote by $u_A$ the mapping of A into $u[A]$ obtained by taking the restriction of u to A. Thus $u_A$ is the mapping of A into $u[A]$ defined by $u_A(x) = u(x)$ for all $x \in A$.

If B is a subset of F, $B \subset F$, we will denote by $u^{-1}[B]$ the inverse image of B by the mapping u:

$$u^{-1}[B] = \{x \in E | u(x) \in B\}.$$

If $B = \{y\}$, the set $u^{-1}[B]$ will be denoted simply by $u^{-1}(y)$. The mapping u of E into F is said to be bijective if $s \neq y \Rightarrow u(x) \neq u(y)$.

The mapping u will be said to be surjective if $u[E] = F$. If u is surjective, then u is said to map E onto F.

If G is a third set, and v is a mapping of F into G, we will denote by vu the composite mapping of E into F defined by $vu(x) = v(u(x))$ for all $x \epsilon E$.

We will designate by $\varepsilon_E$ the identify mapping for the set E: $\varepsilon_E$ is the mapping of E onto E defined by $\varepsilon_E(x) = x$ for all $x \epsilon E$.

If u is a bijective mapping of E onto F, the mapping $u^{-1}$ from F onto E defined by $uu^{-1} = \varepsilon_F$ and $u^{-1}u \neq \varepsilon_E$ will be called the inverse of u.

## 0.2 Permutations

If E designates a set, a permutation of E is any bijective mapping of E onto itself. Every permutation of E has an inverse which in turn is a permutation of E. The composite map $\beta\alpha$ of two permutations $\alpha$ and $\beta$ is again a permutation of E. We will denote by $\alpha^2, \alpha^3, \ldots$ the permutations $\alpha\alpha, \alpha(\alpha\alpha), \ldots$ *The set $G_E$ of all permutations of E with the composition rule $\alpha, \beta \rightarrow \alpha\beta$ is a group. For this reason $\beta\alpha$ is sometimes called the product of $\alpha$ by $\beta$.*

If A is a subset of E and $\alpha$ is a permutation of E such that $\alpha[A] = A$, the restriction $\alpha_A$ of $\alpha$ to A [cf. 0.1] is a permutation of A. If E is a finite set, in order that a mapping $\alpha$ of E into E be bijective, it is necessary and sufficient that $\alpha$ maps E onto E. $\alpha$ is then a permutation of E. ["principe des tiroirs de Dirichlet"] If E is a finite set for which the elements are explicitly given, for example $E = \{m,n,p,q\}$ (set of 4 distinct letters m,n,p.q), a permutation $\alpha$ of E can be explicitly given in the following fashion:

$$\begin{pmatrix} m & n & p & q \\ \alpha(m) & \alpha(n) & \alpha(p) & \alpha(q) \end{pmatrix}.$$

We may thus write

$$\alpha = \begin{pmatrix} m & n & p & q \\ \alpha(m) & \alpha(n) & \alpha(p) & \alpha(q) \end{pmatrix}.$$

0.3 Direct product of sets. Direct produce of permutations.

If E and F designate two sets, we designate by $E \times F$ the set of ordered pairs $(x,y)$, where $x \in E$ and $y \in F$. $E \times F$ is called the direct product of E and F.

If $\alpha$ and $\beta$ designate respectively permutations of E and F, we designate by $\alpha \times \beta$ the permutation of $E \times F$ defined by $\alpha \times \beta\ (x,y) = (\alpha(x), \beta(y))$ for all $x \in E$, $y \in F$. $\alpha \times \beta$ is called the direct product of $\alpha$ and $\beta$.

One can verify the following equations:

(a) $$(\alpha \times \beta)^{-1} = \alpha^{-1} \times \beta^{-1}$$

(b) $$(\alpha \times \beta)\ (\alpha' \times \beta') = (\alpha\alpha' \times \beta\beta')$$

The generalization to the direct product of more than two sets or permutations is straightforward.

0.4 Partition of a set. Permutations compatible with partitions.

If E designates a set, by a partition of E is meant a set P of subsets of E such that:

($P_1$) $$X \in P \text{ and } Y \in P \Rightarrow X = Y \text{ or } X \cap Y = \emptyset.$$

(That is, the elements of P are subsets of E that are pair-wise disjoint.)

($P_2$) $$E = \bigcup_{X \in P} X$$

(That is, every element of E is in some set in P.)

If R is an equivalence relation[1] over E, the set of equivalence classes modulo R is a partition of E called the partition associated with R (or quotient set of E by R).

---

[1] That is, a binary relation which is reflexive, symmetric, and transitive.

We say that the permutation $\alpha$ of E is compatible with a partition P of E if $X \in P \Rightarrow \alpha[X] \in P$. If E is a finite set, one may define a permutation $\bar{\alpha}$ on P by setting $\bar{\alpha}(X) = \alpha[X]$ for all $X \in P$.

### 0.5 *Intransitive classes associated with a permutation

Let E be a finite set and $\alpha$ a permutation of E. The relation R defined by

$$x \sim y \text{ if there is an integer } n \geq 0 \text{ such that } \alpha^n(x) = y$$

is an equivalence relation on E. The equivalence classes defined by this relation are called intransitive classes of $\alpha$.

If X is an intransitive class of $\alpha$ and if x is an arbitrary element of X, then X is identical to the set of all $\alpha^n(x)$, such that n is an integer $\geq 0$.

We say that a permutation $\alpha$ of E is transitive if E itself is the only intransitive class.*

## § 1. ELEMENTARY KINSHIP STRUCTURES

In this first paragraph will be defined the elementary kinship structures and diverse corresponding auxiliary notation (group of kinship terms, regular and irreducible structures, marriage conditions between cross cousins, etc.). After this is done, several important examples for the remainder of the work (nos. 1.4 and 1.9) will be studied.

### 1.1 Definition

By elementary kinship structure of a finite set S will be meant any ordered triplet $(\omega,\mu,\pi)$ of permutations of S satisfying the following axiom:

$$\text{(D)} \quad \pi = \mu\omega$$

$\omega,\mu$, and $\pi$ are called respectively the conjugal function, the maternal function, and the paternal function of the elementary kinship structure $(\omega,\mu,\pi)$.

E 1.1

Preliminary remark on terminology.

The terminology employed in the mathematical model has been chosen in such a way that the notation for the mathematical concepts that have been introduced recalls the ethnological significance of these concepts, so as to facilitate intuitive understanding of the mathematical text. However, it is important to note at this point that, conforming to current mathematical practice, the ethnological terms are simply names and do not modify in any sense the mathematical concepts, which are completely defined by the axioms and definitions.

For example, mathematically, the conjugal function of the elementary kinship structure $(\omega,\mu,\pi)$ does not have, by the term "conjugal", any other sense than that of a mapping of S into S linked to $\mu$ and $\pi$ by the equation $\pi = \mu\omega$.

Or again, when we describe the structure of two exogamous matrilateral moities, the elementary structure of kinship introduced in 1.4 a), we will be defining a precise mathematical concept, but have as interpretation, without affecting the mathematical concept, the ethnographic sense of these terms. Similar remarks apply to the terms "Kariera structure", "Aranda structure", etc.

E 1.1 a

As we have indicated in the introduction, we are interested here in a society of classes. More precisely, a society where the population has been broken up into subpopulations that are pair-wise disjoint that we call matrimonial classes (or more simply, classes) and the only rules for the functioning of the kinship system of this society are in terms of these classes.

These classes are not necessarily localized in space. There are Duponts in Paris and in Bordeaux, etc. (cf. 1, pp. 208-210). We will be able to express this in the mathematical model by means of the concept of the direct product of kinship structures (cf. 2.9).

The format of a society divided into classes is also that utilized by White (cf. 2). Nevertheless, our exposition differs in an essential fashion from his by the fact we define axiomatically the mathematical model, whereas White tries to define the society in an axiomatic fashion (cf. 2, p. 9). Each of the axioms of White (cf. 2, p. 9) can be translated into a property of the mathematical model and some of these properties will be noted in the text.

E 1.1 b

Let us now give ethnological significance to the mathematical terms $S, \omega, \mu$, and $\pi$. First of all, the set S represents the set of all classes of the society under consideration.

Contrary to the identification made by White (cf. [2], p. 9, 1) it is important to not consider S as the set of all individuals. We will see numerous reasons for this in the following.

The conjugal function $\omega$ represents a positive rule that prescribes to a man from class x ($x \in S$) to choose his wife from the class $\omega(x)$ (cf. [2], p. 9, axiom 2, and p. 10; $\omega$ corresponds to the function W of White).

The maternal function $\mu$ represents a positive rule which prescribes that all infants of a woman from class x ($x \in S$) belong to the class $\mu(x)$.

The paternal function $\pi$ similarly represents a positive rule which prescribes that all offspring of a man of class x ($x \in S$) belong to the class $\pi(x)$ (cf. [2], p. 9, axiom 4, and p. 10; $\pi$ corresponds to the function C of White).

Now consider in detail the properties attributed to $\omega, \mu$, and $\pi$ by the axioms.

E 1.1 c

The permutation $\omega$ determines a marriage rule in the society under consideration.

This marriage rule is expressed precisely by the function $\omega$: the class from which a man must choose his wife is completely determined. This translates what is meant by the elementary character (caractère élémentaire) (cf. [1], preface) of the society under consideration. Thus the name "elementary kinship structure" is given to the mathematical model. Note carefully that the elementary character only concerns the functioning of the kinship system in terms of matrimonial classes, and not in terms of actual kinship relations between individuals. Such relations are not formalized in the proposed model. With the Murngin system (cf. 1P 3) there appears a certain amount of freedom (even in terms of classes) in the choice of a spouse: we will try to express the complex nature of the Murngin system in terms of the mathematical model of elementary structures of kinship (cf. no. 3.4).

E 1.1 d

Likewise, the permutations $\mu$ and $\pi$ express a descent rule for the society in question. Axiom (D) implies that $\mu$ and $\pi$ determine the same descent rule, and that this rule is compatible with the marriage rule (or that only licit marriages are considered as being able to be fertile!).

More precisely, the offspring of a man of class x who is married to a woman from class $y = \omega(x)$ will be of class $\pi(x) = \mu(\omega(x))$. In particular, brothers and sisters will always be of the same class.

E 1.1 e

The bijective character (cf. 0.1) of $\omega$ expresses the fact that the rules of marriage require that men belonging to different classes will choose their spouses from different classes (cf. [2], p. 9, axiom 3 of White).

The bijective character of $\pi$ expresses the fact that the offspring of fathers of different classes themselves belong to different classes (cf. [2], p. 9, axiom 5). The same holds true for $\mu$.

Finally, the surjective character of $\omega, \mu$, and $\pi$ expresses the fact that every class of women provides spouses, and offspring can be placed in every class. All classes participate in the functioning of the kinship system, from the viewpoint of marriage and descent.

1.2 *Different methods for defining an elementary kinship structure.

The relation (D): $\pi = \mu\omega$, is equivalent to each of the two following conditions: $\mu = \pi\omega^{-1}$ and $\omega = \mu^{-1}\pi$. Consequently, the elementary kinship structure $(\omega,\mu,\pi)$ over S is well defined if any two of the three permutations $\omega,\mu$, and $\pi$ are specified, and the two that are specified is arbitrary.

For example, if two permutations $\alpha$ and $\beta$ of the set S are given, there is a unique elementary kinship structure $(\omega,\mu,\pi)$ such that $\omega = \alpha$ and $\mu = \beta$; it suffices to set $\pi = \beta\alpha$ for the triplet $(\omega,\mu,\pi)$ to satisfy the axioms. For convenience, we will designate by $(\alpha,\beta)$ the elementary kinship structure defined by setting $\omega = \alpha$ and $\mu = \beta$.

This notation will not cause any inconvenience as long as the context indicates which of the three functions, conjugal, maternal, or paternal, is designatec by $\alpha$ and $\beta$, respectively.

E 1.2

No. 1.2 is a simple mathematical application of axiom (D) to make clear the fact that there are several equivalent ways to define a kinship system. It suffices to give the conjugal function and the maternal function, or the conjugal function and the paternal function, etc.*

1.3 *Group of kinship terms of an elementary kinship structure.

By group of kinship terms for the structure $(\omega,\mu,\pi)$ over the set S is meant the subgroup of $G_S$ [cf. 0.2] generated by the three permutations $\omega,\mu$, and $\pi$. This group will be denoted by $G(\omega,\mu,\pi)$, [or simply by G when no confusion can arrise]. G is the subset of $G_S$ generated by the products $\alpha_1,\alpha_2,\ldots,\alpha_n$, where the $\alpha_1$ are equal to $\omega,\mu,\pi$ or to their inverses.*

E 1.3 b

If $x \in S$, $\mu\pi^{-1}(x) = \mu(\pi^{-1}(x))$ is the class of offspring of the sister of the father of an individual in class x. In particular, $\mu\pi^{-1}(x)$ is the class of patrilateral cross-cousins of a man from class x. We express this by saying that the element $\mu\pi^{-1}$ of G represents the kinship term "patrilateral first cross-cousin".

E 1.3 c

In the same way, $\pi^{-1}\mu(x)$ is the class of offspring of the mother's brother of an individual from class x. We express this by saying that the element $\pi\mu^{-1}$ from G represents the kinship term "matrilateral first cross-cousin".

E 1.3 d

More generally, every element of G has an interpretation in terms of kinship. For example, a list of the 16 kinship terms corresponding to second degree cousins can be found in [2], p. 23.

Note, however, that there is no bijective correspondence between the set of various kinship relations which can link two individuals, and the set G. The same element of G can represent (in the language of classes, not of individuals, remember) different kinship relations. For example, matrilateral and patrilateral parallel first cousins are represented respectively by the elements $\mu\mu^{-1}$ and $\pi\pi^{-1}$ of G and each of these in turn is equal to the same element $\varepsilon_s$. As another example, consider a structure $(\omega,\mu,\pi)$ satisfying the relation $\omega = \pi\mu^{-1}$, that is $\omega\mu = \mu\omega$(cf. no. 1.8). The same element of G represents the kinship term "spouse", and "matrilateral cross-cousin". This only implies that, in the structure under question, the wife of a man and his matrilateral cross-cousin are always in the same class (cf. E 1.8).

E 1.3 e

A study of the structure of the group G can lead to interesting results for the ethnographers. In particular, a classification of the various groups G that are possible yields a classification of the elementary kinship structures for a given set S. Such a classification is studied in [2], p. 26. We will not embark upon this study here (cf. however, the question of isomorphisms of structures, in no. 2.1).*

1.4 Examples: Structures with two exogamous moities.

a) Structure with two exogamous matrilineal moities:

$$S = \{1,2\} \quad \omega = \begin{pmatrix}1,2\\2,1\end{pmatrix} \quad \mu = \varepsilon_s.$$

This structure is well defined by the definition of $\omega$ and $\mu$. The following may be deduced:

$$\pi = \mu\omega = \begin{pmatrix}1,2\\2,1\end{pmatrix} = \omega \text{ and } G(\omega,\mu,\pi) = \{\varepsilon_S,\omega\}.$$

b) Structure with two exogamous patrilineal moities:

$$S = \{1,2\} \quad \omega = \begin{pmatrix}1,2\\2,1\end{pmatrix} \quad \pi = \varepsilon_S.$$

This structure is well defined by the definition of $\omega$ and $\pi$. One may deduce that

$$\mu = \pi\omega^{-1} = \begin{pmatrix}1,2\\2,1\end{pmatrix}.$$

E 1.4

Immediate interpretation: the exogamous character is expressed by the property of $\omega$ exchanging 1 and 2 ($\omega[1] = 2$, $\omega[2] = 1$). $\mu = \varepsilon_S$ (resp. $\pi = \varepsilon_S$) expresses the fact that the offspring are of the same class as the mother (resp. the father), whence the terminology matrilineal (resp. patrilineal).*

1.5 Irreducible structures, reducible structures, induced structures.

An elementary kinship structure $(\omega,\mu,\pi)$ over the set S is said to be irreducible if the only subsets E of S such that

$$\omega[E] = E,\ \mu[E] = E, \text{ and } \pi[E] = E$$

are $\emptyset$ and S.

For example, the structures given in no. 1.4 are irreducible. All structures that are not irreducible are said to be reducible.

If E is a subset of S such that $\omega[E] = E$, $\mu[E] = E$, and $\pi[E] = E$, then $(\omega_E,\mu_E,\pi_E)$ is an elementary kinship structure over E called the structure induced over E by $(\omega,\mu,\pi)$. In the following, we will mainly study irreducible structures. The next two lemmas indicate that the reducible structures can be expressed in terms of irreducible structures (these lemmas will not be used in the following and can be skipped)[1].

---

[1] The proofs of these lemmas are classic and will be left to the reader.

*Lemma 1.1 In order that a kinship structure $(\omega,\mu,\pi)$ over S be irreducible, it is necessary and sufficient that the group $G(\omega,\mu,\pi)$ be transitive in S [that is, for every pair x, y of elements in S, there is an $\alpha\in G$ such that $\alpha(x) = y$].

Lemma 1.2 If $(\omega,\mu,\pi)$ is a kinship structure over S, then the partition P of S defined by the equivalence relation "$x \sim y \iff$ there is an $\alpha\in G(\omega,\mu,\pi)$ such that $\alpha(x) = y$" has the following properties:

a) For all $A\in P$, $\omega[A] = A$, $\mu[A] = A$, and $\pi[A] = A$;

b) For all $A\in P$, $(\omega_A,\mu_A,\pi_A)$ [cf. 0.1] is an irreducible elementary kinship structure over A.*

E 1.5

It is clear that a reducible structure represents a society divided into several subpopulations which have no relations among them other than kinship. The property "for every pair x,y of elements of S there is an $\alpha\in G$ such that $\alpha(x) = y$" is an exact translation of White's Axiom 7 (cf. (2), p. 9). This axiom is therefore equivalent, by virtue of lemma 1, to irreducitiliby.*

## 1.6 *Regular structures.

We will say that the elementary kinship structure $(\omega,\mu,\pi)$ over the set S is regular if $\varepsilon_S$ is the only element in the group $G(\omega,\mu,\pi)$ of kinship terms (cf. 1.3) which leaves invariant at least one element of S [that is, for all $\alpha\in G$ such that $\alpha \neq \varepsilon_S$, $\alpha(x) \neq x$ for all $x\in S$].

*Lemma 1.3 If the structure $(\omega,\mu,\pi)$ over the set S is regular, then the number of elements in the group of kinship terms is less than or equal to n, the number of elements of S.

Proof: Suppose that G has n + 1 distinct elements, say $\alpha_1,\alpha_2,\ldots \alpha_{n+1}$, and $x\in S$. Since S has n elements there are two indices i,j such that $i \neq j$ and $\alpha_i(x) = \alpha_j(x)$. Then $\alpha_j^{-1}\alpha_i(x) = x$, hence $\alpha_j^{-1}\alpha_i = \varepsilon_S$, by the hypothesis of regularity. Therefore $\alpha_i = \alpha_j$, which contradicts the assumption that each of $\alpha_1,\alpha_2,\ldots,\alpha_{n+1}$ are distinct. QED*

E 1.6

The property of regularity expresses, in the mathematical model, the fact that no class in the society is privileged with respect to the other classes in terms of kinship relations. Regularity corresponds exactly to White's Axiom 8 ("Whether two people who are related by marriage or descent links are in the same class depends only on the kind of relationship, not on the class either one belongs to". This axiom is quoted as it is the only one of the eight axioms of White for which the interpretation may be ambiguous.)

It seems reasonable to think that for reasons of symmetry, or rather perhaps for reasons of equality and avoidance of jealousies, natural events will generally lead to kinship systems corresponding to regular structures. However, this does not seem to be a sufficient reason to discard consideration (as White has done) of nonregular structures. We will see that certain nonregular structures shall appear in the decomposition into a direct product of regular structures (for instance, the Aranda system, cf. no. 2.11, below).

Note that we have in the mathematical theory the following result: if a structure $(\omega, \mu, \pi)$ is regular and if $\omega \neq \varepsilon_s$, then $\omega(x) \neq x$ for all $x \in S$. We can deduce from this that for a regular structure, Axiom 6 of White ("A man can never marry a woman of his own class") is equivalent to $\omega = \varepsilon_s$.*

### 1.7 Restricted exchange; generalized exchange.

An elementary kinship structure $(\omega, \mu, \pi)$ over S is said to have restricted exchange if it satisfies the following condition:

$$\text{(RE)} \quad \omega^2 = \varepsilon_s \quad \text{[condition for restricted exchange]}.$$

It is said to have generalized exchange in the contrary case.

If the structure $(\omega, \mu, \pi)$ is regular (1.6), then for there to be restricted exchange, it is necessary and sufficient that there exists an $x \in S$ such that $x = \omega(\omega(x))$.

The structures with two moities presented in 1.4 have restricted exchange. We will see in 1.9 an example of generalized exchange.

E 1.7

The condition of restricted exchange, $\omega^2 = \varepsilon$, expresses the reciprocity of the marriage rule: a man from class x must take a wife from class $\omega(x)$, and reciprocally, a man from class $\omega(x)$ must take a wife from the class $x = \omega(\omega(x))$.

### 1.8 Conditions for marriage between cross-cousins

Let $(\omega,\mu,\pi)$ be an elementary kinship structure over the set S. We will call the condition

$$\text{(MCC)} \quad \omega\mu = \mu\omega$$

the condition of marriageability with the matrilateral cross-cousin. The condition

$$\text{(PCC)} \quad \omega\mu = \mu\omega^{-1}$$

is called the condition of marriageability with the patrilateral cross-cousin.

When the structure $(\omega,\mu,\pi)$ satisfies condition (MCC) (resp. (PCC)) we say that it permits marriage with the matrilateral cross-cousin (resp. patrilateral)).

When the structure $(\omega,\mu,\pi)$ satisfies both (MCC) and (PCC) we say that it permits marriage with the bilateral cross-cousins.

We have, then, the following properties.

Lemma 1.4 If a structure permits marriage with the matrilateral cross-cousin, then in order that it also permit marriage with the patrilateral cross-cousin, it is necessary and sufficient that it has restricted exchange.

Proof: The lemma expresses the following implication:

$$\omega\mu = \mu\omega \Rightarrow \omega\mu = \mu\omega^{-1} \Leftrightarrow \omega^2 = \varepsilon_S,$$

whose verification is immediate.

*Lemma 1.5 In order that a structure $(\omega,\mu,\pi)$ permits marriage with the matrilateral cross-cousin, it is necessary and sufficient that the group $G(\omega,\mu,\pi)$ [3] be commutative.

Proof: The group $G = G(\omega,\mu,\pi)$ is generated by $\omega$ and $\mu$ (by axiom D) and the condition (MCC) simply expresses the fact that $\omega$ and $\mu$ commute.

Lemma 1.6 All regular structures over a set S with four elements permit marriage with the matrilateral cross-cousin.

In particular, there is no regular structure over S permitting marriage with the patrilateral cross-cousin, but not with the matrilateral cross-cousin.

Proof:[1] Let $(\omega,\mu,\pi)$ be a regular structure over S. By lemma 1.3 (no. 1.6) the group $G(\omega,\mu,\pi)$ has, at most four elements, hence is commutative. The remainder of the proof is immediate.*

E 1.8

The condition (MCC) can also be written $\omega = \mu\omega\mu^{-1}$. That is, since $\pi = \mu\omega$, $\omega = \pi\mu^{-1}$, so $\omega(x) = \pi\mu^{-1}(x)$ for all $x\in S$. In other words, the condition $\omega\mu = \mu\omega$ expresses in the mathematical model the fact that the marriage and descent rules permit a man to choose as wife his matrilateral cross-cousin. Note in particular that in such a society marriage with the matrilateral cross-cousin is permitted, but not prescribed (since the class to which first cross-cousins belong can contain other women than these).

Thus, the condition $\omega\mu = \mu\omega$ is compatible with, on the one hand, the rules of marriage and descent formulated in terms of classes; and, on the other hand, a stronger condition (marriage with cross-cousins) formulated in terms of real parents.

The same remarks may be made for the condition (PCC) by replacing matrilateral with patrilateral.

### 1.9 *Examples of elementary kinship structures over a set S with four elements.

[1] Proof due to J. Roubaud.

Let $S = \{p,q,r,s\}$ be a set with four elements.

A) Let

$$\omega_1 = \begin{pmatrix} p & q & r & s \\ r & s & p & q \end{pmatrix}, \mu_1 = \begin{pmatrix} p & q & r & s \\ q & p & s & r \end{pmatrix}, \text{ and } \pi_1 = \mu_1\omega_1 = \begin{pmatrix} p & q & r & s \\ s & r & q & p \end{pmatrix}.$$

It may immediately be verified that $\omega_1^2 = \varepsilon_S$ and $\omega_1\mu_1 = \mu_1\omega_1$ (see fig. 1* in E 1.9). The structure $(\omega_1,\mu_1,\pi_1)$ has restricted exchange and permits marriage with the bilateral cross-cousin [no. 1.8]. It is regular. *The group $G(\omega_1,\mu_1,\pi_1)$ of kinship terms has four elements— $\varepsilon_S,\omega_1$, $\mu_1$, and $\pi_1$— and is the direct product of two groups of two elements.*

The structure $(\omega_1,\mu_1,\pi_1)$ will be called the Kariera structure.

B) Let $\omega_2 = \begin{pmatrix} p & q & r & s \\ q & r & s & p \end{pmatrix}$. The structures $(\omega_2,\omega_2,\omega_2^2)$, $(\omega_2,\omega_2^{-1},\varepsilon_S)$, and $(\omega_2^{-1},\omega_2,\varepsilon_S)$ have generalized exchange. They permit marriage with the matrilateral cross-cousin, but not with the patrilateral cross-cousin [lemma 1.4, no. 1.8]. They are regular and their group of kinship terms is identical to the cyclic group with four terms $\{\varepsilon_S,\omega_2,\omega_2^2,\omega_2^3 = \omega_2^{-1}\}$ generated by $\omega_2$. (See fig. 2 and 3 in E 1.9.)

The structure $(\omega_2,\omega_2,\omega_2^2)$ will be called the theoretical Murngin structure.

C) Let

$$\omega_3 = \omega_2 = \begin{pmatrix} p & q & r & s \\ q & r & s & p \end{pmatrix}, \mu_3 = \begin{pmatrix} p & q & r & s \\ s & r & p & q \end{pmatrix}, \text{ and } \pi_3 = \mu_3\omega_3 = \begin{pmatrix} p & q & r & s \\ r & q & p & s \end{pmatrix},$$

Then $\omega_3\mu_3 = \mu_3\omega_3^{-1}$ and $\omega_3^2 \neq \varepsilon_S$. (See fig. 4 in E 1.9.) The structure $(\omega_3,\mu_3,\pi_3)$ has generalized exchange.

It permits marriage with the patrilateral cross-cousin, but not with the matrilateral cross-cousin.

Contrary to the other examples, this structure is not regular, since $\pi_3(q) = q$, but $\pi_3 \neq \varepsilon_S$ (cf. lemma 1.6).

D) Let

$$\omega_4 = \begin{pmatrix} p & q & r & s \\ p & s & r & q \end{pmatrix}, \mu_4 = \omega_3 = \begin{pmatrix} p & q & r & s \\ q & r & s & p \end{pmatrix}, \text{ and } \pi_4 = \mu_4\omega_4 = \begin{pmatrix} p & q & r & s \\ q & p & r & s \end{pmatrix}.$$

* All figures appear at the end of the text. They are reproduced from the original article with permission of the author and publisher.

Then $\mu_4\omega_4 \neq \omega_4\mu_4$ and $\mu_4\omega_4^{-1} \neq \omega_4\mu_4$, $\omega_4^2 = \varepsilon$. (See fig. 5 in E 1.9.)

The structure $(\omega_4, \mu_4, \pi_4)$ has restricted exchange. It permits marriage neither with the matrilateral cross-cousin nor with the patrilateral cross-cousin. This structure is not regular, since $\omega_4(p) = p$, but $\omega_4 \neq \varepsilon_s$ (see no. 2.11).

E 1.9

The structure $(\omega_1, \mu_1, \pi_1)$ corresponds to the Kariera system (cf. [1], pp. 202, f). We will show below (cf. no. 2.9) that it can be obtained as a direct product.

The structure $(\omega_2, \omega_2, \omega_2^2)$ corresponds to the theoretical Murngin system (cf. [1], pp. 229-232, and pp. 238-242).

The structures $(\omega_2, \omega_2^{-1}, \varepsilon_s)$ and $(\omega_2^{-1}, \omega_2, \varepsilon_s)$ appear in the analysis of the Mara system (cf. [1], p. 248).

Notice that, by virtue of lemma 1.4, for a regular structure over a set S with four elements to permit marriage with the matrilateral cross-cousin, but not with the patrilateral cross-cousin, it is necessary and sufficient that it has generalized exchange.[1]

In systems with four classes, we see that generalized exchange appears as a necessary condition for the dichotomy between matrilateral and patrilateral cross-cousins (cf. [1], p. 229, and p. 271).

Lemma 1.6 allows us to predict that preferential marriage with the matrilateral cross-cousin is going to play a more important role than marriage with the patrilateral cross-cousin, since, according to this lemma, only the former is compatible with regular structures.

The structure $(\omega_3, \omega_3, \omega_3)$ is given as an example of a structure verifying the condition (PCC), but not the condition (MCC).

Finally, the structure $(\omega_4, \omega_4, \omega_4)$ is an example of a structure which does not satisfy either of the conditions (MCC) or (PCC). It will appear in the decomposition of the Aranda structure into a direct product (cf. no. a 2.11).

## 1.10 Pairs, couples, and cycles.

[1] See also E 3.1 b.

Let S be a finite set and let $(\omega,\mu,\pi)$ be an elementary kinship structure over S.

We call, respectively, pair, cycle, and couple, the intransitive classes (see no. 015 above) of the permutations $\omega,\mu$, and $\pi$. Conforming to the results of no. 0.5, the cycle (for example) of an element $a \in S$ is the set of elements a, $\mu(a)$, $\mu^2(a), \ldots, \mu^{n-1}(a)$, where n is the smallest integer k such that $\mu^k(a) = a$. *(If the structure $(\omega,\mu,\pi)$ is regular, note that n is also the smallest integer k such that $\mu^k = \varepsilon_s$).*

In a structure with restricted exchange, all pairs have at most two elements, and exactly two if the structure is regular (no. 1.7).

Conversely, if an elementary kinship structure is such that all pairs have at most two elements, the structure has restricted exchange.

Examples.

In the Kariera structure (no. 1.9 A) it can be seen that there are two pairs, two couples, and two cycles.

Contrarywise, in the theoretical Murngin structure (no. 1.9 B) there are but one pair, a single cycle, and two couples. The concepts of pair, couple, and cycle will be needed below in Nos. 2.11 and 3.1.

E 1.10

The concepts of pair, cycle, and couple introduced in the mathematical model correspond precisely to those used by C. Lévi-Strauss in [1] (see p. 203). Should two classes a and b belong, for example, to the same cycle, it follows that an arbitrary individual from class b will find himself descending by women from a woman belonging to class a, if he traces far enough back in his genealogy.*

## § 2. *TRANSFORMATIONS ON ELEMENTARY KINSHIP STRUCTURES

The transformations of interest here are expressed by the concepts of kinship morphism and isomorphism, of product structure and quotient structure, and are illustrated by analysis of the Kariera and Aranda systems.

### 2.1 Kinship morphisms; isomorphisms.

Let $S_1$ and $S_2$ be two finite sets and let $(\omega_1,\mu_1,\pi_1)$ and $(\omega_2,\mu_2,\pi_2)$ be elementary kinship structures over $S_1$ and $S_2$ respectively. We say that a function $\theta$ mapping $S_1$ into $S_2$ is a kinship morphism[1] (for the structures $(\omega_1,\mu_1,\pi_1)$ and $(\omega_2,\mu_2,\pi_2)$) if

$$\text{(KM)} \quad \theta\omega_1 = \omega_2\theta, \ \theta\mu_1 = \mu_2\theta, \ \theta\pi_1 = \pi_2\theta.$$

If, in addition, $\theta$ is a bijective mapping of $S_1$ onto $S_2$, we will say that is a isomorphism of $S_1$ onto $S_2$ (for the structures $(\omega_1,\mu_1,\pi_1)$ and $(\omega_2,\mu_2,\pi_2)$, respectively).

By virtue of axiom (D) and the associativity of the composition of functions, in order that $\theta$ be a morphism it suffices that two of the three conditions in (KM) be satisfied. For example, if the first two conditions are satisfied, we have:

$$\theta\pi_1 = \theta\mu_1\omega_1 = \mu_2\theta\omega_1 = \mu_2\omega_2\theta = \pi_2\theta.$$

If $\theta$ is an isomorphism for $S_1$ onto $S_2$, then $\theta^{-1}$ is an isomorphism of $S_2$ onto $S_1$.

We say that the elementary kinship structures $(\omega_1,\mu_1,\pi_1)$ and $(\omega_2,\mu_2,\pi_2)$ over $S_1$ and $S_2$, respectively, are isomorphic if there is an isomorphism from $S_1$ onto $S_2$.

If $(\omega_1,\mu_1,\pi_1)$, $(\omega_2,\mu_2,\pi_2)$, and $(\omega_3,\mu_3,\pi_3)$ are elementary kinship structures over $S_1$, $S_2$, and $S_3$, respectively, and if $\theta$ and $\theta'$ are kinship (resp. isomorphisms) of $S_1$ onto $S_2$ and of $S_2$ onto $S_3$ respectively (for the corresponding structures), then $\theta'\theta$ is a kinship morphism (resp. isomorphism) of $S_1$ onto $S_3$.

Note that all the properties of the elementary kinship structures introduced in § 1 are preserved under isomorphisms. That is, if a structure $(\omega,\mu,\pi)$ has one of these properties (for example, generalized exchange), and if the structure $(\omega',\mu',\pi')$ is isomorphic to $(\omega,\mu,\pi)$, then $(\omega',\mu',\pi')$ also has this property. Thus, from the point of view of understanding these properties, two isomorphic structures play the same role.

---

[1] The term "morphism" is part of the current vocabulary for the theory of mathematical structures. See also the comment on the terminology in E 1.1.

This fact justifies the way that we will often manipulate structures that are identical up to isomorphism (see, for example, "identical up to isomorphism" in theorems 1 and 2 in 2.11 and 3.1). All the same, one cannot with complete rigour, identify structures that are "identical up to isomorphism", since one can distinguish structures as mathematical objects even though they are isomorphic. (See the comments in no. E 2.1 b and no. 3.4.)

E 2.1 a

If there is a kinship morphism $\theta$ from $S_1$ onto $S_2$, (for the structures $(\omega_1,\mu_1,\pi_1)$ and $(\omega_2,\mu_2,\pi_2)$, the structure $(\omega_2,\mu_2,\pi_2)$ represents, through $\theta$, an aspect of the functioning of the structure $(\omega_1,\mu_1,\pi_1)$. This interpretation will be made precise when we study the quotient structures (cf. no. 2.4, E 2.4 and E 2.5).

E 2.1 b

As a first example, consider the case where $\theta$ is an isomorphism. For each $x \in S_1$ we have that $\omega_1(x) = \theta^{-1}\omega_2\theta(x)$ and $\mu_1(x) = \theta^{-1}\mu_2\theta(x)$. This shows that knowledge of the structure of $(\omega_2,\mu_2,\pi_2)$ and of $\theta$ completely determines the structure $(\omega_1,\mu_1,\pi_1)$. We also see that the passage from the structure $(\omega_1,\mu_1,\pi_1)$ to the structure $(\omega_2,\mu_2,\pi_2)$ consists in a simple renaming of the elements in the base set ($\theta(x)$ in place of x). Note, however, that isomorphic structures are not any the less distinct mathematical objects. It is important not to confound the two. We will see, for example, in the study of the Murngin system, that the normal and optional structures (cf. [1], p. 220, and no. 3.3) are isomorphic structures over the same set S. However, it is important to distinguish them since the actual functioning of the system consists of an alternation between the two (see the discussion in nos. 3.2 a and e, and also nos. 2.3 D, 3.4, and E 3.2 b).

E 2.1 c

As a second example, suppose that $S_2$ is a structure with two exogamous matrilineal moities, $S_2 = \{1,2\}$ [cf. 1.4 a], and $(\omega_1,\mu_1,\pi_1)$ is a structure over $S_1$ such that there is a homomorphism from $S_1$ onto $S_2$, but otherwise arbitrary.

Let $E_1 = \theta^{-1}(1)$ (resp. $E_2 = \theta^{-1}(2)$) be the set of all $x \in S_1$ such that $\theta(x) = 1$ (resp. $\theta(x) = 2$). The property (KM) implies, since $\omega_2 = \begin{pmatrix} 1,2 \\ 2,1 \end{pmatrix}$ and $\mu_2 = \varepsilon_{S_2}$, that $\omega_1[E_1] = E_2$, $\omega_1[E_2] = E_1$, $\mu[E_1] = E_1$, and $\mu_1[E_2] = E_2$; i.e., $\omega$, maps the elements of $E_1$ into $E_2$ and those of $E_2$ into $E_1$, whereas $\mu_1$ leaves $E_1$ and $E_2$ invariant. That is, the two subpopulations of the population under consideration composed of the individuals belonging to one or the other of the classes $E_1$ or $E_2$, form two exogamous matrilineal moities. This illustrates clearly the interpretation of the morphism $\theta$ from $S_1$ onto $S_2$ as representing the aspect "two exogamous matrilineal moities" of the functioning of $(\omega_1, \mu_1, \pi_1)$.

### 2.2 *Kinship morphisms and kin term group morphisms.

Let $(\omega_1, \mu_1, \pi_1)$ and $(\omega_2, \mu_2, \pi_2)$ be elementary kinship structures over $S_1$ and $S_2$ and let $\theta$ be a kinship morphism of $S_1$ into $S_2$. By virtue of the property (KM), the subset $\theta[S_1]$ of $S_2$ is invariant under $\omega_2, \mu_2$, and $\pi_2$. Therefore if the structure $(\omega_2, \mu_2, \pi_2)$ over $S_2$ is irreducible, then it must follow that $\theta[S_1] = S_2$. Thus we have the result:

Lemma 2.1 If $\theta$ is a kinship morphism of $S_1$ into $S_2$ (for the structures $(\omega_1, \mu_1, \pi_1)$ and $(\omega_2, \mu_2, \pi_2)$ over $S_1$ and $S_2$, respectively), and if the structure $(\omega_2, \mu_2, \pi_2)$ is irreducible, then $\theta[S_1] = S_2$.

By virtue of lemma 2.1, there is no loss of generality by only considering surjective kinship morphisms. We will henceforth say that $\theta$ is a kinship morphism of $S_1$ onto $S_2$.

*Lemma 2.2 Let $(\omega_1, \mu_1, \pi_1)$ and $(\omega_2, \mu_2, \pi_2)$ be kinship structures over $S_1$ and $S_2$, respectively, and let $\theta$ be a kinship morphism from $S_1$ onto $S_2$ ($\theta[S_1] = S_2$). Then there is a homomorphism $\alpha \rightarrow \hat{\alpha}$ (and only one) from the group $G(\omega_1, \mu_1, \pi_1)$ onto the group $G(\omega_2, \mu_2, \pi_2)$ such that $\hat{\alpha}\theta = \theta\alpha$ for all $\alpha \in G(\omega_1, \mu_1, \pi_1)$; further, $\hat{\omega}_1 = \omega_2$, $\hat{\mu}_1 = \mu_1$, and $\hat{\pi}_1 = \pi_2$. In particular, if $\theta$ is an isomorphism, $\alpha \rightarrow \hat{\alpha}$ is an isomorphism of $G(\omega_1, \mu_1, \pi_1)$ onto $G(\omega_2, \mu_2, \pi_2)$.

Proof: Let H be the set of permutations $\alpha$ of $S_1$ for which there is a permutation $\hat{\alpha}$ of $S_2$ such that $\hat{\alpha}\theta = \theta\alpha$ [$\hat{\alpha}$ is well defined by $\alpha$ since $\theta$ is surjective]. One can verify without difficulty that H is a subgroup of $G_{S_1}$, and that $H \supset \{\omega_1, \mu_1, \pi_1\}$. It immediately follows that there exists a function $\alpha \to \hat{\alpha}$ as defined in the lemma. It is easily verified that it is a homomorphism. Finally, if $\theta$ is an isomorphism, we have $\hat{\alpha} = \theta\alpha\theta^{-1}$ and $\alpha = \theta^{-1}\hat{\alpha}\theta$, from which follows the lemma.*

Important Comment.

According to lemma 2.2, if two kinship structures are isomorphic, then the kinship groups are isomorphic as well. But the converse of this property is not true: for a set with two elements and two structures, one patrilineal and the other matrilineal [cf. no. 1.4], it is possible that for both groups the group of kinship terms is the same group of two elements, yet they are not isomorphic.

The structure of the group of kinship terms of an elementary kinship structure is thus not sufficient for characterizing "identical up to isomorphism".

E 2.2

The interpretation of lemma 2.2 is clear, in terms of the function of the morphism $\theta$ (cf. E 2.1 a).

The comment which follows lemma 2.2 shows that knowledge of the group of kinship terms is not sufficient for complete understanding of the functioning of the society under consideration. It is necessary, in addition, to know how the permutations $\omega,\mu$, and $\pi$ are constituted as generators of the group.

2.3 *Examples of morphisms and isomorphisms.

A) Comments on the terminology.

In no. 1.4 and again in no. 1.9 A and B, names have been given (structure with two exogamous moities, Kariera structure, theoretical Murngin structure) to certain structures that have been defined explicitly over certain sets.

These names will be used, in fact, to designate all structures isomorphic to one of these named structures.

Thus, to say: "let $(\omega,\mu,\pi)$ be a Kariera structure over the set S (with four elements)" indicates that the structure $(\omega,\mu,\pi)$ is isomorphic to the structure $(\omega_1,\mu_1,\pi_1)$ introduced in 1.9 A).

B) Let $S_1 = \{p,q,r,s\}$, $(\omega_1,\mu_1,\pi_1)$ be the Kariera structure over $S_1$ introduced in 1.9 A), $S_2 = \{1,2\}$, $\omega_2 = \begin{pmatrix}1,2\\2,1\end{pmatrix}$, $(\omega_2,\varepsilon_{S_2},\omega_2)$ be the structure over $S_2$ with two exogamous matrilineal moities (no. 1.4 a)), and let $\theta$ be the mapping of $S_1$ onto $S_2$ defined by:

$$\theta(p) = \theta(q) = 1, \quad \theta(r) = \theta(s) = 2.$$

is a kinship morphism relative to the structures $(\omega_1,\mu_1,\pi_1)$ and $(\omega_2,\varepsilon_{S_2},\omega_2)$.

C) Let $(\omega_2,\omega_2,\varepsilon_{S_2})$ be the structure over $S_2$ with two exogamous patrilineal moities, and let $\theta'$ be the mapping of $S_1$ onto $S_2$ defined by:

$$\theta'(p) = \theta'(s) = 1, \quad \theta'(q) = \theta'(r) = 2.$$

$\theta'$ is a kinship morphism relative to the structures $(\omega_1,\mu_1,\pi_1)$ and $(\omega_2,\omega_2,\varepsilon_{S_2})$.

D) Let $S_1 = S_2 = S = \{p,q,r,s\}$, and let $\omega_2 = \begin{pmatrix}p & q & r & s\\ q & r & s & p\end{pmatrix}$ [cf. 1.9 B]. The bijective mapping $\theta = \begin{pmatrix}p & q & r & s\\ r & q & p & s\end{pmatrix}$ of S onto S is an isomorphism for the structures $(\omega_2,\omega_2^{-1},\varepsilon_s)$ and $(\omega_2^{-1},\omega_2,\varepsilon_s)$. This may be seen by noting that $\omega_2\theta = \theta\omega_2^{-1}$.

E 2.3

The morphisms $\theta$ and $\theta'$ introduced in 2.3 B and 2.3 C show that the Kariera system represented by the structure $(\omega_1,\mu_1,\pi_1)$ over $S_1$ (cf. 1.9 A) is central to the functioning of both the structure with two exogamous matrilineal moities and that with two exogamous patrilineal moities (cf. [1], p. 203).

We make this observation precise by introducing the notion of direct product [cf. no. 2].

The example 2.3 D shows that two structures over the same set S can be isomorphic without being identical [cf. E 2.1 b].

### 2.4 Quotient structures - definition.

Let $(\omega,\mu,\pi)$ be an elementary kinship structure over the set S.

We say that a partition P of S is compatible with the structure $(\omega,\mu,\pi)$ if each of the permutations $\omega,\mu$, and $\pi$ is compatible with P [that is, cf. no. 0.4, if for all $X \in P$, $\omega[X]$, $\mu[X]$ and $\pi[X]$ belong to P].

By virtue of axiom (D), in order that the partition P of S be compatible with the structure $(\omega,\mu,\pi)$ over S, it suffices that two of the three permutations $\omega,\mu$, and $\pi$ be compatible with P.

Let, therefore, P be a partition of S compatible with the structure $(\omega,\mu,\pi)$. For all $X \in P$, set $\bar{\omega}(X) = \omega[X]$. $\bar{\mu}(X) = \mu[X]$, $\bar{\pi}(X) = \pi[X]$. Then $\bar{\omega},\bar{\mu}$, and $\bar{\pi}$ are permutations of the set P, and $(\bar{\omega},\bar{\mu},\bar{\pi})$ is an elementary kinship structure over the set P. This structure is called the quotient of $(\omega,\mu,\pi)$ by the partition P.

E 2.4

The ethnological interpretation of a partition P compatible with a structure $(\omega,\mu,\pi)$ and of the corresponding quotient structure $(\bar{\omega},\bar{\mu},\bar{\pi})$, is clear: each element of P represents a set of matrimonial classes of the society in question. One can thus consider each element of P as representing a subpopulation. The condition of compatibility of P with the structure $(\omega,\mu,\pi)$ expresses the fact that these subpopulations must also be considered as matrimonial classes, but classes which are nonetheless "much larger" than the primitive classes. To distinguish them from the latter, we will call them P-classes. Thus the quotient expresses, in the mathematical model, that portion of the rules of the kinship system under consideration which can be expressed uniquely in terms of P-classes.

### 2.5 Quotient structures and kinship morphisms.

Lemma 2.3 Let $(\omega_1,\mu_1,\pi_1)$ and $(\omega_2,\mu_2,\pi_2)$ be two elementary kinship structures over the sets $S_1$ and $S_2$, respectively, and let $\theta$ be a kinship morphism of $S_1$ onto $S_2$ [for the structures $(\omega_1,\mu_1,\pi_1)$ and $(\omega_2,\mu_2,\pi_2)$]. The partition $P_1$ of $S_1$ formed by the sets $\theta^{-1}(x_2)(x_2 \in S_2)$ is compatible with the structure $(\omega_1,\mu_1,\pi_1)$, and if $(\bar{\omega}_1,\bar{\mu}_1,\bar{\pi}_1)$ denotes the quotient structure of $(\omega_1,\mu_1,\pi_1)$ by $P_1$, mapping $x_2 \to \theta^{-1}(x_2)$ is an isomorphism of $S_2$ onto $P_1$ for the structures $(\omega_2,\mu_2,\pi_2)$ and $(\bar{\omega}_1,\bar{\mu}_1,\bar{\pi}_1)$.

Proof: Denote by $\bar{\theta}$ the mapping $x_2 \to \theta^{-1}(x_2)$ of $S_2$ onto $P_1$. By its definition, $\bar{\theta}$ is bijective. It remains to show that $\bar{\omega}_1\bar{\theta} = \bar{\theta}\omega_2$ and $\bar{\mu}_1\theta = \bar{\theta}\mu_2$.

We will show, as an example, the first equation. It suffices to verify that $\omega_1[\theta^{-1}(x_2)] = \theta^{-1}[\omega_2(x_2)]$ for all $x_2 \in S_2$; or furthermore that: $\theta^{-1}(x_2) = \omega_1^{-1}[\theta^{-1}[\omega_2(x_2)]]$ for all $x_2 \in S_2$; or finally that: $\theta^{-1}[\omega_2^{-1}(y_2)] = \omega^{-1}[\theta^{-1}(y_2)]$ for all $y_2 \in S_2$. This last equation is equivalent to $(\omega_2\theta)^{-1}(y_2) = (\theta\omega_1)^{-1}(y_2)$ for all $y_2 \in S_2$, which follows from the fact that since $\theta$ is a kinship morphism, $\omega_2\theta = \theta\omega_1$. QED.

Lemma 2.3 has the following reciprocal.

Lemma 2.4 Let $(\omega,\mu,\pi)$ be an elementary kinship structure over the set S, P a partition of S compatible with the structure $(\omega,\mu,\pi)$, and let $(\bar{\omega},\bar{\mu},\bar{\pi})$ be the quotient structure of $(\omega,\mu,\pi)$ by P. For all $x \in S$, let $\theta(x)$ be the unique element X in P such that $x \in X$. The mapping $\theta$ of S onto P thus defined is a kinship morphism for the structures $(\omega,\mu,\pi)$ and $(\bar{\omega},\bar{\mu},\bar{\pi})$.

This can be verified by noting that the equations $\bar{\omega}\theta = \theta\omega$ and $\bar{\mu}\theta = \theta\mu$ are derived directly from the definition of $\theta$ and from the compatibility of the structure $(\omega,\mu,\pi)$ with P.

One can deduce from lemmas 2.3 and 2.4 above the following result: let $(\omega_1,\mu_1,\pi_1)$ and $(\omega_2,\mu_2,\pi_2)$ be two elementary kinship structures over $S_1$ and $S_2$, $\theta$ a kinship morphism of $S_1$ onto $S_2$, and let $P_2$ be a partition of $S_2$ compatible with the structure $(\omega_2,\mu_2,\pi_2)$. The partition $P_1$ of $S_1$ composed of all $\theta^{-1}[X]$ where $X \in P_2$ is compatible with the structure $(\omega_1,\mu_1,\pi_1)$, and the mapping $X \to \theta^{-1}$ is an isomorphism of $P_2$ onto $P_1$ (for the quotient structures of $(\omega_2,\mu_2,\pi_2)$ and $(\omega_1,\mu_1,\pi_1)$ respectively).

In particular, in the case where $\theta$ is an isomorphism, it can be seen that an isomorphism between two structures transforms compatible partitions into compatible partitions and induces an isomorphism between the corresponding quotient structures.

E 2.5

Lemmas 2.3 and 2.4 establish an equivalence, in the mathematical model, between a given partition compatible with a structure, and that of a morphism of this structure onto another structure (which will be isomorphic to the quotient structure).

In terms of the interpretation of quotient structures (cf. E 2.4 above), the lemmas thus precisely indicate the significance of the kinship morphisms noted in E 2.1 a and E 2.1 c.*

2.6 *Quotient structures and groups of kinship terms.

Lemma 2.5 Let $(\omega,\mu,\pi)$ be an elementary kinship structure over the set S, P a partition of S compatible with the structure $(\omega,\mu,\pi)$, and let $(\bar{\omega},\bar{\mu},\bar{\pi})$ be the quotient structure of $(\omega,\mu,\pi)$ by P. Each element $\alpha$ of $G(\omega,\mu,\pi)$ is compatible with the partition P (cf. no. 0.4), and the mapping $\alpha \to \bar{\alpha}$ of $G(\omega,\mu,\pi)$ into $G(\bar{\omega},\bar{\mu},\bar{\pi})$ (defined by $\bar{\alpha}(X) = \alpha[X]$, $\alpha \in G(\omega,\mu,\pi)$, $X \in P$) is a homomorphism of $G(\omega,\mu,\pi)$ onto $G(\bar{\omega},\bar{\mu},\bar{\pi})$.

Corollary. The group $G(\bar{\omega},\bar{\mu},\bar{\pi})$ is isomorphic to the quotient group of $G(\omega,\mu,\pi)$ by the proper subgroup defined by all $\alpha \in G(\omega,\mu,\pi)$ such that $\alpha[X] = X$ for all $X \in P$.

One can verify lemma 2.5 directly, or deduce it from lemmas 2.2 and 2.4 above [no. 2.2 and no. 2.5].*

2.7 *Structures composed of two moities.

We will say that a structure $(\omega,\mu,\pi)$ over the set S is composed of two moities if there is a partition P of S with two elements and compatible with the structure $(\omega,\mu,\pi)$. The quotient structure $(\bar{\omega},\bar{\mu},\bar{\pi})$ of $(\omega,\mu,\pi)$ by P is thus a structure over the set P with two elements. If $M_1$ and $M_2$ are the elements of P, we will say that the structure $(\omega,\mu,\pi)$ is composed of the moities $M_1$ and $M_2$.

If the quotient structure $(\bar{\omega},\bar{\mu},\bar{\pi})$ has two matrilineal exogamous moities (resp. patrilineal), we will say that the structure $(\omega,\mu,\pi)$ is composed of two matrilineal exogamous moities (resp. patrilineal).

If $P = \{M_1,M_2\}$, then $\omega[M_1] = M_2$ and $\omega[M_2] = M_1$; $\mu[M_1] = M_1$ and

$\mu[M_2] = M_2$; and $\pi[M_1] = M_2$ and $\pi[M_2] = M_1$ (in the matrilineal case).

Example for the Kariera structure.

$$S_1 = \{a_1,a_2,b_1,b_2\}; \quad \omega_1 = \begin{pmatrix} a_1 & a_2 & b_1 & b_2 \\ b_2 & b_1 & a_2 & a_1 \end{pmatrix};$$

$$\mu_1 = \begin{pmatrix} a_1 & a_2 & b_1 & b_2 \\ b_1 & b_2 & a_1 & a_2 \end{pmatrix}; \quad \pi_1 = \begin{pmatrix} a_1 & a_2 & b_1 & b_2 \\ a_2 & a_1 & b_2 & b_1 \end{pmatrix}.$$

(See fig. 6.)

$\{a_1,a_2\}$, $\{b_1,b_2\}$ are exogamous patrilineal moities.
$\{a_1,b_1\}$, $\{a_2,b_2\}$ are exogamous matrilineal moities.

Comment.

Note, as is shown by the above example of the Kariera system, a structure can be composed of several distinct compatible partitions with two moities.

E 2.7

Immediate interpretation: for example a structure composed of two exogamous matrilineal moities represents a society divided into two exogamous subpopulations with matrilineal descent.

## 2.8 Quotient structure of a quotient structure, or the refinement of partitions.

Lemma 2.6 Let $(\omega,\mu,\pi)$ be an elementary kinship structure over the set S, $P_1$ and $P_2$ two partitions of S compatible with $(\omega,\mu,\pi)$ and such that $P_2$ is a refinement of $P_1$ (that is, every element of $P_1$ is a union of elements of $P_2$) and let $(\omega_1,\mu_1,\pi_1)$ and $(\omega_2,\mu_2,\pi_2)$ be the quotient structures of $(\omega,\mu,\pi)$ by $P_1$ and $P_2$. If Q is the partition of $P_2$ formed by the sets $\hat{X}_1 = \{X_2 | X_2 \in P_2 \text{ and } X_2 \subset X_1\}$, where the $X_1$ are all of the elements of $P_1$,[1] then Q is compatible with the structure $(\omega_2,\mu_2,\pi_2)$ over $P_2$ and the mapping $X_1 \to \hat{X}_1$ is an isomorphism of $P_1$ onto Q for the structures $(\omega_1,\mu_1,\pi_1)$ and $(\bar{\omega}_2,\bar{\mu}_2,\bar{\pi}_2)$ (where $(\bar{\omega}_2,\bar{\mu}_2,\bar{\pi}_2)$ denotes the quotient structure of $(\omega_2,\mu_2,\pi_2)$ by Q).

Proof: First, it is clear that $X_1 \to \hat{X}_1$ is a bijective mapping of

[1] One could say that Q is the partition quotient of $P_2$ and $P_1$.

$P_1$ onto Q. Second, one can immediately verify that for all $X_1$ $P_1$

$$\bar{\omega}_2(\hat{X}_1) = \omega_2[\hat{X}_1] = \widehat{\omega[X_1]} = \widehat{\omega_1(X_1)}$$

$$\bar{\mu}_2(\hat{X}_1) = \mu_2[\hat{X}_1] = \widehat{\mu[X_1]} = \widehat{\mu_1(X_1)},$$

and the rest of the proof follows immediately.

E 2.8

It is clear that by taking more and more refined compatible partitions, one will obtain quotient structures that more clearly illustrate the functioning of the kinship system under consideration.*

2.9 Direct products of elementary kinship structures.

Let $(\omega_1,\mu_1,\pi_1)$ and $(\omega_2,\mu_2,\pi_2)$ be elementary kinship structures over the sets $S_1$ and $S_2$, respectively. $(\omega_1 \times \omega_2, \mu_1 \times \mu_2, \pi_1 \times \pi_2)$ is an elementary kinship structure over the set $S = S_1 \times S_2$. That is, from equation (b) of no. 0.3 we have:

$$\pi_1 \times \pi_2 = \mu_1\omega_1 \times \mu_2\omega_2 = (\mu_1 \times \mu_2)\ (\omega_1 \times \omega_2).$$

The elementary kinship structure $(\omega_1 \times \omega_2, \mu_1 \times \mu_2, \pi_1 \times \pi_2)$ over the set $S_1 \times S_2$ is called the direct product structure of the structures $(\omega_1,\mu_1,\pi_1)$ and $(\omega_2,\mu_2,\pi_2)$.

Example.

The Kariera structure is the direct product of two structures with two exogamous moities that are, respectively, matrilineal and patrilineal.

Using the notation of nos. 1.4 and 1.9 A, let $(\omega_1,\mu_1,\pi_1)$ be the Kariera structure over $S_1=\{p,q,r,s\}$, and let $(\omega,\varepsilon_s,\omega)$ and $(\omega,\omega,\varepsilon_s)$ be the structures with two exogamous moities that are, respectively, matrilineal and patrilineal over $S = \{1,2\}$ $\left(\omega = \begin{pmatrix}1,2\\2,1\end{pmatrix}\right)$.

Denote by $\theta$ the mapping of $S_1$ onto $S \times S$ defined by

$$\theta(p) = (1,1), \quad \theta(q) = (1,2),$$

$$\theta(r) = (2,2), \quad \theta(s) = (2,1).$$

A simple verification shows that $\theta$ so defined is an isomorphism of $S_1$ onto $S \times S$ for the Kariera structure $(\omega_1,\mu_1,\pi_1)$ over $S_1$, and of the direct product structure of $(\omega,\varepsilon_S,\omega)$ by $(\omega,\omega,\varepsilon_S)$ over $S \times S$.

*Comment.

Note that the group of kinship terms of the direct product structure of two elementary kinship structures is, in general, only a subgroup of the direct product group of the groups of kinship terms of these two structures.*

E 2.9

The isomorphism of the Kariera structure with the direct product of two structures with two exogamous moities that are, respectively, matri- and patrilineal, expresses, in the mathematical model studied here, the intuitive idea that the Kariera system "adds, by the division into two matrilineal moities, a different division perpendicular to the preceding one into patrilineal moities" (see [1], p. 208).

2.10 *Exogamous moities and direct product where one of the factors is a structure with two exogamous moities.

Let $S = \{1,2\}$, $\omega = \begin{pmatrix}1,2\\2,1\end{pmatrix}$, and let $(\omega,\varepsilon_S,\omega)$ be the structure with two matrilineal exogamous moities[1] over S (no. 1.4). Further let S' be a set and let $(\omega',\mu',\pi')$ be an elementary kinship structure over S'.

Denote by $(\hat{\omega},\hat{\mu},\hat{\pi})$ the direct product structure over $S \times S'$ derived from $(\omega,\varepsilon_S,\omega)$ and $(\omega',\mu',\pi')$. Let $\theta$ be the projection of $S \times S'$ onto S ($\theta(i,x) = i$ for all $i \in S$ and $x \in S'$). Set $\theta^{-1}[\{i\}] = E_1$ $(i = 1,2)$. Then $\theta$ is a kinship morphism of $S \times S'$ onto S for the structures $(\hat{\omega},\hat{\mu},\hat{\pi})$ and $(\omega,\mu,\pi)$ (no. 2.1), $E_1$ and $E_2$ are exogamous matrilineal moities for the product structure $(\hat{\omega},\hat{\mu},\hat{\pi})$ (no. 2.7), and the corresponding quotient structure (no. 2.5) is a structure with two exogamous matrilineal moities.

---

[1] One could also begin with the patrilineal structure $(\omega,\omega,\varepsilon_S)$.

These assertions are easily verified. For example, if $i \in \{1,2\}$, and $S \in S'$, then

$$\theta\hat{\omega}(i,x) = \theta(\omega(i),\ \omega'(x)) = \omega(i) = \omega(\theta(i,x)).$$

Hence $\theta\hat{\omega} = \omega\theta$; similarly, $\theta\hat{\mu} = \mu\theta$ and $\mu\hat{\pi} = \pi\theta$.

In addition, by the definition of $(\hat{\omega},\hat{\mu},\hat{\pi})$ and of $\theta$, $\hat{\omega}(i,x) = (\omega(i), \omega'(x))$, hence:

$$\hat{\omega}(i,x) \in E_1 \leftrightarrows \omega(i) = 1 \leftrightarrows i = 2 \leftrightarrows (i,x) \in E_2.$$

Thus $\hat{\omega}[E_2] = E_1$, and similarly, $\hat{\omega}[E_1] = E_2$ and $\hat{\mu}[E_i] = E_i$ $(i = 1,2)$. Note in addition that if one denotes by $\theta'$ the projection of $S \times S'$ onto $S'$, $\theta'$ is a bijective mapping from $E_i$ onto $S'$ and transforms the restriction of $\hat{\mu}$ to $E_i$ $(i = 1,2)$ to $\mu'$ (that is, $\theta'\hat{\mu}(z) = \mu'(\theta'(z))$ for all $z \in S \times S'$).

Conversely, one can ask under what conditions a structure with two exogamous moities is isomorphic to the direct product of a certain structure by a structure with two exogamous moities.

The following lemma provides an answer to this question.

Lemma 2.7 Let $(\hat{\omega},\hat{\mu},\hat{\pi})$ be an elementary kinship structure over the set $\hat{S}$ with 2n elements and composed of two exogamous matrilineal moities over the set $S = \{1,2\}$ (no. 1.4). The following two conditions are equivalent.

1) There is a set $S'$ with n elements, an elementary structure $(\omega',\mu',\pi')$ over $S'$, and a bijective mapping $\theta$ of $\hat{S}$ onto $S \times S'$ such that $\theta^{-1}(\{i\} \times S') = E_1$ $(i = 1,2)$, and $\theta$ is an isomorphism from the structure $(\hat{\omega},\hat{\mu},\hat{\pi})$ over $\hat{S}$ to the direct product structure formed by $(\omega,\varepsilon_s,\omega)$ and $(\omega',\mu',\pi')$ over $S \times S'$.

2) There is a permutation $\sigma$ of $\hat{S}$ such that $\sigma[E_1] = E_2$, $\sigma[E_2] = E_1$, $\sigma^2 = \varepsilon_{\hat{S}}$, $\hat{\omega}\sigma = \sigma\hat{\omega}$, and $\hat{\mu}\sigma = \sigma\hat{\mu}$.

Moreover, if condition 2) is satisfied, the family formed by the sets $\bar{x} = \{x,\sigma(x)\}$, where $x \in \hat{S}$, is a partition of $\hat{S}$ compatible with the structure $(\hat{\omega},\hat{\mu},\hat{\pi})$. One can take for $S'$ this partition; for $(\omega',\mu',\pi')$

the quotient structure of $(\hat{\omega},\hat{\mu},\hat{\pi})$ by S'; and for $\theta$ the mapping defined by

$$\theta(x) = (i,\bar{x}), \text{ if } x \in E_i \ (i = 1,2).$$

Proof: First, if condition 1) is satisfied, it suffices to take for $\sigma$ the mapping of $\hat{S}$ into $\hat{S}$ defined by $\theta\sigma(x) = (2,x')$ if $\theta(x) = (1,x)$ $(x' \in S')$ and $\theta\sigma(x) = (1,x')$ if $\theta(x) = (2,x')$.

Conversely, suppose that condition 2) is satisfied. That the set $S' = \{\bar{x} | x \in \hat{S}\}$ is a partition of $\hat{S}$ follows from the fact that $\sigma^2 = \varepsilon_{\hat{S}}$. We will show that this partition is compatible with the structure $(\hat{\omega},\hat{\mu},\hat{\pi})$.

The equations $\sigma^2 = \varepsilon_{\hat{S}}$ and $\hat{\omega}\sigma = \sigma\hat{\omega}$ imply that $\hat{\omega} = \sigma\hat{\omega}\sigma$. Thus if $x \in E$, then

$$\hat{\omega}[\bar{x}] = \{\hat{\omega}(x),\ \hat{\omega}\sigma(x)\} = \{\sigma(\hat{\omega}\sigma(x)),\ \hat{\omega}\sigma(x)\},$$

since

$$\hat{\omega} = \sigma\hat{\omega}\sigma.$$

Hence

$$\hat{\omega}\ [\bar{x}] \in S'.$$

Similarly,

$$\hat{\mu}\ [\bar{x}] = \{\hat{\mu}(x),\ \hat{\mu}(\sigma(x))\} = \{\hat{\mu}(x),\ \sigma(\hat{\mu}(x))\},$$

since

$$\hat{\mu}\sigma = \sigma\hat{\mu}.$$

Hence

$$\hat{\mu}\ [\bar{x}] \in S'.$$

It only remains to show that $\theta$ is an isomorphism. If $x \in E_i$ and $j \in \{1,2\}$ with $j \neq i$, then

$$\theta\hat{\omega}(x) = (j,\ \overline{\hat{\omega}(x)}),$$

$$= (j,\ \omega'(\bar{x}))$$

$$= (\omega(i),\ \omega'(\bar{x}))$$

$$= (\omega \times \omega')\ \theta\ (x),$$

since $\hat{\omega}(x) \epsilon E_j$, by definition of $\omega'$, and by definition of the product structure. In the same fashion it may be verified that $\theta\hat{\mu} = \hat{\mu}\theta$, which proves the lemma.

*Comment.

The regularity of the structure $(\hat{\omega},\hat{\mu},\hat{\pi})$ combined with the existence of two exogamous matrilineal moities is not a sufficient condition for a decomposition into a direct product (property 1) of lemma 2.7, as is shown by the following example[1] of a structure over a set with eight elements defined by the diagram of figure 7.

It can be seen that $\mu\omega = \omega\mu$, $\omega^2 = \mu$, $\mu^4 = \varepsilon$, and the group of kinship terms reduces to

$$\{\varepsilon,\ \mu,\ \mu^2,\ \mu^3,\ \omega,\ \mu\omega,\ \mu^2\omega,\ \mu^3\omega\},$$

from which it may be deduced that the structure under consideration is regular.

In addition, a simple verification established that there is no permutation having property 2) of lemma 2.7 (where $E_i = \{a_i,b_i,c_i,d_i\}$, $i = 1,2$).*

E 2.10

The operation of forming the direct product of a given structure with the structure composed of two exogamous moities permits the construction of a structure possessing two exogamous moities such that in each one descent rules are as in the given structure.

More exactly, in the case given in 2.10 of a direct product with a structure with two exogamous matrilineal moities, the conjugal function $\hat{\omega}$ of the product structure operates in each of the two moities $E_1$ and $E_2$ in the same fashion as does the conjugal function $\omega'$ of the given

[1] Due to Mme. Dejean.

structure. In particular, the cycles of the product structure are contained in one or the other of the two moities.

## 2.11 *The Aranda structure.

The following general theorem permits the characterization of the Aranda structure.

Theorem 1.

Let n be an integer $\geq 2$ and let S be a set with 2n elements. There exists one and only one elementary kinship structure $(\omega,\mu,\pi)$, unique up to isomorphism, over S composed of two exogamous matrilineal moities which are cycles (no. 1.10) and such that $\omega^2 = \varepsilon_S$ and $\pi^2 = \varepsilon_S$. This structure is regular.

Further, if S' is a set of n elements, the structure $(\omega,\mu,\pi)$ over S is isomorphic to the direct product of the structure with two exogamous matrilineal moities by the structure $(\omega',\mu',\pi')$ over S' defined as follows: $\mu'$ is a transitive permutation of S' (no. 0.5),

$$\omega' = \begin{pmatrix} a'\mu'(a') & \ldots & \mu'^{k}(a') & \ldots & \mu'^{n-1}(a') \\ a'\mu'^{n-1}(a') & \ldots & \mu'^{n-k}(a') & \ldots & \mu'(a') \end{pmatrix},$$

where a' is an element of S', and $\pi' = \mu^{2}{}'\omega'$.[1]

Proof: Let $E_1$ and $E_2$ be two exogamous moities of a structure $(\omega,\mu,\pi)$ that satisfy the conditions of the theorem. $E_1$ and $E_2$ are cycles and as each has n elements, $\mu^n = \varepsilon_S$. Thus $\mu^{-k} = \mu^{n-k}$ $(0 \leq k \leq n)$.

The equation $\pi^2 = \varepsilon_S$ immediately follows, since $\omega^2 = \varepsilon_S$, as $\omega\mu = \mu^{-1}\omega$. It follows, by induction in the integers, that $\omega\mu^k = \mu^{-k}\omega$, hence $\omega\mu^k = \mu^{n-k}$ $(0 \leq k \leq n)$, since $\mu^{-k} = \mu^{n-k}$.

Let a be an element of $E_1$, for example. The elements of $E_1$ can be written in the form $a,\mu(a),\ldots,\mu^{n-1}(a)$, and those of $E_2$ in the form $\omega(a),\omega(\mu(a)) = \mu^{n-1}(\omega(a)),\ldots,\omega(\mu^{n-1}(a)) = \mu(\omega(a))$.

One can deduce from this, without difficulty, the existence and uniqueness of the desired structure. In addition the equations

[1] Note that this structure is not regular.

$$\omega^2 = \varepsilon_s, \quad \mu^n = \varepsilon, \quad \omega\mu^k = \mu^{n-k}\omega \quad (0 \le k \le n)$$

hold by virtue of the fact that the group of kinship terms of $(\omega,\mu,\pi)$ is reduced to the set $\{\varepsilon_s,\mu,\mu^2,\ldots,\mu^{n-1},\omega,\mu\omega,\ldots,\mu^{n-1}\omega\}$. The regularity of the structure $(\omega,\mu,\pi)$ immediately follows.

In order to establish the decomposition of the structure $(\omega,\mu,\pi)$ into a direct product, we will utilize lemma 2.7 (no. 2.10). Define the mapping $\sigma$ of $S$ into $S$ by the equations $\sigma(\mu^k(x)) = \mu^k(\omega(x))$ for $x = a$, $x = \omega(a)$, and for $0 \le k \le n$. $\sigma$ is a permutation of $S$, maps $E_1$ onto $E_2$ and $E_2$ onto $E_1$, and $\sigma^2 = \varepsilon_s$. In addition, $\sigma\mu = \mu\sigma$ and $\sigma\omega = \omega\sigma$. Thus, whether $x = a$ or $x = \omega(a)$, it follows that $\sigma\mu(\mu^k(x)) = \sigma\mu^{k+1}(x) = \mu^{k+1}\omega(x) = \mu(\mu^k(\mu^k(\omega(x)))) = \mu\sigma(\mu^k(x))$, and $\omega\sigma(\mu^k(x)) = \omega\mu^k(\omega(x)) = \mu^{n-k}(x) = \sigma\mu^{n-k}\omega(x) = \sigma\omega(\mu^k(x))$.

Thus $\sigma$ satisfies condition 2) of lemma 2.7. To finish the proof, it suffices to verify that the structure $(\omega',\mu',\pi')$ induced by $(\omega,\mu,\pi)$ over the partition $S' = \{\{x,\sigma(x)\}|x\in S\}$ is of the type described in the theorem. To show this, designate by $\bar{x}$ the class $\{x,\sigma(x)\}$ $(x\in S)$. On the one hand, $\mu'(\overline{\mu^k(a)}) = \overline{\mu^{k+1}(a)}$ $(0 \le k \le n)$, from which follows the transitive character of $\mu'$, since $S' = \{\bar{a},\overline{\mu(a)},\ldots,\overline{\mu^{n-1}(a)}\}$; and on the other hand, $\omega'(\overline{\mu^k(a)}) = \omega[\{\mu^k(a), \sigma\mu^k(a)\}] = \{\omega\mu^k(a), \omega\mu^k\omega(a)\}$

$$= \{\omega\mu^k(a), \mu^{n-k}(a)\} = \{\mu^{n-k}(a), \mu^{n-k}\omega(a)\}$$

$$= \{\mu^{n-k}(a), \sigma\mu^{n-k}(a)\} = \overline{\mu^{n-k}(a)}.$$

This proves theorem 1.

Application to the Aranda structure.

Theorem 1 implies in particular $(n = 4)$ that there is an elementary kinship structure over a set with 8 elements, and only one up to isomorphism, composed of two exogamous matrilineal moities which are cycles and such that the pairs and couples (no. 1.10) each have exactly two elements.

This structure is called the Aranda structure. Lemma 2.7 then implies that this structure can be obtained as a direct product of the structure with two exogamous matrilineal moities by the structure $(\omega_4, \mu_4, \pi_4)$ over the set {p,q,r,s} with 4 elements (no. 1.9 D).

By using this construction, it is easily verified that this structure is regular, but satisfies neither condition (PCC) nor condition (MCC) (no. 1.8).

(See fig. 8.)

E 2.11

Theorem 1 applied to the Aranda structure makes precise in the mathematical model the concepts developed in [1], p. 210 - 213.

Note that theorem 1 permits the characterization of the Aranda structure in a unique fashion, in terms of pairs, couples, and cycles, as the only system with eight classes composed of two exogamous matrilineal moities which are cycles and such that the pairs, as well as the couples, are composed of exactly two classes.

The character of the direct product, as well as the impossibility of marriages between cross-cousins, thus appear as consequences of the properties intervening in this characterization. See for example no. 3.4.

## § 3. *STUDY OF THE MURNGIN SYSTEM

### 3.1 A uniqueness theorem leading from the normal and optional Murngin structures to the theoretical structure.

Theorem 2.

Let S be a set with 4n elements $(n \geq 1)$, $\{E_1, E_2\}$ a partition of S into two subsets with 2n elements, and let $\mu$ be a permutation of S admitting $E_1$ and $E_2$ as intransitive classes (no. 0.5).

In addition, let $\omega_1$ and $\omega_2$ be two permutations of S such that:

(a) $\omega_1 \neq \omega_2$

(b) $\omega_i[E_1] = E_2$ and $\omega_i[E_2] = E_1$ $(i = 1,2)$

(c) $\omega_i^2 = \varepsilon_S$ and $\omega_i\mu = \mu\omega_i$ $(i = 1,2)$

(d) $\omega_1\omega_2 = \omega_2\omega_1$.

Then:

(A) $\omega_1\omega_2 = \mu^n$ and $\omega_1(x) \neq \omega_2(x)$ for all $x \in S$;

(B) If one sets

$$\omega'(x) = \omega_1(x) \text{ and } \omega''(x) = \omega_2(x) \text{ for each } x \in E_1,$$

$$\omega'(x) = \omega_2(x) \text{ and } \omega''(x) = \omega_1(x) \text{ for each } x \in E_2,$$

then $\omega'$ and $\omega''$ so defined are the only permutations $\omega$ of S with the following three properties:

(1) $\omega\mu = \mu\omega$

(2) $\omega^2 \neq \varepsilon_S$

(3) for all $x \in S$, $\omega(x) = \omega_1(x)$ or $\omega(x) = \omega_2(x)$.

In addition, the structures $(\omega', \mu, \mu\omega')$ and $(\omega'', \mu, \mu\omega'')$ admit $E_1$ and $E_2$ as exogamous matrilineal moities.

(C) Suppose, in addition, that n is even, say $n = 2h$.

Let s' denote a set with 2n elements and let $\mu'$ be a transitive permutation of S' (no. 0.5).

Each of the structures $(\omega', \mu, \mu\omega')$ and $(\omega'', \mu, \mu\omega'')$ is then isomorphic to the direct product of the structure with two exogamous matrilineal moities by the structure $(\mu'^h, \mu', \mu'^{h+1})$ over S'. More precisely, if we denote by $\sigma'$ the permutation of S defined by $\sigma'(x) = \mu^{-h}\omega_i(x)$ if $x \in E_i$ $(i = 1,2)$, then $\sigma'$ possesses property (2) of lemma 2.6 (no. 2.10) relative to the structure $(\omega',\mu,\mu\omega')$ over S, and the quotient structure of $(\omega',\mu,\mu\omega')$ by the partition $\{\{x,\sigma'(x)\}|x\epsilon S\}$ associated with $\sigma'$ (lemma 2.6) is isomorphic to $(\mu'^h,\mu',\mu'^{h+1})$. Moreover, an analogous result holds for $\omega''$ if 1 and 2 in the definition of $\sigma'$ are inverted.

Proof:

($\alpha$) Let a be an element of $E_1$. $\mu$ maps $E_1$ and $E_2$ as intransitive classes:

$$E_1 = \{a,\mu(a),\ldots,\mu^{2n-1}(a)\}$$

$$E_2 = \{\omega_1(a),\ \mu(\omega_1(a)),\ldots,\mu^{2n-1}(\omega_1(a))\}.$$

Since $\omega_2(a)\epsilon E_2$, there is an integer $k \leq 2n-1$ (and only one) such that $\omega_2(a) = \mu^k(\omega_1(a))$.

We have then, that $\omega_2(\mu'(a)) = \mu'(\omega_2(a))$ (by (c))

$$= \mu^{k+1}\omega_1(a)$$

$$= \mu^k\mu'\omega_1(a)$$

$$= \mu^k\omega_1(\mu'(a)) \text{ (by (c) for } 0 \leq 1 \leq 2n-1).$$

Thus $\omega_2(x) = \mu^k\omega_1(x)$ for all $x\epsilon E_1$.

Now suppose $y\epsilon E_2$. There is an $x\epsilon E_1$ such that $y = \omega_1(x)$, hence

$$\omega_2(y) = \omega_2\omega_1(x) = \omega_1\omega_2(x) \text{ (by (d))}$$

$$= \omega_1\mu^k\omega_1(x) \text{ (by (c) and the above)}$$

$$= \mu^k(x) = \mu^k\omega_1(y),$$

since $\omega_1^2 = \varepsilon_s$. Thus $\omega_2 = \mu^k\omega_1$. That is, $\omega_1\omega_2 = \mu^k$, but $(\omega_1\omega_2)^{-1} = \omega_1^{-1}\omega_2^{-1} = \omega_1\omega_2$, thus $\mu^{-k} = \mu^k$. Hence $\mu^{2k} = \varepsilon_s$. Since 2n is the smallest integer p such that $\mu^p = \varepsilon_s$ and $k \leq 2n-1$, it follows that $2k = 0$

or $2k = 2n$. This implies that $k = n$, since $k = 0$ implies $\omega_1 = \omega_2$, which contradicts (a).

Finally, $\omega_2(x) \neq \omega_1(x)$ for all $x \in S$, since $\mu^n(x) \neq x$ for all $x \in S$. This proves (A).

(β) It may be verified without difficulty that $\omega'$ and $\omega''$ have properties (1), (2) and (3) and that the structures $(\omega', \mu, \mu\omega')$ and $(\omega'', \mu, \mu\omega'')$ admit $E_1$ and $E_2$ as exogamous matrilineal moities.

Conversely, suppose $\omega$ is a permutation of S having properties (1), (2) and (3).

If $a \in E_i$ $(i = 1,2)$ then by property (3) there is a $j \in \{1,2\}$ such that $\omega(a) = \omega_j(a)$. Property (1) then implies that $\omega(\mu'(a)) = \mu'(\omega(a)) = \mu'(\omega_j(a) = \omega_j(\mu'(a))$ for $0 \leq 1 \leq 2n$. Thus $\omega(x) = \omega_j(x)$ for all $x \in E_i$, since $E_i = \{a, \mu(a), \ldots, \mu^{2n-1}(a)\}$. Hence either $\omega(x) = \omega_1(x)$ for all $x \in E_i$ or $\omega(x) = \omega_2(x)$ for all $x \in E_i$ $(i = 1,2)$.

Property (2) now implies that $\omega = \omega'$ or $\omega = \omega''$, since $\omega_1$ and $\omega_2$ do not have property (2). This establishes (B).

(σ) It may be verified without difficulty that $\sigma'^2 = \varepsilon_s$, $\omega'\sigma' = \sigma'\omega'$, and $\mu\sigma' = \sigma'\mu$. For example, if $x \in E$ it follows that

$$\sigma'^2(x) = \sigma'(\sigma'(x)) = \sigma'(\mu^{-h}\omega_1(x)) \text{ (by definition of } \sigma')$$

$$= \mu^{-h}\omega_2\mu^{-h}\omega_1(x) \text{ (since } \mu^{-h}\omega_1(x) \in E_2)$$

$$= \mu^{-n}\omega_1\omega_2(x) = x \text{ (since } \omega_1\omega_2 = \mu^n).$$

Also

$$\sigma'\omega'(x) = \sigma'(\omega_1(x)) = \mu^{-h}\omega_2\omega_1(x) \text{ (since } \omega_1(x) \in E_2)$$

$$= \omega_2(\mu^{-h}\omega_1(x)) = \omega_2(\sigma'(x))$$

$$= \omega'\sigma'(x).$$

It only remains to show that the quotient structure of $(\omega, \mu, \mu\omega')$ by the compatible partition associated with $\sigma'$ is isomorphic to $(\mu'^h, \mu', \mu'^{h+1})$.

To show this, set $\bar{x} = \{x,\sigma'(x)\}$ for each $x \in S$. It is clear that $\mu$ induces over $\{\bar{x} | x \in S\} = \bar{S}$ a transitive permutation $\bar{\mu}$, since if $a \in E_1$ then the terms $\bar{a}, \bar{\mu}(\bar{a}), \ldots, \bar{\mu}^{2n-1}(\bar{a})$ are all distinct and equal respectively to $\bar{a}, \overline{\mu(a)}, \ldots, \overline{\mu^{2n-1}(a)}$.

Now to show that the permutation $\bar{\omega}'$ induced by $\omega'$ over $\bar{S}$ is identical to $\bar{\mu}^{-h}$. If $x \in E_1$, then if 1 is an integer $\geq 0$, it follows that

$$\bar{\mu}'(\bar{x}) = \overline{\mu'(x)} = \{\mu'(x), \sigma'\mu'(x)\} = \{\mu'(x), \mu'^{-h}\omega_1(x)\}$$

and

$$\bar{\omega}'(\bar{x}) = \overline{\omega'(x)} = \overline{\omega_1(x)} = \{\omega_1(x), \sigma'(\omega_1(x))\}$$

$$= \{\omega_1(x), \mu^{-h}\omega_2\omega_1(x)\} = \{\omega_1(x), \mu^h(x)\},$$

since $\omega_2\omega_1 = \mu^n = \mu^{2h}$. By comparing the expression $\bar{\mu}'(\bar{x})$ with $\bar{\omega}'(\bar{x})$, it can be seen that by setting $1 = h$ we have $\bar{\mu}^h(\bar{x}) = \bar{\omega}'(\bar{x})$, which finishes the proof of theorem 2.

E 3.1 a

Description of theorem 2.

The situation considered in theorem 2 can be described as follows. For a society S composed of 4n matrimonial classes, it is possible to make a partition into two moities $E_1$ and $E_2$ with a matrilineal descent rule (represented by $\mu$) that leaves these two moities invariant.

One can also give two distinct conjugal rules (represented by $\omega_1$ and $\omega_2$) that in conjunction with the descent rule $\mu$ form two kinship systems (the normal and optional Murngin system in the case where n = 2) permitting marriage with bilateral cross-cousins and admitting $E_1$ and $E_2$ as exogamous matrilineal moities.

We now look for a conjugal rule $\omega$ that forms with the descent rule $\mu$ a kinship system having the following two properties. First, marriages are necessarily formed according to one or the other of the two rules $\omega_1$ and $\omega_2$. Second, marriage with the matrilateral cross-cousin is permissible, but not with the patrilateral cross-cousin.

Theorem 2 shows that this problem is uniquely solvable (up to isomorphism, no. 3.2 b) and that the structure which is the unique solution is isomorphic to the direct product of a structure with two exogamous matrilineal moities by a theoretical Murngin structure (or implicit, cf. [1], p. 231 and 238).

E 3.1 b

Theorem 2 also provides a rigorous[1] and simple form for the entire discussion which lead C. Lévi-Strauss to introduce generalized exchange and the implicit Murngin system (cf. [1], p. 216 to 242).

We here note again the important role that lemma 1.4 (no. 1.8) plays in this question. Lemma 1.4 establishes the equivalence, when marriage is authorized with the matrilateral cross-cousin, of generalized exchange and of prohibition of marriage with the patrilateral cross-cousin.

3.2 *Comments on theorem 2.

a) Alternative utilization of the conjugal functions $\omega_1$ and $\omega_2$.

If $x \in E_i$ (i - 1,2), then if $j \in \{1,2\}$ and $j \neq i$, it follows that $\omega'(x) = \omega_i(x)$, $\pi'(x) = \mu\omega'(x) (x \in E_j)$, and $\omega'(\pi'(x)) = \omega_j(\pi(x))$ (verification is straight forward).

b) Among all elementary kinship structures $(\hat{\omega},\hat{\mu},\hat{\pi})$ over S such that $\hat{\mu} = \mu$, and $\hat{\omega}(x) = \omega_1(x)$ or $\hat{\omega}(x) = \omega_2(x)$ for all $x \in S$, there is only one, up to isomorphism, permitting marriage with the matrilateral cross-cousin but not with the patrilateral cross-cousin (no. 1.8, lemma 1.4).

c) The structures $(\omega',\mu,\mu\omega')$ and $(\omega'',\mu,\mu\omega'')$ are both irreducible and regular (nos. 1.5 and 1.6).

d) For each of the structures $(\omega',\mu,\mu\omega')$ and $(\omega'',\mu,\mu\omega'')$, the pairs (no. 1.10) are all subsets of S with 4 elements. In addition, if n = 2, the couples (no. 1.10) are all subsets of S with 2 elements.

To see this, consider the first structure. If $x \in E_i$ (i = 1,2) then

[1] For the case where n = 2.

the pair containing x is the set of all $\omega'^{k}(x)$, where k = 0,1,2,..., so

$$\omega'(x) = \omega_i(x)$$

$$\omega'^2(x) = \omega_j(\omega_i(x)) = \omega_j\omega_i(x) = \mu^n(x) \quad (j\in\{1,2\},\ j \neq i)$$

$$\omega'^3(x) = \omega_i(\mu^n(x)) = \mu^n(\omega_i(x))$$

$$\omega'^4(x) = \omega_j\omega_i(\mu^n(x)) = \mu^{2n}(x) = x.$$

It follows that the pair containing $x\in E_i$ is the following set of 4 elements:

$$\{x,\ \omega_i(x),\ \mu^n(x),\ \omega_i\mu^n(x)\}.$$

Suppose in addition that n = 2. The couple containing $x\in E_i$ is the set of all $(\mu\omega')^k(x)$, where k = 0,1,2,..., so

$$\mu\omega'(x) = \mu\omega_i(x)$$

$$(\mu\omega')^2(x) = \mu\omega_j\mu\omega_i(x) = \mu^2\omega_j\omega_i(x) = x.$$

It follows that the couple containing $x\in E_i$ is the following set with 2 elements:

$$\{x,\ \mu\omega_i(x)\}.$$

(e) The structures $(\omega_1,\mu,\mu\omega_1)$ and $(\omega_2,\mu,\mu\omega_2)$ are isomorphic. Indeed, by setting $\sigma(x) = \mu^h(x)$ when $x\in E_1$ and $\sigma(x) = \mu^{-h}(x)$ when $x\in E_2$, an isormophism $\sigma$ from $(\omega_1,\mu,\mu\omega_1)$ onto $(\omega_2,\mu,\mu\omega_2)$ is defined (it suffices to verify, by virtue of the equation $\omega_2 = \mu^{2h}\omega_1$, that $\sigma^2 = \varepsilon_s$, $\sigma\omega_2 = \omega_1\sigma$, and $\sigma\mu = \mu\sigma$).

E 3.2 a

The structure whose existence is given by theorem 2 would seem by its uniqueness (up to isomorphism) to conveniently represent the actual functioning established empirically by the population to satisfy the exigencies in question (two exogamous matrilineal moities, possible marriage with the matrilateral cross-cousin but not the patrilateral cross-cousin, according to one or the other of the rules $\omega_1$ or $\omega_2$.

The properties listed in no. 3.2 justify this point of view for they are compatible with the observations reported by C. Lévi-Strauss: Property a) translates the heavily stressed fact (cf. [1], p. 223 and 224, for example) of the alternation of the marriage rules that are used: if a man of class x marries according to rule $\omega_1$, his son (of class $\pi(x)$) must marry according to formula $\omega_2$. It is thus clearly established that this alternation is a consequence of other exigencies.

Note that theorem 2 extends, with regard to this point, the results obtained by A Weil (cf. [1], p. 270), who must essentially postulate this alternation (ibid. p. 282, 283).

E 3.2 b

The irreducible character (property C) of the structure that has been obtained seems to contradict the assertion of A Weil (p. 284). This is due, without doubt, in large part to the fact that in the work of Weil the notion of a reducible society is not rigorously defined (nor, moreover, is the mathematical notion of kinship structure). The contradiction would seem to come only from an interpretation of the ambiguity that has been made precise here by theorem 2.

It is clear that this theorem affirms the existence of two proper structures $(\omega', \mu, \mu\omega')$ and $(\omega'', \mu, \mu\omega'')$ whose distinctiveness is not reduced by having an isomorphism between them, and which, in the search for an empirical solution, are very likely to interfere and complicate the true situation. Weil postulates, in order to interpret this double solution, that the society under consideration is in the process of dividing into two parts— the one functioning according to the first system and the other according to the second system. It would appear that this interpretation is not satisfying from the ethnological point of view, for this division has not been noted by the ethnographers (cf. [1], p. 286). We will take up this question of ambiguity below in the conclusion by trying to broach in a more general fashion the functioning of a kinship system in terms of the set of elementary structures compatible with it (cf. no. 3.4).

E 3.2 c

Note that for couples, the length is 4 for the normal and optional Murngin structures, whereas they are of length 2 for the structures $(\omega',\mu,\mu\omega')$ and $(\omega'',\mu,\mu\omega'')$ (no. 3.2 d). It would be interesting to compare these results with ethnographic data.

### 3.3 *Application to the Murngin system.

It suffices to apply theorem 2 in the case where $n = 2$. The structure $(\omega_1,\mu,\mu\omega_1)$ is the normal Murngin structure, and the structure $(\omega_2,\mu,\mu\omega_2)$ is the optional Murngin structure. (See fig. 9.)

The structures $(\omega',\mu,\mu\omega')$ and $(\omega'',\mu,\mu\omega'')$, which are introduced in theorem 2, can be represented, using the same notation as with the normal and optional structures, by the schemas below (fig. 10). The explicit expression in the theorem for $\omega'$ and $\omega''$ can be immediately deduced.

The various properties of these structures already indicated for the general case in no. 3.2 can be verified. These are: irreducibility, regularity, marriage with the matrilateral cross-cousin, generalized exchange, pair and cycles of 4 elements, couples with 2 elements, alternation of marriages of normal and optional type, etc. (See the ethnological commentary in nos. E 3.1 and E 3.2.)

E 3.3

The schemas presented in this paragraph help visualize the results of nos. 3.1 and 3.2 for the case where $n = 2$ in the classical theory of the Murngin system.

## § 4. CONCLUSION AND PERSPECTIVE FOR RESEARCH THE SPECTRE OF A KINSHIP SYSTEM[1]

Theorem 2 is concerned with the search for and study of all elementary kinship structures over a set S admitting a given maternal function $\mu$ com-

[1] This last paragraph contains its own commentaries, as its aim is more to suggest than to develop. It belongs, however, to the mathematical text in the sense that no technical term is employed that has not been mathematically defined.

patible with one or the other of two conjugal functions $\omega_1$ and $\omega_2$, and permitting a dichotomy of the cross-cousins. These conditions are sufficiently restrictive so that, up to isomorphism, there is but one structure with these properties. This uniqueness permits, in the form proposed here, the representation of the "Murngin system" by this single elementary structure.

In reality, this uniqueness is only "up to isomorphism" and theorem 2 affirms the existence of two distinct structures with the required properties. Account must be made of this double solution to get close to the functioning of this complex system.

If, following this idea, one tries to define the Murngin system (obviously from the structural point of view developed here), one is lead to considering it as a set of conditions relative to an elementary kinship structure $(\omega,\mu,\pi)$ over this set S.

These conditions can be of two types: on the one hand, they may be conditions of a general type that do not introduce any constants, such as the condition of generalized exchange or the dichotomy of the cross-cousins. On the other hand, they may be conditions of a particular type that introduce constants, such as being given two moities $E_1$ and $E_2$ (more precisely, subsets of S), being given the maternal function, or being given two permutations $\omega_1$ and $\omega_2$ and requiring that $\omega(x) = \omega_1(x)$ or $\omega(x) = \omega_2(x)$ for all $x \in S$ (as is the case in the Murngin system).

Whatever the conditions may be, let us adopt the following definition of a kinship system.

Definition 1.

A kinship system[1] is a finite set S and a set of conditions of general or particular type $C_i(\omega,\mu,\pi)$ $(1 \leq i \leq n)$ relative to an elementary kinship structure $(\omega,\mu,\pi)$ over this set.

Example 1. Aranda system.

S is a set with 8 elements.

---

[1] In conformation with the terminological format employed in this work (no. E 1.1), the term "kinship system" so defined belongs to the vocabulary of the Mathematical Model.

Condition $C_1$: $(\omega,\mu,\pi)$ comprises two exogamous matrilineal moities which are cycles.

Condition $C_2$: $\omega^2 = \varepsilon_s$.

Condition $C_3$: $\pi^2 = \varepsilon_s$.

The Aranda system has no particular conditions (no constants).

Example 2. Murngin system.

(See no. 3.1.)

$E_1$, $E_2$, $\mu$, $\omega_1$, and $\omega_2$ are the constants of the system.

If a kinship system $(S,C_1,\ldots,C_m)$ is given, then one can attempt to study its functioning by means of the properties of elementary kinship systems which are compatible with it.

Definition 2.

By spectre of the kinship system $(S,C_1,\ldots,C_m)$ is meant the collection of elementary kinship structures $(\omega,\mu,\pi)$ over S which satisfy the conditions $C_1(\omega,\mu,\pi),\ldots,C_m(\omega,\mu,\pi)$.

In theorems 1 and 2 we have above all tried to show that the spectres of the Aranda and Murngin system are composed of structures which are isomorphic among themselves.

In any case, the fact that the spectre of the Murngin system is reduced to two structures $(\omega',\mu,\mu\omega')$ and $(\omega'',\mu,\mu\omega'')$ (see E 3.2 b) appears interesting in and of itself, independent of any isomorphism which links these two structures.

More generally, one could propose a search for means to establish the equations linking the spectre to the functioning of the kinship system under consideration, etc. It would be necessary to begin by defining the mathematical notion of function (obviously in a manner that is in accord with the demographic facts).

BIBLIOGRAPHY

1. C. Lévi-Strauss. Les structures élémentaires de la parenté, P.U.F., Paris, 1949.

2. H. White. An anatomy of kinship, Prentice Hall, Inc., Englewood Cliffs, New Jersey, 1963.

Note: Final copy of the text was typed at the Center for Demographic and Population Genetics of the University of Texas Health Science Center at Houston, by Jeryl Silverman, under grant number GM 19513.

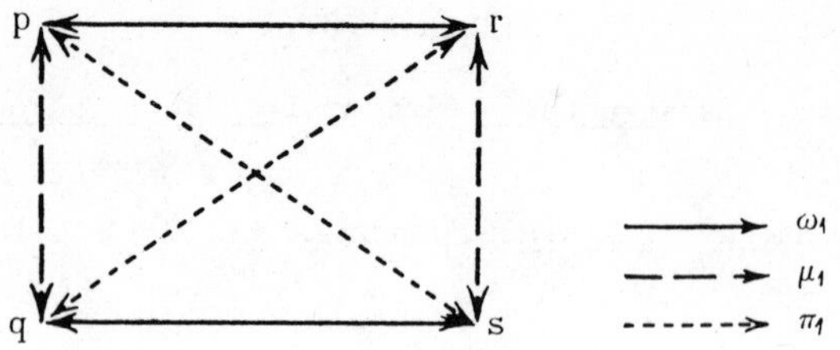


Figure 1

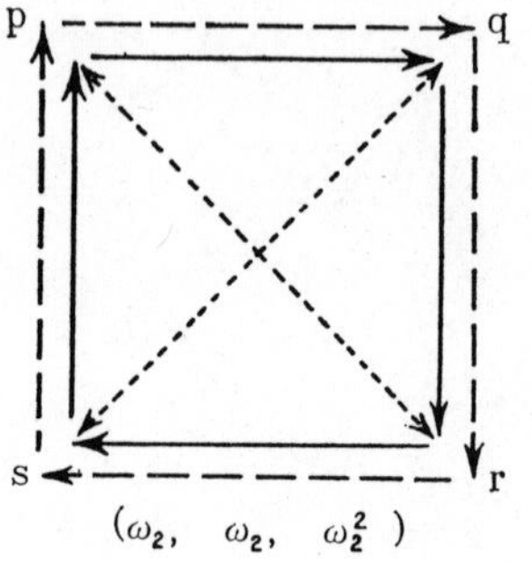


Figure 2

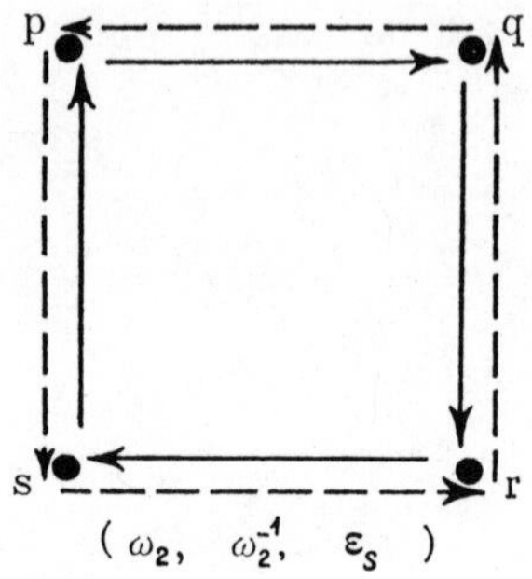


Figure 3

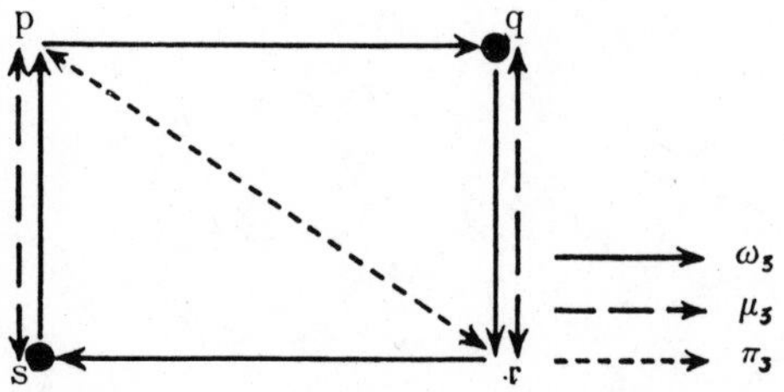


Figure 4

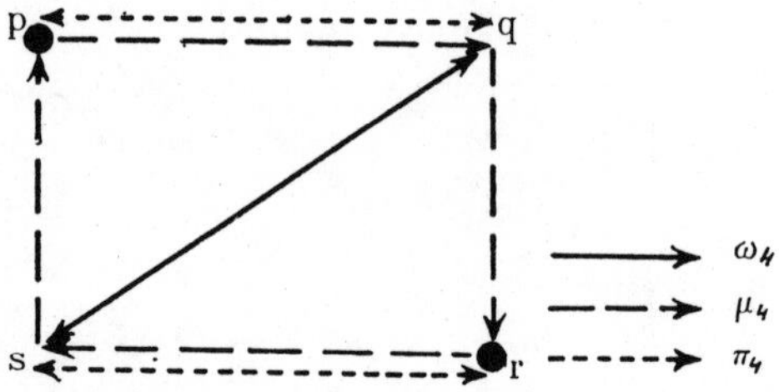


Figure 5

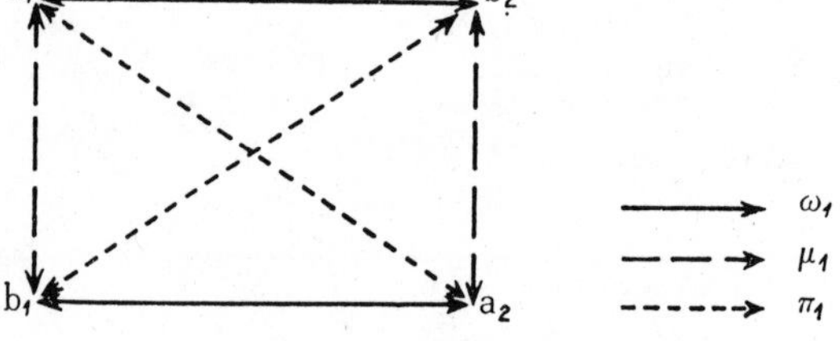


Figure 6

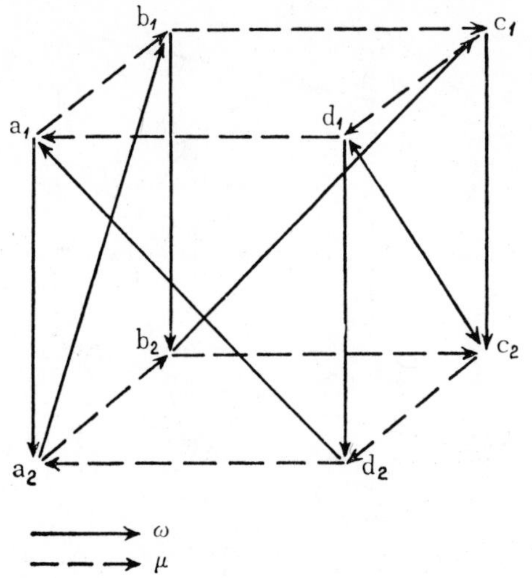


Figure 7

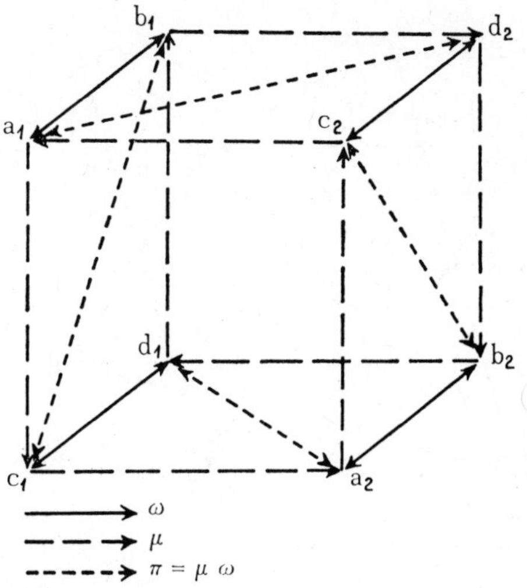


Figure 8. (The notation is that of [1] p. 211.)

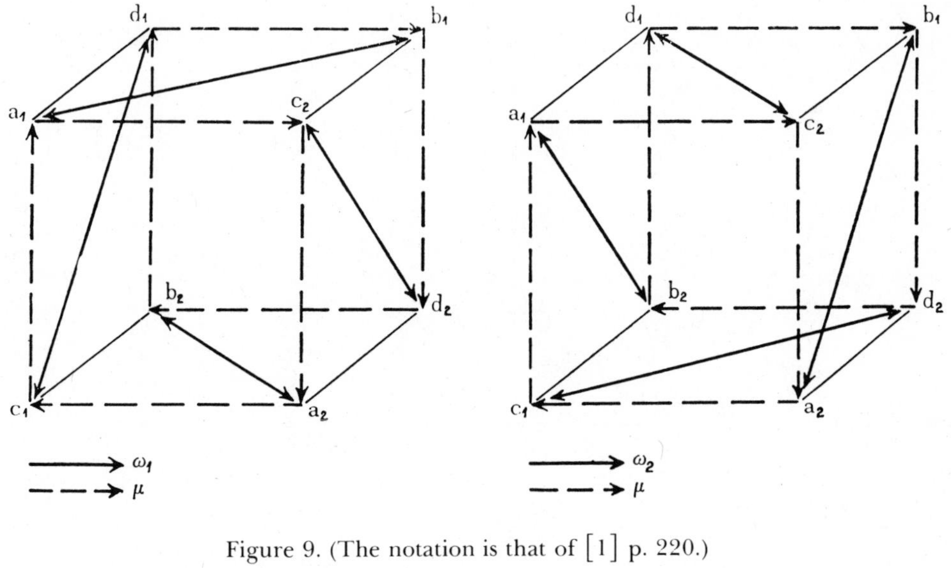


Figure 9. (The notation is that of [1] p. 220.)

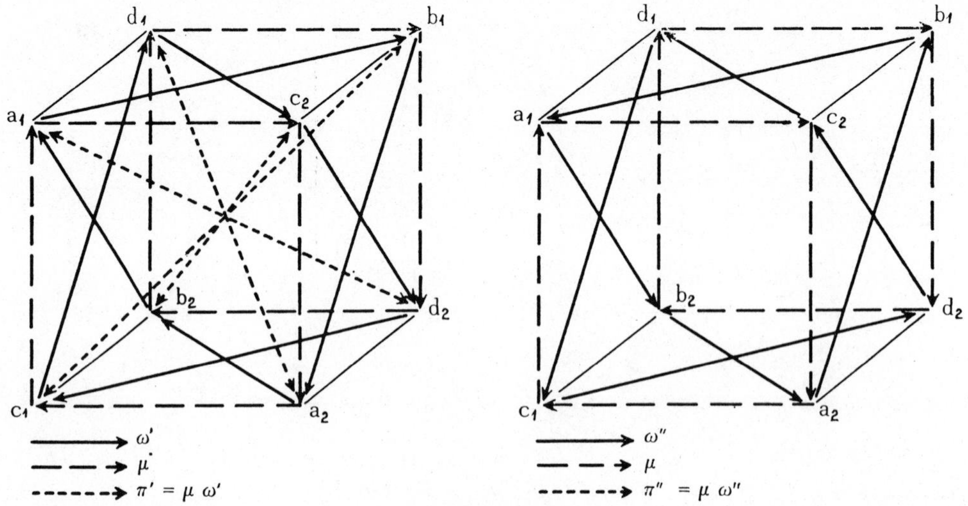


Figure 10

# 14



## Models of Kinship Systems with Prescribed Marriage

**Harrison C. White** *Harvard University*

### 1. Introduction

***Background.*** In 1949 André Weil, in the appendix to Part I of *Elementary Structures of Kinship* by Lévi-Strauss, sketched out one way to analyze in algebraic terms the structures of certain kinship systems. A basic step in his conceptualization is the assignment of all members of the society to a set of a few mutually exclusive and exhaustive marriage types, each husband and wife having the same type. (Each type reflects prescribed marriage of men in one clan to women in another.) He confines himself to systems with bilateral and matrilateral first-cousin marriage which further can be described in terms of cyclic groups. After mentioning that the theory of groups of permutations is applicable, he solves concrete examples through an ingenious and rather specialized use of the addition of $n$-tuples modulo two.

R. R. Bush in an undated mimeo manuscript proposed permutation matrices as a more convenient tool for the analysis. Like Weil, Bush carries out the analysis in terms of marriage types, although the implications for clan membership are always drawn by both authors. Bush suggests that $(M!)^2$ societies with $M$ marriage types are possible, and he works out a few concrete cases for small $M$.

Kemeny, Snell, and Thompson develop the work of Weil using the tool suggested by Bush. In Secs. 10 and 11 of Chapter 5 of *Introduction to Finite Mathematics* they develop the elementary properties of groups and subgroups of permutation matrices; then in Secs. 7 and 8 of Chapter 7 they present a reformulation of the Weil analysis. Although their book can be called an elementary text, these brief sections, with their extensive problem sets, are a major advance over the previous work. The properties of the societies to be investigated are formulated as an integrated set of axioms. The seventh and last axiom, which postulates a kind of homogeneity in the kinship structure, is not foreshadowed in the work of Weil and Bush. Using this plausible axiom, they show there are only a few allowed societies with a given number of marriage types.

***Content.*** In this chapter we systematically derive and describe all distinct kinship structures that satisfy the Kemeny-Snell-Thompson axioms and exhibit one of several kinds of prescribed marriage of particular interest to anthropologists.

A considerable reformulation of the Kemeny-Snell-Thompson approach was desirable. Marriage type is not a concept to be found in the field notes of anthropologists or the thinking of members of the societies. It not only is possible but also proves to be simpler to define the permutation operators in terms of clans. Instead of having one matrix represent the transformation of parents' marriage type into son's type, and another similar matrix represent daughter's marriage type, we deal with one matrix for transforming husband's clan into wife's clan, and another for transforming father's clan into children's clan. Both the formulation and the results are easier to interpret concretely, and the derivations are somewhat simpler.

We show that any abstract group, or equivalently any group of permutations, which can be generated by two elements corresponds to at least one allowed society. Moreover, we find that the regular matrix representation of an abstract group, which is easily written down from the multiplication table, provides a convenient translation of the results of the abstract algebraic derivations into the explicit matrix operators in terms of which the societies can be visualized. However, the same abstract group can represent not only distinct societies but also different types of societies. It is much simpler to consider not groups but pairs of generators of groups in deriving all possible societies of a given kind. Even then it is necessary, after defining what is meant by distinct societies, to show that each allowed pair of generators found yields a distinct society.

Additional tools for analyzing these societies are developed in the remaining sections. In the three earlier works cited, the primary question asked for a given society or type of society was: Which of his female relations can male ego marry? A second question, which can be answered using the results of this chapter, is: What different kinds of relationship can exist between the same two persons in a given type of society? Since there are few clans relative to the number of people, and since everyone in the society is by hypothesis related to some degree, obviously a large number of distinct relationships must relate any pair of people. Two important special cases treated in some detail are the conditions under which (1) two persons can be bilateral cross cousins of the same degree and (2) second- and first-cousin relations of various kinds can coexist between the same two people.

## 2. Axioms

We wish to construct a typology of all prescribed marriage systems that have the following properties:

1. The entire population of the society is divided into mutually exclusive groups, which we call clans. The identification of a person with a clan is permanent. Hereafter $n$ denotes the number of clans.

2. There is a permanent rule fixing the single clan among whose women the men of a given clan must find their wives.
3. By Rule 2, men from two different clans cannot marry women of the same clan.
4. All children of a couple are assigned to a single clan, uniquely determined by the clans of their mother and father.
5. Children whose fathers are in different clans must themselves be in different clans.
6. A man can never marry a woman of his own clan.
7. Every person in the society has some relative by marriage and descent in each other clan: that is, the society is not split into groups not related to each other.
8. Whether two people who are related by marriage and descent links are in the same clan depends only on the kind of relationship, not on the clan either one belongs to.

We also refer to these eight properties as axioms.

### 3. Marriage and Descent Rules as Permutation Matrices

The rule required by Axioms 2 and 3 can be presented in the form of a permutation matrix of side $n$: that is, a square matrix with exactly one entry of unity in each row and column and all other entries zero. Number the clans from 1 to $n$, and let the $i$th row and the $i$th column of the matrix correspond to the $i$th clan. Assume each row of the matrix corresponds to a husband's clan, the wife's clan being identified with the column in which the number 1 appears in that row. Call this matrix $W$. It shows the one clan from whom women of any given clan get their husbands and the one clan from whom the men of any given clan get their wives. Note that polygamy and polyandry are consistent with the axioms, though for simplicity we speak in terms of monogamy.

Since the wife's clan is uniquely determined by her husband's clan, the clan of a couple's children can by Axiom 4 be uniquely specified by the father's clan. Let $C$ be the permutation matrix in which $C_{ij} = 1$ if fathers of clan $i$ have children of clan $j$. $C$ must be of the form of a permutation matrix since, by Axiom 5, children in any given clan have fathers in only one clan, as well as vice versa.

There are $n!$ possible permutation matrices, or all together there are $(n!)^2$ combinations of marriage and descent rules for societies with $n$ clans which have the first five properties. Many of these combinations violate Axioms 6, 7, and 8, and only a small fraction of the valid combinations of rules are structurally distinct. It is necessary to define the latter term precisely in order to count and group distinct structures, but first we must study the implications of Axioms 6, 7, and 8.

$W$ and $C$ not only look like matrices but also can be meaningfully combined by the operation of matrix multiplication. For example, consider the element $(WC)_{ij}$ in the $i$th row and the $j$th column of the matrix $WC$ formed by multiplying the matrices W and C in that order. By the standard definition of matrix multiplication,

$$(WC)_{ij} = \sum_{k=1}^{n} W_{ik} C_{kj}.$$

There is exactly one unity in the $i$th row of $W$: say it occurs in the $p$ column. Similarly, in the $j$th column of $C$ there is only one unity, say in the $q$ row; so the sum on the right in the equation above is zero unless $p = q$, in which case the sum is just unity. In words this means the $(i, j)$ element in $(WC)$ is unity if and only if men in the $i$th clan marry women whose clan brothers are the fathers of children of clan $j$. But there must exist some $j$ such that $C_{pj} = 1$, by Axioms 5 and 4. To put it affirmatively, the matrix $(WC)$ specifies for a man of each clan the clan to which the children of his wife's brother belong. Any ordered series of any powers of $W$ and $C$ when multiplied together will, by the same logic, give a product matrix which is a permutation matrix specifying for each possible clan of a man the clan of a given relative of his.

One possible product matrix is the identity matrix, call it $I$, in which $I_{ij} = 1$ if and only if $i = j$, with all other elements zero. Whatever the clan of a man, any relative of his for whom the product matrix is $I$ will have the same clan that he does. Axiom 6 requires that in the matrix $W$ no diagonal element $W_{ii}$ be unity; certainly $W$ cannot be the identity matrix $I$. Approximately $n!/e$ (where $e = 2.71\ldots$) $n \times n$ permutation matrices satisfy this restriction. $C$ on the other hand can be $I$ when the children of men of any clan are in that clan. If any diagonal element of $C$ is unity, all $C_{ii}$ must be unity, for otherwise some men would be in the same clan as their children but not others, in contradiction of Axiom 8. A parallel argument leads to the conclusion that any product matrix formed from $W$ and $C$ must have no diagonal elements unity or else it must be $I$.

Axiom 8 has further implications. If any man is in the same clan as his own son's son, all men must be; so $C^2 = I$, and so on. If none of the powers $C, C^2, C^3, \ldots, C^{n-1}$ is the identity matrix, then $C^n$ must be. For suppose $C^i \neq I$, $i = 1, \ldots, n$. Then each succeeding generation of sons of sons has a clan different not only from the clan of the man we started with but also from all clans of intermediate ancestors in the male line, given Axiom 8. Thus there must be $n + 1$ clans, in contradiction to our assumption in Axiom 1. So $C^p = I$ for some $1 \leq p \leq n$. Thus any power of $C$ is equal to $C$ to a power between 1 and $p$, inclusive. The same conclusion obviously can be drawn concerning the powers of $W$ and the powers of any product matrix which is made up of $W$ and $C$, since any product matrix corresponds to a relation of a fixed kind.

The inverse of a matrix $M$ is defined by

$$MM^{-1} = M^{-1} M = I.$$

For example, $C^{-1}$ is the matrix that specifies for each clan a son may have what clan his father is in. Thus $(C^{-1})_{ij} = C_{ji}$. Suppose $C$ is of order $p$; that is, $p$ is the lowest integer such that $C^p = I$, where it was shown above that $p \leqslant n$. Since $CC^{p-1} = C^p = I$, the inverse of $C$ can also be written as $C^{p-1}$; and similarly for $W$ and for any product matrix. Thus $C$ and $W$ and their product matrices can be used to describe the change in clan in moving from a given person to his ancestors as well as to his descendants. Thus to every possible relation of a person in the society there corresponds a matrix which is some product of repetitions of $C$ and $W$. We will often use $M$ as a general symbol for such a matrix and call it a *relation matrix*.

There is a final very general restriction implied by Axiom 8 together with Axiom 7. If we calculate the product matrices formed by each of two different sequences of $W$ and $C$ matrices, we often find these two product matrices are equal; that is, the ones and zeros appear in the same places. Let us arbitrarily designate the distinct matrices which result from multiplying $W$ and $C$ in all possible orders and combinations by the symbols $A_1, A_2, A_3, \ldots$. There are exactly $n$ such matrices, as is clear intuitively since one ego has only $n$ essentially different kin relations, one with persons in each of the $n$ clans, and Axiom 8 requires the structure to be homogeneous.

Axiom 7 states that for any pair of clans $k$ and $j$ there is one of the matrices $A_i$ in which the $k, j$ element is unity. A given matrix $A_i$ has only one unity entry in a row $k$; there must be at least $n$ matrices $A_i$. Suppose there is an additional one, $A_{n+1}$. Then in $A_{n+1}$ the $k$th row must be identical with the $k$th row in some $A_i$, $i \leqslant n$. The $k$th row of $A_{n+1}$ can be used to specify what the clan is of some one kind of relative of a person in clan $k$, and similarly for $A_i$. But two persons related in specified ways to a given person also have a specified relation to each other, and here these two persons are in the same clan. It follows that $A_i$ and $A_{n+1}$ must be equal in each row, for otherwise Axiom 8 is violated. In other words, if two of any set of permutation matrices satisfying Axioms 7 and 8 are equal in one element they are identical. There are therefore exactly $n$ distinct permutation matrices generated as products of any $W$ and $C$ matrices for a society with $n$ clans which has properties 1–8.

There is at least one society satisfying all eight axioms for any $n$: that with $C = I$, and $W_{i,i+1} = 1$, $1 \leqslant i < n$, $W_{n,1} = 1$, and all other $W_{i,j} = 0$. The $n$ distinct matrices are $W, W^2, \ldots, W^{n-1}$, and $W^n = I$. Another obvious possibility is a society with $C' = I$, $W'_{i,i+2} = 1$, $1 \leqslant i < n-1$, $W'_{n-1,1} = 1$, and

$W'_{n,2}=1$, and all other $W'_{ij}=0$; again $W'$, $(W')^2, \ldots, (W')^n$ are the distinct matrices. It is intuitively clear that this second society differs from the first only as to the numbering of the clans.

We will use this description of societies by permutation matrices to classify the societies according to what kinds of relatives are allowed to marry. It is natural to say that two societies described by different pairs of $C$ and $W$ matrices are structurally distinct if and only if there is at least one kind of relative who is allowed to marry ego in one society but not the other. Let $M(C, W)$ be any matrix defined as a product of a sequence of powers of $C$ and $W$. Then two societies have equivalent structures when $M(C, W) = W$ in one if and only if $M(C', W') = W'$ in the other where $M$ has the same form in both. In the example above $M(C, W)$ can be written as $W^m$. If $W^m = W$, then $m = jn + 1$ for some integer $j$. But $(W')^n = I$ also; so $(W')^m = W'$ if and only if $W^m = W$.

Very restrictive conditions thus must be satisfied by the $C$ and $W$ matrices. Many fewer than $n!(n!/e)$ pairs of permutation matrices satisfy these restrictions, and even fewer pairs give structurally distinct societies. However, there is no simple way to count the number of distinct societies with properties 1–8. We shall consider only certain general classes of our ideal-type societies. These classes will be defined by the kinds of first-cousin marriages which are allowed in known primitive societies. To simplify the derivations we need to develop a more abstract view of the $A_i$ matrices.

## 4. Groups and Societies

Consider the $n$ distinct relation matrices $A_i$ generated by a $C$ and a $W$ matrix for a society with $n$ clans which satisfies Axioms 1–8. The product of $A_i$ and $A_j$ is some $A_k$ for any $i$ and $j$. We also proved above that one of the $A_i$ is the identity matrix and that for each $A_i$ there is an $A_j$ which is its inverse. Matrix multiplication is associative; that is, $A_i(A_jA_k) = (A_iA_j)A_k$, which is another way of saying, for example, that a man's son's grandson is the same person as his grandson's son. Therefore the set of $A_i$'s constitute a representation of an abstract group. A group is specified by its multiplication table, in which the entry in the $i$th row and $j$th column is $a_k$ when $a_ia_j = a_k$.

In the appendix to this chapter we go a step further and show that a regular representation of *any* abstract group generated by two elements constitutes a set of $A_i$ describing an allowed society. The number of elements in the group is called its *order*. One way to begin classifying societies would be by examining all instances of abstract groups of different orders. There is a well-organized and highly developed literature on the properties of abstract groups in which all groups of order less than, say, 32 are examined in exhaustive detail. Unfortunately this approach is not fruitful, for there are usually numerous pairs of elements in a group which will generate the group. Thus the same group can be isomorphic with the set of distinct relation matrices for two very different societies.

It is more efficient to begin by finding all pairs of abstract group generators $C$ and $W$ which have specified characteristics for a given group size $n$. Then one can construct their regular representation from the multiplication table and diagram each society. These calculations are much simpler, because we can treat $C$ and $W$ and their products as elements in an abstract group algebra rather than as explicit matrices. Once the multiplication table showing all possible products of $C$, $W$, and the other $n - 2$ distinct elements is derived, one can write down a concrete matrix for each element. The $n \times n$ permutation matrix which sends the standard list of elements (the row or column headings of the multiplication table) into the $i$th row of the table is a valid matrix representation of $a_i$. It was proved that there are the same number of clans as elements, and the numbering of the clans is arbitrary; so the matrix representation obtained from the group multiplication table uniquely specifies a society as long as $C$ and $W$ are explicitly identified in the list of elements.

## 5. First-Cousin Marriages

It is both logically and empirically appropriate to classify kinship systems on the basis of the kinds of first cousins allowed to marry. There are four possible kinds of first cousins, if one is male and one female. These can be described most easily by a family tree, in which the symbolic convention is:

△ for male
○ for female
— for sibling relation
| for parent-child relation
= for marriage relation

The arbitrary convention of referring all relations to a male ego will be used throughout. When the girl cousin is the male's father's brother's daughter, the relation graphically is as shown in Fig. 1; this girl cousin is written FBD. When the two siblings who are parents of the first cousins are of the same sex, the latter are termed *parallel* cousins, otherwise *cross* cousins. When the parent of male ego is female, the cousins are said to be *matrilateral* cousins; male ego and the girl are *patrilateral* cousins if it is the father of male ego who is a sibling of one of the girl's parents. In Fig. 1 the boy and girl are patrilateral parallel cousins. When the father of the boy is the brother of the girl's mother and also the mother of the boy is the sister of the girl's father, they are said to be *bilateral* cross cousins.

A fundamental question to be answered is: What kinds of relations can marry in a society specified by given $C$ and $W$ matrices? Begin with Fig. 1. The matrix $C^{-1}C$ specifies in a row $i$ by a unity entry the column for the clan of the girl who is a patrilateral parallel cousin of male ego in clan $i$. The clan of the boy's father, say $j$, is specified in the $i$th row of $C^{-1}$, the father's brother is in the same clan, and the latter's children are in the clan specified in the $j$th row of $C$, that is in the clan specified in the $i$th row of $(C^{-1}C)$. In this case it is easy to see directly from Axioms 2–5 required for all societies considered here that the boy and girl cannot marry, and $CC^{-1} = I$, as it should.

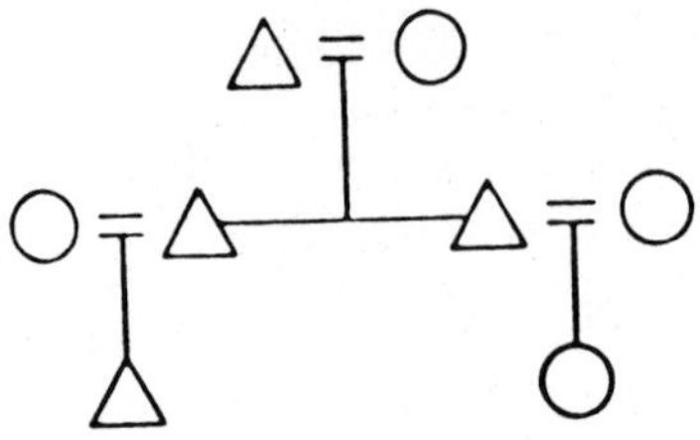

**Fig. 1.** Patrilateral parallel cousins.

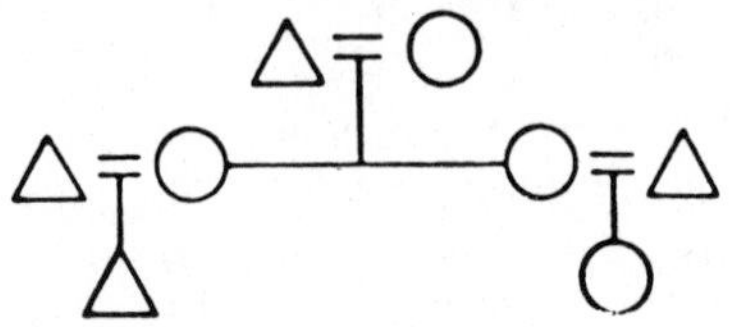

$M_2 = C^{-1}WW^{-1}C$

**Fig. 2.** Matrilateral parallel cousins. M is the relation matrix, in which the girl's clan is specified by the column in which a unity entry appears in the row specified by the boy's clan.

If a boy can marry a girl, her clan must be that indicated by a unity entry in a row of $W$ corresponding to the boy's clan. In symbolic terms, if $M$ is the matrix describing the clan of a girl relation of male ego (for Fig. 1, $M = C^{-1}C$), then

$$M = W$$

is the condition that must be satisfied if the girl is to be a legitimate marriage partner of the boy, whatever his clan. By Axiom 6, $W \neq I$, ever, so in no society satisfying Axioms 1–8 can a boy marry his female patrilateral parallel cousin.

Marriage between matrilateral parallel cousins (see Fig. 2) is also forbidden, since there

$$M = C^{-1}WW^{-1}C$$

or $M = I$, using the associative law. Just as the conversion of husband's to wife's clan is specified by $W$, the conversion of wife's clan to husband's clan is given by $W^{-1}$.

The other two kinds of cousins, matrilateral and patrilateral cross cousins, can marry for some $W$ and $C$ matrices. (See Figs. 3 and 4.) Matrilateral cross cousins can marry if and only if

$$W = M = C^{-1}WC.$$

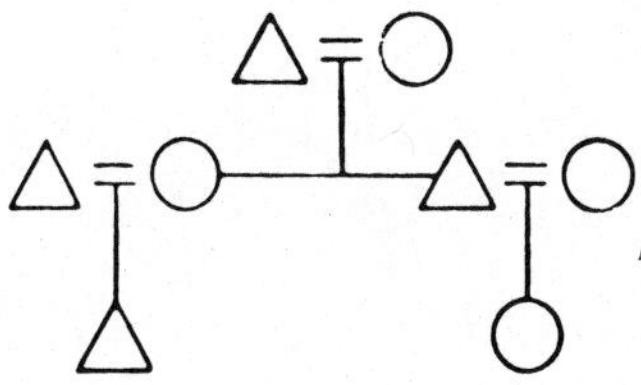

$M_3 = M_4^{-1}$

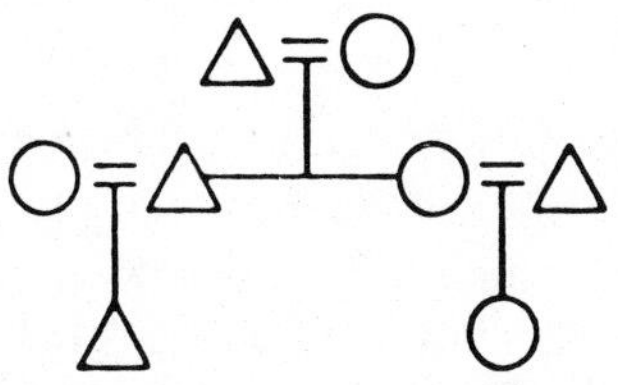

$M_3 = C^{-1}WC$

**Fig. 3.** Matrilateral cross cousins.

$M_4 = C^{-1}W^{-1}C$

**Fig. 4.** Patrilateral cross cousins.

If we premultiply both sides of the equation by $C$, we have $CW = (CC^{-1})WC$, or

$$CW = WC$$

as the necessary and sufficient condition. In other words, the order in which $W$ and $C$ are multiplied does not affect the product matrix; that is, $W$ and $C$ commute. Since all relation matrices are generated as products of $W$ and $C$, it follows at once that all $n$ distinct matrices $A_i$ for a given society must commute with each other if matrilateral cross-cousin marriage is allowed.

Male ego is allowed to marry many other kinds of relations than matrilateral cross cousins in a society described by a commutative group of matrices, and there are several different subtypes of such societies. The first major goal will be to specify and analyze all such societies. It proves more orderly to define subtypes by simple algebraic conditions on $C$ and $W$ than by what other types of relatives marry.

From Fig. 4 it can be seen that patrilateral cross-cousin marriage is allowed if and only if

$$W = M = C^{-1}W^{-1}C.$$

Again premultiplying both sides of the equation by $C$, we have

$$CW = W^{-1}C$$

for the necessary and sufficient condition. This can be called the semicommutative condition. An alternative form is obtained by first postmultiplying both sides of this equation by $W^{-1}$:

$$C(WW^{-1}) = W^{-1}CW^{-1} \quad \text{or} \quad C = W^{-1}CW^{-1},$$

and then premultiplying both sides by $W$:

$$WC = CW^{-1}.$$

The second major goal is to identify and describe all societies in which this equation holds, and for each to find what kinds of relatives can marry.

There is an ambiguity in our categories of cross-cousin relations. Cousins may be bilateral; if they are cross cousins, then both Fig. 3 and Fig. 4 describe the relation of male ego to his girl cousin. In a society in which bilateral cross cousins may marry, both $WC = CW$ and $CW = W^{-1}C$ must apply. But then

$$W^{-1}C = WC$$

if we combine the two equations. If we postmultiply by $C^{-1}$, then

$$W = W^{-1}$$

is a necessary condition for bilateral cross-cousin marriage. An alternative form is

$$W^2 = I;$$

that is, the order of $W$ must be two.

Furthermore, bilateral cross cousins cannot exist in a society unless $W^2 = I$. From Fig. 3,

$$M = C^{-1}WC$$

must describe the girl's clan by columns for the boy's clan by rows; but by Fig. 4,

$$M = C^{-1}W^{-1}C$$

must also describe this transformation of clans. A given girl can be in only one clan; so

$$C^{-1}W^{-1}C = C^{-1}WC$$

is required if there is not to be a contradiction. But this can be reduced to the equation $W = W^{-1}$.

On the other hand, in any society there can be bilateral parallel cousins. In both Fig. 1 and Fig. 2 the relation matrix between the boy and girl cousins is just the identity. Bilateral parallel cousins can never marry but can exist in any society; bilateral cross cousins exist only in societies in which $W^2 = I$ and can marry if and only if in addition $WC = CW$.

The condition $W^2 = I$ has a very simple interpretation: there must be an even number of clans in the society, and each clan must swap women as wives with another clan. The basic typology of societies will be

I. *Bilateral marriage*, in which $W^2 = I$ and $WC = CW$.
II. *Matrilateral marriage*, where $WC = CW$, but $W^2 \neq I$.
III. *Patrilateral marriage*, where $WC = CW^{-1}$, but $W^2 \neq I$.
IV. *Paired clans*, where $W^2 = I$, $WC \neq CW$.
V. *Residual.*

In the first three names "cross cousin" is omitted, since parallel first cousins can never marry. Only in I and IV societies can there exist bilateral cross cousins. Observe that any two of the three conditions imply the other: for example, if

$$WC = CW^{-1} \quad \text{and} \quad W^2 = I$$

then

$$W = W^{-1} \quad \text{and} \quad WC = CW.$$

# 15



Reprinted from *VIIIth Congr. Anthropol. Ethnol. Sci., Vol. II: Ethnology*, 90–92 (1968)

# Formal Analysis of Prescriptive Marriage System: The Murngin Case

PIN-HSIUNG LIU

In 1949 André Weil proposed a unique concept of marriage types for the analysis of section systems. He wished to prove that the theory of groups of permutations is applicable to the study of prescribed marriage systems. Weil's idea was developed by a group of mathematicians, though sporadically, and thus a kind of new mathematical approach to the kinship structure has been established. Following Weil, Robert Bush in his undated mimeo manuscript introduced permutation matrices as a more effective tool for analysis [see Bush (1963)]. Kemeny, Snell, and Thompson (1956) first systematized the properties of the prescriptive marriage systems as an integrated set of axioms. All distinct kinship structures that satisfy the abovementioned axioms are systematically derived and described by White (1963), who adopted more practical "generators" to set structural analysis on a more concrete basis.

In the past two decades kinship mathematics as developed by mathematicians has made remarkable progress, but there still remain controversial problems, and the effective range of the mathematical approach is obviously limited. White himself recognized the failure of his structural analysis of those societies, such as Murngin or Purum, which practice unilateral cross-cousin marriage. Many reasons for this failure may exist, as discussed by White and by Reid (1967), but it is time to question whether the method they applied is sound or not.

If we examine the writings of Weil and his successors, the fatal fact emerges that their method does not allow one to distinguish oblique marriage from cross-cousin marriage. This defect is caused by the fact that the mathematicians neglect important anthropological phenomena, as, for example, the regulation of marriage among the descent groups.

Here I propose the following working hypotheses for a new mathematical approach to the prescriptive marriage systems.

1. The matrilateral cross-cousin marriage system is described by an $n$-generation cycle, where the marriage cycle (or circulation connubium) is derived from $n$ hordes or clans, the minimum number for $n$ being 3.

2. Under the same conditions, the patrilateral cross-cousin marriage system is described by a two-generation cycle, the cycle of alternating generations.

3. The bilateral cross-cousin marriage system is described by the above-mentioned two principles, the $n$-generation and 2-generation cycles, but in this case $n$ may be 2 or more.

The present study is designed to examine only the first hypothesis. The Murngin case is chosen. Methodologically, the new mathematical approach proposed by Harvey and Liu (1967) is adopted. In that approach the kin relationships are all reduced to the two basic units of parent and child. Kinship category is determined by the products of the two units as generators. The kinship categories represented by the numerical notation system are computable, and mathematically they are regulated by algebraic theory.

The model of genealogical space composed by societies of the section system is completely different from that composed by societies characterized by kinship categories or bilateral systems. Therefore, the numerical kinship notation system is not applicable to the former. New generators appropriate to the section system must be selected. For this purpose I propose here two basic units, father–child link and mother–child link. The former is represented by $m$ and the latter by $f$, which are used as generators for analysis of the Murngin system.

It is an established theory that at least three descent groups are necessary to form a marriage cycle for the practice of the "remote exchanges" characterized by unilateral cross-cousin marriages. The Murngin's sixty-odd patrilineal clan-hordes, which are the minimum unit of an exogamous group, are divided into two moieties which intermarry. These are known as *Dua* and *Yiritcha*. Each moiety is further subdivided into four sections, making a total of eight sections. For *Dua* the sections are *Buralang* (O), *Warmut* (P), *Balang* (Q), and *Karmarung* (R). For *Yiritcha* they are *Bulain* (S), *Kaijark* (T), *Ngarit* (U), and *Bangardi* (V). The symbols used in this paper for the respective sections are given in parentheses.

According to the Murngin law a marriage among the clans of the same moiety is prohibited, so the number of clans or hordes required to organize the marriage cycle must be even, and the minimum number is four. Figure 1 depicts the marriage cycle of the matrilateral cross-cousin marriage system composed by four clans which are represented by a patriline. The alphabetical order represents four patriclans and the numbers four matriclans. Solid lines stand for patrilines and dotted lines for matrilines.

The diagram shows hypothesis 1 working: each descent line is regulated by a 4-generation cycle; thus the four patrilines and the four matrilines intersect each other to produce 16 segments. In Figure 2 the general symbols used in Fig. 1 are replaced by the specific section symbols used for the Murngin system. Clans belonging to the same moiety are distinguished by subscripts.

The Murngin system is further characterized by the extraordinary rule of pairing two marriage classes, regular and alternate, for each section. These characteristics are best shown by a Cayley diagram, Figure 3, where each segment is represented by the accumulation of generators $m$ and $f$. Both generators are determined modulo 4, and $O_1$ is chosen as origin or for the position of ego (I). Arrows on the matrilineal path indicate the direction of transit from ascending to descending generation. The lines of

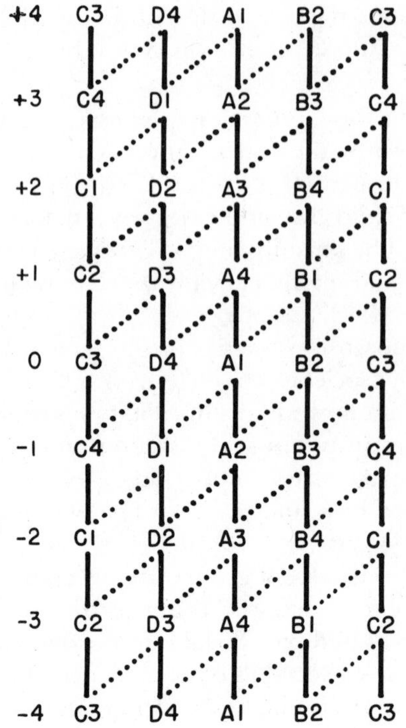


Fig. 1

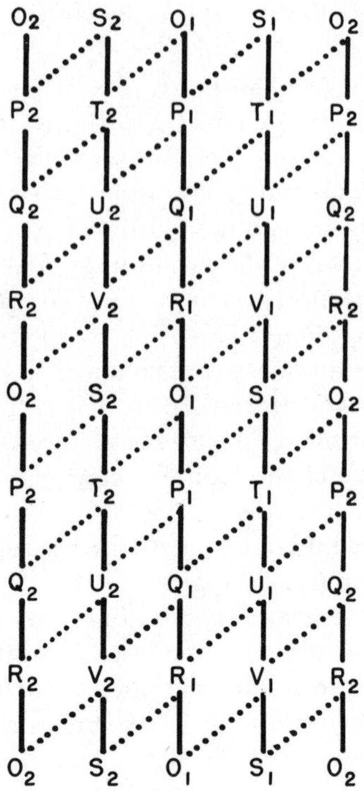


Fig. 2

descent are constant and are not affected by the type of marriage. The patrilineal path is reversible. In the case of regular marriage it proceeds in the direction of alphabetical order, and for alternate marriage in reverse order.

Each segment plays multiple roles and expresses multiple combinations of generators. For example,

$$R_2 = mf^2, f^2m, fmf, mf^{-2}, f^{-2}m, f^{-1}mf^{-1}, m^{-3}f^2, m^{-3}f^{-2}, \ldots$$

Among them the form $f^i m^j$ is chosen as a representative for each segment. To express this a new numerical notation system is given. We adopt a two-place system and assign the first place to the exponent of $f$ and the second place to the exponent of $m$. For example, $f^2m$ ($R_2$) is represented as 21. Now, the Murngin's basic marriage cycle, composed of four clan-hordes ($G_4$: superscripts indicate the generation cycle and subscripts indicate the number of hordes), can be expressed as a set of 16 segments:

$$G_4^4 = \left\{ \begin{array}{llll} 00\ (O_1) & 10\ (T_2) & 20\ (Q_2) & 30\ (V_1) \\ 01\ (P_1) & 11\ (U_2) & 21\ (R_2) & 31\ (S_1) \\ 02\ (Q_1) & 12\ (V_2) & 22\ (O_2) & 32\ (T_1) \\ 03\ (R_1) & 13\ (S_2) & 23\ (P_2) & 33\ (U_1) \end{array} \right\} \qquad \begin{array}{l} \underline{m}\text{: mod} = 4 \\ \underline{f}\text{: mod} = 4 \end{array}$$

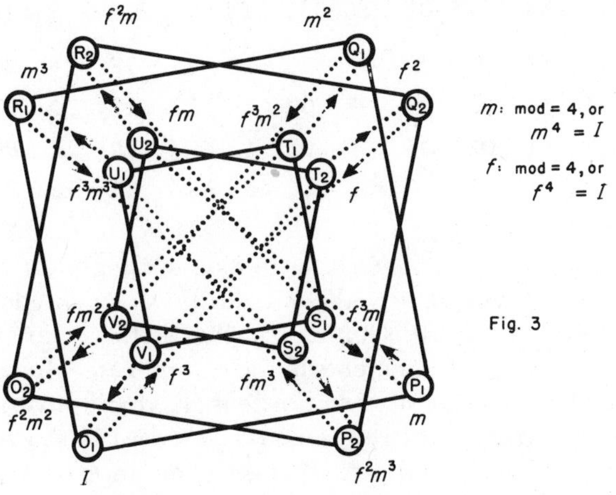


Fig. 3

This segment diagram shows an algebraic group characterized by the following mathematical properties:

1. Binary operation: The generations of the same descent line (or the exponents of the same generators) are computable. The formula is $f^p m^q \cdot f^r m^s = f^{p+r} m^{q+s}$. For example, 23 · 31 = 10 (mod $m, f$ = 4).

2. Identity: 00 is the identity segment. For example, 00 · 21 = 21 · 00 = 21.

3. Associativity: This property is satisfied as (21 · 13) · 02 = 21 · (13 · 02) = 32.

4. Inverses: Each element or segment has its inverse in the set. The formula is $(f^p m^q)^{-1} = f^{4-p} m^{4-q}$. For example, $(21)^{-1} = 23$, or 21 · 23 = 00.

5. Commutative: This property is shown as 13 · 22 = 22 · 13 = 31.

Because it is commutative, and because of the finite number of its elements, this group is called a "finite Abelian group." In the segment diagram each column represents a patrilineal subgroup. Each row represents a matrilineal subgroup. The multiplication table of this group is

| · | 00 | 01 | 02 | 03 | 10 | 11 | 12 | 13 | 20 | 21 | 22 | 23 | 30 | 31 | 32 | 33 |
|---|---|---|---|---|---|---|---|---|---|---|---|---|---|---|---|---|
| 00 | 00 | 01 | 02 | 03 | 10 | 11 | 12 | 13 | 20 | 21 | 22 | 23 | 30 | 31 | 32 | 33 |
| 01 | 01 | 02 | 03 | 00 | 11 | 12 | 13 | 10 | 21 | 22 | 23 | 20 | 31 | 32 | 33 | 30 |
| 02 | 02 | 03 | 00 | 01 | 12 | 13 | 10 | 11 | 22 | 23 | 20 | 21 | 32 | 33 | 30 | 31 |
| 03 | 03 | 00 | 01 | 02 | 13 | 10 | 11 | 12 | 23 | 20 | 21 | 22 | 33 | 30 | 31 | 32 |
| 10 | 10 | 11 | 12 | 13 | 20 | 21 | 22 | 23 | 30 | 31 | 32 | 33 | 00 | 01 | 02 | 03 |
| 11 | 11 | 12 | 13 | 10 | 21 | 22 | 23 | 20 | 31 | 32 | 33 | 30 | 01 | 02 | 03 | 00 |
| 12 | 12 | 13 | 10 | 11 | 22 | 23 | 20 | 21 | 32 | 33 | 30 | 31 | 02 | 03 | 00 | 01 |
| 13 | 13 | 10 | 11 | 12 | 23 | 20 | 21 | 22 | 33 | 30 | 31 | 32 | 03 | 00 | 01 | 02 |
| 20 | 20 | 21 | 22 | 23 | 30 | 31 | 32 | 33 | 00 | 01 | 02 | 03 | 10 | 11 | 12 | 13 |
| 21 | 21 | 22 | 23 | 20 | 31 | 32 | 33 | 30 | 01 | 02 | 03 | 00 | 11 | 12 | 13 | 10 |
| 22 | 22 | 23 | 20 | 21 | 32 | 33 | 30 | 31 | 02 | 03 | 00 | 01 | 12 | 13 | 10 | 11 |
| 23 | 23 | 20 | 21 | 22 | 33 | 30 | 31 | 32 | 03 | 00 | 01 | 02 | 13 | 10 | 11 | 12 |

**FORMAL ANALYSIS OF PRESCRIPTIVE MARRIAGE SYSTEM**

| | | | | | | | | | | | | | | | | |
|---|---|---|---|---|---|---|---|---|---|---|---|---|---|---|---|---|
| 30 | 30 | 31 | 32 | 33 | 00 | 01 | 02 | 03 | 10 | 11 | 12 | 13 | 20 | 21 | 22 | 23 |
| 31 | 31 | 32 | 33 | 30 | 01 | 02 | 03 | 00 | 11 | 12 | 13 | 10 | 21 | 22 | 23 | 20 |
| 32 | 32 | 33 | 30 | 31 | 02 | 03 | 00 | 01 | 12 | 13 | 10 | 11 | 22 | 23 | 20 | 21 |
| 33 | 33 | 30 | 31 | 32 | 03 | 00 | 01 | 02 | 13 | 10 | 11 | 12 | 23 | 20 | 21 | 22 |

Next, let us expand the marriage cycles to have six, eight, or more clans participating. According to our working hypothesis 1, such systems should be regulated by a 6-generation cycle and produce $6^2$ segments in the first case, or regulated by an 8-generation cycle and produce $8^2$ segments in the second case. However, in the Murngin system the 4-generation cycle of each horde is fixed and invariable regardless of the number of hordes participating in the marriage cycle. This is not fully consistent with our working hypothesis, which states that the generation cycle should be equal to the number of clans or hordes. The hypothesis must be revised to make it consistent with the properties of Murngin system.

Figures 4 and 5 show diagrams of matrilateral cross-cousin marriage organized from six or eight patriclans or hordes. They produce 4 × 6 and 4 × 8 segments, re-

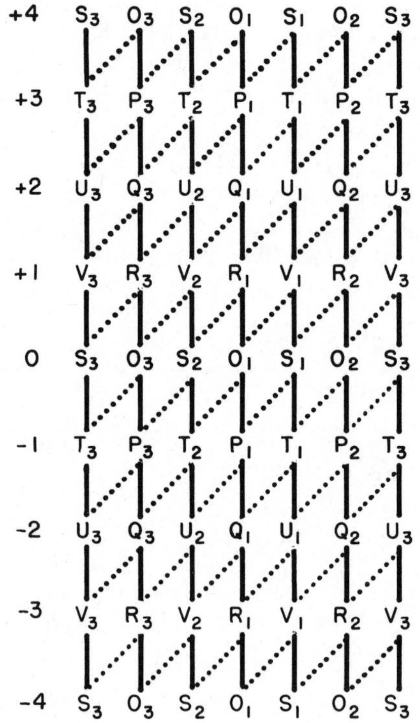


Fig. 4

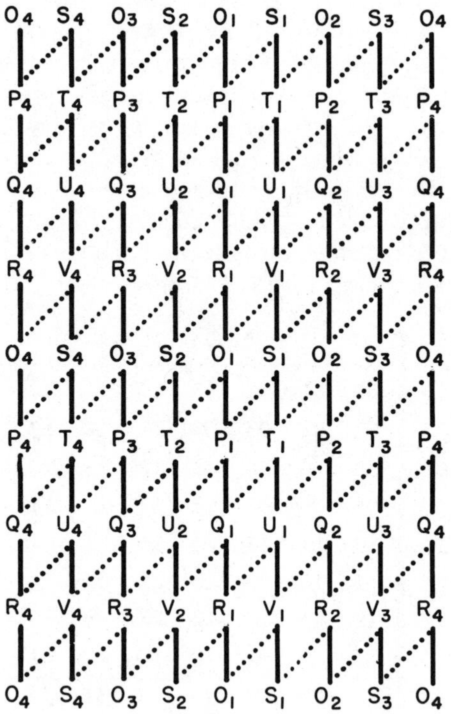


Fig. 5

spectively. The generation cycles of the matrilines are not four. Now they are determined by the least common multiple (LCM) of the generation cycle of patrilines and the number of patrilines. Then the number of matrilines is determined by the quotient of the total number of segments and the generation cycle of the matriline. Thus in the six-patriline case, the matriline is two and is regulated by a 12-generation cycle; in the eight patriline case, the matriline is four and is regulated by a 8-generation cycle. These groups are expressible in the following segment diagrams, respectively:

$$G_6^4 = \left\{ \begin{array}{cccccc} 00\ (O_1) & 10\ (T_2) & 20\ (Q_3) & 30\ (V_3) & 40\ (O_2) & 50\ (T_1) \\ 01\ (P_1) & 11\ (U_2) & 21\ (R_3) & 31\ (S_3) & 41\ (P_2) & 51\ (U_1) \\ 02\ (Q_1) & 12\ (V_2) & 22\ (O_3) & 32\ (T_3) & 42\ (Q_2) & 52\ (V_1) \\ 03\ (R_1) & 13\ (S_2) & 23\ (P_3) & 33\ (U_3) & 43\ (R_2) & 53\ (S_1) \end{array} \right\} \quad \begin{array}{l} \underline{m}: \text{mod} = 4 \\ \underline{f}: \text{mod} = 12 \end{array}$$

$$G_8^4 = \left\{ \begin{array}{cccccccc} 00\ (O_1) & 10\ (T_2) & 20\ (Q_3) & 30\ (V_4) & 40\ (O_4) & 50\ (T_3) & 60\ (Q_2) & 70\ (V_1) \\ 01\ (P_1) & 11\ (U_2) & 21\ (R_3) & 31\ (S_4) & 41\ (P_4) & 51\ (U_3) & 61\ (R_2) & 71\ (S_1) \\ 02\ (Q_1) & 12\ (V_2) & 22\ (O_3) & 32\ (T_4) & 42\ (Q_4) & 52\ (V_3) & 62\ (O_2) & 72\ (T_1) \\ 03\ (R_1) & 13\ (S_2) & 23\ (P_3) & 33\ (U_4) & 43\ (R_4) & 53\ (S_3) & 63\ (P_2) & 73\ (U_1) \end{array} \right\}$$

$$\underline{m}: \text{mod} = 4, \quad \underline{f}: \text{mod} = 8$$

The general form of the Murngin marriage cycle can be given in a segment diagram as follows:

$$G_{2n}^4 = \left\{ \begin{array}{ccccccc} 00 & 10 & 20 & - & - & - & (2n-2)0 \quad (2n-1)0 \\ 01 & 11 & 21 & - & - & - & (2n-2)1 \quad (2n-1)1 \\ 02 & 12 & 22 & - & - & - & (2n-2)2 \quad (2n-1)2 \\ 03 & 13 & 23 & - & - & - & (2n-2)3 \quad (2n-1)3 \end{array} \right\} \quad \begin{array}{l} \underline{m}: \text{mod} = 4 \\ \underline{f}: \text{mod} = \text{LCM of } 4 \\ \qquad \text{and } 2n \end{array}$$

The compound groups of circulation connubium or marriage cycle of the Murngin system (G) can be expressed in the canonical form.

$$G = G_4^4 + G_6^4 + G_8^4 + \cdots + G_{2n}^4.$$

## Conclusion

In this study a new numerical kinship notation system for section systems is presented. We adopt two basic kinship units, father–child link and mother–child link as generators for the analysis of the Murngin system. The groups of circulation connubium or marriage cycle are demonstrated in a segment diagram, and their mathematical properties are characterized by the fact that they are finite Abelian groups. Meanwhile, the working hypothesis proposed for the analysis of matrilateral cross-cousin marriage system has been shown to be incomplete. It needs to be revised as follows, in order to make it consistent with the properties of the Murngin system.

The matrilateral cross-cousin marriage system is described by an $n$-generation cycle, where the marriage cycle (or circulation connubium) is derived from $n$ hordes or clans, the minimum number for $n$ being 3. When the generation cycle governing the horde or clan is fixed as $p$, and $p \geqslant 1$, then the other descent line (if the former is patrilineal, the latter is matrilineal, or vice versa) will be regulated by a $q$-generation cycle, with $q$ being the least common multiple of $p$ and $n$.

## References

Bush, Robert R. 1963. An algebraic treatment of rules of marriage and descent. Appendix II of H. C. White, *An Anatomy of Kinship.* Englewood Cliffs, N.J., Prentice-Hall, Inc., pp. 159–172.

Harvey, J. H. T. and Pin-hsiung Liu. 1967. Numerical kinship notation system: mathematical model of genealogical space. *Bulletin of the Institute of Ethnology, Academia Sinica,* 23:1–22.

Kemeny, J. G., J. L. Snell, and G. L. Thompson, 1956. *Introduction to Finite Mathematics.* Englewood Cliffs, N.J., Prentice-Hall, Inc., pp. 343–353.

Reid, R. M. 1967. Marriage systems and algebraic group theory: a critique of White's *An Anatomy of Kinship. Amer. Anthropol.* 69:171–178.

Weil, André. 1949. Sur l'étude algébrique de certain types de lois de marriage (système Murngin). In Appendice à la Première Partie, *Les Structures élementaire de la parenté.* Claude Lévi-Strauss, Paris, Presses Universitaires de France, pp. 278–285. Also translated as Appendix I in H. C. White, *An Anatomy of Kinship.* Englewood Cliffs, N. J., Prentice-Hall, Inc.

White, H. C. 1963. *An Anatomy of Kinship: Mathematical Models for Structures of Cumulated Roles.* Englewood Cliffs, N.J., Prentice-Hall, Inc.

# 16



Reprinted by permission of Cambridge University Press from *Ann. Human Genet.*, **35**, 179–196 (1971)

## Analysis of population structure

### II. Two-dimensional stepping stone models of finite length and other geographically structured populations*

By TAKEO MARUYAMA

*National Institute of Genetics, Mishima, Japan*

#### 1. INTRODUCTION AND MODELS

Many human populations as well as other animal and plant populations are divided into colonies (villages). These colonies are usually distributed geographically on a plane, and the size of a colony may be small in the sense that the random drift may cause appreciable variation in the gene frequency among colonies. Usually these colonies constituting a population are not completely separated but there are some exchanges among them, and the closer two colonies are geographically the more exchanges there are. Therefore genetical similarity between colonies is a function of their distance and, of course, of the migration rate. Strictly speaking real population structure may be too complicated to be handled mathematically. However, the stepping stone model of a population structure proposed by Kimura (1953) is a mathematically tractable approximation to real situations. The population consists of colonies, each located at a grid-point of 2-dimensional integer lattice. All colonies have equal and finite size ($N$) which does not vary in time. We use $(i, j)$ to denote the position of a colony in the habitat and let $n_k$ be the number of colonies along the $k$th co-ordinate axis, i.e. $1 \leqslant i_k \leqslant n_k$. There are $n_1 n_2$ colonies altogether. We assume that migration occurs only between geographically adjacent colonies at a constant rate which depends on the direction but not on the location of a colony, except at the boundary of the habitat. This model of restricted migration between geographically adjacent colonies is the opposite of Wright's 'island model' where migration is independent of the distance. Let $m_k$ be the migration rate along the $k$th co-ordinate axis, i.e. a colony at $(i, j)$ receives immigration from the colonies at $(i-1, j)$ and $(i+1, j)$ at the rate $\frac{1}{2}m_1$, provided these colonies exist. In addition to the migration between adjacent colonies, every colony (except possibly a boundary colony) receives immigration from an outside world at the rate $m_\infty$. The meaning of this will be given later. When a colony located at the boundary of the habitat receives, in addition to $m_\infty$, the same amount, $\frac{1}{2}m^k$, of the outside world immigration as of that from its neighbouring colony, we call the boundary an absorbing type. And when a boundary colony receives no more than $m_\infty$ from the outside world, it is called a reflecting type. This is illustrated in Fig. 1. Fig. 2 shows the six possibilities of combinations of boundaries which will be analysed in §2 below.

Considering a single locus at which neutral alleles are segregating, we treat the two situations (1) the number of possible alleles at this locus is finite and (2) it is infinite (or very large) so that every mutation creates an allele which does not concurrently exist in the population. For (1), considering a particular allele, we obtain the variance of the gene frequency of that allele and the correlation between the frequencies in different colonies. We also obtain the probability that two homologous genes sampled from specified colonies (or a single colony) are the same allele

* Contribution no. 808 from the National Institute of Genetics, Mishima, Shizuoka-ken, 411 Japan. Aided in part by a Grant-in-Aid from the Ministry of Education, Japan.

(the probability of identity). In (2), two homologous genes can be the same allele only if they are identical by descent. Thus, for (2) assuming equal mutation rate for all genes, we obtain the probability that two genes are identical by descent.

For both situations, (1) and (2), except for colonies at an absorbing boundary, the migration from the outside world can be considered as mutation. This can also correspond to real migration from an outside world in which the gene frequency is constant for situation (1), and all genes are different alleles for (2). Therefore a habitat with reflecting boundary may be considered as a closed population. To a colony at an absorbing boundary, this migration must come from a real outside world of large size.

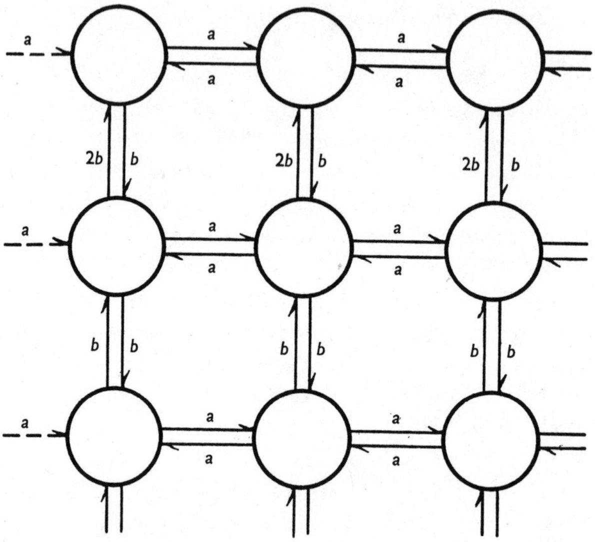


Fig. 1. Illustration of the boundary of a habitat. The circles indicate colonies, the solid arrows indicate the migration between colonies, and the broken arrows indicate the migration from an outside world. In the figure, the left vertical boundary is of the absorbing type and the upper horizontal boundary is of the reflecting type. $a = \frac{1}{2}m_1$, $b = \frac{1}{2}m_2$.

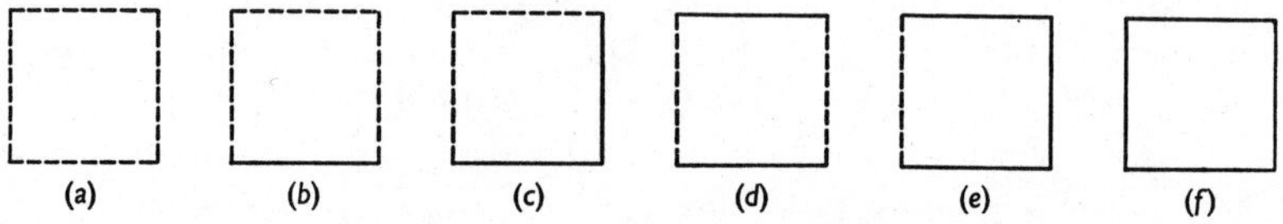


Fig. 2. Various combinations of absorbing and reflecting boundaries. Each square indicates a habitat. A side indicated by a broken line means an absorbing boundary and a side indicated by a solid line means a reflecting boundary.

## 2. ANALYSIS FOR THE VARIANCE AND CORRELATION OF THE GENE FREQUENCY

Let $\bar{q}$ be the expected gene frequency of a particular allele under consideration, i.e. $\bar{q}$ is equal to the gene frequency of the outside world. In particular, if there are $K$ alleles with equal mutation rate, $\bar{q} = 1/K$, and if the mutation rate from any one to any other is $u$, $m_\infty = uK/(K-1)$.

Let $p_{ij}+\bar{q}$, $i = 1, 2, \ldots, n_1$ and $j = 1, 2, \ldots, n_2$ be the gene frequency in the colony located at $(i, j)$ and let $\mathbf{P}$ be the matrix $[p_{ij}]_{n_1 . n_2}$.

In order to express the recurrence relation for the matrix $\mathbf{P}$, we define the following square matrices:

$$\mathbf{A}_n \equiv \begin{bmatrix} 0 & 1 & & & & \\ 1 & 0 & 1 & & & \\ & 1 & 0 & 1 & & \\ & & \cdot & \cdot & \cdot & \\ & & & 1 & 0 & 1 \\ & & & & 1 & 0 \end{bmatrix}_{n.n}, \quad \mathbf{B}_n \equiv \begin{bmatrix} & 2 & & & & \\ 1 & 0 & 1 & & & \\ & \cdot & \cdot & \cdot & & \\ & & \cdot & \cdot & \cdot & \\ & & & \cdot & \cdot & \cdot \\ & & & 1 & 0 & 1 \\ & & & & 2 & 0 \end{bmatrix}_{n.n},$$

$$\mathbf{C}_n \equiv \begin{bmatrix} 0 & 2 & & & & \\ 1 & 0 & 1 & & & \\ & 1 & \cdot & \cdot & & \\ & & \cdot & \cdot & \cdot & \\ & & & 1 & 0 & 1 \\ & & & & 1 & 0 \end{bmatrix}_{n.n}.$$

(The suffix $n.n$ indicates that each is an $n \times n$ matrix, and all unspecified elements are zero.) Also, for $i = 1, 2$,

$$\mathbf{U}_i \equiv (1-m_i)\,\mathbf{I}_{n_i} + (\tfrac{1}{2}m_i)\,\mathbf{A}_{n_i}, \quad \mathbf{V}_i \equiv (1-m_i)\,\mathbf{I}_{n_i} + (\tfrac{1}{2}m_i)\,\mathbf{B}_{n_i}$$

and $\quad \mathbf{W}_i \equiv (1-m_i)\,\mathbf{I}_{n_i} + (\tfrac{1}{2}m_i)\mathbf{C}_{n_i}$,

where $\mathbf{I}_{n_i}$ is the identity matrix of order $n_i$.

If all the four boundaries are of absorbing type (case (*a*) of Fig. 2) the recurrence equation for the $\mathbf{P}$ is

$$\mathbf{P}' = (1-m_\infty)\,\mathbf{U}_1\mathbf{P}\mathbf{U}_2 + \mathbf{\Phi}, \tag{2·1}$$

where the prime on $\mathbf{P}$ indicates its value in the next generation and the element $\xi_{ij}$ of $\mathbf{\Phi} = [\xi_{ij}]$ is the random variable due to the sampling of gametes. If all the four boundaries are of reflecting type (case (*f*) of Fig. 2), the recurrence equation is

$$\mathbf{P}' = (1-m_\infty)\,\mathbf{V}_1\mathbf{P}\mathbf{V}_2^T + \mathbf{\Phi}. \tag{2·2}$$

For case (*b*) of Fig. 2,

$$\mathbf{P}' = \mathbf{U}_1\mathbf{P}\mathbf{W}_2^T + \mathbf{\Phi}. \tag{2·3}$$

Equations (2·1) to (2·3) should be sufficient to suggest the recurrence equation for other possible combinations of boundaries (see Fig. 2).

For an arbitrary matrix $\mathbf{M} = [\mathbf{c}_1, \mathbf{c}_2, \ldots, \mathbf{c}_n]$, where $\mathbf{c}_i$ indicates the $i$th column, let us write the column vector

$$\begin{bmatrix} \mathbf{c}_1 \\ \mathbf{c}_2 \\ \cdot \\ \cdot \\ \cdot \\ \mathbf{c}_n \end{bmatrix}$$

as $\hat{\mathbf{M}}$. Thus $\hat{\mathbf{M}}$ is a single column of numbers. With this notation equations (2·1) to (2·3) can be rewritten respectively as

$$\hat{\mathbf{P}}' = (1-m_\infty)\,\mathbf{U}_1 \otimes \mathbf{U}_2^T\hat{\mathbf{P}}+\hat{\mathbf{\Phi}}, \qquad (2\cdot1')$$

$$\mathbf{P}' = (1-m_\infty)\,\mathbf{V}_1 \otimes \mathbf{V}_2^T\,\hat{\mathbf{P}}+\hat{\mathbf{\Phi}}, \qquad (2\cdot2')$$

and

$$\mathbf{P}' = (1-m_\infty)\,\mathbf{U}_1 \otimes \mathbf{W}_2^T\,\hat{\mathbf{P}}+\hat{\mathbf{\Phi}}, \qquad (2\cdot3')$$

where $\otimes$ indicates the Kronecker product of two matrices, i.e. if

$$\mathbf{X} = [x_{ij}]_{m.n} \quad \text{and} \quad \mathbf{Y} = [y_{ij}]_{p.q}$$

then

$$\mathbf{X}\otimes\mathbf{Y} \equiv \begin{bmatrix} \mathbf{X}y_{11}, & \mathbf{X}y_{12}, & \ldots, & \mathbf{X}y_{1n} \\ \vdots & & & \\ \mathbf{X}y_{m1}, & \mathbf{X}y_{m2}, & \ldots, & \mathbf{X}y_{mn} \end{bmatrix}_{mp.nq}.$$

Define

$$\mathbf{Q} \equiv E[\hat{\mathbf{P}}\hat{\mathbf{P}}^T] \equiv [C_{iji'j'}] \qquad (2\cdot4)$$

where $E$ stands for the expectation. Since the $\xi_{ij}$ has mean 0, the $\xi_{ij}$'s are uncorrelated, and the $p_{ij}$'s and the $\xi_{ij}$'s are uncorrelated, we have, from (2·1′) to (2·3′), the following recurrence equations for $\mathbf{Q}$ respectively,

$$\mathbf{Q}' = (1-m_\infty)^2\{\mathbf{U}_1\otimes\mathbf{U}_2^T\}\,\mathbf{Q}\{\mathbf{U}_2\otimes\mathbf{U}_1^T\}+\{\bar{q}(1-\bar{q})\,\mathbf{I}-\mathbf{D}\}/2N, \qquad (2\cdot1'')$$

$$\mathbf{Q}' = (1-m_\infty)^2\{\mathbf{U}_1\otimes\mathbf{V}_2^T\}\,\mathbf{Q}\{\mathbf{V}_2\otimes\mathbf{U}_1^T\}+\{\bar{q}(1-\bar{q})\,\mathbf{I}-\mathbf{D}\}/2N, \qquad (2\cdot2'')$$

$$\mathbf{Q}' = (1-m_\infty)^2\{\mathbf{U}_1\otimes\mathbf{W}_2^T\}\,\mathbf{Q}\{\mathbf{W}_2\otimes\mathbf{U}_1^T\}+\{\bar{q}(1-\bar{q})\,\mathbf{I}-\mathbf{D}\}/2N, \qquad (2\cdot3'')$$

in which $\mathbf{D}$ is a diagonal matrix consisting of the diagonal elements of $\mathbf{Q}$, and $\mathbf{I}$ is the identity matrix of order $n_1n_2$. Equations (2·1″) to (2·3″), in which $\mathbf{Q}'$ is replaced by $\mathbf{Q}$ (at equilibrium), are the equations to be solved. For example, (2·1″) becomes

$$\mathbf{Q}-(1-m_\infty)^2\{\mathbf{U}_1\otimes\mathbf{U}_2^T\}\,\mathbf{Q}\{\mathbf{U}_2\otimes\mathbf{U}_1^T\} = \{\bar{q}(1-\bar{q})\,\mathbf{I}-\mathbf{D}\}/2N. \qquad (2\cdot1''')$$

With fixed values of $m_\infty$, $m_1$ and $m_2$, the left side of (2·1‴) can be considered as a linear transformation on the set of all matrices of order $n_1n_2$. Eigenvectors of the Kronecker product of two matrices are the Kronecker products of eigenvectors of each matrix, and a similar relationship holds also for eigenvalues. For these properties, see a book on linear algebra, for example, Jacobson (1953), ch. VII. Since the complete spectral analysis is possible for the matrix $\mathbf{A}_n$, we can obtain all the eigenvectors and eigenvalues of the left side of (2·1‴). Let

$$\mathbf{s}_k^{(n)} = \left(\sin\frac{\pi k}{n+1},\ \sin\frac{2\pi k}{n+1},\ldots,\ \sin\frac{n\pi k}{n+1}\right)^T,$$

$$\mathbf{s}_{kl} \equiv \mathbf{s}_k^{(n_1)}\otimes\mathbf{s}_l^{(n_2)}$$

and

$$\mathbf{E}_{klk'l'} \equiv \mathbf{s}_{kl}\,\mathbf{s}_{k'l'}^T.$$

Notice that $\mathbf{s}_k^{(n_i)}$ is an eigenvector of $\mathbf{U}_i$ and the corresponding eigenvalue is

$$1-m_i\{1-\cos(\pi k/(n_i+1))\}.$$

Since an eigenvector of the direct product of two matrices is the direct product of two eigenvectors of each matrix, and the eigenvalue is the product of the corresponding eigenvalues, $\mathbf{s}_{kl}$ is an eigenvector of $\mathbf{U}_1\otimes\mathbf{U}_2^T$, and the corresponding eigenvalue is

$$\left\{1-m_1\left(1-\cos\frac{\pi k}{n_1+1}\right)\right\}\left\{1-m_2\left(1-\cos\frac{\pi l}{n_2+1}\right)\right\}.$$

By similar reasoning, $\mathbf{E}_{klk'l'}$ is the eigenvector (matrix) of the operator

$$\{\mathbf{U}_1 \otimes \mathbf{U}_2^T\}\{.\}\{\mathbf{U}_2 \otimes \mathbf{U}_1^T\}.$$

Therefore we have

$$\mathbf{E}_{klk'l'} - (1-m_\infty)^2\{\mathbf{U}_1 \otimes \mathbf{U}_2^T\}\,\mathbf{E}_{klk'l'}\{\mathbf{U}_2 \otimes \mathbf{U}_1^T\} = \lambda_{klk'l'}\,\mathbf{E}_{klk'l'} \qquad (2\cdot1'''')$$

and

$$\langle \mathbf{E}_{klk'l'} | \mathbf{E}_{iji'j'} \rangle = \delta_{ki}\delta_{lj}\delta_{k'i'}\delta_{l'j'},$$

where $\langle | \rangle$ indicates the innerproduct of two matrices, the sum of products of corresponding elements, and $\delta_{ij}$ is the Kronecker delta. The eigenvalue in (2·1'''') is

$$\lambda_{klk'l'} = 1-(1-m_\infty)^2\left\{1-m_1\left(1-\cos\frac{\pi k}{n_1+1}\right)\right\}\left\{1-m_2\left(1-\cos\frac{\pi l}{n_2+1}\right)\right\}$$
$$\times\left\{1-m_1\left(1-\cos\frac{\pi k'}{n_1+1}\right)\right\}\left\{1-m_2\left(1-\cos\frac{\pi l'}{n_2+1}\right)\right\}. \qquad (2\cdot5)$$

This follows from the theorem that if $P(x)$ is a polynomial in $x$ and $\mathbf{A}$ is a matrix with eigenvalue $\lambda$, then $P(\lambda)$ is an eigenvalue of the matrix $P(\mathbf{A})$. For these properties of matrix algebra, see Jacobson (1953).

If we let $\mathbf{Q} \equiv \Sigma c_{klk'l'}\mathbf{E}_{klk'l'}$, substituting this $\mathbf{Q}$ into (2·1''') and using the orthogonality of the $\mathbf{E}$'s, we have

$$c_{klk'l'} = 4\langle \mathbf{E}_{klk'l'} | \bar{q}(1-\bar{q})\,\mathbf{I}-\mathbf{D}\rangle / 2Nn_1n_2\lambda_{klk'l'}.$$

If we know the diagonal elements of the $\mathbf{Q}$, we can determine the $c$'s explicitly. However, notice that the value of $\langle \mathbf{E}_{klk'l'} | \bar{q}(1-\bar{q})\,\mathbf{I}-\mathbf{D}\rangle$ with $k \neq k'$ or $l \neq l'$ is much smaller than that with $k = k'$ and $l = l'$. In fact, it is zero if either $k+k'$ or $l+l'$ is an odd integer. Therefore, for the time being, ignoring those with $k \neq k'$ or $l \neq l'$, we have

$$\mathbf{Q} \propto \sum_{k=1}^{n_1}\sum_{l=1}^{n_2}\mathbf{E}_{klkl}/\lambda_{klkl}$$

or

$$C_{iji'j'} \approx c\sum_{k=1}^{n_1}\sum_{l=1}^{n_2}\sin\frac{i\pi k}{n_1+1}\sin\frac{j\pi l}{n_2+1}\sin\frac{i'\pi k}{n_1+1}\sin\frac{j'\pi l}{n_2+1}\Big/\lambda_{klkl}. \qquad (2\cdot6)$$

Thus the correlation of the $p_{ij}$'s between colonies is given by

$$r_{iji'j'} = C_{iji'j'}/\sqrt{(C_{ijij}C_{i'j'i'j'})}. \qquad (2\cdot7)$$

The variance of the $p_{ij}$'s should be recalculated, because the differences among the variances were ignored in the derivation of (2·6). Let $v_{ij}$ be the variance of the $p_{ij}$. Then using the $r_{iji'j'}$'s of (2·7) and the recurrence equation (2·1'''), we have

$$v_{ij} \approx \frac{\bar{q}(1-\bar{q})}{1+2N\{1-(1-m_\infty)^2B\}},$$

where $$B = (1-m_1)^2(1-m_2)^2 + m_1(r_{i,j,i-1,j}+r_{i,j,i+1,j}) + m_2(r_{i,j,i,j-1}+r_{i,j,i,j+1}), \qquad (2\cdot8)$$

in which the value of the $r_{ijij}$ with any subscript exceeding the range $[1, n_1]$ or $[1, n_2]$ is zero.

The six possibilities illustrated in Fig. 2 can be solved by similar methods (cf. Maruyama, 1970*a*, *b*). We use the following abbreviations:

$s_{ijk} \equiv \sin(j\pi k/(n_i+1))$, $t_{ijk} \equiv \sin(j\pi k/2n_i)$, $u_{ijk} \equiv \cos((j-1)\pi(k-1)/(n_i-1))$,

$\sigma_{ij} \equiv 1-m_i\{1-\cos(\pi j/(n_i+1))\}$, $\tau_{ij} \equiv 1-m_i\{1-\cos(\pi j/2n_i)\}$,

$v_{ij} \equiv 1-m_i\{1-\cos(\pi(j-1)/(n_i-1))\}$, $\Delta_i = 1$ if $i = 1$ or $i = n_i$ (or $i = n_2$),

$\Delta_i = 2$ otherwise, $\theta_{11} = \theta_{1n_1} = \theta_{21} = \theta_{2n_2} = \frac{1}{2}$ and

$\theta_{ij} = (2n_i-4)/(2n_i-3)$ otherwise, $\alpha \equiv (1-m_\infty)^2$.

Then the formulas are

$$C_{iji'j'} = c\left[\sum_{k=1}^{n_1} s_{1ik}s_{1i'k}\left\{\sum_{l=1}^{2n_2-1}\frac{t_{2jl}t_{2j'l}}{1-\alpha\sigma_{1k}^2\tau_{2l}^2}+\frac{1}{2n_2-2}\sum_{\substack{l\neq l'\\ l,l'=\text{odd}}}\frac{(-1)^{(\frac{1}{2}(l+l'))-1}t_{2jl}t_{2j'l'}}{1-\alpha\sigma_{1k}^2\tau_{2l}\tau_{2l'}}\right\}\right] \quad \text{for } (b), \quad (2\cdot9)$$

$$C_{iji'j'} = c\left[\sum_{k=1}^{2n_1-1} t_{1ik}t_{1i'k}\left\{\sum_{l=1}^{2n_2-1}\frac{t_{2jl}t_{2j'l}}{1-\alpha\tau_{1k}^2\tau_{2l}^2}+\frac{1}{2n_2-2}\sum_{\substack{l\neq l'\\ l,l'=\text{odd}}}\frac{(-1)^{(\frac{1}{2}(l+l'))-1}t_{2jl}t_{2j'l'}}{1-\alpha\tau_{1k}^2\tau_{2l}\tau_{2l'}}\right\}\right.$$

$$\left.+\frac{1}{2n_1-2}\sum_{\substack{k\neq k'\\ k,k'=\text{odd}}}(-1)^{(\frac{1}{2}(l+l'))-1}t_{1ik}t_{1i'k'}\left\{\sum_{l=1}^{2n_2-1}\frac{t_{2jl}t_{2j'l}}{1-\alpha\tau_{1k}\tau_{1k'}\tau_{2l}^2}-\frac{1}{2n_1}\sum_{\substack{l\neq l'\\ l,l'=\text{odd}}}\frac{(-1)^{(\frac{1}{2}(l+l'))-1}t_{2jl}t_{2j'l'}}{1-\alpha\tau_{1k}\tau_{1k'}\tau_{2l}\tau_{2l'}}\right\}\right] \quad \text{for } (c) \quad (2\cdot10)$$

$$C_{iji'j'} = c\left[\sum_{k=1}^{n_1} s_{1ik}s_{1i'k}\left\{\sum_{l=1}^{n_2}\frac{\theta_{2l}u_{2jl}u_{2j'l}}{1-\alpha\sigma_{1k}^2v_{2l}^2}-\frac{2}{2n_2-3}\sum_{\substack{l\neq l'\\ l+l=\text{even}}}\frac{\Delta_l\Delta_{l'}u_{2jl}u_{2jl'}}{1-\alpha\sigma_{1k}^2u_{2l}v_{2l'}}\right\}\right] \quad \text{for } (d), \quad (2\cdot11)$$

$$C_{iji'j'} = c\left[\sum_{k=1}^{2n_1} t_{1ik}t_{1i'k}\left\{\sum_{l=1}^{n_2}\frac{\theta_{2l}u_{2jl}u_{2j'l}}{1-\alpha\tau_{1k}^2v_{2l}^2}-\frac{2}{2n_2-3}\sum_{\substack{l\neq l'\\ l+l'=\text{even}}}\frac{\Delta_l\Delta_{l'}u_{2jl}u_{2j'l'}}{1-\alpha\tau_{1k}^2v_{2l}v_{2l'}}\right\}\right.$$

$$\left.-\frac{1}{2n_1}\sum_{\substack{k\neq k,\\ k,k'=\text{odd}}}t_{1ik}t_{1i'k'}\left\{\sum_{l=1}^{n_2}\frac{\theta_{2l}u_{2jl}u_{2j'l}}{1-\alpha\tau_{1k}\tau_{1k'}v_{2l}^2}-\frac{2}{2n_2-3}\sum_{\substack{l\neq l'\\ l+l'=\text{even}}}\frac{\Delta_l\Delta_{l'}u_{2jl}u_{2j'l'}}{1-\alpha\tau_{1k}\tau_{1k'}v_{2l}v_{2l'}}\right\}\right] \quad \text{for } (e), \quad (2\cdot12)$$

$$C_{iji'j'} = c\left[\sum_{k=1}^{n_1}\theta_{1k}u_{1ik}u_{1i'k}\left\{\sum_{l=1}^{n_2}\frac{\theta_{2l}u_{2jl}u_{2j'l}}{1-\alpha v_{1k}^2v_{2l}^2}-\frac{2}{2n_2-3}\sum_{\substack{l\neq l'\\ l+l'=\text{even}}}\frac{\Delta_l\Delta_{l'}u_{2jl}u_{2j'l'}}{1-\alpha v_{1k}^2v_{2l}v_{2l'}}\right\}\right.$$

$$\left.-\frac{2}{2n_1-3}\sum_{\substack{k\neq k'\\ k+k'=\text{even}}}\Delta_k\Delta_{k'}u_{1ik}u_{1i'k'}\left\{\sum_{l=1}^{n_2}\frac{\theta_{2l}u_{2jl}u_{2j'l'}}{1-\alpha v_{1k}v_{1k'}v_{2l}^2}-\frac{2}{2n_2-3}\sum_{\substack{l\neq l'\\ l+l'=\text{even}}}\frac{\Delta_l\Delta_{l'}u_{2jl}u_{2j'l'}}{1-\alpha v_{1k}v_{1k'}v_{2l}v_{2l'}}\right\}\right] \quad \text{for } (f), \quad (2.13)$$

Using the values of the $C_{iji'j'}$ given in (2·9) to (2·13), the correlations $(r_{iji'j'})$ can be calculated by formula (2·7). The variances $(v_{ij})$ are given by formula (2·8), there

$$r_{ojkl} = r_{iokl} = \dots = 0$$

if the boundary is absorbing type and $r_{ojkl} = r_{2jkl}$, $r_{iokl} = r_{i2kl}$, etc., if it is reflecting type.

## 3. PROBABILITY OF IDENTITY

In the previous section, we have dealt with a particular allele and obtained the variance and correlation. A quantity related to these is 'the probability of identity', i.e. the probability that two homologous genes are identical in constitution. For this, we assume that there are $K$ possible alleles at a locus under consideration, and that the mutation rate from one specified allele to another is $u/(K-1)$. In terms of the previous section, the expected frequency of each allele among the migration from the outside world is $1/K = \bar{q}$, and the relationship between $u$ and $m_\infty$ is $m_\infty = uK/(K-1)$. Then the probability $(f_{ijij})$, that two randomly chosen homologous genes (without replacement) from the colony at $(i, j)$ are the same allele, is given by

$$f_{ijij} = K\{v_{ij} + K^{-2}\}, \quad (3\cdot1)$$

where $v_{ij}$ is given in (2·8). And the probability for two homologous genes chosen from two different colonies is

$$f_{iji'j'} = K\{r_{iji'j'}\sqrt{(v_{ij}v_{i'j'})} + K^{-2}\}, \quad (3\cdot2)$$

where $r_{iji'j'}$ is given in (2·7).

With infinitely many possible alleles at the locus, the formula corresponding to (3·1) becomes

$$f_{ijij} = v_{ij}, \tag{3·3}$$

where $v_{ij}$ is the same as that in (2·8), except that $\bar{q}(1-\bar{q})$ is replaced by 1, and the formula corresponding to (3·2) is

$$f_{iji'j'} = r_{iji'j'}\sqrt{(v_{ij}v_{i'j'})}, \tag{3·4}$$

where $v_{ij}$ is the same as that of (3·3).

The probability of identity is based on the gene frequencies. Thus it can be measured by a sample from the population, while the actual number of alleles maintained in a population can be measured only if the entire population is surveyed. The effective number of alleles ($n_e$) defined by Kimura & Crow (1964) is the reciprocal of the average allelism probability.

Although formulas (3·1) to (3·4) are obtained from the correlation analysis, we can give a direct analysis for the probability of identity. The following analysis given for a special habitat is to show an alternative and direct method, and to find a relationship between the homozygous frequency and the total genetic variability. We consider a torus-like habitat formed by the direct product of two circles of $n_1$ and $n_2$ points. Thus there are $n_1 n_2$ colonies, each of which has $N$ individuals. Let $f_{ij}$ be the probability that two randomly chosen homologous genes, one from each of colonies separated $i$ steps along the first co-ordinate axis and $j$ steps along the second, are identical. Then the recurrence equation for the $f_{ij}$'s are (ignoring the terms involving $m_1^2$, $m_2^2$, or $u^2$)

$$\begin{aligned} f_{ij}' = (1-u)^2[&(1-m_1-m_2)^2 f_{ij} + m_1(1-m_1-m_2)\{f_{i-1,j}+f_{i+1,j}\} + m_2(1-m_1-m_2)\{f_{i,j-1}+f_{i,j+1}\}] \\ &+(2u/K)[(1-m_1-m_2)^2(1-f_{ij}) + m_1(1-m_1-m_2)\{(1-f_{i-1,j})+(1-f_{i+1,j})\} \\ &+m_2(1-m_1-m_2)\{(1-f_{i,j-1})+(1-f_{i,j+1})\}] \end{aligned} \tag{3·5}$$

for $2 \leqslant i \leqslant n_1-2$ and $2 \leqslant j \leqslant n_2-2$. With the possibility of two genes coming from the same colony in the previous generation, the recurrence equation must be changed accordingly; for example (again ignoring high order terms of the migration rates),

$$\begin{aligned} f_{00}' = (1-u)^2&\left[(1-m_1-m_2)^2\left\{\left(1-\frac{1}{2N}\right)f_{00}+\frac{1}{2N}\right\}+2m_1(1-m_1-m_2)f_{10}+2m_2(1-m_1-m_2)f_{01}\right] \\ +\frac{2u}{K-1}&\left[(1-m_1-m_2)^2\left(1-\frac{1}{2N}\right)(1-f_{00})+2m_1(1-m_1-m_2)(1-f_{10})+2m_2(1-m_1-m_2)(1-f_{01})\right]. \end{aligned}$$

These recurrence equations can be expressed neatly by matrices. Let $\mathbf{F} \equiv [f_{ij}]$, $\mathbf{I}_n \equiv$ identity matrix of order $n$, $\mathbf{J} \equiv$ the $n_1 \times n_2$ matrix all of whose entries are 1, $\mathbf{G} \equiv$ the $n_1 \times n_2$ matrix whose (1, 1) entry is $1-f_{00}$ and all the others are 0, and $\mathbf{C}_n \equiv$ circular matrix of order $n$, i.e. a matrix whose $(n, 1)$ and $(i, i+1)$, $i = 1, 2, \ldots, n-1$, entries are 1 and all the other entries are 0. Let

$$\mathbf{M}_i \equiv \left\{(1-m_i)\,\mathbf{I}_{n_i} + \frac{m_i}{2}(C_{n_i} + \mathbf{C}_{n_i}^{-1})\right\}^2.$$

Then

$$\mathbf{F}' = (1-u)^2\mathbf{M}_1\mathbf{F}\mathbf{M}_2 + \frac{2u}{K-1}\mathbf{M}_1(\mathbf{J}-\mathbf{F})\,\mathbf{M}_2 + \frac{1}{2N}\left\{1-\frac{2uK}{K-1}\right\}\mathbf{M}_1\,\mathbf{G}\mathbf{M}_2. \tag{3·6}$$

Equation (3·6) for equilibrium $\mathbf{F}$ can be solved by the same method as that used in the previous section. The final formulas are

$$f_{ij} = \frac{1}{K} + \frac{\left\{1-\dfrac{2uK}{K-1}\right\}(1-f_{00})}{2Nn_1n_2}\sum_{k=0}^{[\frac{1}{2}n_1]}\sum_{l=0}^{[\frac{1}{2}n_2]}\frac{\Delta_k\Delta_l\xi_{kl}\cos\dfrac{i\pi k}{n_1}\cos\dfrac{j\pi l}{n_2}}{1-\left\{1-\dfrac{2uK}{K-1}\right\}\xi_{kl}} \quad (i = 0, 1, \ldots, n_1, j = 0, 1, \ldots, n_2), \tag{3·7}$$

13-2

where $$\xi_{kl} = \left\{1 - m_1\left(1 - \cos\frac{\pi k}{n_1}\right)\right\}^2 \left\{1 - m_2\left(1 - \cos\frac{\pi l}{n_2}\right)\right\}^2,$$

$\Delta_i = 1$ if $i = 0$ or $i = \frac{1}{2}n_1$ (or $\frac{1}{2}n_2$) and $\Delta_i = 2$ otherwise, and $[x]$ indicates the largest integer $\leqslant x$. And

$$f_{00} = \left[\frac{1}{K} + \frac{\left\{1 - \frac{2uK}{K-1}\right\} W}{2Nn_1 n_2}\right] \Big/ \left[1 + \frac{\left\{1 - \frac{2uK}{K-1}\right\} W}{2Nn_1 n_2}\right], \tag{3·8}$$

where $$W = \sum_{k=0}^{[\frac{1}{2}n_1]} \sum_{l=0}^{[\frac{1}{2}n_2]} \Delta_k \Delta_l \xi_{kl} \Big/ \left[1 - \left\{1 - \frac{2uK}{K-1}\right\} \xi_{kl}\right].$$

The factor $\xi_{kl}$, on the numerator of (3·7), which does not appear in formulas (2·6), (2·9) to (2·13), is due to the migration, in this model, taking place after the sampling of gametes, whereas the order is reversed in the previous model, and this makes very little difference.

The average of the $f_{ij}$'s over the whole population is

$$\bar{f} = \frac{1}{n_1 n_2} \sum_{k=0}^{n_1-1} \sum_{l=0}^{n_2-1} f_{ij} = \frac{\frac{4Nn_1 n_2 u}{K-1} + (1 - f_{00})\left(1 - \frac{2uK}{K-1}\right)}{4Nn_1 n_2 u \left(\frac{K}{K-1}\right)}. \tag{3·9}$$

It is important that formula (3·9) does not involve the migration rate. As $K$ becomes large, (3·9) is simply $(1 - f_{00})/4Nn_1 n_2 u$.

So far the migration is restricted to four geographically neighbouring colonies. We can remove this restriction and impose a general migration pattern on the model. Let $m_{xy}$ be the migration rate at which an individual moves $x$ steps along the first co-ordinate axis and $y$ steps along the second, i.e. the colony at $(i, j)$ receives migration from the colony at $(i-x, j-y)$ at the rate $\frac{1}{4}m_{xy}$ and similarly from the colonies at $(i-x, j+y)$, $(i+x, j-y)$ and $(i+x, j+y)$. In particular, $m_{10}$ and $m_{01}$ correspond respectively to $m_1$ and $m_2$ in the above analysis. With the general migration pattern, the only modification needed in formulas (3·7) ~ (3·9) is to change the $\xi_{kl}$ to

$$\xi_{kl} = \prod_{x,y} \left\{1 - m_{xy}\left(1 - \cos\frac{\pi xk}{n_1}\cos\frac{\pi yl}{n_2}\right)\right\}^2,$$

where $\prod_{x,y}$ indicates the multiplication in the all possible combinations of $x$ and $y$.

## 4. GENERALIZATION OF (3·9)

In the previous sections, we assumed equal colony size and a regular migration pattern in order to simplify the problem. However, we often encounter situations where we would like to remove these assumptions, for example see Smith (1969). In this section, we deal with a population subdivided into partially isolated colonies of arbitrary size and show that an analogous formula to (3·9) exists for such a population. Here the dimension of the habitat is irrelevant. We define the following quantities: $L \equiv$ the number of colonies; $N_i \equiv$ the number of individuals in colony $i$; $N_T \equiv \sum_{i=1} N_i =$ the total population size; $K \equiv$ the number of possible alleles; $f_{ii} \equiv$ the probability that two different homologous genes, chosen at random from the same colony $i$, are identical; $f_{ij} \equiv$ the probability that two homologous genes, one from colony $i$ and the other from $j$, are identical; $u \equiv$ total mutation rate, $u/(K-1) =$ mutation rate from one specified allele to another; $m_{ij} \equiv$ the probability that an individual born in colony $i$ migrates to $j$.

We assume

$$\sum_{j=1}^{L} m_{ij} = 1 \quad \text{for all } i \tag{4·1}$$

and

$$\sum_{i=1}^{L} m_{ij} N_i = N_j \quad \text{for all } j. \tag{4·2}$$

If we let $\alpha_{ij} \equiv m_{ij} N_i / N_j$, we have

$$f'_{ij} = (1-u)^2 \left\{ \sum_{k=1}^{L} \sum_{l=1}^{L} \alpha_{ki}\alpha_{lj} f_{kl} + \sum_{k=1}^{L} \left(\frac{1-f_{kk}}{2N_k}\right) \alpha_{ki}\alpha_{kj} \right\} + \left\{ \frac{2u(1-u)}{K} + \frac{u^2(K-2)}{(K-1)^2} \right\}$$
$$\times \left\{ \sum_{k=1}^{L} \sum_{l=1}^{L} \alpha_{ki}\alpha_{lj}(1-f_{kl}) - \sum_{k=1}^{L} \alpha_{ki}\alpha_{kj} \left(\frac{1-f_{kk}}{2N_k}\right) \right\}.$$

Let $\mathbf{M} \equiv [\alpha_{ij}]$, $\mathbf{F} \equiv [f_{ij}]$, $\mathbf{J} \equiv$ a matrix of order $L$ all of whose elements are 1, and $\mathbf{G}_0$ be a diagonal matrix consisting of $(1-f_{11})/2N_1$, $(1-f_{22})/2N_2$, ..., $(1-f_{LL})/2N_L$. Then the $\mathbf{F}$ at equilibrium satisfies

$$\mathbf{F} = \{(1-u)^2 - v\} \mathbf{M}^T\mathbf{F}\mathbf{M} + v\mathbf{M}^T\mathbf{J}\mathbf{M} + \{(1-u)^2 - v\} \mathbf{M}\mathbf{G}_0\mathbf{M}, \tag{4·3}$$

where $v = 2u(1-u)/K - u^2(K-2)/(K-1)^2$.

For an $L \times L$ matrix $\mathbf{A} = [a_{ij}]$ with $a_{ij} \geq 0$ for all $i$ and $j$, let

$$\|\mathbf{A}\| \equiv \frac{1}{N_T^2} \sum_{i=1}^{L} \sum_{j=1}^{L} N_i N_j a_{ij}.$$

Then notice that, for such matrices $\mathbf{A}$ and $\mathbf{B}$ $\|\mathbf{A}+\mathbf{B}\| = \|\mathbf{A}\| + \|\mathbf{B}\|$ and that $\|\mathbf{M}^T\mathbf{A}\| = \|\mathbf{A}\|$ and $\|\mathbf{A}\mathbf{M}\| = \|\mathbf{A}\|$. Applying these properties to (4·3), we have

$$\|\mathbf{F}\| = \{(1-u)^2 - v\} \|\mathbf{F}\| + v + \{(1-u)^2 - v\}(1-\bar{f}_0),$$

where

$$\bar{f}_0 = \sum_{i=1}^{L} f_{ii} N_i / N_T.$$

Since $\|\mathbf{F}\|$ is the weighted average of the $f_{ij}$'s, $\bar{f} = \|\mathbf{F}\|$, we have

$$\left.\begin{aligned} \bar{f} &= \frac{(1-\bar{f}_0)\left[1-(2u-u^2)\dfrac{K}{K-1} - \dfrac{u^2}{(K-1)^2}\right] + \dfrac{4N_T u(1-u)}{K-1} + \dfrac{4N_T(K-2)u^2}{(K-1)^2}}{2N_T\left[(2u-u^2)\dfrac{K}{K-1} - \dfrac{u^2}{K-1}\right]} \\ &\left(\bar{f} = \frac{(1-\bar{f}_0)(1-u)^2}{4Nu}, \quad \text{if} \quad K = \infty\right), \end{aligned}\right\} \tag{4·4}$$

which agrees with (3·9) if we ignore terms of order $u^2$. (4·1) and (4·2) are a necessary and sufficient condition for this formula to be valid. Formula (4·4) holds also for a continuously distributed population (see 6·8). See also Crow and Maruyama (1971).

## 5. ANALYSIS FOR NON-EQUILIBRIUM STATE

In the previous sections, we considered only the equilibrium state, but here we treat the rate of decay in existing genetic variability on the assumption of no mutation. Let $f_{ij}$, $\mathbf{M}$, $\| \; \|$, $N_i$, $L$, $\bar{f}_0$, $\mathbf{G}_0$ and $N_T$ stand for the same quantities as in §4. Let $\mathbf{H} \equiv [1-f_{ij}]$. Without mutation, the recurrence equation for the matrix $\mathbf{H}$ is

$$\mathbf{H}' = \mathbf{M}^T\mathbf{H}\mathbf{M} - \mathbf{M}^T\mathbf{G}_0\mathbf{M}. \tag{5·1}$$

Therefore

$$\|\mathbf{H}'\| = \|\mathbf{H}\| - \|\mathbf{G}_0\| = \|\mathbf{H}\| - \frac{1}{2N_T^2}\sum_{i=1}^{L}(1-f_{ii})N_i$$
$$= \|\mathbf{H}\| - (1-\bar{f}_0)/2N_T \tag{5·2}$$

and
$$\lambda_t = \frac{\|\mathbf{H}_{t+1}\|}{\|\mathbf{H}_t\|} = 1 - \frac{1-\bar{f}_0^{(t)}}{2N_T\|\mathbf{H}_t\|} \quad \text{or} \quad 1-\lambda_t = \frac{1-\bar{f}_0^{(t)}}{2N_T\|\mathbf{H}_t\|}, \tag{5·3}$$

where $t$ indicates time in generation. (5·3) means that the rate of decay in the weighted mean heterozygosity, $\|\mathbf{H}_t\|$, is reduced by the factor of the local heterozygosity divided by the global heterozygosity times the population size. This formula was first obtained by Robertson (1964) by a different argument.

Formula (5·3) is true for all $t$, but the most important case is the rate for large $t$, for which $\lambda_\infty$ is called the largest eigenvalue. To find the largest eigenvalue for an arbitrary case is very difficult. With some restrictions, we can obtain a rather useful lower bound for the $\lambda_\infty$. We assume that all $N_i$'s are equal to $N$ = a constant and for each $j$,

$$\sum_{i=1, i\neq j}^{L} m_{ij}$$

is equal to $m_T$ = a constant. Since at steady decay, the form of the $\mathbf{H}$ does not change in time, unless it is multiplied by the factor $\lambda_\infty$, from (5·1) we have

$$\|\mathbf{G}_0\| + \left(1-\frac{1}{2N}\right)\sum_{i=1}^{L}\left\{\sum_{\substack{j,k \\ i\neq k \text{ or } i\neq j}} m_{ij}m_{ik}(1-f_{ii})\right\}/L^2 = \sum_{i=1}^{L}\left\{\sum_{\substack{j,k \\ i\neq k \text{ or } i\neq j}} m_{ji}m_{ki}(1-f_{jk})\right\}/L^2.$$

Notice,
$$\|\mathbf{G}_0\| \leqslant \sum_{i=1}^{L}\left\{\sum_{\substack{j,k \\ i\neq k \text{ or } i\neq j}} m_{ji}m_{ki}(1-f_{jk})\right\}/L^2$$

and
$$\sum_{\substack{j,k \\ i\neq k \text{ or } i\neq j}} m_{ji}m_{ki}(1-f_{jk}) \leqslant \|\mathbf{H}\| \sum_{\substack{j,k \\ i\neq j \text{ or } i\neq k}} m_{ji}m_{ki}/L.$$

Therefore
$$\|\mathbf{G}_0\| \leqslant \|\mathbf{H}\| \sum_{i=1}^{L}\left\{\sum_{j,k} m_{ji}m_{ki}\right\}/L^2.$$

Ignoring higher-order term of the $m_T$, we have

$$\|\mathbf{G}_0\| \leqslant 2\|\mathbf{H}\|m_T/L^2.$$

Substituting this inequality into (5·2), we have

$$\lambda_\infty = \|\mathbf{H}'\|/\|\mathbf{H}\| \geqslant 1 - 2m_T/L \quad \text{or} \quad 1-\lambda_\infty \leqslant 2m_T/L. \tag{5·4}$$

Exact values of the $\lambda_\infty$ for special cases of geographically structured populations were investigated in Maruyama (1970*c*, *d*). For one-dimensional stepping stone model, we have

$$1-\lambda_\infty \approx 10m/L^2 \quad \text{(circular habitat)}, \tag{5·5}$$
$$\approx 5m/L^2 \quad \text{(linear habitat)},$$

if $2mN < L/10$ (circular), or $2mN < L/5$ (linear), and

$$1-\lambda_\infty \approx 1/2NL \tag{5·6}$$

if the above inequality is reversed. More generally, with migration allowing more than one step away, $m$ in the above formulas should be replaced by the variance

$$\sigma^2 = \sum_k k^2 m_k,$$

where $m_k$ is the migration rate going $k$ steps away in one generation. For a two-dimensional square habitat of the type (*f*) of Fig. 2, we have the following three asymptotic limits (Maruyama, unpublished),

$$m_T = m_1 + m_2 \quad \text{and} \quad m_1 = m_2,$$

$$\lim_{m_T \to 0} 1 - \lambda_\infty = 2m_T/L, \tag{5·7}$$

$$1 - \lambda_\infty \approx 1/2NL, \quad \text{if} \quad m_T N > 2, \tag{5·8}$$

and

$$1 - \lambda_\infty \approx m_T/4L, \quad \text{if} \quad m_T N < 2 \tag{5·9}$$

$$\approx m_T/2L \quad \text{(torus-like space)}$$

(5·9) holds only for a large value of $L$.

Among the quantities determined by the largest eigenvalue, a very important one is the time required for a polymorphic population to become monomorphic and genetic variability to cease to be maintained in a population. The average time required for a 'lucky' mutant gene to become fixed in a population is approximately equal to

$$2/(1 - \lambda_\infty) \tag{5·10}$$

and the second moment is

$$6/(1 - \lambda_\infty)^2 \tag{5·11}$$

(cf. Maruyama 1971).

It is worth noting that, for a large two-dimensional habitat, $1 - \lambda_\infty$ is proportional to the reciprocal of $L$ (the number of the colonies) whereas for a large one-dimensional habitat it is proportional to the reciprocal of the square of $L$. This suggests that a mutant gene spreads faster in a one-dimensional habitat than in a two-dimensional habitat, and therefore that a one-dimensional population maintains more genetic variability than a two-dimensional population of the same size. This point is illustrated in terms of local differentiation of the gene frequencies in Table 6. See also Kimura & Maruyama (1971) and Crow & Maruyama (1971).

## 6. Continuous model

We treat briefly a population occupying a continuous habitat for two reasons: (1) there are such populations in nature and (2) in a continuous model we use migration distance, which is the data given in many circumstances, instead of migration rate. The method of solving a continuous model is similar to that used for the stepping-stone models. Here we will discuss only a torus-like space. We assume also that the number of possible alleles is so large that every mutation creates an allele which is new to the population. The two-dimensional surface of a torus does not have an obvious counterpart in nature. However, since it has an exact solution, it points the way to an approximation to a more realistic model.

The space considered here is the direct product of two circles of circumference $L_1$ and $L_2$. Let $D$ be the number of individuals in a unit area on the habitat. For notational convenience we use a Cartesian co-ordinate system to designate a position on the surface in the space. Let $f(t, x, y)$ be the probability that two homologous genes, separated distance $x$ along the first co-ordinate axis and distance $y$ along the second co-ordinate axis at time $t$, are the same allele. Extend $f(t, x, y)$ on to the whole $(x, y)$ plane as a doubly periodic function of period $L_1$ and $L_2$. Let $m(\Delta t, x, y)$ be the migration function that gives the probability that an individual moves distance $x$ and $y$ in the directions of the first and second co-ordinate axes in the time interval $\Delta t$. Let

$$r(\Delta t, x, y) = \int_{-\infty}^{\infty}\int_{-\infty}^{\infty} m(\Delta t, x-\xi, y-\eta)\, m(\Delta t, \xi, \eta)\, d\xi\, d\eta.$$

Then the $f(t, x, y)$ satisfies

$$f(t+\Delta t, x, y) = (1-u\Delta t)^2 \int_{-\infty}^{\infty}\int_{-\infty}^{\infty} r(\Delta t, \xi, \eta)\Big\{f(t, x-\xi, y-\eta) + \frac{\Delta t\,\delta(x-\xi|\mathrm{mod}\,L_1)\,\delta(y-\eta|\mathrm{mod}\,L_2)\,(1-f(t, x-\xi, y-\eta))}{2D}\Big\} d\xi\, d\eta + o(\Delta t), \tag{6·1}$$

where $u$ is the mutation rate, $x-\xi|\mathrm{mod}\,L_1$ means that the value of $x-\xi$ is reduced by the modulo of $L_1$, and $\delta(.)$ is Dirac's delta function. (6·1) is the same as that given in Malécot (1948), except that his is for an infinite habitat.

We now assume

$$\int_{-a}^{a} x m(\Delta t, x, y)\, dx = \int_{-a}^{a} y m(\Delta t, x, y)\, dy = 0 \quad \text{for all } a \text{ and } \Delta t, \tag{6·2}$$

$$\lim_{\Delta t \downarrow 0} \frac{1}{\Delta t}\int_{|x|>\delta} m(\Delta t, x, y)\, dx = \lim_{\Delta t \downarrow 0} \frac{1}{\Delta t}\int_{|y|>\delta} m(\Delta t, x, y)\, dy = 0 \quad \text{for every } \delta > 0 \tag{6·3}$$

and

$$\left.\begin{aligned} \lim_{\Delta t \downarrow 0} \frac{1}{\Delta t}\int_{-\infty}^{\infty}\int_{-\infty}^{\infty} x^2 m(\Delta t, x, y)\, dx\, dy = \sigma_x^2 < \infty, \\ \lim_{\Delta t \downarrow 0} \frac{1}{\Delta t}\int_{-\infty}^{\infty}\int_{-\infty}^{\infty} y^2 m(\Delta t, x, y)\, dy\, dx = \sigma_y^2 < \infty. \end{aligned}\right\} \tag{6·4}$$

From (6·1), together with assumptions (6·2) to (6·4), we have the following equation which equilibrium $f(t, x, y)$ satisfies, $f \equiv f(x, y) = f(t, x, y)$

$$\sigma_x^2 \frac{\partial^2 f}{\partial x^2} + \sigma_y^2 \frac{\partial^2 f}{\partial y^2} - 2uf + \frac{\delta(x)\,\delta(y)}{2D}(1-f) = 0. \tag{6·5}$$

The solution of (6·5) which we seek is a periodic even function in both variables. It turns out to be

$$f(x, y) = \frac{L_1 L_2 (1-f(0, 0))}{D} \sum_{k=0}^{\infty}\sum_{l=0}^{\infty} \frac{\cos(2\pi kx)/L_1 \cos(2\pi ly)/L_2}{\{\pi^2(L_2^2 k^2 \sigma_x^2 + L_1^2 l^2 \sigma_y^2) + uL_1^2 L_2^2\}\,\Delta_k \Delta} \tag{6·6}$$

and

$$f(0, 0) = \frac{W}{4D+W} \tag{6·7}$$

in which

$$W = L_1 L_2 \sum_{k=0}^{\infty}\sum_{l=0}^{\infty} \frac{1}{\{\pi^2(L_2^2 k^2 \sigma_x^2 + L_1^2 l^2 \sigma_y^2) + uL_1^2 L_2^2\}\,\Delta_k \Delta},$$

where $\Delta_0 = 2$ and $\Delta_i = 1$ if $i \neq 0$.

The average $f(x, y)$ over the whole population is

$$\bar{f} = \frac{1}{L_1 L_2}\int_0^{L_1}\int_0^{L_2} f(x, y)\, dx\, dy = \frac{1-f(0, 0)}{4L_1 L_2 D u}$$
$$= \frac{1-f(0, 0)}{4N_T u}, \tag{6·8}$$

which is analogous to (4·4).

Although the analysis is given for a torus-like space, asymptotically this space becomes a plane. For a sufficiently large and closed habitat, formulas (6·6) to (6·8) can be applied, even if it is a rectangular habitat.

## 7. SIMULATIONS AND NUMERICAL EXAMPLES

In order to check the validity of the preceding analyses, I have performed a number of simulations by computer. Sampling of gametes and migration were done probabilistically by

Table 1. *Comparison of the results of simulations on the correlation and the variance of the gene frequencies with their theoretical expectations, formulas* (2·6) *to* (2·13)

(In all the cases below, $n_1 = n_3 (n_1 \times n_3$ = number of colonies) = 5, $2N$ (colony size) = 100, $m_1 = m_2$ (migration rate) = 0·1, $m_\infty$ (migration rate from outside world) = 0·01, and $\bar{q}$ (expected gene frequency) = 0·5.)

| | Correlation coefficient of the gene frequencies between the colonies specified | | | | | Variance of the gene frequencies in the colony | Model and formulas |
|---|---|---|---|---|---|---|---|
| Colony | (1, 2) | (1, 3) | (1, 5) | (2, 2) | (4, 4) | (1, 1) | |
| (1, 1) | 0·2469 | 0·0924 | −0·0094 | 0·1422 | 0·0296 | 0·00736 (S.) | (*a*) (2·6), (2·7), (2·8) |
| | 0·2556 | 0·0856 | 0·0112 | 0·1564 | 0·0164 | 0·00789 (T.) | |
| (1, 1) | 0·3626 | 0·1656 | 0·0365 | 0·2716 | 0·0466 | 0·00930 (S.) | (*d*) (2·11), (2·7), (2·8) |
| | 0·3683 | 0·1843 | 0·0455 | 0·2809 | 0·0532 | 0·00993 (T.) | |
| (1, 1) | 0·4119 | 0·1716 | 0·0502 | 0·3989 | 0·0486 | 0·00922 (S.) | (*e*) (2·12), (2·7), (2·8) |
| | 0·4101 | 0·1870 | 0·0654 | 0·4002 | 0·0730 | 0·01100 (T.) | |
| (1, 1) | 0·6084 | 0·4137 | 0·2388 | 0·5219 | 0·1936 | 0·01941 (S.) | (*f*) (2·13), (2·7), (2·8) |
| | 0·6273 | 0·4171 | 0·2506 | 0·5356 | 0·2192 | 0·02065 (T.) | |

S., Simulation; T., theoretical.

Table 2. *Comparison of the results of simulations on the probability of identity* ($f_{iji'j'}$) *and the genetic variability with their theoretical expectations, formulas* (3·3) *and* (3·4)

(In all the cases below, $n_1 = n_2 = 5$, $2N = 40$, $u$ (mutation rate) = 0·001, $K$ (number of possible alleles) = ∞, and the model used is of type ($f$) of Fig. 2.)

| | Probability of identity between colonies specified | | | | | | Effective no. of alleles ($n_e$) in the entire population and homozygote frequency | | | |
|---|---|---|---|---|---|---|---|---|---|---|
| Colony | (1, 1) | (1, 2) | (1, 3) | (1, 5) | (2, 2) | (5, 5) | $n_e = 1/\bar{f}$ | $f_0$ | | |
| (1, 1) | 0·7940 | 0·3669 | 0·1681 | 0·0572 | 0·3308 | 0·0172 | 7·331 | 0·7648 | (S.) | $m_1 = m_2 = 0·0025$ |
| | 0·7715 | 0·3489 | 0·1441 | 0·0412 | 0·2352 | 0·0159 | 7·541 | 0·7582 | (T.) | |
| (1, 1) | 0·7001 | 0·4020 | 0·2203 | 0·0765 | 0·3149 | 0·0455 | 5·701 | 0·6823 | (S.) | $m_1 = m_2 = 0·005$ |
| | 0·7028 | 0·3917 | 0·2096 | 0·0945 | 0·2992 | 0·0546 | 5·491 | 0·6808 | (T.) | |
| (1, 1) | 0·5242 | 0·4173 | 0·3297 | 0·2730 | 0·3770 | 0·2513 | 3·315 | 0·4958 | (S.) | $m_1 = m_2 = 0·025$ |
| | 0·5594 | 0·4308 | 0·3429 | 0·2781 | 0·3888 | 0·2447 | 3·148 | 0·5266 | (T.) | |

S., Simulation; T., theoretical.

generating pseudo-random numbers uniformly distributed on (0, 1). The correlation, the variance, the probability of being identical alleles, genetic variability etc. were calculated and compared with the theoretical expectations. Agreement between them was good in all the cases compared. In Table 1 the correlation and variance of a particular allele are compared. They are to show the validity of the formulas developed in § 2. In Table 2 the results of simulations on the probability of being identical alleles are compared with their theoretical expectation. In this table, the total variability measured by the effective number of alleles is also presented.

In Table 3 the rate of decay which was calculated from the data of a simulation by using formula (5·3), is compared with the exact rate.

Table 3. *The rate $(1-\lambda_t)$ of decay of genetic variability calculated from observed local- and global-heterozygosities by using formula* (5–3)

(The parameters in the simulation are $n = 40$, $2N = 2$, $m = 0.2$ and exact value of $(1-\lambda_\infty) = 0.00125$.)

$$\left(\bar{h}_0^{(t)} = \sum_{i=1}^{n} h_{ii}^{(t)}/n, \quad \|\mathbf{H}_t\| = \sum_{i,j=1}^{n} h_{ij}^{(t)}/n^2.\right)$$

| | | | | | | |
|---|---|---|---|---|---|---|
| $(1-\lambda_\infty)t$ | 0·25 | 0·50 | 0·75 | 1·00 | 1·25 | 1·50 |
| $\bar{h}_0^{(t)}/[2Nn\|\mathbf{H}_t\|]$ | 0·00136 | 0·00123 | 0·00125 | 0·00126 | 0·00113 | 0·00155 |
| $(1-\lambda_\infty)t$ | 1·75 | 2·00 | 2·25 | 2·50 | 2·75 | 3·00 |
| $\bar{h}_0^{(t)}/[2Nn\|\mathbf{H}_t\|]$ | 0·00115 | 0·00125 | 0·00149 | 0·00123 | 0·00151 | 0·00113 |

Average of the 12 observations: 0·00129.

Using the torus-like population structure of § 3 and the recurrence relation (5·1), the asymptotic rates of decay in heterozygosity were calculated numerically by computer. They are compared with the values given by formulas (5·7) to (5·9) in Table 4.

Table 4. *The rate of decay of genetic variability in a two-dimensional square habitat with the colonies arranged on the surface of a torus*

(The approximation, (5·7), (5·8) or (5·9), is compared with the exact value of the dominant eigenvalue obtained by a matrix iteration method.)

| No. of colonies | Migration rate ($m_T$) | Colony size ($2N$) | $m_T N$ | Dominant eigenvalue $(1-\lambda_\infty)$ | | Formula |
|---|---|---|---|---|---|---|
| | | | | Exact | Theoretical approximation | |
| 96 × 96 | 0·2 | 2 | 0·2 | $1.19 \times 10^{-5}$ | $1.09 \times 10^{-5}$ | (5·9) |
| 100 × 100 | 0·5 | 2 | 0·5 | $2.33 \times 10^{-5}$ | $2.50 \times 10^{-5}$ | (5·9) |
| 150 × 150 | 0·4 | 2 | 0·4 | $9.36 \times 10^{-6}$ | $8.89 \times 10^{-6}$ | (5·9) |
| 190 × 190 | 0·1 | 1 | 0·1 | $1.48 \times 10^{-6}$ | $1.39 \times 10^{-6}$ | (5·9) |
| 190 × 190 | 0·2 | 4 | 0·4 | $2.40 \times 10^{-6}$ | $2.77 \times 10^{-6}$ | (5·9) |
| 10 × 10 | 0·2 | 40 | 4·0 | $2.30 \times 10^{-4}$ | $2.50 \times 10^{-4}$ | (5·8) |
| 30 × 30 | 0·3 | 20 | 3·0 | $4.79 \times 10^{-5}$ | $5.55 \times 10^{-5}$ | (5·8) |
| 96 × 96 | 0·01 | 400 | 2·0 | $2.28 \times 10^{-7}$ | $2.71 \times 10^{-7}$ | (5·8) |
| 96 × 96 | 0·4 | 50 | 10·0 | $2.05 \times 10^{-6}$ | $2.17 \times 10^{-6}$ | (5·8) |
| 10 × 10 | 0·00001 | 2 | 0·00001 | $1.86 \times 10^{-7}$ | $2.00 \times 10^{-7}$ | (5·7) |
| 96 × 96 | 0·001 | 2 | 0·001 | $1.63 \times 10^{-7}$ | $2.17 \times 10^{-7}$ | (5·7) |

At the last part of § 5 I have stated that the time required for a mutant gene to become fixed in a population has the mean $2/(1-\lambda_\infty)$ and the second moment $6/(1-\lambda_\infty)^2$. Using a one-dimensional circular model, I have performed several simulations. Their means and second moments are compared with the theoretical expectations in Table 5.

A large amount of genetic variability at protein level (isozymes) has been found by Harris (1966), Lewontin & Hubby (1966), and many others, and it is of considerable interest to know whether the variability is due to natural selection or to the random drift of the gene frequency.

Table 5. *Time until a single mutant gene reaches fixation in a population occupying a circular habitat*

(The results of simulations are compared with the theoretical expectations, (5·10) and (5·11). The values of $\lambda_\infty$ in the table are given by (5·5) or (5·6).)

| No. of colonies ($n$) | Colony size $2N$ | Migration rate | | Variance $\sigma^2$ | Fixation time: Mean | | Fixation time: 2nd moment | | No. of simulations |
|---|---|---|---|---|---|---|---|---|---|
| | | $m_1$ | $m_2$ | | Simulation | Theoretical $2/(1-\lambda_\infty)$ | Simulation | Theoretical $6/(1-\lambda_\infty)^2$ | |
| 20 | 2 | 0·05 | 0 | 0·05 | 1333·80 | 1600 | $2{\cdot}34\times10^6$ | $3{\cdot}84\times10^6$ | 35 |
| 20 | 2 | 0·05 | 0·0125 | 0·1 | 785·81 | 800 | $7{\cdot}47\times10^5$ | $9{\cdot}60\times10^5$ | 91 |
| 30 | 2 | 0·1 | 0 | 0·1 | 2005·15 | 1800 | $5{\cdot}77\times10^6$ | $4{\cdot}80\times10^6$ | 40 |
| 30 | 2 | 0·1 | 0·025 | 0·2 | 876·77 | 900 | $1{\cdot}15\times10^6$ | $1{\cdot}20\times10^6$ | 35 |
| 5 | 50 | 0·1 | 0·1 | 0·5 | 490·45 | 500 | $3{\cdot}33\times10^5$ | $3{\cdot}75\times10^5$ | 40 |
| 10 | 25 | 0·2 | 0·1 | 0·6 | 462·56 | 500 | $3{\cdot}01\times10^5$ | $3{\cdot}75\times10^5$ | 48 |

The often used criterion is to investigate the gene frequencies at different localities and if they are all similar we tend to attribute the polymorphism to natural selection and if there is a geographical cline in the frequencies we tend to think there is geographically differentiated selection. Here I want to show that both phenomena can arise entirely by the random drift of selectively neutral alleles.

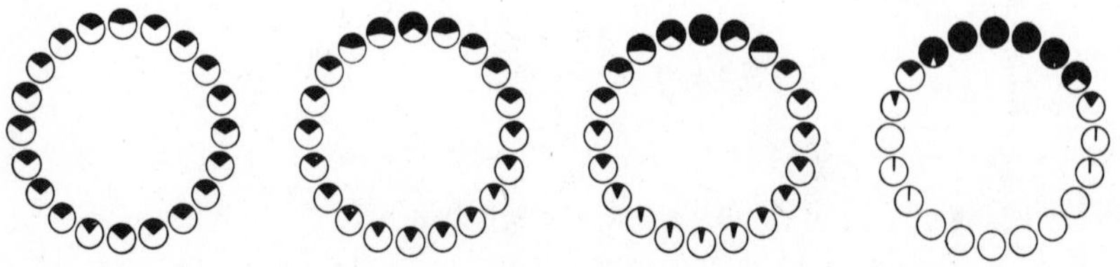

(a) $2N=100$, $m_1=0{\cdot}4$, $m_2=0{\cdot}2$, $\sigma^2=1{\cdot}2$ (b) $2N=100$, $m_1=0{\cdot}05$, $m_2=0{\cdot}0125$, $\sigma^2=0{\cdot}1$ (c) $2N=20$, $m_1=0{\cdot}05$, $m_2=0{\cdot}0125$, $\sigma^2=0{\cdot}1$ (d) $2N=20$, $m_1=0{\cdot}01$, $\sigma^2=0{\cdot}01$

Fig. 3. Local differentiation of gene frequencies (simulation). The circles indicate colonies and the dark part of each circle indicates the frequency of the mutant gene in the corresponding colony. The gene frequency in the entire population (20 colonies) is between 0·30 and 0·31. (*a*) represents a situation in which $2\sigma^2N \gg \frac{1}{10}n$, (*d*) represents a situation in which $2\sigma^2N \ll \frac{1}{10}n$, and (*b*) and (*c*) are intermediate situations.

Table 6. *Examples of numerical values of several statistics* ($f_0$, $\bar{f}$, etc.) *for continuous one- and two-dimensional models, assuming* $K = \infty$

| Habitat size | Density ($D$) | Variance of dispersion ($\sigma^2$) | $D\sigma^2$ | Mutation rate ($u$) | $f_0$ | $\bar{f}$ | $f_{min}$ | $(1-f_0)/(1-\bar{f})$ |
|---|---|---|---|---|---|---|---|---|
| 200 × 200 | 100 | 0·001 | 0·1 | $10^{-7}$ | 0·883 | 0·073 | 0·066 | 0·126 |
| 200 × 200 | 100 | 1·0 | 100·0 | $10^{-7}$ | 0·386 | 0·384 | 0·383 | 0·996 |
| 200 × 200 | 1000 | 0·001 | 1·0 | $10^{-7}$ | 0·430 | 0·636 | 0·033 | 0·590 |
| 500 × 500 | 100 | 0·001 | 0·1 | $10^{-7}$ | 0·866 | 0·013 | 0·013 | 0·136 |
| 500 × 500 | 100 | 0·01 | 1·0 | $10^{-7}$ | 0·444 | 0·056 | 0·51 | 0·588 |
| 500 × 500 | 100 | 1·0 | 100·0 | $10^{-7}$ | 0·095 | 0·090 | 0·090 | 0·995 |
| 500 × 500 | 1000 | 0·01 | 10·0 | $10^{-7}$ | 0·074 | 0·009 | 0·009 | 0·935 |
| 1000 × 1000 | 100 | 0·01 | 1·0 | $10^{-8}$ | 0·487 | 0·128 | 0·123 | 0·589 |
| 1000 × 1000 | 10 | 0·01 | 0·1 | $10^{-7}$ | 0·874 | 0·031 | 0·028 | 0·130 |
| 1000 × 1000 | 1 | 100·0 | 100·0 | $10^{-7}$ | 0·962 | 0·962 | 0·962 | 0·998 |
| 1000 × 1000 | 1 | 0·1 | 0·1 | $10^{-7}$ | 0·969 | 0·779 | 0·775 | 0·141 |
| 1000 × 100 | 10 | 0·1 | 1·0 | $10^{-7}$ | 0·787 | 0·534 | 0·456 | 0·458 |
| 1000 × 10 | 10 | 0·1 | 1·0 | $10^{-6}$ | 0·890 | 0·275 | 0·074 | 0·152 |
| 1000 × 10 | 1 | 1·0 | 1·0 | $10^{-6}$ | 0·970 | 0·738 | 0·628 | 0·113 |
| 1000 × 1 | 10 | 0·01 | 0·1 | $10^{-6}$ | 0·994 | 0·141 | 0·002 | 0·007 |
| 1000 × 1 | 10 | 0·1 | 1·0 | $10^{-6}$ | 0·982 | 0·429 | 0·208 | 0·030 |
| 1000 × 1 | 100 | 1·0 | 100·0 | $10^{-6}$ | 0·744 | 0·640 | 0·590 | 0·713 |
| 2000 × 1 | 100 | 0·1 | 10·0 | $10^{-6}$ | 0·848 | 0·190 | 0·019 | 0·187 |

The preceding analyses suggest that when the dominant eigenvalue ($\lambda_\infty$) is given by (5·5), (5·7) or (5·9) there will be a tendency to geographical cline in gene frequencies; on the other hand, when it is given by (5·6) or (5·8) the gene frequencies at different localities tend to be all similar. Using the circular model the following simulations are performed. We place a single mutant gene in the population and follow the frequency of this gene. If it becomes extinct from the population or fixed in it, we place a new mutant gene, and repeat this process many times. But except on these occasions no mutant gene is introduced into the population. When and only when the gene frequency of a mutant gene in the whole population is between 0·30 and 0·31, we record the gene frequency of each colony. Among such records, cases in which colony 1 had the highest frequency were averaged for each colony. They are presented in Fig. 3. It is clear from Fig. 3 that both geographical cline and uniformity in the gene frequencies at different localities can be attributed entirely to the random drift.

Finally, using the formulas at the bottom of page 218 of Maruyama (1970*a*), and (6·6) and (6·7) of this paper, I want to compare numerically the local differentiation in terms of the probability of identity for continuously distributed large but finite populations of different shapes; linear, long but narrow, and square habitats. We compare the following four quantities: the homozygous probability, $f(0)$ or $f(0, 0)$; the average probability of identity, $\bar{f}$; the minimum value of the $f(x)$ or $f(x, y)$, $f_{\min}$ – that is, the probability of identity between individuals farthest separated in the population; and $(1-f(0))/(1-\bar{f})$ or $(1-f(0\ 0))/(1-\bar{f})$, the factor by which the dominant eigenvalue of a structured population differs from that of a panmictic population of the same size, see (5·3). These quantities give a good indication of how a population is locally differentiated. They are given in Table 6. It is worth noting that the local differentiation is very strong in a linear or narrow habitat, whereas it is rather weak in a truly two-dimensional habitat. See also Kimura & Maruyama (1971).

## SUMMARY

Two-dimensional stepping-stone models of finite length and other population models were analysed with special reference to the variance of gene frequency, the correlation of gene frequency between colonies, the probabilities of identity within a colony and between colonies, and the genetic variability maintained in the entire population. Each of these quantities is given explicitly as a function of the colony size, the location of colony, the migration rate and the mutation rate.

It was shown that for a closed population the total genetic variability is known from the local homozygosity without knowing the population structure.

Under the assumption of no mutation, the decreasing rate of genetic variability was investigated. The rate was expressed by a ratio of the local and global heterozygosity. Some limiting cases and one upper bound for the asymptotic rate were given.

I would like to acknowledge the encouragement and help offered, at all times, by Dr Motoo Kimura, who has also suggested the subject of this paper to me. I am also grateful to Professor J. F. Crow, to whom I owe a number of substantial improvements in this paper.

## REFERENCES

Crow, J. F. & Maruyama, T. (1971). The number of neutral alleles maintained in a finite, geographically structured population. *Theoretical Population Biology* (in the press).

Harris, H. (1966). Enzyme polymorphisms in man. *Proceedings of the Royal Society* B **164**, 298–310.

Jacobson, N. (1953). *Lectures in Abstract Algebra*. Princeton: Van Nostrand.

Kimura, M. (1953). 'Stepping stone' model of population. *Annual Report of National Institute of Genetics, Japan* **3**, 62–3.

Kimura, M. & Crow, J. F. (1964). The number of alleles that can be maintained in a finite population. *Genetics* **49**, 725–38.

Kimura, M. & Maruyama, T. (1971). Pattern of neutral polymorphism in a geographically structured population. *Genetical Research* (in the press).

Lewontin, R. C. & Hubby, J. L. (1966). A molecular approach to the story of genic heterozygosity in natural populations. II. Amount of variation and degree of heterozygosity in natural populations of *Drosophila pseudoobscura*. *Genetics* **54**, 595–609.

Malécot, G. (1948). *Les mathématiques de l'hérédité*. Paris: Masson.

Maruyama, T. (1970*a*). Analysis of population structure. I. One-dimensional stepping-stone models of finite length. *Annals of Human Genetics* **34**, 201–19.

Maruyama, T. (1970*b*). Stepping stone models of finite length. *Advances in Applied Probability* **2**, 229–258.

Maruyama, T. (1970*c*). On the rate of decrease of heterozygosity in circular stepping stone models of populations. *Theoretical Population Biology* **1**, 101–19.

Maruyama, T. (1970*d*). Rate of decrease of genetic variability in a subdivided population. *Biometrika* **57**, 299–311.

Maruyama, T. (1971). Speed of gene substitution in a geographically structured population. *American Naturalist* **105**, 253–65.

Robertson, A. (1964). The effect of nonrandom mating within inbred lines on the rate of inbreeding. *Genetical Research* **5**, 164–7.

Smith, C. A. B. (1969). Local fluctuation in gene frequencies. *Annals of Human Genetics* **32**, 251–60.

# VI

# The Current State of Unsolved Problems

# Editor's Comments on Paper 17

17 **Haldane and Jayakar:** An Enumeration of Some Human Relationships
*J. Genet.*, **58**, No. 1, 81–107 (1962)

One of the original motivations for collecting the papers in this volume was a surprise to the editor similar to that expressed by this paper in its introductory paragraph. On first sight, one might expect there are only a few types of kin-based human relationships, and that therefore all the important ones (and perhaps a few that are not important), would have already been studied, their genetic effects and sociological importance known. Part of the answer is provided by this paper; there are a huge number of even simple relationships, and even these may be interpreted differently in various societies.

It is useful to note that in approaching what they apparently conceived of as a new topic, the authors devised techniques founded on the logical ideas of Carnap and Russell (see the bibliography of the paper), much as Cotterman had done. It is also useful to recognize that the resulting technique closely resembles that of Macfarlane. This leaves one with the nearly correct impression that some areas of progress were remiss for the intervening 80 years. However, the main progress was in other areas, as discussed in Part V, in social theory. Apart from a basically intuitive structuralism such as that which apparently guided Wright, genetics has been more concerned with the demographic than the structural. Dahlberg (1929, 1948) explored certain relationships between family sizes and numbers of marriages between cousins of arbitrary degree of removal. Nei and Imaizumi (1962) provided a formula for number of marriages between relatives of arbitrary degree, while Nei et al. (1970) carried out an empirical study. For other references on this topic, see the paper of Cavalli-Sforza et al. in this volume (Paper 19). In addition, recent work by Cockerham (1970) needs to be considered.

It is impossible, and perhaps irresponsible, to include a paper by J. B. S. Haldane without some comment on his other work. With specific reference to this paper and the topic of this volume, see the work by Haldane and Moshinsky (1939), which may be viewed as a theoretical foundation for the genetic techniques of the present paper. Bartlett and Haldane (1935) is one of the earliest papers to use matrix techniques, hence has been slightly ignored in my historic treatment of Etherington. This is justified by the fact that Haldane in that paper was concerned with problems other than mathematical foundations per se. Haldane (1948) is a still fundamental paper on the techniques and areas of concern of human population genetics.

For a more general summary of Haldane's theoretical work, see Haldane (1931), which has an appendix on his basic mathematical theory. For a nearly complete bibliography and an interesting biography of his personal life and development, see Clark (1969).

# 17



Reprinted from *J. Genet.*, **58**, No. 1, 81–107 (1962)

# AN ENUMERATION OF SOME HUMAN RELATIONSHIPS

BY J. B. S. HALDANE AND S. D. JAYAKAR
*Genetics & Biometry Research Unit, C.S.I.R.*

## INTRODUCTION

In view of the importance attached to relationship in most human cultures, religions, and legal codes, it might be thought that at least the simpler ones, for example those not more distant than that of uncle, would have been enumerated, and a terminology adopted for them, either in colloquial or legal language, or in that of anthropologists. This is not the case. For example if we say that *B* is *A*'s first cousin we shall show that this may mean any of 48 relationships if the sexes of *A* and *B* are not specified, and 192 if they are specified. Only a few of these are given special names in any language known to us. All but 6 of the 48 involve at least one multiple marriage, or its biological equivalent, among the grandparents of *A* and *B*. However even when simultaneous polygamy and divorce are both illegal, remarriage after the death of a spouse is permitted and may be enjoined, and its frequency is shown by the existence of such words as "stepmother" and "half-sister". With increasing study of the extent and effects of human inbreeding an exact terminology is desirable, though we do not suppose that our own cannot be improved. We venture to hope that our enumeration may be of value to anthropologists.

(*a*) As we are concerned with biological relationships, it is irrelevant whether the parents of any member of the pedigree were married when he or she was born. Nevertheless we shall frequently use the word "marriage" to mean "marriage or its biological equivalent".

(*b*) We confine ourselves to cognate or "blood" relationships, that is to say relationships between two persons who have at least one known latest common ancestor in common, or of whom one is the ancestor of the other. By "latest common ancestor" is meant the last-born common ancestor. Thus two sons of the same man by different women have two grandparents in common, but their father is their latest common ancestor.

(*c*) We also confine ourselves to relationships of *B* to *A* where no latest common ancestor is more remote than a grandparent of *A* or *B*.

(*d*) We further neglect the possibility of any inbreeding in the ancestry of *A* or *B* (though of course their offspring, if any, are inbred). For it can be seen that by condition (*c*) this would imply a fertile sexual union between parent and offspring or between whole or half brother and sister. A union between grandparent and grandchild, uncle and niece, or aunt and nephew would violate condition (*c*). Thus if *A* is the child of a man and his niece, this means that the same person is *A*'s grandparent and great-grandparent. Incestuous unions between parent and offspring, or between sibs, rarely occur, and are still more rarely discovered with certainty. We are aware that in Bali (*fide* Karve, 1953) twins of opposite sex were compelled to marry. A consideration of such possibilities would greatly extend our analysis. We do however consider unions, such as that of a man with a woman and her daughter by another man, which are forbidden in many cultures. Such unions do not involve inbreeding.

(*e*) Finally we adopt the convention that the sexes of the relatives *A* and *B* are irrelevant to the relationship. This convention reduces the number of relationships listed to one quarter of the possible number, and greatly simplifies the symbolism. However it involves some complication in the consideration of sex-linked genes. The structure of most languages is illogical. Thus a man or a woman usually uses the same word for his or her brother, and another for his or her sister, though in India different words of older and younger brothers, and for older and younger sisters, are usual. However a man and a woman almost always use the same words. Thus the description of the relationship of *B* to *A* depends on the sex of *B* and not that of *A*. But according to Karve (1953) the Nambudri Brahmins of Keraḷa are more logical. A woman calls her elder brother "*oppa*", a man "*jyesṭhan*". A woman calls her elder sister "*chettati*" or "*jyəsṭhati*", a man calls her "*oppol*". The terms for younger sibs do not appear to depend on *A*'s sex.

Ignoring the sexes of *A* and *B* we have to enumerate 129 distinct relationships, or 516 if the sexes of *A* and *B* are considered. This large number is mainly due to the possibilities of polygamy, remarriage, and their biological equivalents. If no member of the pedigree has a child by more than one spouse or sexual partner, these numbers are reduced to 17 and 68.

Monozygotic twinning leads to a slight complication. Monozygotic twins have the same genotype apart from mutation, and their children presumably resemble one another as closely as those of the same individual. The effect of such twinning is considered in a special section.

The classification adopted in Roman law (see Morton, 1961 for diagrams) is excellent so far as it goes, and will be adopted here. The degree of consanguinity is the number of "steps" between two relatives, the parent-child relationship being regarded as a step. Thus parents are relatives of degree 1, sibs of degree 2. This system does not take cognizance of half-relationships, and even with the restrictions which we have adopted, *B* can be simultaneously related to *A* in two different degrees.

A geneticist asks three main types of question about a given relationship:

(1) What correlations may be expected between the phenotypic characters of *A* and *B* as the result of their relationship ?

(2) If *A* and *B* are of different sexes, and have children, how may these children be expected to differ from the general population ?

(3) What are the frequencies in a population studied, of marriages, or their biological equivalent, between relatives of various different kinds, and of progeny of such marriages ? These two frequencies are not quite the same if inbreeding affects the net fertility of a marriage, or the viability of its children.

The first two questions can be answered with the aid of coefficients of relationship which are here calculated. It must however be stated that these coefficients may be misleading for at least five reasons. The population studied may be divided into more or less endogamous groups, based either on geographical isolation, or on religious, occupational, or caste differences. In some societies spouses are positively correlated for certain phenotypic characters, presumably as a result of choice. Selective deaths may alter correlations; thus maternal-foetal incompatibility must raise the correlation between mothers and surviving children with respect to some antigens. There is presumably a tendency for relationships to accumulate. Thus those who are first cousins are probably more often also second or third cousins than are persons unrelated in the fourth degree. Finally the environments of relatives are correlated. All these causes must tend to increase correlations.

All the relationships which we shall consider are irreflexive in the terminology of logic. If *B* is *A*'s paternal half-brother, he cannot be identical with *A*. So we must be careful not to define *A*'s paternal half-brother as *A*'s father's son, and if we are using symbolic logic, to add the non-identity symbol *J* where it is needed to ensure irreflexivity (see Carnap, 1958, pp. 117, 223). None of the relations in our list is transitive, like "ancestor" and "descendant". Some are intransitive, others non-transitive, for example the relation "whole sib". The symmetry or otherwise of relationships is more important. It is clear that some relationships are symmetric, for example if *B* is *A*'s mother's whole sister's child, *A* is *B*'s mother's whole sister's child. Others are asymmetric. Thus if *B* is *A*'s mother's whole brother's child, *A* is *B*'s father's whole sister's child. Such asymmetric relations evidently occur in pairs, each member of a pair being the converse of the other. A third class of relations, the non-symmetric, also exists. If *B* has a relation *R* of this class to *A*, then *A* may or may not have it to *B*. Thus if *B* is *A*'s sister, *A* may be *B*'s sister or her brother. None of the relations in our classification fall into this class, as they would do if *B*'s sex were specified while *A*'s was not. We avoid this class either by specifying the sex of neither *A* nor *B*, or by specifying the sex of both. We can thus classify all our 129 relationships as symmetrical or fully asymmetrical. In fact 37 are symmetrical, and the other 92 occur in converse pairs.

## COEFFICIENTS OF RELATIONSHIP

The terminology of coefficients of relationship is unfortunately imprecise. Wright (1922) used the lower case letter $f$ for his coefficient, which is much the most important of the group. Recent authors have however used $F$. Haldane and Moshinsky (1939) used $f'$ for sex-linked loci. This had however been used by Wright for $f_{n-1}$ when $f$ denoted $f_n$. We suggest that the Greek letter $\phi$ be used. $f$ and $\phi$ enable us to make statements about the gametes of $B$, given information about a gamete of $A$. But we also require information about the diploid genotype of $B$, given that of $A$. This information may be given by two more coefficients, $F$ for autosomal loci, and $\Phi$ for sex-linked loci in members of the homogametic sex. A similar coefficient for $Y$-linked genes would be unity for males with a latest common male ancestor connected to each by a series of males, (e.g. if $B$ is $A$'s father's paternal half-brother) and otherwise zero.

The values of $f$ and $F$ are independent of the sexes of $A$ and $B$, so no difficulty arises from grouping our relations in fours. But this is not so for $\phi$, which may be defined as the limiting probability, when a sex-linked gene is very rare, that if it is present in a gamete of $A$, it will also be found in the first gamete of $B$ examined. This probability differs with the sexes of $A$ and $B$ and may assume four different values. By listing these, we have been able to reduce our list of relationships from 516 to 83, and we consider that this compensates for a possible lack of logicality. Our definitions, which follow, are based on Malécot's (1948) useful fiction of an ancestral population in which any given allele was present at only one locus. This greatly facilitates the calculation of the coefficients in finite pedigrees. We shall call such a gene 'rare'. From these definitions one can readily calculate probabilities when a gene is not rare.

$f$ symbolizes Wright's coefficient of relationship or of inbreeding. It is more precisely described as the coefficient of single autosomal relationship. The rules for its calculation are well-known. $f_{AB}=f_{BA}$ is the probability that if $A$ has produced a gamete carrying a rare gene, the first tested gamete of $B$ will carry the same gene. When the gene has a frequency $q$ in the population this probability becomes $q+f\,(1-q)$. If there is no dominance, the somatic resemblance between $A$ and $B$ for any character is given by the correlation

$$\rho_{AB} = \frac{2\sigma^2 f_{AB}}{\sigma^2+\epsilon^2}.$$

$$= 2h^2 f_{AB}.$$

where $\sigma^2$ is the part of the variance of the character concerned due to additive gene effects, $\epsilon^2$ the part due to inhomogeneity of the environment, and $h^2$ its heritability. (Here and later we neglect maternal effects, environmental correlation between relatives, epistasis, and other complicating factors listed by Lerner (1950), Kempthorne (1957) and others.) However dominance introduces a complication. If $A$ and $B$ are related through both parents of each, then there is a finite probability $F_{AB}$, when neither $A$ nor $B$ is inbred (i.e. $f_A=f_B=0$) that if $A$ is homozygous for a rare gene, $B$ will also be homozygous. This, as Malécot showed, can be calculated as follows. If $P$ and $Q$ are the parents of $A$, $R$ and $S$ those of $B$, then

$$F_{AB}=f_{PR}\,f_{QS}+f_{PS}\,f_{QR},$$

$$\text{while } f_{AB}=\tfrac{1}{4}(f_{PR}+f_{QS}+f_{PS}+f_{QR}).$$

If now $F_{AB}$ is not zero, their somatic resemblance is given by

$$\rho_{AB}=\frac{2\sigma^2 f_{AB}+\tau^2 F_{AB}}{\sigma^2+\tau^2+\epsilon^2}$$

where $\tau^2$ is the part of the variance of the character concerned due to "dominance deviations",

Fisher (1918) first explained the higher correlations between sibs than between parent and offspring on these lines. $F$ is usually zero, for example for ordinary first cousins, but it is not zero for double half first cousins, though a large sample would be needed to verify the increased correlation. If $p+q=1$, the array of $B$, given that $A$ is **gg,** and **G** stands for any allelomorph of **g,** is

$$(1-4f+F)p^2\,\mathbf{GG}+2p\,[q+2f\,(p-q)-Fp]\,\mathbf{Gg}+(q^2+4fpq+Fp^2)\,\mathbf{gg}.$$

The value of $F_{AB}$ is irrelevant to the offspring of $A$ and $B$.

$\phi_{AB}$ is similarly definable as the probability that if $A$ has produced a gamete carrying a rare sex-linked gene, the first tested gamete of $B$ will carry the same gene. $\phi$ is defined as their coefficient of single sex-linked relationship; and it is easily seen that $\phi_{BA}=\phi_{AB}$. If $A$ and $B$ have a child $C$, $\phi_C$ is not defined if $C$ is a male. If $C$ is a female $\phi_C=\phi_{AB}$ is her coefficient of sex-linked inbreeding. When the gene **s** is not rare but has a frequency $q$, and $p+q=1$, then the probable genotypes of $B$, if $A$ has produced a gamete carrying **s**, are:—

$$♂\ (1-f)p\mathbf{S}+(q+fp)\mathbf{s}$$
$$♀\ (1-2f)p^2\mathbf{SS}+2\ (pq+fp^2-fpq)\ \mathbf{Ss}+(q^2+2fpq)\ \mathbf{ss}$$

where **S** stands for all alleles of **s.**

Haldane and Moshinsky (1939) gave rules for calculating $\phi_{AB}$ which are correct if none of their latest common ancestors are inbred, as in the pedigrees here considered. The amended rules, valid for a finite pedigree, are as follows. List all paths by which $A$ and $B$ are connected to latest common ancestors. Ignore any path passing through two males in succession. Count the steps in the remainder with the following conventions. The step from mother to daughter counts as one. That from father to daughter counts as zero, so does that from mother to son, except that where a woman is a latest common ancestor the steps between two of her sons count as one. If $n_i$ is the number of steps so counted in the $i$-th path from $A$ to $B$, and $\phi_i$ is the coefficient of sex-linked inbreeding of the latest common ancestor (if a woman) on this path, then

$$\phi_{AB}=\Sigma 2^{-n_i-1}(1+\phi_i).$$

Similarly if $P$ and $Q$ are the parents of $A$, $R$ and $S$ those of $B$,

$$\Phi_{AB}=\phi_{PR}\,\phi_{QS}+\phi_{PS}\,\phi_{QR},$$

with the further convention that if $P\equiv R$, $\phi_{PR}=1$ if $P$ is a male, and $\phi_{PR}=\frac{1}{2}$ if $P$ is a female. This is the probability that if $A$ is **ss,** $B$ will also be **ss.** In practice it is simpler to calculate these values directly, adopting Malécot's fiction, for the 18 fundamental relationships from which all the others in Table 1 except $F$ and $\Phi$ can be built up by addition.

## THE LOGICAL STRUCTURE OF HUMAN RELATIONSHIPS

Carnap (1958) describes human relationships in terms of symbolic logic in his sections 15c, 17b, 30c, 31b, 54a, 54b, and 55c. He is able to derive all cognate relationships from the relation *Par* (Parent of) and the class *Ml* (Male), though in this system $x$ is only said to be the husband of $y$ if he has begotten a child by her. Further primitive signs are needed for legal relationships such as "Husband of", but they do not concern us here. Logicians have been more interested in defining relationships for which names already exist than in enumeration. In practice we gain in symmetry by adopting two primitive signs $M$ and $P$ for relations, rather than one. Let $M$ designate the maternal relationship, i.e. $M(x,y)$ means "$x$ is the mother of $y$", and let $P$ similarly designate the paternal relationship. Instead of $M^{-1}$ and $P^{-1}$ for the converse relationships we shall use Russell's (1903) sign $\breve{M}$ and $\breve{P}$. $\breve{M}\ (x,y)$ means that $x$ is a child born of $y$, or $y$ the mother of $x$. $M$ and $P$ are one-valued, $\breve{M}$ and $\breve{P}$ are many-valued.

The logical products of all these four symbols are null for the human species. Thus $M.P$ would mean that $x$ was both mother and father of $y$, which is only possible in self-fertilized organisms. However their relative products are meaningful. Thus $M|M$ means maternal grandmother, and $M|P$ paternal grandmother (mother of father). Cognate relationships involving a common ancestor are symbolized by a series of one or more inverse signs followed by one or more direct signs. The first direct sign must be the converse of the last converse sign. Thus $M|P$ is null. It means that $x$'s mother is $y$'s father. $M|M$ means that $x$ and $y$ have the same mother. To ensure that $x$ is not identical with $y$ we must use the non-identity sign $\mathcal{J}$. Thus $M|M.\ \mathcal{J}$ means that $x$ has the same mother as $y$ but is not identical. $x$ and $y$ often have several latest common ancestors. Thus $M|M.\ \breve{P}|P.\ \mathcal{J}$ means that $x$ has the same mother as $y$, and also the same father, and is not identical; hence $x$ is $y$'s whole sister or brother. This is however a somewhat cumbrous expression, which we replace by a single letter. However we are aware that logicians may prefer a symbolism which fits into the corpus of symbolic logic. And it may also readily be translated into the symbols of binary arithmetic. Thus if $x$ is the child of $y$'s mother's father and of $y$'s father's whole sister, a relation which we symbolize by $\left.\begin{matrix} wH \\ hMw \end{matrix}\right\}$, the relationship in the terminology of this section is $(\breve{P}|P.\mathcal{J})\ |M{\cdot}\breve{M}|\ (M|M.\ \breve{P}|P.\mathcal{J})\ |P$. If we write *0* for $\breve{M}$ and $M$ and 1 for $\breve{P}$ and $P$, using a decimal point for the latest common ancestor, with the convention that children of the same parent are not identical, this becomes $1{\cdot}10\frown00{\cdot}01\frown01{\cdot}11$, using Russell's sign for logical product. It is instructive to note why we could not use symbols for "Son of" and "Daughter of" as our primitive signs.

## Symbolism

In the interests of brevity we use the following symbolism. Each letter stands for a human being. *w* means a wife, woman, Weib, etc. *h* means a man (husband, homme, Herr, homo, etc.). The symbols *m* (which may mean male, man, maschio, mother, Mutter, mā (Hindi) etc.) and *f* (which may mean father, female, femme, etc.) are liable to be misleading. Any set of letters (not more than three in this paper) symbolises a relationship of *B* to *A* through a common ancestor, beginning with *A*'s parent. The common ancestor is written with a capital letter, and letters subsequent to it represent descendants of the common ancestor, and ancestors of *B*. Thus *wH* means that *A*'s mother *w* was a daughter of *H*, the father of *B*. Thus *B* is *A*'s maternal half-uncle. Similarly *hWh* means that *B* is the child of *A*'s maternal half-brother. If there are several common ancestors we use a bracket. Thus $\left.\begin{matrix} hH \\ wWw \end{matrix}\right\}$ means that *A*'s paternal grandfather *H* had a child *B* by the maternal half sister of *A*'s mother. That is to say *H* married his son's wife's maternal half-sister. Thus both *H* and *W* must have married twice (or had children by two sexual partners). Where a pair of common ancestors are married, we use the symbol *M*. Thus *hM* replaces $\left.\begin{matrix} hW \\ hH \end{matrix}\right\}$ and means that *B* is a whole sib of *A*'s father *h*, that is to say a paternal aunt or uncle of *A*.

We are aware that we have no symbols for the relationships in a direct ancestral line, namely parent and child, and the symbols are ambiguous for grandparent, and grandchild, which come within our scope. But here unambiguous words or phrases are available in most languages. The Indian languages are richer than the European in terms for relationships. Thus in Hindi "brother" is "*bhai*" and "sister" "*bahin*" or "*bahan*". The four types of male first cousin, namely the sons of *A*'s mother's sister, mother's brother, father's sister, and father's brother (in our symbolism *wMw*, *wMh*, *hMw*, and *hMh*) are called *mausera-bhai*, *mamera-bhai*, *phuphera-bhai*, and *chachera-bhai*, respectively, their sisters being called *mauseri-bahan*, etc. These are derived from the names of the four types of aunt and uncle. But although remarriage of widowers is normal in northern India, and polygyny was not rare, there are no special words for *wHw*, etc. There is therefore a need for a symbolism.

In translating our symbolism into that of symbolic logic, the order should be reversed. *w* before the capital becomes $\breve{M}$, after it $M$. *h* before the capital becomes $\breve{P}$, after it $P$. $W$ becomes $\breve{M}|M.\,J$; $H$ becomes $\breve{P}|P.\,J$, and $M$ becomes $(\breve{M}|M.\ \breve{P}|P.\,J)$. Inclusion in a bracket is replaced by the full stop or other sign for a logical product or intersection class.

## The Enumeration

Table 1 gives our enumeration. The third column gives the converse of each relationship. *S* denotes that a relationship is symmetrical, and is therefore its own converse. There are only 17 relationships which do not involve at least one remarriage. They are the six relations between ancestor and descendant which head our list, and the following:—

*M*, *wM*, *Mw*, *hM*, *Mh*, *wMw*, *wMh*, *hMw*, *hMh*, $\left.\begin{matrix} wMw \\ hMh \end{matrix}\right\}$, and $\left.\begin{matrix} wMh \\ hMw \end{matrix}\right\}$. In Indian communities where widows do not remarry, relationships containing the letter *W* do not occur, or are not recognized.

Some of the multiple relationships involving remarriage or polygamy are no doubt bizarre, and several of them can only occur after marriages to agnates (spouse's blood-relatives) which are forbidden in some cultures. However they are often encouraged in others. Thus where a man marries two full sisters, simultaneously or successively, their children will be in the relation $\left.\begin{matrix} H \\ wMw \end{matrix}\right\}$ (paternal half sib and full cousin) to one another. Again in some polygamous cultures, and particularly in ruling families, a man might inherit his father's wives and concubines, and have access to all of them but his own mother. The child of a woman by her first husband and by his son by another wife are in the relationship $\left.\begin{matrix} W \\ Hh \end{matrix}\right\}$, that is to say *B* is *A*'s maternal half-brother and paternal nephew. Mythology furnishes examples of still stranger relationships. Thus in the Mahabharata the children of the brothers Yudhiṣṭhira and Arjuna by their co-wife Draupadi were legally $\left.\begin{matrix} W \\ hMh \end{matrix}\right\}$; but since according to the epic Yudhiṣṭhira

Table 1. *Relationships and coefficients of relationship*

| | Relation | Symbol | Converse | $f$ | $F$ | $\phi_{11}$ | $\phi_{12}$ | $\phi_{21}$ | $\phi_{22}$ | $\Phi$ |
|---|---|---|---|---|---|---|---|---|---|---|
| | Degree 1 | | | | | | | | | |
| 1, 2 | Parent | — | Child | $\frac{1}{4}$ | 0 | 0 | $\frac{1}{2}$ | $\frac{1}{2}$ | $\frac{1}{4}$ | 0 |
| | Degree 2 | | | | | | | | | |
| 1, 2 | Mother's Parent | — | Daughter's child | $\frac{1}{8}$ | 0 | $\frac{1}{2}$ | $\frac{1}{4}$ | $\frac{1}{4}$ | $\frac{1}{8}$ | 0 |
| 3, 4 | Father's Parent | — | Son's child | ,, | ,, | 0 | 0 | 0 | $\frac{1}{4}$ | 0 |
| 5 | Maternal half sib | *W* | *S* | $\frac{1}{8}$ | 0 | $\frac{1}{2}$ | $\frac{1}{4}$ | $\frac{1}{4}$ | $\frac{1}{8}$ | 0 |
| 6 | Paternal half sib | *H* | *S* | ,, | ,, | 0 | 0 | 0 | $\frac{1}{4}$ | 0 |
| 7 | Full sib | *M* | *S* | $\frac{1}{4}$ | $\frac{1}{4}$ | $\frac{1}{2}$ | $\frac{1}{4}$ | $\frac{1}{4}$ | $\frac{3}{8}$ | $\frac{1}{2}$ |
| | Degree 3 | | | | | | | | | |
| 1, 2 | Half aunt or uncle | *wW* | *Ww* | $\frac{1}{16}$ | 0 | $\frac{1}{4}$ | $\frac{1}{8}$ | $\frac{1}{8}$ | $\frac{1}{16}$ | 0 |
| 3, 4 | ,, ,, ,, | *wH* | *Hw* | ,, | ,, | 0 | $\frac{1}{4}$ | 0 | $\frac{1}{8}$ | ,, |
| 5, 6 | ,, ,, ,, | *hW* | *Wh* | ,, | ,, | 0 | 0 | $\frac{1}{4}$ | $\frac{1}{8}$ | ,, |
| 7, 8 | ,, ,, ,, | *hH* | *Hh* | ,, | ,, | 0 | 0 | 0 | 0 | ,, |
| 9, 10 | Aunt or uncle | *wM* | *Mw* | $\frac{1}{8}$ | 0 | $\frac{1}{4}$ | $\frac{3}{8}$ | $\frac{1}{8}$ | $\frac{3}{16}$ | 0 |
| 11, 12 | ,, ,, ,, | *hM* | *Mh* | ,, | ,, | 0 | 0 | $\frac{1}{4}$ | $\frac{1}{8}$ | 0 |
| 13, 14 | Double half aunt or uncle (Fig. 1) | *wW* } *hH* | *Ww* } *Hh* | $\frac{1}{8}$ | $\frac{1}{16}$ | $\frac{1}{4}$ | $\frac{1}{8}$ | $\frac{1}{8}$ | $\frac{1}{16}$ | 0 |
| 15, 16 | ,, ,, ,, | *wH* } *hW* | *Hw* } *Wh* | ,, | ,, | 0 | $\frac{1}{4}$ | $\frac{1}{4}$ | $\frac{1}{4}$ | $\frac{1}{4}$ |
| 17, 18 | Half aunt or uncle and half niece or nephew (Fig. 2) | *wW* } *Hh* | *Ww* } *hH* | $\frac{1}{8}$ | $\frac{1}{16}$ | $\frac{1}{4}$ | $\frac{1}{8}$ | $\frac{1}{8}$ | $\frac{1}{16}$ | 0 |
| 19 | ,, ,, ,, ,, ,, ,, | *wH* } *Hw* | *S* | ,, | ,, | 0 | $\frac{1}{4}$ | $\frac{1}{4}$ | $\frac{1}{4}$ | $\frac{1}{4}$ |
| 20 | ,, ,, ,, ,, ,, ,, | *hW* } *Wh* | *S* | ,, | ,, | 0 | $\frac{1}{4}$ | $\frac{1}{4}$ | $\frac{1}{4}$ | $\frac{1}{4}$ |
| | Degree 4 | | | | | | | | | |
| I | One common grandparent | | | | | | | | | |
| 1 | Half first cousin Fig .3 | *wWw* | *S* | $\frac{1}{32}$ | 0 | $\frac{1}{8}$ | $\frac{1}{16}$ | $\frac{1}{16}$ | $\frac{1}{32}$ | 0 |
| 2 | ,, ,, ,, | *wHw* | *S* | ,, | ,, | $\frac{1}{4}$ | $\frac{1}{8}$ | $\frac{1}{8}$ | $\frac{1}{16}$ | ,, |
| 3, 4 | ,, ,, ,, | *wWh* | *hWw* | ,, | ,, | 0 | $\frac{1}{8}$ | 0 | $\frac{1}{16}$ | ,, |
| 5, 6 | ,, ,, ,, | *wHh* | *hHw* | ,, | ,, | 0 | 0 | 0 | 0 | ,, |
| 7 | ,, ,, ,, | *hWh* | *S* | ,, | ,, | 0 | 0 | 0 | $\frac{1}{8}$ | ,, |
| 8 | ,, ,, ,, | *hHh* | *S* | ,, | ,, | 0 | 0 | 0 | 0 | ,, |
| II | Two common grandparents | | | | | | | | | |
| A | No remarriage | | | | | | | | | |
| 9 | Full first cousin (Fig. 4) | *wMw* | *S* | $\frac{1}{16}$ | 0 | $\frac{3}{8}$ | $\frac{3}{16}$ | $\frac{3}{16}$ | $\frac{3}{32}$ | 0 |
| 10, 11 | ,, ,, ,, | *wMh* | *hMw* | ,, | ,, | 0 | $\frac{1}{8}$ | 0 | $\frac{1}{16}$ | ,, |
| 12 | ,, ,, ,, | *hMh* | *S* | ,, | ,, | 0 | 0 | 0 | $\frac{1}{8}$ | ,, |
| B | Two remarriages | | | | | | | | | |
| 13 | Double half first cousin (Fig. 5) | *wWw* } *hWh* | *S* | $\frac{1}{16}$ | $\frac{1}{64}$ | $\frac{1}{8}$ | $\frac{1}{16}$ | $\frac{1}{16}$ | $\frac{5}{32}$ | $\frac{1}{16}$ |
| 14 | ,, ,, ,, ,, | *wWw* } *hHh* | *S* | ,, | ,, | $\frac{1}{8}$ | $\frac{1}{16}$ | $\frac{1}{16}$ | $\frac{1}{32}$ | 0 |
| 15 | ,, ,, ,, ,, | *wHw* } *hWh* | *S* | ,, | ,, | $\frac{1}{4}$ | $\frac{1}{8}$ | $\frac{1}{8}$ | $\frac{3}{16}$ | $\frac{1}{8}$ |

Table 1. *Relationships and coefficients of relationship*—(Contd.)

| | Relation | Symbol | Converse | $f$ | $F$ | $\phi_{11}$ | $\phi_{12}$ | $\phi_{21}$ | $\phi_{22}$ | $\Phi$ |
|---|---|---|---|---|---|---|---|---|---|---|
| B | Two remarriages | | | | | | | | | |
| 16 | Double half first cousin | *wHw*<br>*hHh* | *S* | $\frac{1}{16}$ | $\frac{1}{64}$ | $\frac{1}{4}$ | $\frac{1}{8}$ | $\frac{1}{8}$ | $\frac{1}{16}$ | 0 |
| 17 | ,, ,, ,, ,, | *wWh*<br>*hWw* | *S* | ,, | ,, | 0 | $\frac{1}{8}$ | $\frac{1}{8}$ | $\frac{1}{8}$ | $\frac{1}{16}$ |
| 18 | ,, ,, ,, ,, | *wHh*<br>*hHw* | *S* | ,, | ,, | 0 | 0 | 0 | 0 | 0 |
| 19, 20 | ,, ,, ,, ,, | *wWh*<br>*hHw* | *wHh*<br>*hWw* | ,, | ,, | 0 | $\frac{1}{8}$ | 0 | $\frac{1}{16}$ | 0 |
| C | Double remarriage | | | | | | | | | |
| 21, 22 | Double half first cousin (Fig. 6) | *wWw*<br>*wHh* | *wWw*<br>*hHw* | $\frac{1}{16}$ | 0 | $\frac{1}{8}$ | $\frac{1}{16}$ | $\frac{1}{16}$ | $\frac{1}{32}$ | 0 |
| 23, 24 | ,, ,, ,, ,, | *wWh*<br>*wHw* | *wHw*<br>*hWw* | ,, | ,, | $\frac{1}{4}$ | $\frac{1}{4}$ | $\frac{1}{8}$ | $\frac{1}{8}$ | 0 |
| 25, 26 | ,, ,, ,, ,, | *wWh*<br>*hHh* | *hWw*<br>*hHh* | ,, | ,, | 0 | $\frac{1}{8}$ | 0 | $\frac{1}{16}$ | 0 |
| 27, 28 | ,, ,, ,, ,, | *wHh*<br>*hWh* | *hHw*<br>*hWh* | ,, | ,, | 0 | 0 | 0 | $\frac{1}{8}$ | 0 |
| III | Three common grand-parents | | | | | | | | | |
| A | One remarriage | | | | | | | | | |
| 29 | Full and half first cousin (Fig. 7) | *wMw*<br>*hWh* | *S* | $\frac{3}{32}$ | $\frac{1}{32}$ | $\frac{3}{8}$ | $\frac{3}{16}$ | $\frac{3}{16}$ | $\frac{7}{32}$ | $\frac{3}{16}$ |
| 30 | ,, ,, ,, ,, | *wMw*<br>*hHh* | *S* | ,, | ,, | $\frac{3}{8}$ | $\frac{3}{16}$ | $\frac{3}{16}$ | $\frac{3}{32}$ | 0 |
| 31, 32 | ,, ,, ,, ,, | *wMh*<br>*hWw* | *hMw*<br>*wWh* | ,, | ,, | 0 | $\frac{1}{8}$ | $\frac{1}{8}$ | $\frac{1}{16}$ | 0 |
| 33, 34 | ,, ,, ,, ,, | *wMh*<br>*hHw* | *hMw*<br>*wHh* | ,, | ,, | 0 | $\frac{1}{8}$ | 0 | $\frac{1}{16}$ | 0 |
| 35 | ,, ,, ,, ,, | *hMh*<br>*wWw* | *S* | ,, | ,, | $\frac{1}{8}$ | $\frac{1}{16}$ | $\frac{1}{16}$ | $\frac{5}{32}$ | $\frac{1}{16}$ |
| 36 | ,, ,, ,, ,, | *hMh*<br>*wHw* | *S* | ,, | ,, | $\frac{1}{4}$ | $\frac{1}{8}$ | $\frac{1}{8}$ | $\frac{3}{16}$ | $\frac{1}{8}$ |
| B | Triple remarriage | | | | | | | | | |
| 37 | Triple half first cousin (Fig. 8) | *wWh*<br>*hWw*<br>*wHw* | *S* | $\frac{3}{32}$ | $\frac{1}{64}$ | $\frac{1}{4}$ | $\frac{1}{4}$ | $\frac{1}{4}$ | $\frac{3}{16}$ | $\frac{1}{16}$ |
| 38, 39 | ,, ,, ,, | *wWw*<br>*wHh*<br>*hWh* | *wWw*<br>*hHw*<br>*hWh* | ,, | ,, | $\frac{1}{8}$ | $\frac{1}{16}$ | $\frac{1}{16}$ | $\frac{5}{32}$ | $\frac{1}{16}$ |
| 40 | ,, ,, ,, | *wWw*<br>*wHh*<br>*hHw* | *S* | ,, | ,, | $\frac{1}{8}$ | $\frac{1}{16}$ | $\frac{1}{16}$ | $\frac{1}{32}$ | 0 |
| 41 | ,, ,, ,, | *wWh*<br>*hHh*<br>*hWw* | *S* | ,, | ,, | 0 | $\frac{1}{8}$ | $\frac{1}{8}$ | $\frac{1}{8}$ | $\frac{1}{16}$ |
| 42, 43 | ,, ,, ,, | *wHw*<br>*wWh*<br>*hHh* | *wHw*<br>*hWw*<br>*hHh* | ,, | ,, | $\frac{1}{4}$ | $\frac{1}{4}$ | $\frac{1}{8}$ | $\frac{1}{8}$ | 0 |
| 44 | ,, ,, ,, | *wHh*<br>*hHw*<br>*hWh* | *S* | ,, | ,, | 0 | 0 | 0 | $\frac{1}{8}$ | 0 |

Table 1. *Relationships and coefficients of relationship*—(Contd.)

| | Relation | Symbol | Converse | $f$ | $F$ | $\phi_{11}$ | $\phi_{12}$ | $\phi_{21}$ | $\phi_{22}$ | $\Phi$ |
|---|---|---|---|---|---|---|---|---|---|---|
| IV | Four common grandparents | | | | | | | | | |
| A | Two marriages | | | | | | | | | |
| 45 | Double first cousin (Fig. 9) | *wMw* } *hMh* | *S* | $\frac{1}{8}$ | $\frac{1}{16}$ | $\frac{3}{8}$ | $\frac{3}{16}$ | $\frac{3}{16}$ | $\frac{7}{32}$ | $\frac{3}{16}$ |
| 46 | ,, ,, ,, | *wMh* } *hMw* | *S* | ,, | ,, | 0 | $\frac{1}{8}$ | $\frac{1}{8}$ | $\frac{1}{8}$ | $\frac{1}{16}$ |
| B | Cyclical remarriage | | | | | | | | | |
| 47 | Quadruple half first cousin (Fig. 10) | *wWw* *wHh* *hWh* *hHw* } | *S* | ,, | $\frac{1}{32}$ | $\frac{1}{8}$ | $\frac{1}{16}$ | $\frac{1}{16}$ | $\frac{5}{32}$ | $\frac{1}{16}$ |
| 48 | ,, ,, ,, | *wWh* *wHw* *hWw* *hHh* } | *S* | ,, | ,, | $\frac{1}{4}$ | $\frac{1}{4}$ | $\frac{1}{4}$ | $\frac{3}{16}$ | $\frac{1}{16}$ |
| | Degrees 2 and 3 | | | | | | | | | |
| 1, 2 | Half sib and half aunt or uncle (Fig. 11) | *W* } *hH* | *W* } *Hh* | $\frac{3}{16}$ | $\frac{1}{8}$ | $\frac{1}{2}$ | $\frac{1}{4}$ | $\frac{1}{4}$ | $\frac{1}{8}$ | 0 |
| 3, 4 | ,, ,, ,, | *H* } *wW* | *H* } *Ww* | ,, | ,, | $\frac{1}{4}$ | $\frac{1}{8}$ | $\frac{1}{8}$ | $\frac{5}{16}$ | $\frac{1}{4}$ |
| | Degrees 2 and 4 | | | | | | | | | |
| I | Two remarriages | | | | | | | | | |
| 1 | Half sib and half first cousin (Fig. 12) | *W* } *hWh* | *S* | $\frac{5}{32}$ | $\frac{1}{16}$ | $\frac{1}{2}$ | $\frac{1}{4}$ | $\frac{1}{4}$ | $\frac{1}{4}$ | $\frac{1}{4}$ |
| 2 | ,, ,, ,, | *W* } *hHh* | *S* | ,, | ,, | $\frac{1}{2}$ | $\frac{1}{4}$ | $\frac{1}{4}$ | $\frac{1}{8}$ | 0 |
| 3 | ,, ,, ,, | *H* } *wWw* | *S* | ,, | ,, | $\frac{1}{8}$ | $\frac{1}{16}$ | $\frac{1}{16}$ | $\frac{9}{32}$ | $\frac{1}{8}$ |
| 4 | ,, ,, ,, | *H* } *wHw* | *S* | ,, | ,, | $\frac{1}{4}$ | $\frac{1}{8}$ | $\frac{1}{8}$ | $\frac{5}{16}$ | $\frac{1}{4}$ |
| II | One remarriage | | | | | | | | | |
| 5 | Half sib and full cousin (Fig. 13) | *W* } *hMh* | *S* | $\frac{3}{16}$ | $\frac{1}{8}$ | $\frac{1}{2}$ | $\frac{1}{4}$ | $\frac{1}{4}$ | $\frac{1}{4}$ | $\frac{1}{4}$ |
| 6 | ,, ,, ,, | *H* } *wMw* | *S* | ,, | ,, | $\frac{3}{8}$ | $\frac{3}{16}$ | $\frac{3}{16}$ | $\frac{11}{32}$ | $\frac{3}{8}$ |
| | Degrees 3 and 4 | | | | | | | | | |
| I A | One remarriage | | | | | | | | | |
| 1, 2 | Half aunt or uncle and half cousin (Fig. 14) | *wW* } *wHh* | *Ww* } *hHw* | $\frac{3}{32}$ | 0 | $\frac{1}{4}$ | $\frac{1}{8}$ | $\frac{1}{8}$ | $\frac{1}{16}$ | 0 |
| 3, 4 | ,, ,, ,, | *wH* } *wWw* | *Hw* } *wWw* | ,, | ,, | $\frac{1}{8}$ | $\frac{5}{16}$ | $\frac{1}{16}$ | $\frac{5}{32}$ | 0 |
| 5, 6 | ,, ,, ,, | *hW* } *hHh* | *Wh* } *hHh* | ,, | ,, | 0 | 0 | $\frac{1}{4}$ | $\frac{1}{8}$ | 0 |
| 7, 8 | ,, ,, ,, | *hH* } *hWw* | *Hh* } *wWh* | ,, | ,, | 0 | 0 | $\frac{1}{8}$ | $\frac{1}{16}$ | 0 |

Table 1. *Relationships and coefficients of relationship*—(Contd.)

| | Relation | Symbol | Converse | $f$ | $F$ | $\phi_{11}$ | $\phi_{12}$ | $\phi_{21}$ | $\phi_{22}$ | $\Phi$ |
|---|---|---|---|---|---|---|---|---|---|---|
| B | Two remarriages | | | | | | | | | |
| 9, 10 | Half aunt or uncle and half cousin (Fig. 15) | *wW*<br>*hWh* | *Ww*<br>*hWh* | $\frac{3}{32}$ | $\frac{1}{32}$ | $\frac{1}{4}$ | $\frac{1}{8}$ | $\frac{1}{8}$ | $\frac{3}{16}$ | $\frac{1}{8}$ |
| 11, 12 | ,, ,, ,, | *wW*<br>*hHh* | *Ww*<br>*hHh* | ,, | ,, | $\frac{1}{4}$ | $\frac{1}{8}$ | $\frac{1}{8}$ | $\frac{1}{16}$ | 0 |
| 13, 14 | ,, ,, ,, | *wH*<br>*hWw* | *Hw*<br>*wWh* | ,, | ,, | 0 | $\frac{1}{4}$ | $\frac{1}{8}$ | $\frac{3}{16}$ | $\frac{1}{8}$ |
| 15, 16 | ,, ,, ,, | *wH*<br>*hWw* | *Hw*<br>*wHh* | ,, | ,, | 0 | $\frac{1}{4}$ | 0 | $\frac{1}{8}$ | 0 |
| 17, 18 | ,, ,, ,, | *hW*<br>*wWh* | *Wh*<br>*hWw* | ,, | ,, | 0 | $\frac{1}{8}$ | $\frac{1}{4}$ | $\frac{3}{16}$ | $\frac{1}{8}$ |
| 19, 20 | ,, ,, ,, | *hW*<br>*wHh* | *Wh*<br>*hHw* | ,, | ,, | 0 | $\frac{1}{8}$ | $\frac{1}{4}$ | $\frac{1}{8}$ | 0 |
| 21, 22 | ,, ,, ,, | *hH*<br>*wWw* | *Hh*<br>*wWw* | ,, | ,, | $\frac{1}{8}$ | $\frac{1}{16}$ | $\frac{1}{16}$ | $\frac{1}{32}$ | 0 |
| 23, 24 | ,, ,, ,, | *hH*<br>*wHw* | *Hh*<br>*wHw* | ,, | ,, | $\frac{1}{4}$ | $\frac{1}{8}$ | $\frac{1}{8}$ | $\frac{1}{16}$ | 0 |
| II A | Double remarriage | | | | | | | | | |
| 25, 26 | Half aunt or uncle and double half cousin (Fig. 16) | *wW*<br>*wHh*<br>*hWh* | *Ww*<br>*hHw*<br>*hWh* | $\frac{1}{8}$ | $\frac{1}{32}$ | $\frac{1}{4}$ | $\frac{1}{8}$ | $\frac{1}{8}$ | $\frac{3}{16}$ | $\frac{1}{8}$ |
| 27, 28 | ,, ,, ,, | *wH*<br>*wWw*<br>*hHw* | *Hw*<br>*wWw*<br>*wHh* | ,, | ,, | $\frac{1}{8}$ | $\frac{5}{16}$ | $\frac{1}{16}$ | $\frac{5}{32}$ | 0 |
| 29, 30 | ,, ,, ,, | *hW*<br>*hHh*<br>*wWh* | *Wh*<br>*hHh*<br>*hWw* | ,, | ,, | 0 | $\frac{1}{4}$ | $\frac{1}{8}$ | $\frac{3}{16}$ | $\frac{1}{8}$ |
| 31, 32 | ,, ,, ,, | *hH*<br>*hWw*<br>*wHw* | *Hh*<br>*wWh*<br>*wHw* | ,, | ,, | $\frac{1}{4}$ | $\frac{1}{8}$ | $\frac{1}{4}$ | $\frac{1}{8}$ | 0 |
| II B | One remarriage | | | | | | | | | |
| 33, 34 | Half aunt or uncle and cousin (Fig. 17) | *wW*<br>*hMh* | *Ww*<br>*hMh* | $\frac{1}{8}$ | $\frac{1}{16}$ | $\frac{1}{4}$ | $\frac{1}{8}$ | $\frac{1}{8}$ | $\frac{3}{16}$ | $\frac{1}{8}$ |
| 35, 36 | ,, ,, ,, | *wH*<br>*hMw* | *Hw*<br>*wMh* | ,, | ,, | 0 | $\frac{1}{4}$ | $\frac{1}{8}$ | $\frac{3}{16}$ | $\frac{1}{8}$ |
| 37, 38 | ,, ,, ,, | *hW*<br>*wMh* | *Wh*<br>*hMw* | ,, | ,, | 0 | $\frac{1}{8}$ | $\frac{1}{4}$ | $\frac{3}{16}$ | $\frac{1}{8}$ |
| 39, 40 | ,, ,, ,, | *hH*<br>*wMw* | *Hh*<br>*wMw* | ,, | ,, | $\frac{3}{8}$ | $\frac{3}{16}$ | $\frac{3}{16}$ | $\frac{3}{82}$ | 0 |
| III | Two remarriages | | | | | | | | | |
| 41 | Half aunt or uncle, nalf nephew or niece, and half cousin (Fig. 18) | *wH*<br>*Hw*<br>*wWw* | *S* | $\frac{5}{32}$ | $\frac{1}{16}$ | $\frac{1}{8}$ | $\frac{5}{16}$ | $\frac{5}{16}$ | $\frac{9}{32}$ | $\frac{1}{4}$ |
| 42 | ,, ,, ,, | *hW*<br>*Wh*<br>*hHh* | *S* | ,, | ,, | 0 | $\frac{1}{4}$ | $\frac{1}{4}$ | $\frac{1}{4}$ | $\frac{1}{4}$ |

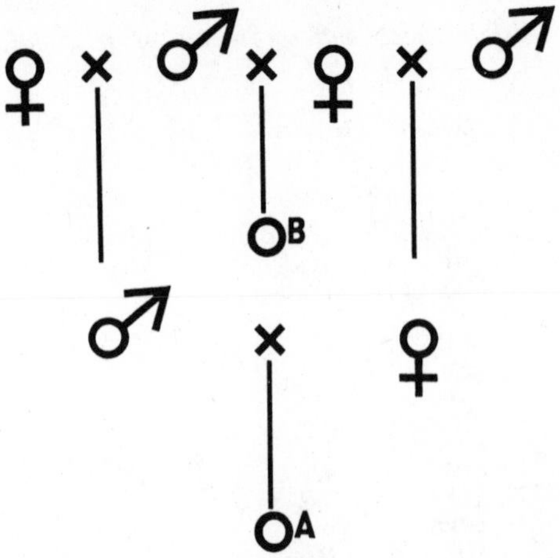


Fig. 1 The relationship symbolised by $\left\{\begin{matrix} wW \\ h\,H \end{matrix}\right.$

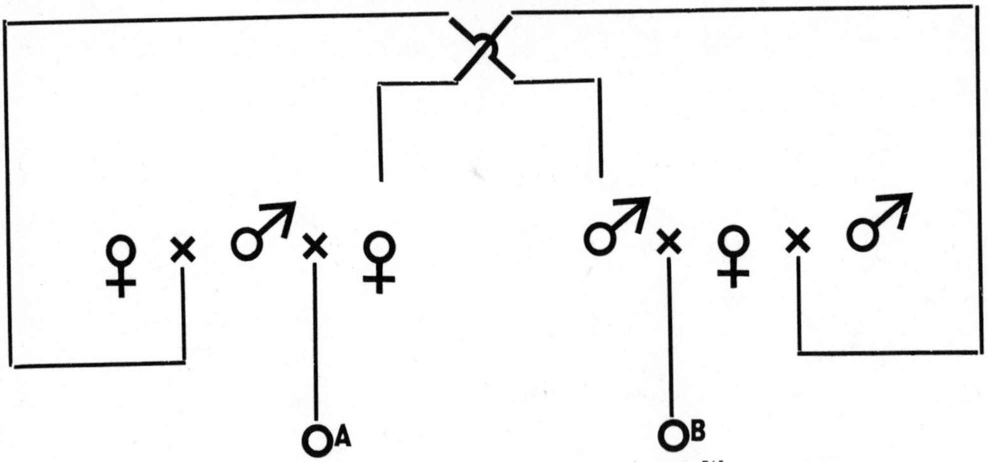


Fig. 2 The relationship symbolised by $\left\{\begin{matrix} wW \\ H\,h \end{matrix}\right.$

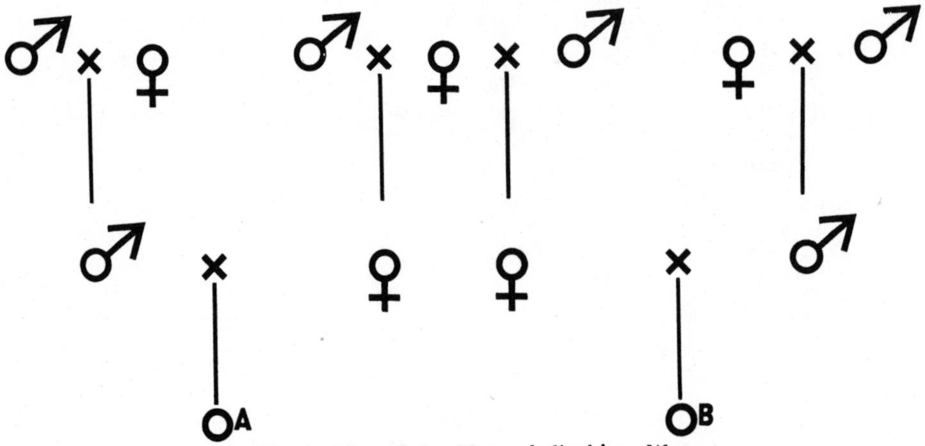


Fig. 3 The relationship symbolised by wWw

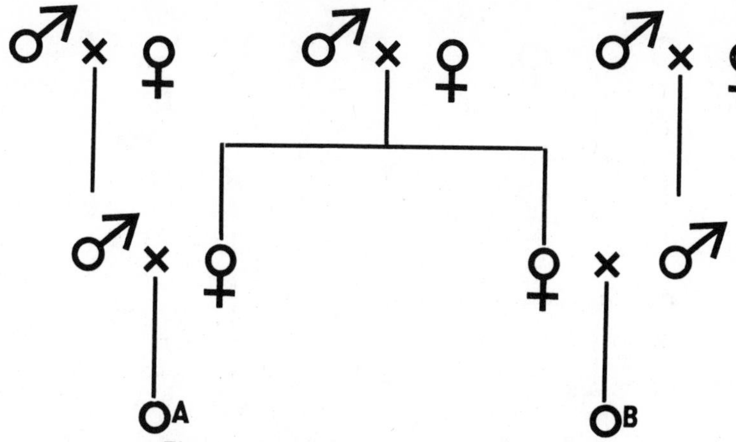


Fig. 4 The relationship symbolised by wMw

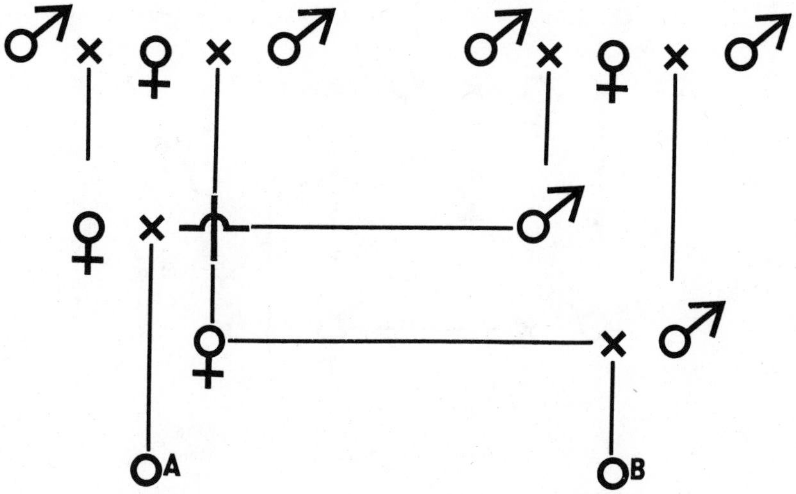


Fig. 5 The relationship symbolised by $\begin{cases}\text{wWw}\\ \text{hWh}\end{cases}$

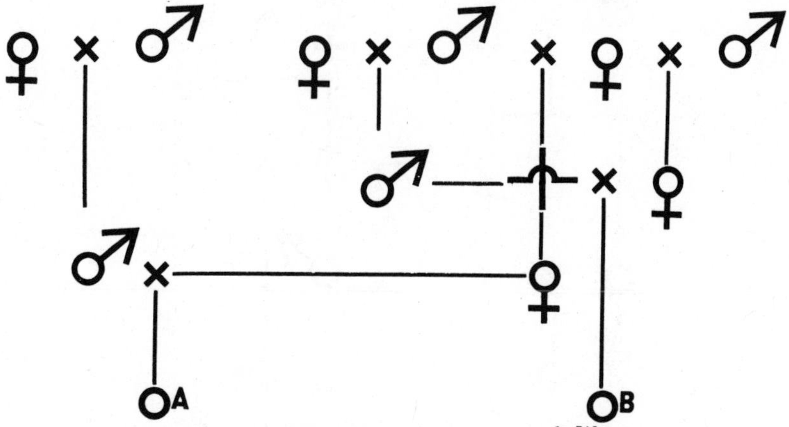


Fig. 6 The relationship symbolised by $\begin{cases}\text{wWw}\\ \text{wHh}\end{cases}$

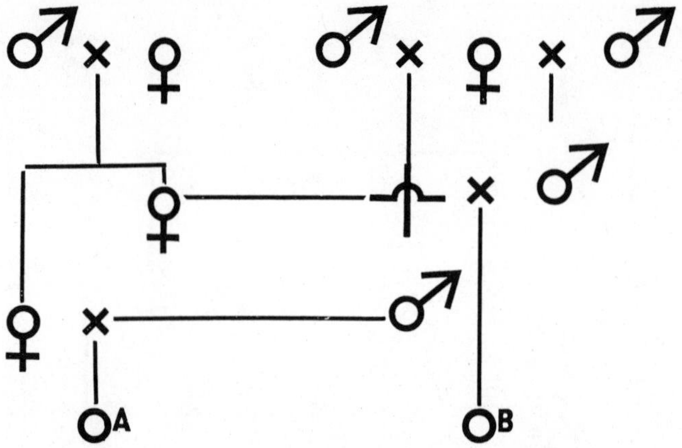


Fig. 7 The relationship symbolised by {wMw / hWh}

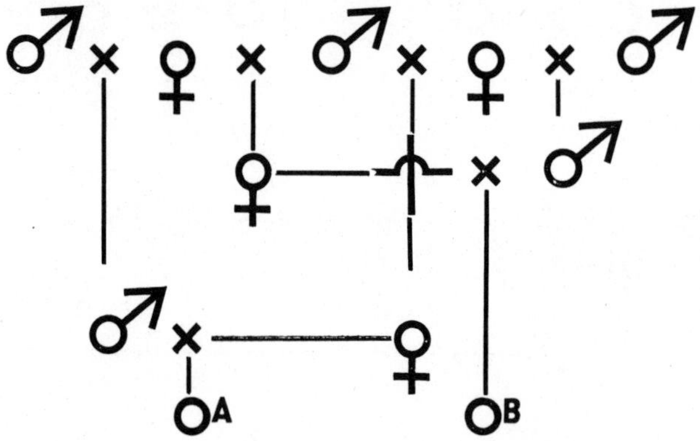


Fig. 8 The relationship symbolised by {wWw / hWh / wHw}

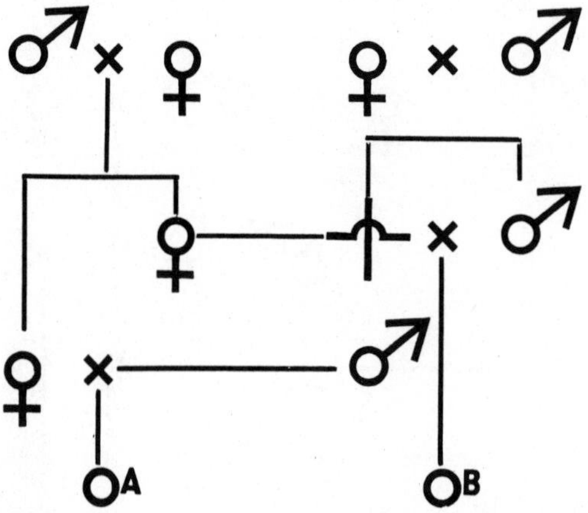


Fig. 9 The relationship symbolised by {wMw / hMh}

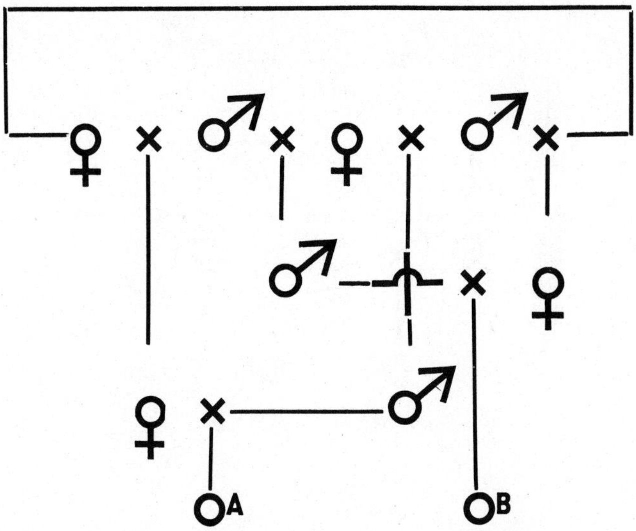


Fig. 10 The relationship symbolised by $\begin{cases} \text{wWw} \\ \text{wHh} \\ \text{hWh} \\ \text{hHw} \end{cases}$

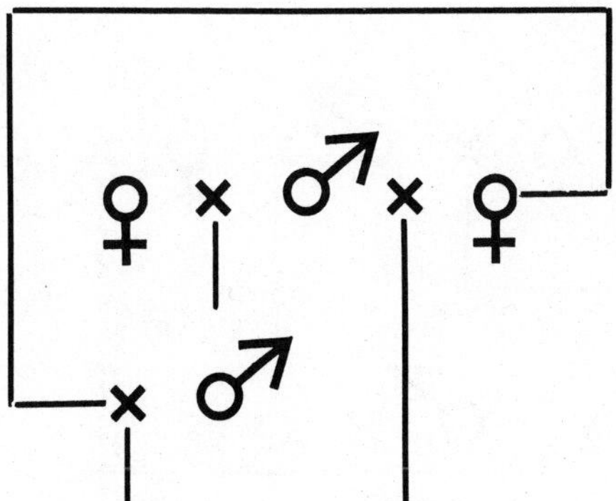


Fig. 11 The relationship symbolised by $\begin{cases} \text{W} \\ \text{hH} \end{cases}$

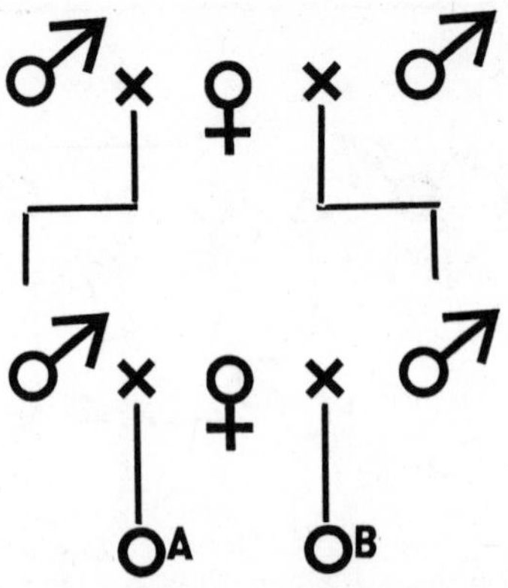


Fig. 12 The relationship symbolised by $\left\{\begin{matrix} W \\ hWh \end{matrix}\right.$

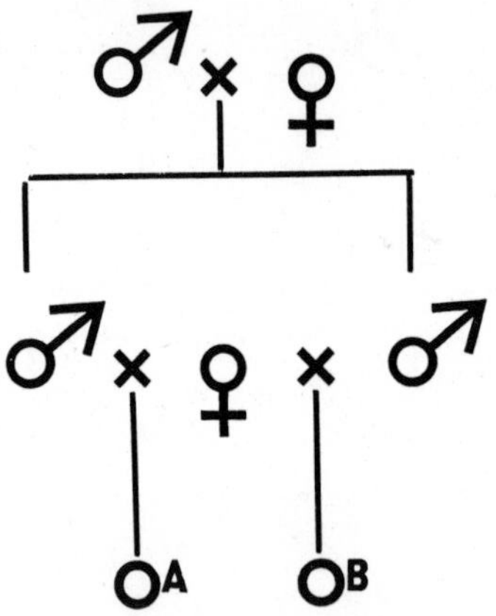


Fig. 13 The relationship symbolised by $\left\{\begin{matrix} W \\ hMh \end{matrix}\right.$

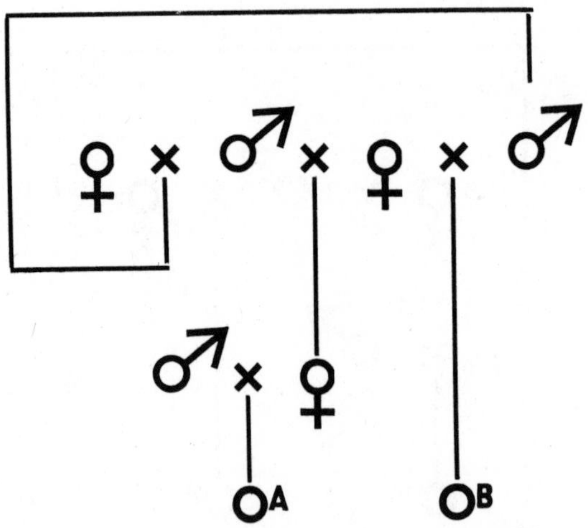


Fig. 14 The relationship symbolised by $\left\{\begin{matrix} wW \\ wHh \end{matrix}\right.$

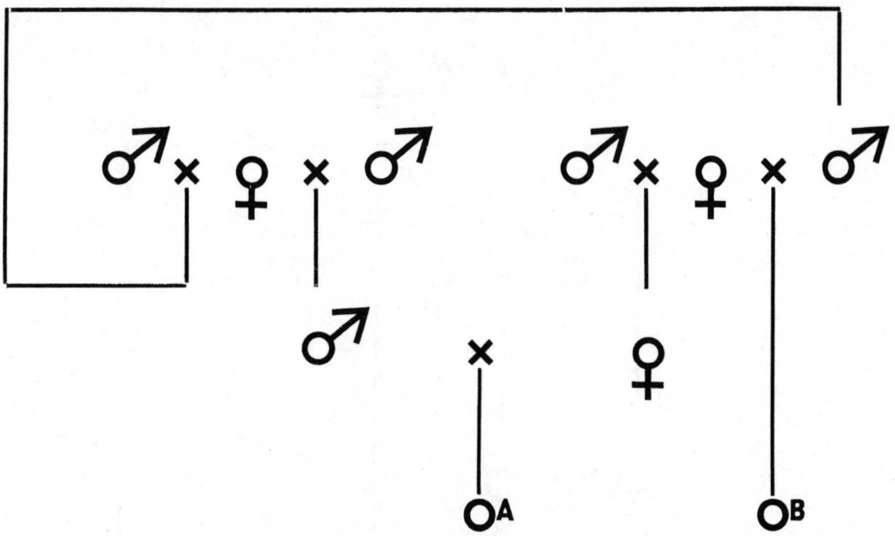


Fig. 15 The relationship symbolised by $\begin{cases} \text{wW} \\ \text{hWh} \end{cases}$

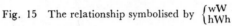

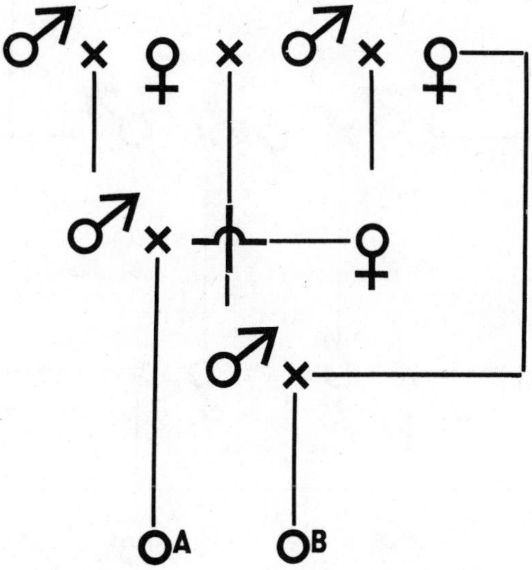


Fig. 16 The relationship symbolised by $\begin{cases} \text{wW} \\ \text{wHh} \\ \text{hWh} \end{cases}$

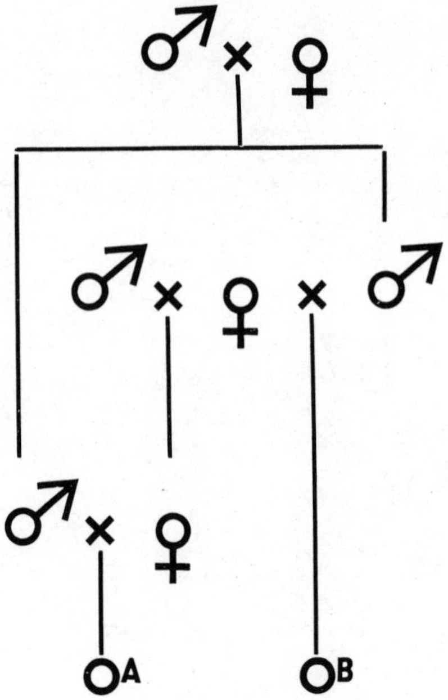


Fig. 17 The relationship symbolised by $\begin{cases} \text{wW} \\ \text{hMh} \end{cases}$

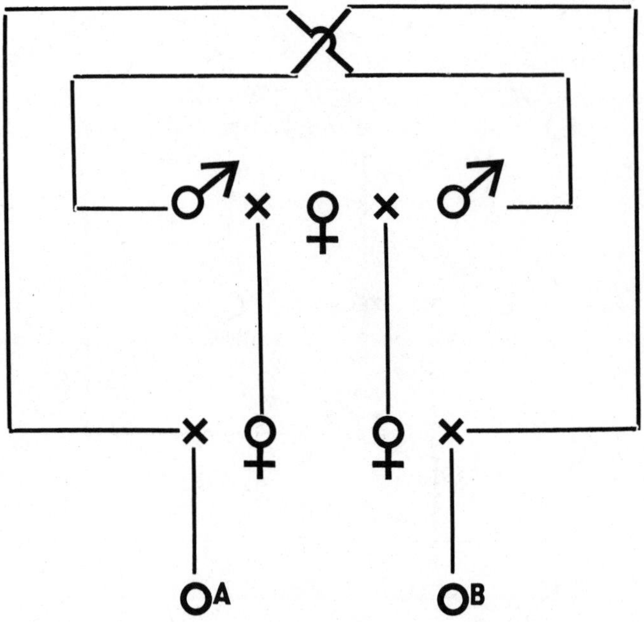


Fig. 18 The relationship symbolised by $\begin{cases} \text{wH} \\ \text{Hw} \\ \text{wWw} \end{cases}$

and Arjuna, though legally sons of Pandu, were in fact children of their mother Kunti by different males, their relationship was $\left.\begin{matrix} W \\ hWh \end{matrix}\right\}$. To take a still more startling relationship, according to some Greek mythologists, after Odysseus' death his son Telemachas by Penelope married Circe, his son Telegonos by Circe married Penelope. The children of these second marriages were in the relation $\left.\begin{matrix} Wh \\ hW \\ hHh \end{matrix}\right\}$. Each, if a male, was half-uncle, half-nephew, and half-first cousin of the other. It is perhaps unlikely that this relationship, which closes our list, has occurred outside mythology.

While relationships whose symbol contains a single $H$ or a single $W$ are not rare, those with several such are much rarer, and those involving a triple remarriage of five persons must be very rare, even though such a chain of marriages could occur without divorce. On the other hand a cyclical remarriage (Fig. 10) must be rare, though it occurs in communities where divorce is easy and spouses exchanged. But the sequel of two marriages between the unrelated progeny of the cyclical remarriage must be very rare indeed, whereas a marriage between two brothers and two sisters, which yields similar coefficients of relationship, is not unusual.

The relationships fall into 24 groups, in each of which the structure, as shown by the pedigree, is the same, apart from differences of sex. Hence in each group the values of $f$ and $F$, which refer to autosomal genes, and are thus not affected by sex, are constant. The groups are separated in Table 1. The values of $f$ and $F$ may of course be the same in two groups of different structure, for example the grandparent and grandchild group and the half sib group. There is no difficulty in being sure that one has enumerated all the possible relationships in such a group. One has merely to permute the sexes of all relevant ancestors of $A$ and $B$ in every way consistent with the condition that the parents of any individual are of different sexes. The number of relationships in a group then ranges from 1 to 16 (subject to multiplication by 4 if the sexes of $A$ and $B$ are taken into account). It is a little harder to lay down rules to ensure that the enumeration of the isomorphic groups is exhaustive.

The following argument shows that there are 48 and only 48 types of first cousin relationship. $A$ has four grandparents, his mother's mother $C_2$, his father's mother $C_1$, his mother's father $\gamma_2$, his father's father $\gamma_1$. Similarly $B$'s four grandparents may be labelled $C_2'$, $C_1'$, $\gamma_2'$, $\gamma_1'$. $A$ and $B$ may have 1, 2, 3, or 4 grandparents in common.

First consider in how many ways they may have one in common. We can choose any of $A$'s four grandparents, say $C_1$. This grandparent must be a female grandparent of $B$, and can be assigned to $B$ in two ways, either as $C_1'$ or $C_2'$. Thus there are 8 different relationships. We can choose two like-sexed grandparents of $A$ in 2 ways, and assign them to $B$ in 2 ways. So they can have two like-sexed grandparents in 4 ways. We can choose two unlike-sexed grandparents of $A$ in 4 ways, and assign them in 4 ways. So $A$ and $B$ can have 2 unlike-sexed grandparents in 16 ways, and 2 grandparents in 20 ways. We can choose 3 grandparents of $A$ in 4 ways, and assign them to $B$ in 4 ways, making 16 possibilities. Finally if they have all 4 grandparents in common the grandmothers may be assigned to $B$ in 2 ways, and the grandfathers in 2 ways, making 4 in all. This agrees with the figures 8, 20, 16, and 4, given in Table 1.

However we actually used a method which can be illustrated from the relations between first cousins. We write down the 8 single path relationships:—

*wWw*, *wWh*, *hWw*, *hWh*, *wHw*, *wHh*, *hHw*, and *hHh*.

Any one of these is compatible with any other differing from it by two or more steps. A capital letter represents two steps. Thus *wWw* is compatible with all the others except *wWh* and *hWw*. Thus 8 of the 28 possible pairs of paths are excluded because they involve incest, e.g. $\left.\begin{matrix} wWw \\ wWh \end{matrix}\right\}$. This leaves 20 admissible relationships with 2 grandparents in common. A third path may be added to each pair provided it is not incompatible with either of the first two. The triad of paths must include two with the same common capital letter. There are only 4 such pairs, and each such pair can be associated with any path with the other capital letter. For example $\left.\begin{matrix} wWh \\ hWw \end{matrix}\right\}$ can be associated with any path including $H$. Thus there are 16 relationships involving 3 grandparents. Similarly each of the two compatible pairs including $W$ can be associated with either of the two including $H$, giving 4 quadruple relationships. In fact we represented each single path by a point, joined the compatible points by 20 lines, and then picked out triangles and quadrilaterals with both diagonals. Two paths differing only in their capital letters are condensed, e.g. $\left.\begin{matrix} wWw \\ wHw \end{matrix}\right\}$ to *wMw*. The same principles can be applied to the relationships of mixed degree, though here some multiple relationships are excluded because they involve more than 3 generations.

The calculation of coefficients of relationship is not difficult. $B$ may be related to $A$ by at most 4 single paths; and since a gene found in $A$ and $B$ can only be inherited from one of their common ancestors, the contributions of these paths to $f$ and $\phi$ are additive. The single paths are symbolized by $W$, $H$, numbers 1 to 8 of degree 3, and numbers 1 to 8 of degree 4. Thus having calculated the values of $f$ and $\phi$ for these paths we can obtain them for all the rest by addition. Thus the values for $\left.\begin{matrix} wMw \\ hWh \end{matrix}\right\}$ are obtained by adding those for *wWw*, *wHw*, and *hWh*. The values of $F$ and $\Phi$ are sums of two products of $f$ and $\phi$ values which may be looked up in the earlier parts of Table 1, and of the values when $A \equiv B$, which are given in the first line of Table 2.

The most important of the coefficients is Wright's $f$, which among other things, determines the biological dangers of inbreeding, and its biological advantage in reducing the risk of maternal-foetal incompatibility. As the latter, though it has probably been detected by Goldschmidt (1961), has not been calculated, we discuss it in the next section. The prohibitions against marriage with cognate relatives, if, as has been suggested, they are based on the observation that inbreeding produces more defective children than outbreeding, are not very logical, as has often been pointed out. Marriage between half brother and half sister is generally prohibited, and a union between them is criminal in many legal codes. However, by the standard of $f$, they are no more closely related than niece and uncle or aunt and nephew, whose union is legal in some codes, and criminal in few or none, double first cousins (who may marry where uncle and niece may not, as in Britain) and several complex relations. In British law a man may not marry his half-niece, who is less closely related ($f=\frac{1}{16}$) than his double first cousin.

## THE EFFECT OF INBREEDING ON DISEASE AND DEATH DUE TO MATERNAL-FOETAL INCOMPATIBILITY

Consider an autosomal locus where a gene **A** determines the production of an antigen such that an **Aa** child may immunize an **aa** mother, and cause its own death, or more usually that of a subsequent **Aa** child. The symbol **A** is used both for genes such a $\mathbf{A}_1$ which appear often to immunize a mother to her first **Aa** child, and those such as **D** which rarely immunize the mother during the first pregnancy.

If $p$ is the frequency of **A**, $q$ that of **a**, i.e. of all alleles not producing the antigen in question, the frequency of mothers lacking the antigen is $q^2$. In a random mating population a fraction $p$ of their children, or $pq^2$ of all children, is liable to be immunized. Of these $p^2q^2$ are in families consisting of **Aa** children only, $pq^3$ in families consisting of **Aa** and **aa.** In the case of genes such as **D** the former are at a considerably greater risk. If the father is a relative of the mother with given values of $f$ and $F$, the frequency of **Aa** among the children of **aa** mothers is reduced from $p$ to $(1-2f)p$. The fraction belonging to all-**Aa** families is reduced from $p^2$ to $(1-4f+F)p^2$, while that belonging to mixed families is changed from $pq$ to $p[q+2f(p-q)-Fp]$, which may be an increase or decrease according to the value of $p$. If $K$ of the **Aa** children of homozygous, and $k$ of the children of heterozygous fathers are killed by incompatibility, the death-rate from this cause is $pq^2\,(Kp+kq)$ when parents are unrelated, and

$$pq^2[Kp+kq-f\,(2Kp-kp+kq)+F(K-k)p]$$

when they are related. Thus inbreeding saves the lives of a fraction

$$f+\frac{(f-F)\,(K-k)\,p}{Kp+kq}$$

of the babies which would otherwise die from the effect of incompatibility. $f>F$ and $K>k$, so the fraction exceeds $f$, but cannot exceed $2f-F$. This however is the fraction saved at any one locus. If the number of independent loci concerned is large enough, it is conceivable that the death-rate may be reduced quite considerably by cousin marriage.

Sex-linked genes responsible for human antigens are not yet known.* Still less is there any evidence that such genes are ever responsible for incompatibility between mother and foetus. However it is interesting to calculate what effect they have if they exist. Let **S** and **s** be a pair of sex-linked allelomorphs, or groups of allelomorphs, such that **S** determines the production of an antigen capable of immunizing **ss** mothers, while **s** does not. A **ss** mother cannot bear **S** sons, but may bear **Ss** daughters, who may be eliminated. Let $p$ and $q$ be the frequencies of **S** and **s,** and $k$ the fraction of **Ss** daughters of **ss** mothers eliminated. The fraction of daughters eliminated if parents

* We learn that one has at last been discovered.

are unrelated is clearly $kpq^2$, and it can be seen that there is unstable equilibrium when $p=q=\frac{1}{2}$, as in the autosomal case. The frequency of males among the surviving offspring would thus be $\frac{1}{2-kpq^2}$ if no other causes were disturbing the sex ratio.

If the parents are related, the frequency of **Ss** among the daughters of **ss** mothers is reduced from $p$ to $(1-\phi)p$ where $\phi$ is the coefficient of relationship of the parents. Thus the frequency of males would be reduced from $(2-kpq^2)^{-1}$ to $[2-k(1-\phi)pq^2]^{-1}$. Goldschmidt however found more males when the parents were related. So incompatibility, if it accounts for a lower abortion rate, cannot account for a higher frequency of males when parents are related. If $p=\frac{1}{3}$, which gives the maximal elimination, and $k=0{\cdot}1$, which is very high, the frequency of males would only be 50·373%, and if the coefficient of sex-linked relationship of parents were 1/8, it would only be reduced to 50·326%, which could hardly be verified on a sample less than a million. There might be several such loci on the $X$ chromosome, but a high mortality rate and a high frequency of the rarer allele could seldom coexist. Even if, between them, they accounted for the whole excess of males observed, inbreeding would rarely reduce the ratio by even one eighth of the way towards equality. We conclude that such effects, if they occur, are unimportant.

The effect of inbreeding in raising the male frequency, if it is confirmed, is perhaps more likely to be due to the same group of causes which eliminate male mammals, and in general members of the heterogametic sex, in interspecific hybrids (Haldane, 1922).

### The Effect of Monozygotic Twinning

If $B$ is $A$'s monozygotic twin, the effect on the coefficients of relationship is as shown in the first line of Table 2. If a pair of like-sexed sibs are monozygotic twins they are believed to be genetically identical. If any pair of "paths" connecting $B$ to $A$ passes through a pair of whole brothers or sisters, these may be monozygotic twins. In double first cousinship of the first type, either the sisters, the brothers, or both, may be monozygotic twins. The relationships which can be transformed by monozygotic twinning are listed in Table 2. The first one can of course only be transformed if $A$ and $B$ are of like sex, the second if $B$ is a woman, the third if $A$ is a woman, the fourth if $B$ is a man and the fifth if $A$ is a man. Altogether 18 relationships can be so transformed. The second column gives the original value of $f$, the fourth and later those of the transformed relationship. The value of $f$ is always increased, but never more than doubled. That of $F$ can be increased up to four times. The marriage of two pairs of monozygotic twins may not be as rare as might be thought. A biometric study of the resemblance between the children of monozygotic twins would be well worth carrying out.

### Relationships when Relatives are Inbred

We shall not consider these in any detail, but it may be instructive to work out the number of first cousinships possible if $A$, $B$, or both are inbred. $A$ may have only 3

Table 2. *Enhancement of relationship by monozygotic twinning*

| Original relationship | $f$ | Enhanced relationship | $f$ | $F$ | $\phi_{11}$ | $\phi_{12}$ | $\phi_{21}$ | $\phi_{22}$ | $\Phi$ |
|---|---|---|---|---|---|---|---|---|---|
| *M* | $\frac{1}{4}$ | Identity | $\frac{1}{2}$ | 1 | 1 | .. | .. | $\frac{1}{2}$ | 1 |
| *wM* | $\frac{1}{8}$ | Mother | $\frac{1}{4}$ | 0 | .. | $\frac{1}{2}$ | .. | $\frac{1}{4}$ | 0 |
| *Mw* | ,, | Woman's child | ,, | ,, | .. | .. | $\frac{1}{2}$ | $\frac{1}{4}$ | 0 |
| *hM* | ,, | Father | ,, | ,, | 0 | .. | $\frac{1}{2}$ | .. | .. |
| *Mh* | ,, | Man's child | ,, | ,, | 0 | $\frac{1}{2}$ | .. | .. | .. |
| *wMw* | $\frac{1}{16}$ | *W* | $\frac{1}{8}$ | 0 | $\frac{1}{2}$ | $\frac{1}{4}$ | $\frac{1}{4}$ | $\frac{1}{8}$ | 0 |
| *hMh* | ,, | *H* | ,, | ,, | 0 | 0 | 0 | $\frac{1}{4}$ | 0 |
| *wMw*<br>*hWh* | $\frac{3}{32}$ | *W*<br>*hWh* | $\frac{5}{32}$ | $\frac{1}{16}$ | $\frac{1}{2}$ | $\frac{1}{4}$ | $\frac{1}{4}$ | $\frac{1}{4}$ | $\frac{1}{4}$ |
| *wMw*<br>*hHh* | ,, | *W*<br>*hHh* | ,, | ,, | $\frac{1}{2}$ | $\frac{1}{4}$ | $\frac{1}{4}$ | $\frac{1}{8}$ | 0 |
| *hMh*<br>*wWw* | ,, | *H*<br>*wWw* | ,, | ,, | $\frac{1}{8}$ | $\frac{1}{16}$ | $\frac{1}{16}$ | $\frac{9}{32}$ | $\frac{1}{8}$ |
| *hMh*<br>*wHw* | ,, | *H*<br>*wHw* | ,, | ,, | $\frac{1}{4}$ | $\frac{1}{8}$ | $\frac{1}{8}$ | $\frac{5}{16}$ | $\frac{1}{4}$ |
| *wMw*<br>*hMh* | $\frac{1}{8}$ | *W*<br>*hMh* | $\frac{3}{16}$ | $\frac{1}{8}$ | $\frac{1}{2}$ | $\frac{1}{4}$ | $\frac{1}{4}$ | $\frac{1}{4}$ | $\frac{1}{4}$ |
| ,, | ,, | *H*<br>*wMw* | ,, | ,, | $\frac{3}{8}$ | $\frac{3}{16}$ | $\frac{3}{16}$ | $\frac{11}{32}$ | $\frac{3}{8}$ |
| ,, | ,, | *M* | $\frac{1}{4}$ | $\frac{1}{4}$ | $\frac{1}{2}$ | $\frac{1}{4}$ | $\frac{1}{4}$ | $\frac{3}{8}$ | $\frac{1}{2}$ |
| *W*<br>*hMh* | $\frac{3}{16}$ | *M* | ,, | ,, | ,, | ,, | ,, | ,, | ,, |
| *H*<br>*wMw* | ,, | *M* | ,, | ,, | ,, | ,, | ,, | ,, | ,, |
| *wW*<br>*hMh* | $\frac{1}{8}$ | *H*<br>*wW* | $\frac{3}{16}$ | $\frac{1}{8}$ | $\frac{1}{4}$ | $\frac{1}{8}$ | $\frac{1}{8}$ | $\frac{5}{16}$ | $\frac{1}{4}$ |
| *Ww*<br>*hMh* | ,, | *H*<br>*Ww* | ,, | ,, | $\frac{1}{4}$ | $\frac{1}{8}$ | $\frac{1}{8}$ | $\frac{5}{16}$ | $\frac{1}{4}$ |
| *hH*<br>*wMw* | ,, | *W*<br>*hH* | ,, | ,, | $\frac{1}{2}$ | $\frac{1}{4}$ | $\frac{1}{4}$ | $\frac{1}{8}$ | 0 |
| *Hh*<br>*wMw* | ,, | *W*<br>*Hh* | ,, | ,, | $\frac{1}{2}$ | $\frac{1}{4}$ | $\frac{1}{4}$ | $\frac{1}{8}$ | 0 |

grandparents in two ways. His double grandfather had a son and daughter by different women, and $A$ is the offspring of their union. Or he may have only one grandmother as a result of a union between her offspring by two men. In either case the three grandparents can be specified uniquely. $A$ can have only two grandparents in three ways. His grandfather may have united with a daughter, and is therefore $A$'s father and grandfather. Or his grandmother may have united with a son. We do not consider these possibilities further, as they give rise to mixed relationships of degree 3 and 4. Finally $A$ may be the offspring of the union of a whole brother and sister.

7

This possibility will be considered. All these possibilities are of course realised in animal breeding.

Table 3. *Summary of Tables* 1 *and* 2

| Degree of relationship | Symmetrical relationships | Pairs of converse relationships | Number enhanced by monozygosity |
|---|---|---|---|
| 1 | 0 | 1 | 0 |
| 2 | 3 | 2 | 1 |
| 3 | 2 | 9 | 4 |
| 4 | 24 | 12 | 7 |
| 2+3 | 0 | 2 | 0 |
| 2+4 | 6 | 0 | 2 |
| 3+4 | 2 | 20 | 4 |
| | 37 | 46 | 18 |

It is convenient to denote the relationships where $A$ has $m$ grandparents and $B$ has $n$ by $(m, n)$. We have shown that there are 48 relationships of class (4, 4). Let us now find the number in class (3, 4). First consider the relationships with 1 grandparent in common. If $A$ has only one grandfather he can be assigned to $B$ in 2 ways. But $A$'s grandmother can be chosen in 2 ways and assigned in 2. So they may have one grandparent in common in 6 ways, with 6 more when $A$ has only one grandmother, making 12. Again if $A$ has only one grandfather we can choose two grandmothers in one way and assign them to $B$ in 2 ways; we can choose a grandfather and grandmother in 2 ways and assign them in 4. Similarly if $A$ has only one grandmother. Thus $A$ and $B$ can have 2 common grandparents in 20 ways. Similarly $A$'s 3 grandparents may be assigned in 4 ways if he has one grandmother, or 8 in all. The total number of cousinships of class (3, 4) is thus $12+20+8$, or 40.

Similar calculations may be made for other classes, and we find the following numbers

| | |
|---|---|
| (4, 4) | 48 |
| (3, 4) and (4, 3) | 80 |
| (2, 4) and (4, 2) | 16 |
| (3, 3) | 40 |
| (2, 3) and (3, 2) | 20 |
| (2, 2) | 3 |
| | 207 |

Thus the number of possible first cousin-ships has been increased from 48 to 207, or 828 if the sexes of $A$ and $B$ are considered. The grand total of all relationships would be considerably increased, since new relationships of mixed degree are possible, for example $A$ may be $B$'s parent and grandparent, parent and half-sib, and so on. Our $h$, $w$ symbolism is not adapted to describe relationships involving inbreeding of either relative, that is to say relation through a line of intermediaries which forks.

## Relationships of Higher Degree

It is quite possible to enumerate relationships of higher degree, and to calculate their coefficients. We have not attempted this task for two reasons. The first is its gigantic character. As will be seen, a list of relationships between second cousins on the lines of Table 1 would occupy several volumes. The second is that as soon as we reach asymmetrical relations of degree 4 such as that of great-uncle, or relations of higher degree such as first cousin once removed, it becomes possible for $A$ or $B$ to be inbred, and related through chains of relatives which split. We shall merely calculate the number of distinct relationships between second cousins which do not involve inbreeding, so that both $A$ and $B$ have 8 distinct grandparents. Clearly they are all built up of 32 constituents such as $whWhh$, though at most 8 of these can be combined, since there are at most 8 common grandparents. Let the great-grandmothers of $A$ be $C_1$, $C_2$, $C_3$, $C_4$, the great-grandfathers $\gamma_1$, $\gamma_2$, $\gamma_3$, $\gamma_4$, each defined by her or his relation to $A$. Thus $C_2$ might be $A$'s mother's father's mother. Let $C'_1$, $C'_2$, $C'_3$, $C'_4$, $\gamma'_1$, $\gamma'_2$, $\gamma'_3$, $\gamma'_4$ be the great grandparents of $B$, similarly defined. From 1 to 8 of these may be common great grandparents of both. Consider the grandmothers. If a single one is common to $A$ and $B$ this can occur in $4^2$ or 16 ways, for that of $A$ can be any of $C_1$, $C_2$, $C_3$, and $C_4$, and that of $B$ can be any one of $C'_1$, $C'_2$, $C'_3$ and $C'_4$. Thus the same woman could be $A$'s mother's father's mother and $B$'s father's father's mother. Similarly a second identical pair can be chosen in $3^2=9$ ways, but since the order of the pairs is irrelevant we have $16\times 9\div 2$ or 72 possibilities. Similarly 3 identities can be chosen in $4^2.3^2.2^2\div 3!$, or 96 ways, and four identities in $4^2.3^2.2^2.1^2\div 4!$, or 24 ways. In fact the identity of $i$ great grandmothers allows of $\dfrac{(4!)^2}{[(4-i)!]^2 i!}$ possibilities. The same is clearly true for the great grandfathers. Now suppose $A$ and $B$ have 5 great-grandparents in common, 4 may be great grandmothers and 1 great-grandfather, or 1 great-grandmother and 4 great-grandfathers, each giving $24\times 16$, or 384 possibilities. Or they may have 3 great-grandmothers and 2 great-grandfathers, or 2 and 3, in common, each giving $72\times 96$, or 6912 possibilities. We thus have the possibilities given in Table 4.

There are thus 43,680 or $2^5.3.5.7.13$ distinct relationships when neither $A$ nor $B$ is inbred. But $A$'s or $B$'s parents may, without incest, be any one of the 48 types of first cousin, descended from only 7 to 4 great-grandparents. When allowance is made for this there may be about 9 million types of second cousin.

## Discussion

Our main results are summarized in Table 3. Perhaps the most unexpected is the large number of possible relationships of mixed degrees 3 and 4. Some of these may

Table 4. *Possible relations of outbred second cousins*

| Number of common greatgrandparents | Partitions | Possible relationships | Total |
|---|---|---|---|
| 1 | 0+1 | $2\times16$ | 32 |
| 2 | 0+2, 1+1 | $2\times72+16^2$ | 400 |
| 3 | 0+3, 1+2 | $2\times96+2\times16\times72$ | 2,496 |
| 4 | 0+4, 1+3, 2+2 | $2\times24+2\times16\times96+72^2$ | 8,304 |
| 5 | 1+4, 2+3 | $2\times16\times24+2\times72\times96$ | 14,592 |
| 6 | 2+4, 3+3 | $2\times72\times24+96^2$ | 12,672 |
| 7 | 3+4 | $2\times96\times24$ | 4,608 |
| 8 | 4+4 | $24^2$ | 576 |
| | | | 43,680 |

not be very rare. Where it is permitted, there is nothing surprising in a widower marrying a woman, and his son by another marriage marrying her younger sister. The children of these marriages are in the relationship symbolized by $\left.\begin{matrix} Hh \\ wMw \end{matrix}\right\}$.

We are aware that we may have made errors both by omitting relationships permitted by our hypothesis, by miscalculation, or by the overlooking of misprints. We shall be thankful to receive corrections, and to publish them if they are accepted. Our symbolism may be found unacceptable. We have given reasons for preferring it to a bulkier if more logical symbolism, and for using *w* and *h* rather than *f* and *m*, *m* and *f*, or *m* and *p*. It has the grave demerits of not covering ancestral relationships or their converses, relations beyond the fourth degree, or relationships involving incest. It has however the merit that it can readily be extended to cover the distinction between older and younger brothers or sisters where this is sociologically important. If by putting a sign above or below that of the common ancestor or ancestors we denote younger and elder sibs, while an asterisk denotes twins, this need can be met. Thus $\overline{W}$ would denote a younger maternal half sib, $\underline{Mw}$ the child of an elder full sister, $wM^*$ a twin of *A*'s mother, and so on.

There is a very serious gap between genetics and anthropology, and anything which can help to bridge that gap is worth attempting.

## Summary

The logical structure of human relationships is discussed. When the relatives are not inbred their genetical implications can be summarized by four coefficients of relationship, namely Wright's coefficient of single autosomal relationship, a coefficient of double autosomal relationship, a coefficient of single sex-linked relationship, and one of double sex-linked relationship between women only. A simple symbolism for human relationships is developed. There are 516 possible relationships between $A$ and $B$, not involving an ancestor of either more remote than a grandparent, and not involving incest. Table 1 gives the coefficients of relationship for all of them. 192 are relationships between first cousins. In 66 cases the coefficients may be raised by monozygotic twinning. The effect of relationship on incompatibility between mother and foetus is described. There is a brief discussion of relationships involving incest, and those between second cousins.

## REFERENCES


CARNAP, R. (1958). *Introduction to symbolic logic and its applications.* Dover, New York.

FISHER, R. A. (1918). The correlation between relatives on the supposition of Mendelian inheritance. *Trans. Roy. Soc. Ed.*, **52,** II, no. 15.

GOLDSCHMIDT, E. (1961). Viability studies in Jews from Kurdistan. Conference on Human Population Genetics in Israel, Abstracts of Communications.

HALDANE, J. B. S. (1922). Sex ratio and unisexual sterility in hybrid animals. *J. Genet.*, **12,** 101-109.

HALDANE, J. B. S. AND MOSHINSKY, P. (1939). Inbreeding in Mendelian populations with special reference to human cousin marriage. *Ann. Eug.*, **9,** 321-340.

KARVE, I. (1953). Kinship organisation in India. *Deccan College Monograph Series*, 11.

KEMPTHORNE, O. (1957). *Introduction to Genetic Statistics.* John Wiley & Sons, Inc., New York.

LERNER, I. M. (1950). *Population Genetics and Animal Improvement.* Cambridge University Press.

MALÉCOT, G. (1948). *Les mathématiques de l'Hérédité.* Masson, Paris.

MORTON, N. E. (1961). Morbidity of children from consanguineous marriages. *Progress in medical genetics.* Vol. **1,** 261-291. Grune and Stratton, New York.

RUSSELL, B. (1903). *The principles of mathematics.* Cambridge.

WRIGHT, S. (1922). Coefficients of inbreeding and relationship. *Amer. Nat.*, **56,** 330-339.


## ERRATUM

An enumeration of some human relationships by J. B. S. Haldane and S. D. Jayakar, P. 94, Fig. 8

for $\begin{cases} \text{wWw} \\ \text{hWh} \\ \text{wHw} \end{cases}$ please read $\begin{cases} \text{wWh} \\ \text{hWw} \\ \text{wHw} \end{cases}$

# Editor's Comments on Paper 18

18 **Wright:** The Interpretation of Population Structure by F-Statistics with Special Regard to Systems of Mating
*Evolution*, **19**, 395–420 (Sept. 1965)

It is difficult both in reality and in theory to clearly isolate the effects of finite population size, breeding system of a local group, and migration between local groups seen as having independent breeding systems. This difficulty is reflected in the somewhat artificial separation of the referenced bibliography in this introduction and in the introductions to papers by Cavalli-Sforza et al. and by T. Maruyama. Here I shall mention certain papers on finite size or on comparison of theoretical structures, since those are the principal importances of the present paper.

A critical and still useful paper on population size is that by Crow (1954) which summarizes formulas on effective breeding size of a population and its relation to the mean and variance of family size. But in contrast with Wright's paper, this reference does not emphasize the actual breeding structure or mating system of a population. In connection with the articles in Wright's bibliography, one should also see the paper of Alan Robertson (1964), who concludes in part that "circular mating systems are thus a special type of sublining in which the mating system follows the same plan in each generation" (1964, p. 167). This argues in effect that these particular systems are examples of marriage systems in the sense of Part V, and therefore should be eventually treatable by those techniques. [Note also that all such systems are "1-stable minimal sequences" in the sense of Ballonoff (1974b) and that techniques for generalizing such structurally dependent techniques to systems that do not identically repeat themselves may be possible through graph theory methods.]

Thus, in addition to comparing the differences between probability and correlational interpretations of inbreeding, Wright is comparing the effects of two different regular systems of mating, in finite populations. The alternative theoretical technique appears to be to ignore or minimize concern with the particular regular system of mating, or existing mix of systems of mating, with the at least nominal purpose of studying broader evolutionary problems. This emphasis allows one to assume either large populations or large numbers of smaller populations, so that the mathematics of continuous processes becomes applicable. An early attempt on this problem related to methods used in Crow's paper mentioned above was by Crow and Morton (1955), but, for example, Moran (1958) was already more involved with the solution to these continuous variable models [compare to Watterson (1959), who used markovian, hence discrete, techniques in another comparative paper]. However, the most complete treatment was carried out by Kimura (1964) in a small monograph.

Perhaps the clearest contrast between the two major viewpoints that must be separated in population genetics can be seen by reading Kimura and Ohta (1971). This work is essentially a summary of the continuous process approach to population genetics, touching only in a few places on the problems raised here, and even then (principally in Chapter 8) emphasizing continuous process models where possible.

This is not wrong in itself, but is somewhat like trying to treat physics as statistical mechanics, with only passing reference to quantum models. It works well but has its price; the very results which motivate that chapter (and Wright's paper here) were obtained by direct concern with finitely structured breeding systems.

# 18

Reprinted from *Evolution*, **19**, 395–420 (Sept. 1965)

# THE INTERPRETATION OF POPULATION STRUCTURE BY F-STATISTICS WITH SPECIAL REGARD TO SYSTEMS OF MATING

Sewall Wright

*Department of Genetics, University of Wisconsin, Madison, Wisconsin*[1, 2]

Accepted May 15, 1965

Kimura and Crow (1963b) have recently made an interesting comparison between two classes of systems of mating within populations of constant size: ones in which there is maximum avoidance of consanguine mating and ones in which all matings are between close relatives around an unbroken circle. These are illustrated in Figs. 1 and 2 in populations of eight. The rate of decrease of heterozygosis in the former class had, as they note, been found long before to approach $1/(4N)$ asymptotically with increasing size of population, $N$ (Wright, 1921, 1933a). Two cases with patterns of mating similar to those of Kimura and Crow's second class, except that the matings were between neighbors along infinitely extended lines instead of around a circle, had also been considered in these papers. These systems consisted of exclusive mating of half-sibs or of first cousins, otherwise with a minimum of relationship. It was found that there is no equilibrium in either case short of complete fixation locally, in spite of the linear increase in number of different ancestors with increasing number of ancestral generations. This was in contrast to systems (half first cousin or second cousin) in which this increase is more than linear and a steady state is rapidly attained with respect to heterozygosis.

Kimura and Crow were surprised to find that the limiting rates of decrease of heterozygosis in their circular systems are much less than under maximum avoidance approaching $[\pi/(2N+4)]^2$ in the case of half-sib matings and $[\pi/(N+12)]^2$ under first-cousin matings with large $N$. Maximum avoidance delays the onset of the decrease in heterozygosis but after a great many generations the lower rates under the circular systems cause these to fall below the systems of maximum avoidance of corresponding population number, in progress toward fixation.

The authors' surprise was occasioned by the fact that maximum avoidance had been reported in my early papers as approximately halving the rate of decrease in heterozygosis found under random mating, in populations of the same size. A later demonstration (Wright, 1938c, 1939) that the effective size of populations is approximately doubled by using exactly two offspring per parent as parents in the next generation in maintaining population size (assumed in the regular systems of mating) instead of by drawing offspring at random, was unfortunately not applied to this case until its implication was noted by Kimura and Crow. Actually, the limiting rate of decrease of heterozygosis is very slightly *greater* under maximum avoidance than under random mating if exactly two offspring are used from each parent in both cases.

I fully agree with the interpretation of the paradox given by Kimura and Crow. It is, however, I think, instructive to apply to these systems the set of $F$-coefficients devised for describing population structure in breeds of livestock and in natural populations (Wright, 1943a, 1946, 1951). The following discussion is divided into three parts: first, the justification for using the theoretical correlation coefficients between gametes as the basis for description of population structure; second, a review of previous applications of a set of such correlations; and third, the application to the

[1] Paper No. 988 from the Department of Genetics, University of Wisconsin.

[2] This study was conducted under grant No. GB 1317 from the National Science Foundation.

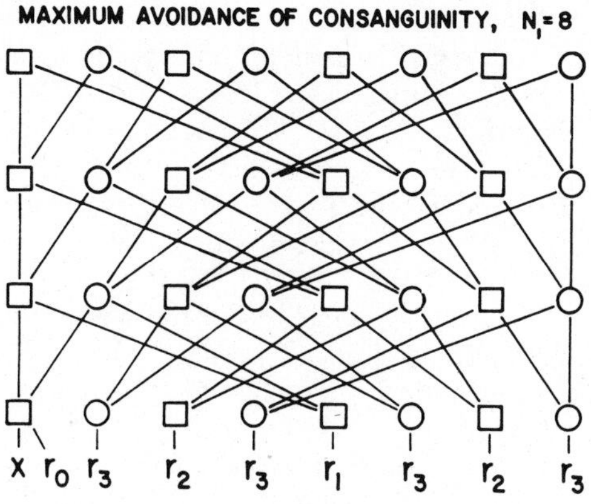


FIG. 1.

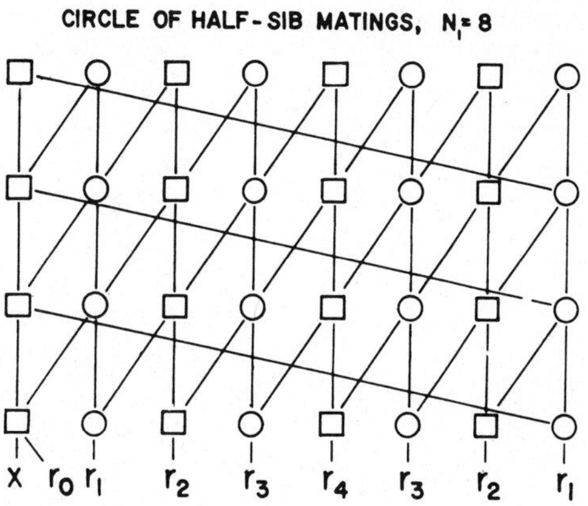


FIG. 2.

systems of maximum avoidance and of circular mating.

## PART I. THE FIXATION INDEX $F$ AS THE CORRELATION BETWEEN UNITING GAMETES

The basic coefficient in the set of statistics referred to above is a coefficient $F$ ($f$ in early papers) defined in the 1921 paper as the correlation between homologous genes of uniting gametes under a given mating pattern, relative to the total array of these in random derivatives of the foundation stock. While in most of the cases considered, the basis for the correlation was relationship, formulae were also obtained where the basis was phenotypic similarity in assortative mating. In a paper in the next year (1922a) this coefficient was designated the coefficient of inbreeding in cases in which it is based on relationship of the parents. It should be emphasized that this is a narrower concept than that of the 1921 paper. Later (Wright, 1951), $F$ in the broader context was designated the fixation index. The quantity $P$ ($= 1 - F$), which is often more convenient than $F$, was designated the panmictic index. It is to be noted that an individual, as representative of a certain mating system, or a population may have more than one fixation index. The most important as a rule, is the inbreeding coefficient pertaining to neutral autosomal disomic loci but there are other inbreeding coefficients pertaining to neutral sex linked (Wright, 1933a, 1951) and polysomic loci (1938b, 1951). There may also be fixation indices relating to loci subject to assortative mating or even to differential selection.

The value of $F$ in terms of that in preceding generations was determined by the form of correlation analysis, designated path analysis. It is desirable to sketch this briefly for comparison with alternative methods.

Path analysis (Wright, 1954, 1963a) is based essentially on the algebraic manipulation of standardized partial regression coefficients (unidirectional path coefficients) in systems of variables, measurable or hypothetical, in which each one that is not treated as an ultimate factor is represented, usually by an arrow diagram, as completely determined by certain others. This requires as a rule an independent residual factor to give the postulated completeness. These determining factors are often represented as determined similarly by more remote ones, and so on, until all lines of determination end in ultimate factors which must be assumed to be correlated with each other. These correlations are indicated in a diagram by bidirectional arrows, if the factors are not known to be independent, in order that the system may be a completely self-contained one.

If all relations are linear, it holds rigorously that the correlation between any two variables is the sum of contributions from all legitimate paths through the system that tend to contribute to it, the value of the contribution being the product of the path coefficients for the component paths, of which one may be a correlation coefficient (bidirectional) the rest all unidirectional. In the most condensed form, the correlation between two variables, $X$ and $Y$, may be expressed as $r_{XY} = \Sigma p_{Xi} r_{Yi}$ where $p_{Xi}$ is the unidirectional path coefficient relating $X$ to one its immediate determiners and $r_{Yi}$ is the correlation between this and $Y$ (which reduces to 1 if $Y$ is itself an immediate determiner). This expression may be expanded as far as the diagram permits, by application of this sort of equation to $r_{Yi}$ itself. The self-correlations yield very useful equations of the type, $r_{XX} = \Sigma p_{Xi} r_{Xi} = 1$, in complete systems. In tracing connecting paths through a diagram one must never go forward along an arrow (including an end of a bidirectional one) and then back along another. At least two critics have considered this to be an arbitrary rule but common "descendants" do not tend to contribute to correlation between "ancestors."

A pedigree is a system of the sort described above. Under disomic autosomal heredity there are two sorts of elementary paths. One is for a path from zygote ($Z$) to a determining gamete ($G$) and has the value $a$ $(= p_{ZG}) = \sqrt{1/[2(1+F)]}$, which follows at once from the equation expressing complete and equal determination by the pertinent genes of the two uniting gametes, $r_{ZZ} = 2a(a + aF) = 1$. The other is for a path from gamete to parental zygote ($Z'$) with the value $b$ $(= p_{GZ'}) = \sqrt{\tfrac{1}{2}[1 + F']}$, where the prime indicates preceding generation, derivable at once from the fact that the correlation $r_{GZ'}$ which equals $b$ because there is only one connecting path, must equal the correlation $r_{Z'G'} = a'(1 + F')$, if there is no intervening selection. Note that there is no implicit assumption in either of these equations with respect to number or frequency of alleles or to values attributed to them. It is, however, assumed that there are no selective differences. In the case of sex-linked heredity (♀$XX$, ♂$XY$) the path coefficient relating a male zygote to the determining egg has the value 1 there being no path to the sperm with no $X$ chromosome) and the path coefficient relating an $X$ bearing sperm to the male zygote that produced it is also 1 because of complete determination. In polysomic systems, the analysis must be in terms of pairs of genes that are brought together, not to the sets of alleles of gametes.

In practice, most use is made of two compound path coefficients: a unidirectional one relating a gamete of one generation to one of the two back of it in the preceding generation, which under disomic autosomal heredity always has the value $ba' = \tfrac{1}{2}$, and the bidirectional correlation between two random gametes from the same individual with value $b^2 = \tfrac{1}{2}(1 + F')$. Thus the correlation between two gametes (whether uniting ones or not) has the value $\Sigma[(\tfrac{1}{2})^n(1 + F_A)]$ where $n$ is the number of individuals along a path that contributes to it, there being $n - 1$ unidirectional compound components with the value ½ each, and one bidirectional one, connecting two gametes of the common ancestor, $A$, with the value $\tfrac{1}{2}(1 + F_A)$ where $F_A$ is the inbreeding coefficient of this ancestor. This formula is the usual one for use in calculating inbreeding coefficients and correlations between gametes other than uniting ones, where there is only sporadic inbreeding. Closely related is the correlation between two zygotes $Z_1$ and $Z_2$ as sums of contributions from the uniting gametes, $Z_1 = G_1 + G_2$, $Z_2 = G_3 + G_4$ (Wright, 1922a).

$$r_{Z_1Z_2} = \sum \frac{(\tfrac{1}{2})^{n-1}(1 + F_A)}{\sqrt{(1 + F_{Z_1})(1 + F_{Z_2})}}$$

If there is systematic inbreeding, it is more convenient to trace the paths back of the gametes under consideration for only one gamete-to-gamete generation in each case before completing them by correla-

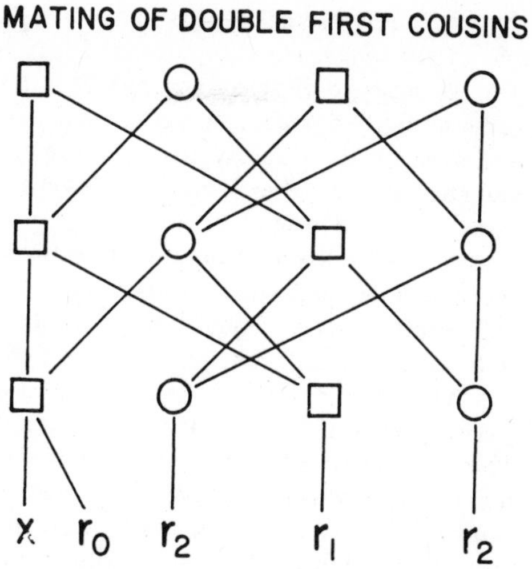


FIG. 3. Gametic correlations with x.

tional paths. Thus if one traces $G_1$ to $G_5$ and $G_6$, and $G_2$ to $G_7$ and $G_8$, we have $r_{12} = \frac{1}{4}[r_{57} + r_{58} + r_{67} + r_{68}]$.

As a simple example, consider the case of mating of double first cousins (Fig. 3), which comes under the heads of both maximum avoidance and of circular first-cousin mating. There are three kinds of correlations among gametes of the same generation: between gametes of the same individual ($r_0$), between gametes produced by sibs ($r_1$), and between ones produced by first cousins ($r_2$). The last is also the inbreeding coefficient, $F$. These may be expressed in terms of $r$'s of the preceding generation by inspection (Fig. 3).

$$r_0 = \tfrac{1}{2}(1 + r_2')$$
$$r_1 = \tfrac{1}{4}(2r_0' + 2r_2')$$
$$r_2 = \tfrac{1}{4}(2r_1' + 2r_2')$$

The desired expression for $F$ is obtained by substitution in the third. Using numbers of primes to indicate the numbers of preceding generation $F = \frac{1}{2}F' + \frac{1}{4}F'' + \frac{1}{8}F''' + \frac{1}{8}$. There is usually simplification by substitution of $1 - P$ for $F$.

$$P = \tfrac{1}{2}P' + \tfrac{1}{4}P'' + \tfrac{1}{8}P'''$$

On putting $\lambda = P/P' = P'/P'' = P''/P'''$ to obtain the limiting ratio of $P$ to $P'$, expected here because of the constant size of population.

$$8\lambda^3 - 4\lambda^2 - 2\lambda - 1 = 0, \qquad \lambda = 0.91964$$

The most important property of $P$, assuming that the pedigrees trace to a random breeding foundation stock, is that it gives the amount of heterozygosis ($y$) relative to that under random mating, $y_0 = 2q(1-q)$, where $q$ is gene frequency. By constructing a $2 \times 2$ correlation array of uniting alleles in the two allele case, it comes out at once that $F = 1 - (y/y_0)$ and thus that $P = y/y_0$. From this the zygotic composition of the population with respect to the pair of alleles $[qA + (1-q)a]$ can be written at once in terms of deviations from complete fixation, $[q - Pq(1-q)]AA + [2Pq(1-q)]Aa + [(1-q) - Pq(1-q)]aa$ (Wright, 1922b) or in two algebraically equivalent ways: as deviations from panmixia, $[q^2 + Fq(1-q)]AA + [2q(1-q) - 2Fq(1-q)]\ Aa + [(1-q)^2 + Fq(1-q)]$ $aa$ or as the weighted average of panmictic and fixed components, $[Pq^2 + Fq]AA + [2Pq(1-q)]Aa + [P(1-q)^2 + F(1-q)]$ $aa$. A coefficient $\alpha$ proposed by Bernstein (1930) in this connection is the same as $F$.

These formulae can be extended to any number of alleles by the principle that any group of alleles e.g. $(q_iA_i + q_jA_j)$, can be treated formally as if one, $(q_i + q_j)A_{ij}$ (Wright, 1931, 1951). Thus the frequency of any homozygote $A_iA_i$ (opposed to all of its alleles collectively) is of the type $[Pq_i^2 + Fq_i]$ in which $F$ is obtained by path analysis, and that of any collective pair, $A_{ij}$, as opposed to all of the rest collectively, is of the type $[P(q_i + q_j)^2 + F(q_i + q_j)]$ by the same analysis, whence by subtraction, that of any heterozygote, $A_iA_j$, is of the type $2Pq_iq_j$. The ratio of the total amount of heterozygosis, $2\Sigma Pq_iq_j$, $(i < j)$, to that in the randombred foundation stock, $2\Sigma q_iq_j$, is thus $P(= 1 - F)$ irrespective of number, frequencies, or values of alleles. $F$, as calculated by path analysis, is thus a rather unusual sort of correlation coefficient, but the principle may easily be verified by actually setting up the correlation array with arbitrary frequencies and values for each allele in a multiple system and calculating the correlation coefficient by the usual formula (Wright, 1963a).

If the average inbreeding coefficient is calculated from pedigrees that trace to a foundation stock that is itself inbred, it gives the correlation between gametes with histories of the sorts indicated, relative to the gene frequencies of this foundation stock, not to those of any more remote foundation stock to which the latter traces.

Thus the inbreeding coefficient (or more generally the fixation index), $F$, and the panmictic index, $P$, must always be specified as relative to the particular foundation stock to have any meaning.

It is not necessary here to discuss the alternative method of calculating the decline in heterozygosis under systems of mating, based on types of matings instead of gametic unions. This method, as developed by Bartlett and Haldane (1934, 1935) gives a more complete account of the consequences of systems of mating but is much more cumbersome than path analysis in arriving at the most significant parameter and does not permit generalization from small to large populations as readily. Thus the analysis of double first-cousin matings which requires a $12 \times 12$ matrix and thus solution of a characteristic equation of the 12th degree by the method of Bartlett and Haldane (Fisher, 1949) required only solution of a cubic equation by path analysis (as indicated above) and could easily be generalized to any population of size $2^m$ with maximum avoidance of consanguine mating (Wright, 1921, 1933a). The equations of path analysis are also expressible in matrix form but this has been needed only in exceptionally complicated cases (Wright, 1963a).

Another approach to such problems was that of Malécot (1948). He introduced the interpretation of the inbreeding coefficient $F$ as the probability of identity of representatives of a locus, due to descent from a common origin. Since the probability that an autosomal gene is identical with one of the two a generation back of it is $\frac{1}{2}$ (in the absence of selection) and thus the same as the compound path coefficient $ba'$, and since the probability that two random gametes of the same individual have the same gene by origin is $\frac{1}{2}(1 + F')$, and thus equal to the compound bidirectional coefficient $b^2$, and since probabilities, like path coefficients, compound by multiplication along paths, the resulting formula $F = \Sigma[(\frac{1}{2})^n(1 + F_A)]$ is identical with the basic formula of path analysis in this case. There is similar agreement in the cases of sex linkage and polysomy. This mathematical identity of the methods also applies if probabilities are traced only to the preceding gamete generation before combining with the probabilities of identity by origin of the two gametes to which these lead.

This method is therefore identical as far as calculation is concerned with path analysis as applied to gametes. The only difference is in interpretation. The same relation can be interpreted either as the correlation between two gametes, relative to the foundation stock, or as the probability of identity by origin from this foundation stock.

The concepts are thus very closely related. The principle of equal probabilities at segregation is indeed directly involved in assigning the value of the elementary path coefficient $b$. The principles of probability were also explicitly involved in the application of path analysis to assortative mating (Wright, 1921) and to all applications to random breeding populations (*e.g.*, ones with $N_m$ males and $N_f$ females, Wright, 1931).

An especially important example of the use of principles of probability in calculating correlations has been in determining the phenotypic correlation between relatives with respect to loci involving dominance (Malécot, 1948) and factor interaction (Kempthorne, 1954; Cockerham, 1954). Path analysis was developed as a method of dealing rigorously with correlations in systems of multiple variables connected exclusively by linear relations and as such did not seem competent to deal with the non-linear relations usually involved in phenotypic correlations (except in cases in which there could be no correlation between the dominance or interaction deviations of the relatives in question).

It is possible, however, to extend path analysis to include contributions to correlation from parallel all-or-none joint contributions of two or more variables to the variances of the two non-inbred relatives. The value of the compound joint path coefficient is the product of all of the elementary coefficients in all of the parallel paths and is thus zero if any one of these is zero. With this addition, the formulae for the effects of dominance and the various types of factor interaction agree with those of the above authors (Wright, 1963b).

Returning to relations between gametes, the significance of the inbreeding coefficient and of the similar parameter for pairs of gametes in general, is undoubtedly much enhanced by the fact that they can always be interpreted either as correlation coefficients or as probabilities of identity of origin, in both cases relative to a specified foundation stock.

There is, however, an important difference between those interpretations in applicability in the broader context of a system of parameters useful in concise discrimination among population structures. Correlation coefficients vary between −1 and +1 while probabilities can only vary between 0 and +1. A correlation between gametes as calculated from a pedigree cannot be negative (as may be seen from the general formula $F = \Sigma[(\frac{1}{2})^n(1 + F_A)]$ and thus can always be identified with a probability. It may, however, be useful to find the correlation between such gametes relative to the array of gametes of their own generation. Such a correlation (not capable of direct calculation from the pedigree) may be negative. Thus it is negative by definition, if there is maximum avoidance of consanguinity, which is one of the classes of mating systems with which this paper is especially concerned.

There may also be a possibility of negative correlation between uniting gametes (and thus more heterozygosis than under random mating) in the broader context of $F$ as a fixation index in cases in which there are other reasons for correlation than consanguinity. The effects of assortative mating, based on a phenotypic correlation, $r$, between mates, or genotypic correlation $m = h^2r$ where $h^2$ is the heritability of the character, affected by a system of multiple ($n$) pairs of alleles with equal frequencies, no dominance and equivalent effects, with respect to which the assortative mating occurs, were dealt with by path analysis at the same time as the effects of systems of inbreeding (Wright, 1921). The phenotypic correlation was taken as constant. Less restrictive postulates may be made but this case suffices for the present purpose.

The genotypic correlation, $m$, may be expected in general to reach a steady state, $\hat{m}$, at less than 1. The correlation between uniting gametes (with respect to the array of loci involved) was represented by $f$; that between uniting genes of a single locus by $f_u$. The latter, however, may be considered a case of the fixation index since the amount of heterozygosis at each locus relative to that is the random breeding foundation stock is given by $1 - \hat{f}_u$.

$$F = \hat{f}_u = \hat{m}/(2n - 2n\hat{m} + \hat{m})$$

$$\hat{f} = \hat{m}/(2 - \hat{m})$$

These are positive if $m$ is positive but negative if $m$ is negative. In the extreme case of perfect disassortative mating ($r = -1$) and complete heritability ($h^2 = 1$, $m = -1$), $F = -1/(4n - 1)$ and $\hat{f} = -\frac{1}{3}$. The amount of heterozygosis relative to that in the foundation stock ($\frac{1}{2}$ under the assumptions) is given by $P = 1 - F = 4n/(4n - 1)$ and thus is greater than 1 in this case. In absolute terms, $y = 2n/(4n - 1)$ under the assumption that gene frequency is $\frac{1}{2}$.

The case of a locus in which there is a steady state because of selection against both homozygotes in favor of the heterozygotes similarly gives a negative fixation index.

| | Frequency | $w$ | |
|---|---|---|---|
| $AA$ | $q^2$ | $1-s$ | $\bar{w}=1-sq^2-t(1-q)^2$ |
| $Aa$ | $2q(1-q)$ | 1 | $\Delta q=-(s+t)q(1-q)\times[q-\hat{q}],\ \hat{q}=t/(s+t)$ |
| $aa$ | $(1-q)^2$ | $1-t$ | $\hat{w}=1-[st/(s+t)]$ |

Calculation of the correlation between uniting gametes gives the negative fixation index $F=-st/[s+t-st]$, and $P=(s+t)/(s+t-st)$ is in excess of 1.

## Part II. The *F*-Statistics

In studying the history of the British Shorthorn cattle from the herdbook records (Wright, 1923a, 1923b; McPhee and Wright, 1925, 1926) it became obvious that mere specification of the average inbreeding coefficient at successive periods was not adequate. It was found desirable to supplement this in ways which will be discussed briefly later.

A system was developed from this start (Wright, 1943a, 1946, 1951) for describing the properties of hierarchically subdivided natural populations. Three parameters were proposed in the 1951 paper in terms of a total population ($T$), subdivisions ($S$), and individuals ($I$). $F_{IT}$ is the correlation between gametes that unite to produce the individuals, relative to the gametes of the total population. $F_{IS}$ is the average over all subdivisions of the correlation between uniting gametes relative to those of their own subdivision. $F_{ST}$ is the correlation between random gametes within subdivisions, relative to gametes of the total population. The list can be extended if there are further subdivisions.

The above three $F$-statistics are not independent. One of two demonstrations of their interrelation, given in the 1943 paper, is repeated below in the later symbolism. The effects of accidents of sampling are here ignored.

The amount of heterozygosis ($y_T$) in the total population, whatever its structure, is as follows in terms of total gene frequency $q_T$, as brought out earlier.

$$y_T=2q_T(1-q_T)(1-F_{IT})$$

Assume first that the total is divided into many ($n$) random breeding demes ($D$) with varying gene frequencies, $q_D$, and amounts of heterozygosis, $y_D=2q_D(1-q_D)$. The total amount of heterozygosis is the average of these.

$$y_T=\frac{2}{n}\sum^n q_D(1-q_D)=2\left[q_T-\frac{1}{n}\sum^n q_D^2\right]$$

The value of the term $\frac{1}{n}\sum^n q_D^2$ may be obtained from the variance of gene frequencies of demes within the total, $\sigma_{q(DT)}{}^2$:

$$\sigma_{q(DT)}{}^2=\frac{1}{n}\sum^n(q_D-q_T)^2=\frac{1}{n}\sum^n q_D^2-q_T^2.$$

Thus $y_T=2[q_T(1-q_T)-\sigma_{q(DT)}{}^2]$.

This formula was enunciated first by Wahlund (1928). On equating the two expressions for $y_T$,

$$\sigma_{q(DT)}{}^2=q_T(1-q_T)F_{IT}.$$

It should be noted that the correlation between random gametes drawn from demes ($F_{DT}$) is the same as $F_{IT}$ if there is random mating within the deme.

Consider next, division of the total into subdivisions that are themselves inbred. As before, $y_T=2q_T(1-q_T)(1-F_{IT})$ but as an average $y_T=\frac{2}{n}\sum^n[q_S(1-q_S)](1-F_{IS})$ $=2(1-\bar{F}_{IS})\left(q_T-\frac{1}{n}\sum q_S^2\right)$ assuming that $F_{IS}$ and $q_S$ are independent. Again,

$$\sigma_{q(ST)}{}^2=\frac{1}{n}\sum^n(q_S-q_T)^2=\frac{1}{n}\sum^n q_S^2-q_T^2.$$

$$y_T=2[(1-\bar{F}_{IS})q_T(1-q_T)-\sigma_{q(ST)}{}^2].$$

Thus $\sigma_{q(ST)}{}^2=q_T(1-q_T)[F_{IT}-F_{IS}]/[1-F_{IS}]$, dropping the bar over $F_{IS}$.

If now, completely random mating were instituted in the subdivisions there would be no change in their gene frequencies and hence none in $\sigma_{q(ST)}^2$, but now $\sigma_{q(ST)}^2 = q_T(1-q_T)F_{ST}$ where $F_{ST}$ is defined as the correlation between random gametes of the subdivision.

Thus $F_{ST} = (F_{IT} - F_{IS})/(1-F_{IS})$.

This is simplified if expressed in terms of $P$'s.

$$P_{IT} = P_{IS}P_{ST}$$

If there are secondary subdivisions into local races ($R$) which may themselves be inbred ($F_{IR} \neq 0$), $P_{IS} = P_{IR}P_{RS}$ and $P_{IT} = P_{IR}P_{RS}P_{ST}$.

Such analysis may be continued as far as there is hierarchic subdivision.

These equations add another interpretation of the $F$-statistics to the three already treated (as correlations, as functions of the relative amount of heterozygosis, and, in some cases, as probabilities of identity by origin). $F_{IT}$ gives the ratio of the variance $\sigma_{q(DT)}^2$ of gene frequencies of random breeding subdivisions ($D$) (if these occur) to its maximum possible value $q_T(1-q_T)$, expected if the subdivisions are completely isolated and each completely fixed, thus forming the array $q_T AA + (1-q_T)aa$. $F_{IS} = \frac{1}{m}\sum^{m}\sigma_{q(DS)}^2/q_S(1-q_S)$ gives the average of such ratios among subdivisions. Most importantly, however, $F_{ST}$ is the ratio of the actual variance of gene frequencies of subdivisions to its limiting value, irrespective of their own structures. $F_{ST}$ is thus necessarily positive. $F_{IS}$, while usually positive, is negative if there is systematic avoidance of consanguine mating within the subdivisions. $F_{IT}$ is positive if there is systematic subdivision, whether into demes ($F_{IS} = 0$, $F_{IT} = F_{ST}$) or into inbred groups, but can be negative if there is no systematic subdivision and there is prevailing avoidance of consanguine mating.

If pedigrees can be traced some distance back, a pedigree $F$ can be obtained which differs from $F_{IT}$ above in relating to a still more comprehensive total than currently exists, the total of all hypothetically similar populations, derivable from the foundation stock of the earlier period. This, as noted, is necessarily positive and interpretable as the probability of identity of origin of the uniting gametes in the current total populations.

In addition to the above properties, the $F$-statistics have important relations to the statistics of quantitatively varying characters. If the effects of genes are completely additive (semidominance and no factor interactions) the mean of the character in the total population or its subdivision depends only on the gene frequencies, unaffected by the $F$-statistics. The variances are, however, dependent on the latter. The variance of individuals in the total population, $\sigma_{IT}^2$, is analyzed below into the variance of subdivision means, $\sigma_{ST}^2$, and the mean of the variances within subdivisions, $\sigma_{IS}^2$. The general formula does not seem to have been given before. That for randombred subdivisions has long been known (Wright, 1921, 1943b).

The variance that would be found with the same gene frequencies but random mating throughout is represented by $\sigma_{IT(0)}^2$.

| | General | Randombred subdivisions ($F_{IS} = 0$, $F_{ST} = F_{IT}$) |
|---|---|---|
| $\sigma_{ST}^2$ | $2F_{ST}\sigma_{IT(0)}^2$ | $2F_{IT}\sigma_{IT(0)}^2$ |
| $\sigma_{IS}^2$ | $(1 + F_{IT} - 2F_{ST})\sigma_{IT(0)}^2$ | $(1 - F_{IT})\sigma_{IT(0)}^2$ |
| $\sigma_{IT}^2$ | $(1 + F_{IT})\sigma_{IT(0)}^2$ | $(1 + F_{IT})\sigma_{IT(0)}^2$ |

If there is other than semidominance, the mean is affected in a way that is responsible for the well-known effects of inbreeding. At a given gene frequency, the mean with a given value of $F_{IT}$, represented by $M(F_{IT})$, is related to the means under random mating, $M(0)$, and under complete fixation of subdivisions, $M(1)$, by the formula (Wright, 1922b).

$$M(F_{IT}) = M(0) + F_{IT}[M(1) - M(0)]$$

This applies to subgroups on replacing $F_{IT}$ by $F_{IS}$.

The total variance under inbreeding, $\sigma_{IT(F)}^2$, is related to that under random mating throughout by the formula (Wright, 1951).

$$\sigma_{IT(F)}^2 = (1 - F_{IT})\sigma_{IT(0)}^2 + F\sigma_{IT(1)}^2 + F_{IT}(1 - F_{IT})[M(1) - M(0)]^2$$

The analysis of this into the components, $\sigma_{ST}^2$ and $\sigma_{IS}^2$, requires third and fourth moments of the distribution of gene frequencies among subdivisions. The somewhat surprising results for dominant characters under progressive inbreeding of completely isolated subdivisions have been presented by Alan Robertson (1952). The somewhat different ones where there is a steady state because of a balance between local inbreeding and immigration have also been presented (Wright, 1952).

The theory of the distribution of gene frequencies under joint action of systematic and random processes deals with aspects of population structure that are not in general amenable to path analysis because of the non-linear action of selection. A connection can be established, however, in cases in which there is linearity. The general formula for the steady state distribution at a single locus is as follows letting $\Delta q$ represent the change per generation that systematic factors tend to produce and $\sigma_{\Delta q}^2$, the contribution to variance from random ones (Wright, 1938a).

$$\phi(q) = (C/\sigma_{\Delta q}^2) \exp\left[2\int(\Delta q/\sigma_{\Delta q}^2)\, dq\right]$$

Immigration is responsible for a linear pressure on gene frequency $\Delta q_S = -m(q_S - q_T)$ in which $m$ is the amount of replacement of local genes (frequency $q_S$) by immigrant ones (frequency $q_T$). The sampling variance is $\sigma_{\Delta q}^2 = q_S(1-q_S)/2N$. Substitution yields:

$$\phi(q) = \frac{\Gamma(4Nm)}{\Gamma(4Nmq_T)\Gamma[4Nm(1-q_T)]} \times q_S^{4Nmq_T-1}(1-q_S)^{4Nm(1-q_T)-1}.$$

This formula was obtained in a different way earlier (Wright, 1931).

$$\bar{q}_S = \int_0^1 q_S\phi(q_S)\, dq_S = q_T$$

$$\sigma_{q(ST)}^2 = \int_0^1 (q_S - q_T)^2\phi(q_S)\, dq_S = q_T(1-q_T)/[4Nm+1]$$

Thus $F_{ST} = \dfrac{1}{4Nm+1}$ in this case (Wright, 1931).

By path analysis $F_{ST} = (1-m)^2\{1/(2N) + [1-1/(2N)]F_{ST}'\}$.

At equilibrium $F_{ST}' = F_{ST}$, giving (Wright, 1943a, 1951) $F_{ST} = (1-m)^2/[2N-(2N-1)(1-m)^2] \approx 1/(4Nm+1)$.

The two approximate determinations of $F_{ST}$ thus agree. More generally $F_{ST}$ in the broad sense can always be obtained, at least empirically, for the variance of distribution of gene frequencies even in cases involving selection, from the formula $F_{ST} = \sigma_{q(ST)}^2/q_T(1-q_T)$. The results, of course, apply only to the particular loci in question.

A more direct determination of $F_{ST}$ is possible in the important case in which there is a balance between local selection pressure and immigration, that varies among the subdivisions. In the simplest case (additive gene effects, $s$ always less than $m$ in absolute value).

$$\Delta q_S = sq_S(1-q_S) - m(q_S - q_T)$$

$$\hat{q}_S = q_T + \frac{s}{m}q_T(1-q_T) \text{ approximately}$$

(Wright, 1931)

$$\sigma_{\hat{q}_S}^2 = \sigma_{q(ST)}^2 = \sigma_{(s/m)}^2 q_T^2(1-q_T)^2$$
$$F_{ST} = \sigma_{(s/m)}^2 q_T(1-q_T)$$

The situation, if $s$ varies from positive values greater than $m$ to negative ones also greater in absolute value, is more complicated but it would be possible to calculate $F_{ST}$ for the locus in question, again as a fixation index, not an inbreeding coefficient. The point here is that many different aspects of population structure can be brought under a common viewpoint by means of the $F$-statistics.

There was nothing in the derivation of the three basic $F$-statistics that implies anything about the degree of isolation of the subdivisions or their arrangement in space. They may be completely isolated, at one extreme, or merely arbitrarily bounded portions of a continuum, at the other. Their gene frequencies may be distributed at random in the total population or in an orderly cline. The full account of an actual population obviously requires a map, and detailed accounts of the various regions within it and their relations. There are, however, additional general aspects of

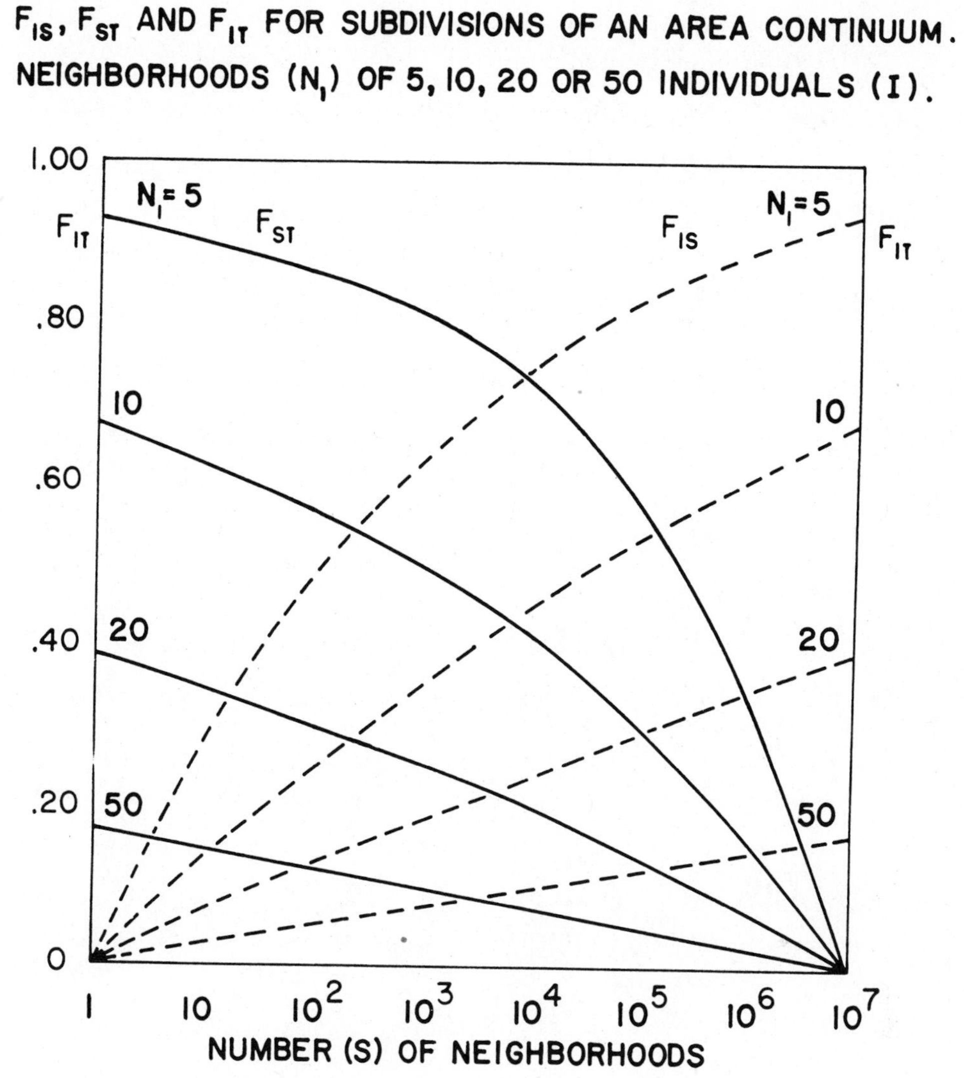


FIG. 4.

the structure than can be brought out by extension of the hierarchic pattern. This can be illustrated by considering the application of $F$-statistics to the properties of a continuum in which there is some degree of local isolation because of restricted dispersion (Wright, 1940, 1943a, 1946, 1951).

The most important unit is the "neighborhood," the size of population ($N_1$) from which the parents may be treated as if drawn at random. The properties of the system depend largely on $N_1$. If the distribution of birthplaces of parents are distributed normally relative to those of offspring in the case of a linear continuum, such as a shoreline, effective $N_1$ is equivalent to the number of reproducing individuals along a strip $2\sqrt{\pi\sigma}$ long. If the distribution is normal in all directions in an area continuum, effective $N_1$ is equivalent to the number in a circle of radius $2\sigma$ (Wright, 1946).

Arbitrary subdivisions ($S$) may be of any size from that of the neighborhood up to the total ($T$) under consideration. Various systems of mating have been considered (Wright, 1946). It will suffice here to review only the simplest case, that of random union of gametes in a monoecious population since other cases do not differ very much. It will be convenient to use $x$ for a varying number of ancestral generations.

In the case of an area continuum, ancestors of generation $x$ are drawn from an effective population of size $xN_1$. It was shown that for a subdivision ($S$ generations):

$$F_{IS} = \sum_1^{S-1} t_x \Big/ \left[2 - \sum_1^{S-1} t_x\right], \quad [t_1 = 1/N_1]$$

$$t_x = \frac{(x-1) - (1/N_1)}{x} t_{x-1}.$$

$\Sigma t_x$ can be calculated as the sum of such terms, or approximated by an integration formula or, as pointed out by D. J. Hooton, by the formula $\sum^{S-1} t_x = 1 - St_S$.

$F_{IT}$ is analogous for the total ($T$ generations).

$$F_{ST} = (F_{IT} - F_{IS})/(1 - F_{IS})$$

In the case of a linear continuum, ancestors of generation $x$ are drawn from an effective population of size $\sqrt{x}\,N_1$. If $F_{IS}$ refers to a strip containing $\sqrt{S}$ neighborhoods:

$$F_{IS} = \sum^{S-1} t_x/[2 - \sum^{S-1} t_x], \quad [t_1 = 1/N_1]$$

$$t_x = \frac{[\sqrt{x-1} - (1/N_1)]}{\sqrt{x}} t_{x-1}$$

$$\sum^{S-1} t_x = 1 - \sqrt{S}\, t_S.$$

The population structure may best be indicated (as in Fig. 4) by plotting $F_{IS}$ $(= \sigma_{q(IS)}{}^2/q_S(1-q_S))$ and $F_{ST}$ $(= \sigma_{q(ST)}{}^2/q_T(1-q_T))$ against log $S$ over the range 0 to log $T$. $F_{IT}$ is the last value of the former and first of the latter. The interpretation of $F_{IS}$ as a measure of the variability among neighborhoods within areas of increasing size suffers from the disadvantage that its denominator increases with increasing $S$, though less rapidly than the numerator. $F_{ST}$, however, is without this drawback and shows how differentiation of neighborhoods builds up decreasing amounts of differentiation in areas of larger size.

The comparison of the curves for different sizes of neighborhood shows, however, that $N_1$ must be very small (in the case of an area continuum) for an appreciable differentiation even of neighborhoods. This is not the case with a linear continuum.

It should be added that reversible mutation ($u$, $v$) or a small amount of universal dispersion ($m$) imposes a limit beyond which $F_{IS}$ cannot increase with increasing $S$. This can be estimated by an integration formula but more easily for an area continuum by a formula suggested by Alan Robertson $\Sigma t = 1 - [1 - (1-m)^2]^{1/N_1}$. Fig. 5 shows how $F_{ST}$, ($N_1 = 20$) is affected by increasing $m$. The rates of mutation to ($v$) and from ($u$) the gene may be incorporated into $m$. Balanced universal selection has somewhat similar effects.

If the distributions of parental distances are not normal, the forms of these curves are somewhat modified. The distributions of ancestral populations approach normality. With leptokurtic parental distributions, which seem to be most usual (Bateman, 1950), there is little modification in the case of areas but considerably more damping in the case of linear ranges.

This type of theory has been extended to the important model of population structure in which there are clusters of high density distributed uniformly over the range of the species. The size of population of such clusters may be large but the rise in $F_{IS}$ depends on the small increments to the size of the ancestral populations due to small amounts of dispersion per generation in the population as a whole. If this increment is treated as the significant "$N_1$," the rise of $F_{IS}$ from zero for the cluster is substantially the same irrespective of the cluster size (Wright, 1951).

Malécot (1948) has attacked the problem of isolation by distance in a very dif-

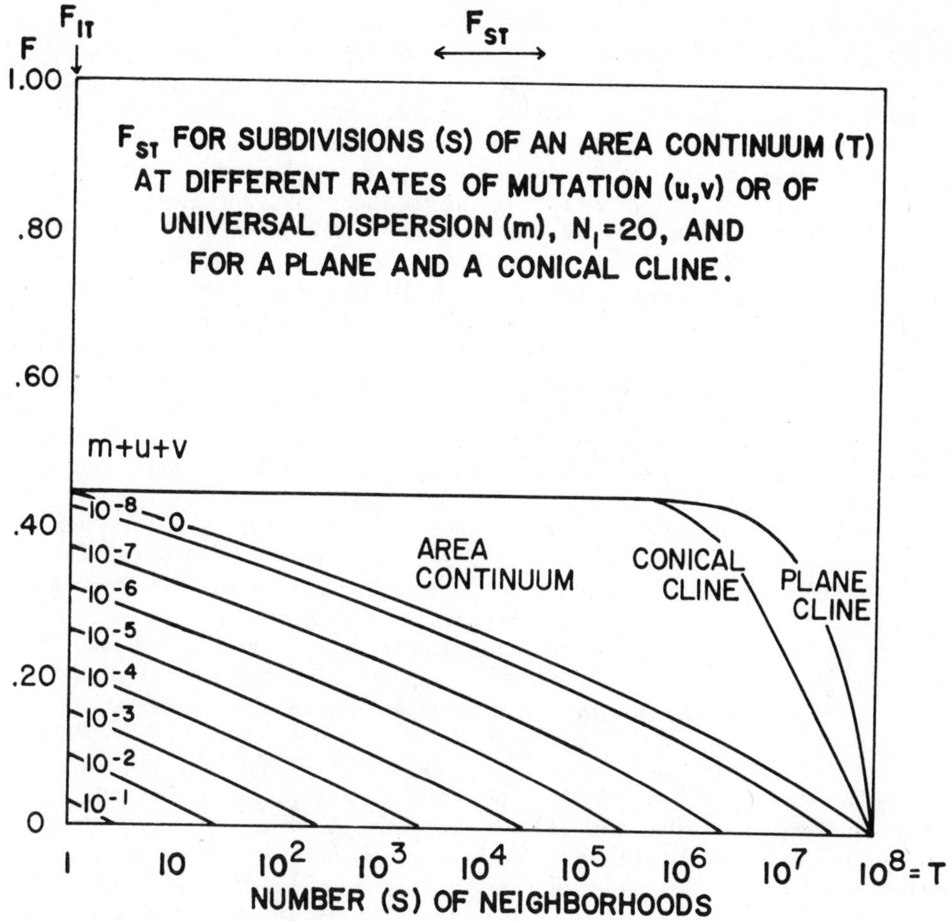


FIG. 5.

ferent way. His primary objective was the correlation between regions as a function of their distance apart. Kimura (1953) studied what he called the stepping stone model which was similar to a cluster model in one dimension. Kimura and Weiss (1964) have extended this to two and three dimensions. They also dealt primarily with the correlation between regions as functions of distance but also considered variances. Points of contact have been established between these models and those of the *F*-statistics but no full comparison has yet been made because of differences in mathematical form.

If, as assumed above, local differentiation is due only to the balancing of local inbreeding and dispersion, $F_{ST}$ for an area continuum falls off much more rapidly with increasing $S$ than it does in the case of a cline, based on orderly regional changes in the conditions of selection. Consider the case of a uniform gradient in one direction over a square total area (a plane cline). If this area is divided into $n$ squares, the variance is only $1/n$ less than for very small subdivisions, corresponding to neighborhoods in the preceding case. The curve for $F_{ST}$ is virtually level up to large fractions of the total (Fig. 5).

In the case of a conical cline, with uniform change of average in all directions from a high (or low) center, there is more rapid falling off of amount of differentiation with increasing size of the areas under consideration than in the case of a plane cline, but still much less than in the chaotic pattern of differentiation due to balancing between inbreeding and dispersion. With seven equal areas (a central one surrounded by six others), the variance of area averages is 51.6 per cent of that of very small subdivisions. With 19 such areas in two concentric zones about a central one, the corresponding figure is 78.9 per cent. With 37 such areas in three concentric zones about a central one, it is 88.7 per cent and in the case of 61, it is 93 per cent. The curve for $F_{ST}$ is still virtually level over most of the range of values of $S$ in a large total population (Fig. 5).

We return here to the very different sort of application of the theory that has been made for the study of breed history. In this case, pedigree $F$, treated as $F_{IT}$ measures inbreeding relative to the foundation stock and thus to a hypothetical contemporary total ($T$). The contemporary breed of any period under study is treated as a subdivision, $S$, of this total. $F_{ST}$ is the correlation between random gametes from the parental generation relative to the same total. This can be calculated from random pedigrees by random association of sires and dams, while $F_{IS}$, the correlation between mating gametes within the contemporary array of gametes cannot be calculated directly from pedigrees but must be calculated from the formula $F_{IS} = (F_{IT} - F_{ST})/(1 - F_{ST})$. It is negative if $F_{ST}$ is greater than $F_{IT}$.

The method actually used in studying the history of the British Shorthorns (McPhee and Wright, 1925, 1926) was to select a considerable number of bulls and cows at random from Coates' herdbook of a given year, and construct random samples of the pedigrees of the sire and dam of each (Wright and McPhee, 1925). At first, single random lines, the sequence of sires and dams being determined by coin tossing, were carried back of all four grandparents but later it was recognized that it was preferable to trace only two such lines, one back of each parent. The average value of $F_{IT}$ for the breed at the time is given by $(k/2m)(1 + \bar{F}_A)$ if there are $k$ cases of a common entry among $m$ such pairs of pedigree lines. This approaches 1 if nearly every pair shows a common entry and the average inbreeding coefficient of this animal itself ($A$) approaches 1. In this form a standard error is easily calculated. The calculation of $F_{ST}$ is exactly the same except that the pedigree samples of sires and dams of different animals are matched at random for common entries.

In the original study, the calculation of $F_{IT}$ was supplemented by calculations of the correlation between random animals of each period and two bulls, Charles Colling's Favourite born in 1793, about which the breed was formed, and Amos Cuickshank's Champion of England, born 1859 about which it was largely reformed nearly a century later. These correlations were compared with the correlation between random animals of the same period. The latter is closely related to $F_{ST}$ which as "the amount of inbreeding due to random mating" was calculated but not wholly correctly. Table 1 and Fig. 6 show $F_{IT}$ separately for males and females as well as the average revised figures for $F_{ST}$, $F_{IS}$, and the other correlations referred to.

This table and figure show that a very substantial amount of inbreeding was built up in the course of a century; $F_{ST} = 0.249$ in 1920. This was possible because of relatively small effective numbers, especially in early years.

It was noted (Wright, 1931) that the effective number, $N_e$, may differ from the number of mature individuals for at least three reasons: an extreme sex ratio, differences in productivity beyond expectation under random sampling, and passage through a bottleneck of small size. The effect of sex ratio was indicated by the formula $N_e = 4N_mN_f/(N_m + N_f)$ which approaches $4N_m$ if there are many more females than males. The effect of differential productivity is given in a monoecious population with random union of gametes

TABLE 1. *F-statistics and coefficients of relationship for the British Shorthorn cattle at successive periods.*

| | $F_{IT}$ | | | $F_{ST}$ | $F_{IS}$ | Relationship | | |
|---|---|---|---|---|---|---|---|---|
| | ♂ | ♀ | avg. | | | Favourite | C. of E. | Interse |
| 1780 | 0 | 0 | 0 | – | – | – | – | – |
| 1810 | 19.1 ± 1.5 | 14.3 ± 1.3 | 16.6 | 12.8 | +4.4 | 44.3 | 26.3 | 22.0 |
| 1825 | 21.9 ± 2.1 | 16.1 ± 3.7 | 19.9 | 16.1 | +4.5 | 51.3 | 29.9 | 26.8 |
| 1850 | 18.4 ± 1.9 | 16.9 ± 3.7 | 18.0 | 20.0 | –2.5 | 50.1 | 26.1 | 33.9 |
| 1875 | 25.9 ± 2.1 | 29.0 ± 3.8 | 27.4 | 24.1 | +4.3 | 57.6 | 32.9 | 37.8 |
| 1900 | 23.2 ± 2.1 | 22.6 ± 3.8 | 22.9 | 24.1 | –1.6 | 52.1 | 39.2 | 39.3 |
| 1920 | 25.4 ± 4.0 | 26.7 ± 4.1 | 26.0 | 24.9 | +1.5 | 55.2 | 45.5 | 39.5 |
| Dairy | – | 27.5 ± 2.1 | – | 25.3 | +2.9 | 56.1 | 42.1 | 39.7 |

by the formula $N_e = (4N - 2)/(2 + \sigma_k^2)$ where $\sigma_k^2$ is the variance of number of offspring that reach maturity about the mean $\bar{k} = 2$ in a population of constant size (Wright, 1939). Kimura and Crow (1963a) arrived at $(4N - 4)/(2 + \sigma_k^2)$ as the effective number with $N/2$ males and $N/2$ females to which, as in all cases with separate sexes 0.5 must be added for comparison with the case of random union of gametes. Finally the effective number is the harmonic mean of the varying numbers over a number of generations and thus is dominated by the numbers at bottlenecks of population size (Wright, 1938c, 1939).

All of these played roles in the Shorthorn breed. The total numbers were relatively small in the foundation period which was thus something of a bottleneck. The number of sires was much less than the number of dams and the effect of this was exaggerated by wide differences in the productivity of the sires that were used, especially with respect to sons that became sires. The breed was essentially founded by intensive inbreeding to the bull Favourite to which the breed as a whole came to have a relationship somewhat closer than between parent and offspring. This was made possible to a large extent by the building up of strains such as Thomas Bates' Duchesses with average inbreeding coefficient of 0.409 over eight generations (64 cows) (Fig. 6) and average relationship to Favourite of 0.656 (Wright, 1923b).

It may be seen that in the early years, the bulls were of more inbred origin than the cows. In the later years, increasing relationship was built up to the bull Champion of England until this reached 0.455 by 1920 with the diffusion through the breed of the beefy Scotch strain, centering in this bull.

In the earlier years and one later period there was excess contemporary inbreeding ($F_{IS} > +0.043$) indicating the building up of somewhat inbred strains such as the Duchesses. In two periods, however, $F_{IS}$ was actually negative, indicating a slight excess tendency toward crossing differentiated strains. Thus the relative steadiness in the rise of $F_{ST}$ contrasts with wide fluctuations in $F_{IT}$.

A phenotypically somewhat differentiated group, those with registered milking records (McPhee and Wright, 1926) showed little differentiation in pedigree from the contemporary breed (1920) but showed relatively high $F_{IS}$ suggesting a slight tendency toward segregation, substantiated by a relatively low relationship to Champion of England.

## PART III. APPLICATION OF $F$-STATISTICS TO SYSTEMS OF MATING

In applying the set of $F$-statistics to systems of mating, the closed lines are to be considered as subdivisions ($S$) of an indefinitely large random array of such lines ($T$), derived similarly from a random breeding foundation stock. The inbreeding coefficients are thus of the type $F_{IT}$. $F_{ST}$ is the average of the correlations among all gametes of a line relative to this total. It

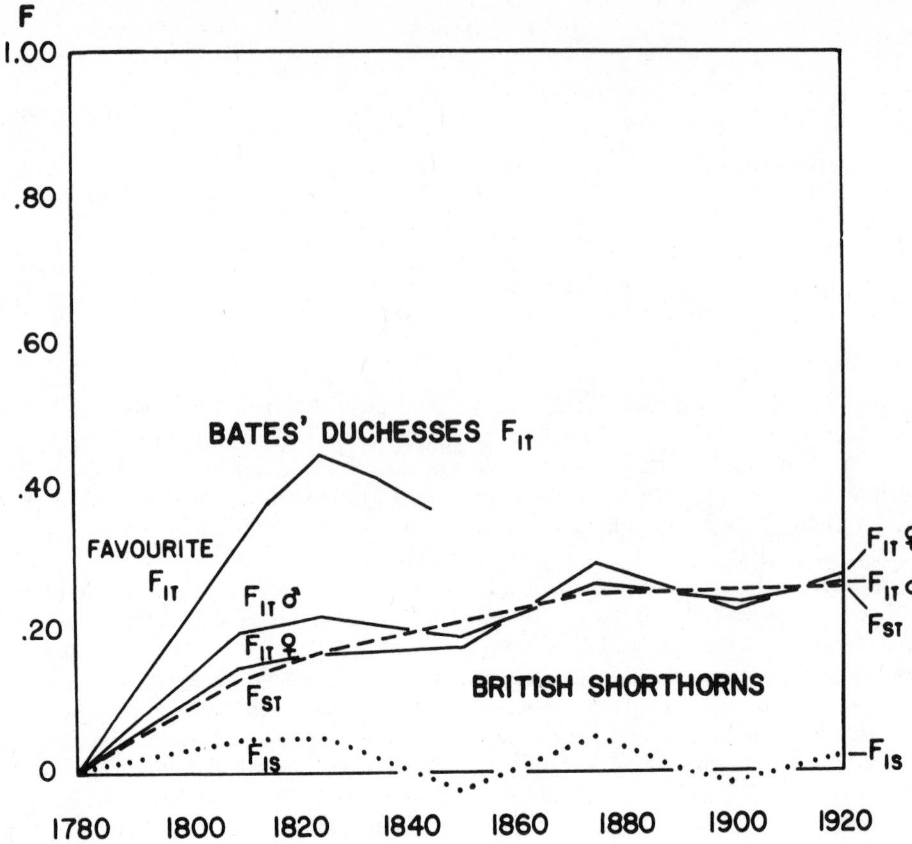


FIG. 6.

might seem that with separate sexes, one should average only the correlations between eggs and sperms, but if a very large random breeding population is to be produced in a given generation, the gametes of the $N$ zygotes should contribute equally. It is the average correlation among gametes of such a random breeding strain from each line, relative to the total of all possible lines that constitutes $F_{ST}$. $F_{IS}$ is the average correlation between uniting gametes relative to the gene frequencies of their own lines.

It will be convenient to begin with the extreme case of sib-mating. There are two kinds of correlations in a generation: between gametes of the same individual ($r_0$) and between gametes of different individuals ($r_1 = F_{IT}$).

$$
\begin{aligned}
r_0 &= \tfrac{1}{2}(1 + r_1') \\
F_{IT} = r_1 &= \tfrac{1}{2}(r_0' + r_1') \\
&= \tfrac{1}{4}[2F' + F'' + 1]. \\
P_{IT} &= \tfrac{1}{2}P_{IT}' + \tfrac{1}{4}P_{IT}''
\end{aligned}
$$

Since these types are equally numerous

$$
\begin{aligned}
F_{ST} &= \tfrac{1}{2}(r_0 + r_1) = \tfrac{1}{4}[2F + F' + 1] \\
P_{ST} &= \tfrac{1}{2}P_{IT} + \tfrac{1}{4}P_{IT}' \\
P_{IS} &= P_{IT}/P_{ST} = P_{ST}'/P_{ST}
\end{aligned}
$$

The values of $F_{IT}$, $F_{ST}$, and $F_{IS}$ in seven generations of sib mating are given in Table 2.

The correlation ($F_{IT}$) between uniting gametes, relative to the total, lags a generation behind that ($F_{ST}$) between random gametes, also relative to the total and is smaller by a ratio that oscillates about $\lambda$ $(= \tfrac{1}{4}[1 + \sqrt{5}] = 0.80902)$. The average

TABLE 2. *Values of* $F_{IT}$, $F_{ST}$, *and* $F_{IS} = (F_{IT} - F_{IS})/(1 - F_{IS})$ *by generation.*

| | 0 | 1 | 2 | 3 | 4 | 5 | 6 | 7 |
|---|---|---|---|---|---|---|---|---|
| $F_{IT}$ | 0 | 0.2500 | 0.3750 | 0.5000 | 0.5937 | 0.6719 | 0.7344 | 0.7852 |
| $F_{ST}$ | 0.2500 | 0.3750 | 0.5000 | 0.5937 | 0.6719 | 0.7344 | 0.7852 | 0.8262 |
| $F_{IS}$ | −0.3333 | −0.2000 | −0.2500 | −0.2308 | −0.2381 | −0.2353 | −0.2364 | −0.2360 |

correlation between uniting gametes ($F_{IS}$) relative to its line, is thus negative and rapidly approaches $(\lambda - 1)/\lambda = -0.23607$ in oscillatory fashion.

The question may arise as to whether $F_{IS}$ has any meaning in such small populations. In populations consisting of one male and one female, the correlation coefficient between egg and sperm is indeterminate (0/0), unless both parents are heterozygous, in which case, it is zero. If the correlation arrays are made diagonally symmetrical by tabulating gamete against gamete, irrespective of kind, the coefficient is indeterminate only if both parents are homozygous in the same allele; but this is a class of matings that approaches 100 per cent in frequency as the inbreeding proceeds.

This difficulty disappears, however, if the various types of mating are weighted by their variances in the direct calculations of $F_{IS}$. The essentially different kinds of mating, their correlation arrays, their variances, and correlation coefficients are as follows, noting that $A$ and $a$ may be exchanged with no essential change in kind.

| $AA \times AA$ | $AA \times aa$ | $AA \times Aa$ | $Aa \times Aa$ |
|---|---|---|---|
| 0 1 | 0.50 0 | 0.25 0.50 | 0.25 0.25 |
| 0 0 | 0 0.50 | 0 0.25 | 0.25 0.25 |
| 0 1 | 0.50 0.50 | 0.25 0.75 | 0.50 0.50 |
| $\sigma^2 = 0$ | $\sigma^2 = 0.25$ | $\sigma^2 = 0.1875$ | $\sigma^2 = 0.25$ |
| $r = 0/0$ | $r = -1$ | $r = -0.3333$ | $r = 0$ |

The frequencies of these types in seven successive generations of sib mating and the weighted averages of the coefficients are given in Table 3.

It may be seen that the actual weighted averages of the correlations between uniting gametes agree exactly with the theoretical values. $F_{IS}$ can thus be given a concrete meaning even in this very extreme case. It cannot, of course, be interpreted as a probability because of its negative value.

In maximum avoidance systems with population number $N = 2^m$ (illustrated in Fig. 1 for $N = 8$), there are $m + 1$ different kinds of correlations between gametes of the same generation. Labeling these $r_0, r_1 \text{---} r_m$ in order of remoteness, $F_{IT} = r_m = \tfrac{1}{2}(r_{m-1}' + r_m')$ and $r_0$, which as always is $\tfrac{1}{2}(1 + F_{IT}')$, can be written $\tfrac{1}{2} \times (1 + r_m')$. All of the others are of the form $r_x = \tfrac{1}{2}(r_{x-1}' + r_m')$. $F_{IT} = \tfrac{1}{2}F_{IT}' + \tfrac{1}{4}F_{IT}'' \text{ - - - } \left(\frac{1}{2^{m+1}}\right)F_{IT}^{(m+1)\,primes}$ or, after starting, $F_{IT}' - [1/(4N)]F_{IT}^{(m+2)\,primes}$. $P_{IT} = P_{IT}' - [1/(4N)]P^{(m+2)\,primes}$. Thus $k$ $(= 1 - \lambda)$ approaches 1 (4$N$) asymptotically (Wright, 1933a). A closer approximation is given by the equation $k = 1/[4N(1-k)]^{m+1}$ or approximately $k \approx [1/(4N)][1 + k]^{m+1} \approx [1/(4N)][1 + (m+1)k]$, $k = 1 - \lambda \approx 1/[4N - (m+1)]$ (communicated by Alan Robertson).

The average correlation between gametes is given by $F_{ST} = (1/N)[r_0 + r_1 + 2r_2 + 4r_3 \text{---} 2^{m-1}r_m]$.

All of the $r$'s can readily be expressed in terms of $F_{IT}$ $(= r_m)$ of the same and following generations ($r_{m-1}' = 2r_m - r_m'$ etc.) and similarly with the $P$'s. Thus in the case of 16-fold fourth cousins $P_{ST} = \tfrac{1}{128}[64P_{IT} + 32P_{IT}' + 15P_{IT}'' + 7P_{IT}''' + 3P_{IT}^{IV} + P_{IT}^{V}]$ and $P_{IS} = P_{IT}/P_{IS}$: limiting value, $128\lambda^5/[64\lambda^5 + 32\lambda^4 + 15\lambda^3 + 7\lambda^2 + 3\lambda + 1] = 1.04207$. $F_{IS} = 1 - P_{IS}$: limiting value, −0.04207. Values of $1 - \lambda$ are given in Table 4 and of $F_{IS}$ in Table 5 for various values of $N$.

In the case of circular mating of half-sibs, population number $N = 2^m$ (illustrated in Fig. 2 for $N = 8$), there are $[(N/2) + 1]$ different kinds of correlations. Letting $n = (N/2)$ and numbering

TABLE 3. *Frequencies of types of matings by generations.*

| | $r\ (= F_{IS})$ | $\sigma^2$ | 0 | 1 | 2 | 3 | 4 | 5 | 6 | 7 |
|---|---|---|---|---|---|---|---|---|---|---|
| $AA \times AA$, $aa \times aa$ | 0/0 | 0 | 0.1250 | 0.2812 | 0.4141 | 0.5254 | 0.6157 | 0.6891 | 0.7484 | 0.7965 |
| $AA \times aa$ | −1 | 0.2500 | 0.1250 | 0.0313 | 0.0391 | 0.0254 | 0.0220 | 0.0172 | 0.0141 | 0.0113 |
| $AA \times Aa$, $aa \times Aa$ | −0.3333 | 0.1875 | 0.5000 | 0.3750 | 0.3437 | 0.2734 | 0.2246 | 0.1812 | 0.1469 | 0.1187 |
| $Aa \times Aa$ | 0 | 0.2500 | 0.2500 | 0.3125 | 0.2031 | 0.1758 | 0.1377 | 0.1125 | 0.0906 | 0.0734 |
| | | | 1.0000 | 1.0000 | 1.0000 | 1.0000 | 1.0000 | 1.0000 | 1.0000 | 1.0000 |
| $\bar{F}_{IS} = \Sigma r\sigma^2 f / \Sigma \sigma^2 f$ | | | −0.3333 | −0.2000 | −0.2500 | −0.2308 | −0.2381 | −0.2353 | −0.2364 | −0.2360 |

these from $r_0$ to $r_n$, $F_{IT} = r_1$ and $r_0 = \frac{1}{2}(1 + r_1')$. The smallest of these correlations is $r_n = \frac{1}{2}(r_{n-1}' + r_n')$. All the rest are of the type $r_x = \frac{1}{4}[r_{x-1}' + 2r_x' + r_{x+1}']$. On making the substitution $r_x = 1 - P_x$, $P_0 = \frac{1}{2}P_1'$, and the others are as above with replacement of $r$'s by $P$'s. The values can be calculated generation after generation. The limiting ratio $\lambda = P/P'$ was obtained by Kimura and Crow by arranging equivalents of these $P$-equations in a matrix. The characteristic equation could then be expressed as a function of a known determinant (Wolstenholme's) on making the substitution $\lambda = \frac{1}{2}(1 + \cos\theta)$. This yielded the solution $\sin\theta = \cot n\theta$.

It is not, however, necessary to set up the matrix and characteristic equation, given this substitution. The equation $P_n = \frac{1}{2}(P_{n-1}' + P_n')$ can be written $P_{n-1}' = P_n'(2\lambda - 1) = P_n' \cos\theta$. The equation $P_{n-1} = \frac{1}{4}[P_{n-2}' + 2P_{n-1}' + P_n']$ can be written $P_{n-2}' = P_n'[2\cos^2\theta - 1] = P_n' \cos 2\theta$. The equation $P_{n-2} = \frac{1}{4}[P_{n-3}' + 2P_{n-2}' + P_{n-1}']$ can be written $2P_n' \cos 2\theta \cos\theta = P_{n-3}' + P_n' \cos(2\theta - \theta)$ giving $P_{n-3}' = P_n'(\cos 2\theta \cos\theta - \sin 2\theta \sin\theta) = P_n' \cos 3\theta$. In general $P_x = P_n \cos[(n - x)\theta]$, including $P_0 = P_n \cos n\theta$. Also, $P_0 = \frac{1}{2}P_1' = P_n \cos[(n - 1)\theta]/(1 + \cos\theta)$.

$\operatorname{Cos} n\theta(1 + \cos\theta) = \cos[(n - 1)\theta] = \cos n\theta \cos\theta + \sin n\theta \sin\theta$; $\sin\theta = \cos n\theta / \sin n\theta$ as given by Kimura and Crow.

With large $n$, $\theta$ becomes small, $\sin\theta$ and $\cos n\theta$ approach $\theta$, and $\sin n\theta$ approaches 1. Thus $\theta$ approaches $\pi/[2(n + 1)]$, $\cos\theta$ approaches $\sqrt{1 - \theta^2} \approx 1 - \theta^2/2$ and $1 - \lambda = \frac{1}{2}[1 - \cos\theta]$ approaches $[\pi/4(n + 1)]^2$ or $[\pi/(2N + 4)]^2$, again as given by Kimura and Crow.

All of the correlations appear twice around the circle except $r_0$ and $r_n$ since $r_{N-x} = r_x$. Thus $F_{ST} = (1/N)[r_0 + 2\sum^{n-1} r_x + r_n]$. $P_{IS}$ approaches a limiting value that can be obtained as the limiting value of $NP_1/[P_0 + 2\sum^{n-1} P_x + P_0]$ in terms of the above cosine formula, after determining the value of $\theta$. Values of $1 - \lambda$ and $F_{IS}$ are given in Tables 4 and 5 respectively.

In the case of circular mating of first cousins, $N = 2^m$, there are $(N/4) + 2$ different kinds of correlations between gametes. Letting $n = N/4$ in this case, and numbering the correlations according to remoteness, $F_{IT} = r_2$ and thus $r_0 = \frac{1}{2}(1 + r_2')$. The correlation between gametes of sibs is $r_1 = \frac{1}{2}(r_0' + r_2')$. The correlation between the most remote gametes is $r_{n+1} = \frac{1}{2}[r_n' + r_{n-1}']$. All of the others are of the type $r_x = \frac{1}{4}[r_{x-1}' + 2r_x' + r_{x+1}']$. Replacing $r_x$ by $1 - P_x$, $P_0 = \frac{1}{2}P_2'$ and in the others merely replace $r$'s by corresponding $P$'s. Kimura and Crow again solved for the limiting ratio $\lambda = P/P'$ by setting up the $P$-matrix and making the same substitution, $\lambda = \frac{1}{2}(1 + \cos\theta)$, as in the preceding case.

TABLE 4. *The limiting rates of decrease of heterozygosis ($1-\lambda$) under six closed systems of mating at different sizes of population (N). In the first two systems, the parents are drawn at random ($\sigma_k^2 = 2$); in the others just two are used from each parent of the preceding generation ($\sigma_k^2 = 0$).*

| $N$ | 1 ♂ ($N-1$) ♀'s $\sigma_k^2 = 2$ | $N/2$ ♂'s $N/2$ ♀'s $\sigma_k^2 = 2$ | Maximum avoidance ($\sigma_k^2 = 0$) | $N/2$ ♂'s $N/2$ ♀'s ($\sigma_k^2 = 0$) | Circle of first cousins ($\sigma_k^2 = 0$) | Circle of half sibs ($\sigma_k^2 = 0$) |
|---|---|---|---|---|---|---|
| 2 | 0.1910 | 0.1910 | 0.1910 | 0.1910 | — | — |
| 4 | 0.1396 | 0.1096 | 0.0804 | 0.0764 | 0.0804 | 0.0727 |
| 8 | 0.1228 | 0.0586 | 0.0362 | 0.0344 | 0.0347 | 0.0249 |
| 16 | 0.1159 | 0.0303 | 0.0170 | 0.0164 | 0.0142 | 0.0076 |
| 32 | 0.1120 | 0.0154 | 0.0082 | 0.0080 | 0.0053 | 0.0021 |
| Large $N$ | 0.1096 | $1/[2N+1]$ | $1/[4N-(m+1)]$ | $1/[4N-3]$ | $[\pi/(N+12)]^2$ | $[\pi/(2N+4)]^2$ |

By the same sort of reasoning as before, $P_n = P_{n-1}\cos\theta$ and in general $P_x = P_{n-1}\cos[(n+1-x)\theta]$ except for $P_0$ for which there are two equations to be solved.

$$P_0 = P_{n-1}\cos[(n-1)\theta]/(1+\cos\theta)$$
$$P_0 = 2\lambda P_1 - P_2 = P_{n-1}(1+\cos\theta)\cos n\theta - \cos[(n-1)\theta]$$

From these $\sin\theta[2+\cos\theta] = \cos n\theta/\sin n\theta$ as given by Kimura and Crow.

This requires that $\theta$ be approximately $\pi/2(n+3)$, leading to the approximate value, $1-\lambda = \theta^2/4 = [\pi/4(n+3)]^2 = [\pi/(N+12)]^2$ as given by Kimura and Crow.

The correlations $r_0$ and $r_1$ appear only once around the circle, $r_{n+1}$ appears twice while all of the others appear four times.

$$F_{ST} = \frac{1}{N}[r_0 + r_1 + 4\sum_2^n r_x + 2r_{n+1}]$$

$P_{IS}$ approaches a limiting value that can readily be found by the limiting cosine formulae for the $P$'s. Values of $1-\lambda$ and $F_{IS}$ are again given in Tables 4 and 5 respectively.

TABLE 5. *The limiting correlation between uniting gametes relative to the array of gametes of the same line ($F_{IS}$) in systems in which there is not random mating.*

| $N$ | Maximum avoidance | Circle of first cousins | Circle of half sibs |
|---|---|---|---|
| 2 | −0.2361 | — | — |
| 4 | −0.1824 | −0.1824 | −0.0784 |
| 8 | −0.1170 | −0.0733 | +0.2221 |
| 16 | −0.0711 | +0.1011 | +0.5160 |
| 32 | −0.0421 | +0.3293 | +0.7274 |

Fig. 7 compares the modes of approach toward fixation ($F_{IT}$) under various systems of mating in populations of eight individuals, starting in all cases from the first generation that shows an increase and thus not allowing for lags on starting from a random breeding stock. The most rapid of these is that with one male and seven females with replacement by a random male and seven random females. The next is similar except that there are four males and four females. In terms of $N_m$ males, $N_f$ females (Wright, 1931).

$$F = F' + \tfrac{1}{2}N_e(1-2F'+F''),$$
$$N_e = \frac{4N_mN_f}{N_m+N_f}$$
$$(1-\lambda) = \tfrac{1}{2}\left[1+\frac{1}{N_e}-\sqrt{1+\frac{1}{N_e^2}}\right]$$
$$\approx \frac{1}{2N_e+1}$$

If $N_m = 1$, $N_f = 7$, $(1-\lambda) = \tfrac{1}{14}[9-\sqrt{53}] = 0.1228$

If $N_m = 4$, $N_f = 4$, $(1-\lambda) = \tfrac{1}{16}[9-\sqrt{65}] = 0.0586$

The maximum avoidance system (quadruple second cousins) comes next ($1-\lambda = 0.03622$), but differs little from the following two: circular mating of first cousins ($1-\lambda = 0.03475$) and mating of four males and four females with replacement by selection of just two offspring per parent in all three. In the last case, the effective number in relation to the system of completely random replacement is given by Kimura and Crow (1963a) as $(4N-4)/(2+\sigma_k^2)$ which with $\sigma_k^2 = 0$ give $2N-2$

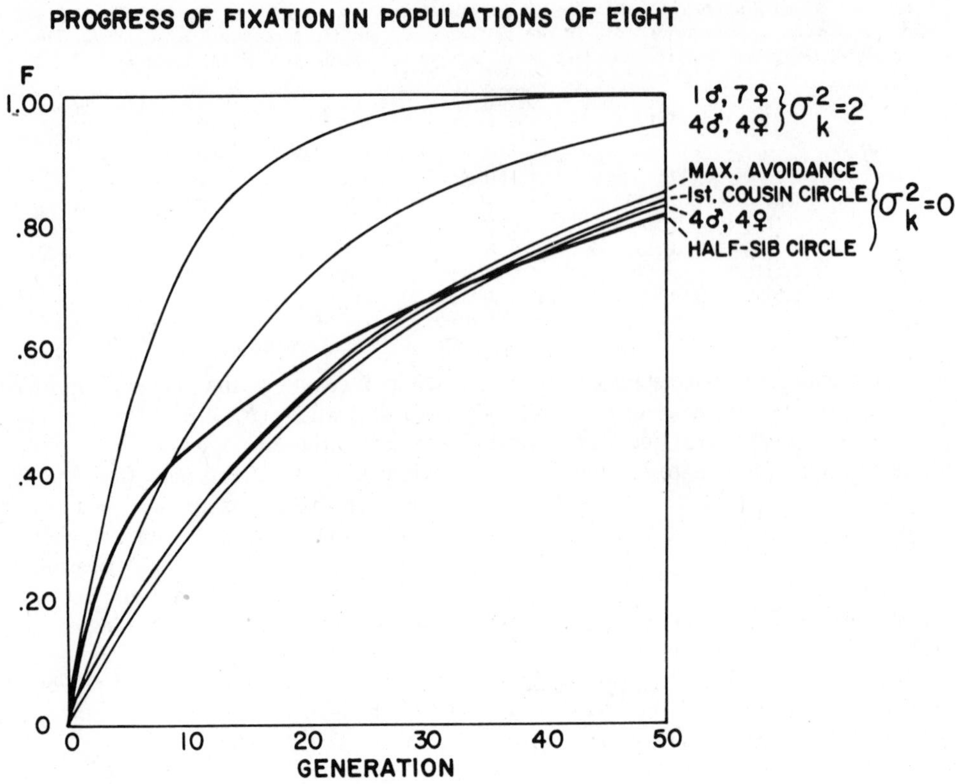


FIG. 7.

and thus 14 if $N = 8$. This replaces $N_e$ in the formula above by 14.

$$1 - \lambda = \tfrac{1}{28}[15 - \sqrt{197}] = 0.0344$$

This differs little from $1/(4N-3)$ which $1-\lambda$ approaches with large $N$.

Progress under circular mating of half-sibs follows a very different course from any of the others. There is almost as rapid an early rise as with one male, seven females followed by crossing of all of the other lines because of much the lowest ultimate rate (0.0249).

Table 4 compares the limiting rates of decrease of heterozygosis, $1-\lambda$ in various sizes of population. This is always slightly greater under maximum avoidance ($N > 2$) than under random mating with equal numbers of males and females and selection of just two offspring per parent in both. Circular mating of first cousins is the same system as maximum avoidance if $N = 4$ but has a slight lower rate if $N = 8$. It has a very slightly higher rate (if $N = 8$) than under the above random system. With $N = 16$ or more the rate becomes less than either of the preceding. The limiting rate under circular half-sib mating is always the lowest of those considered here and becomes very much so as $N$ is increased. The greater tendency under this system to maintain heterozygosis than under maximum avoidance was the paradox discussed by Kimura and Crow (1963b).

Table 5 gives the correlation, $F_{IS}$, between uniting gametes relative to their own lines in the three systems in which there is not random mating. $F_{IS}$ is, of course, always negative under maximum avoidance but it rapidly approaches zero as $N$ is increased.

Circular first cousin mating passes from

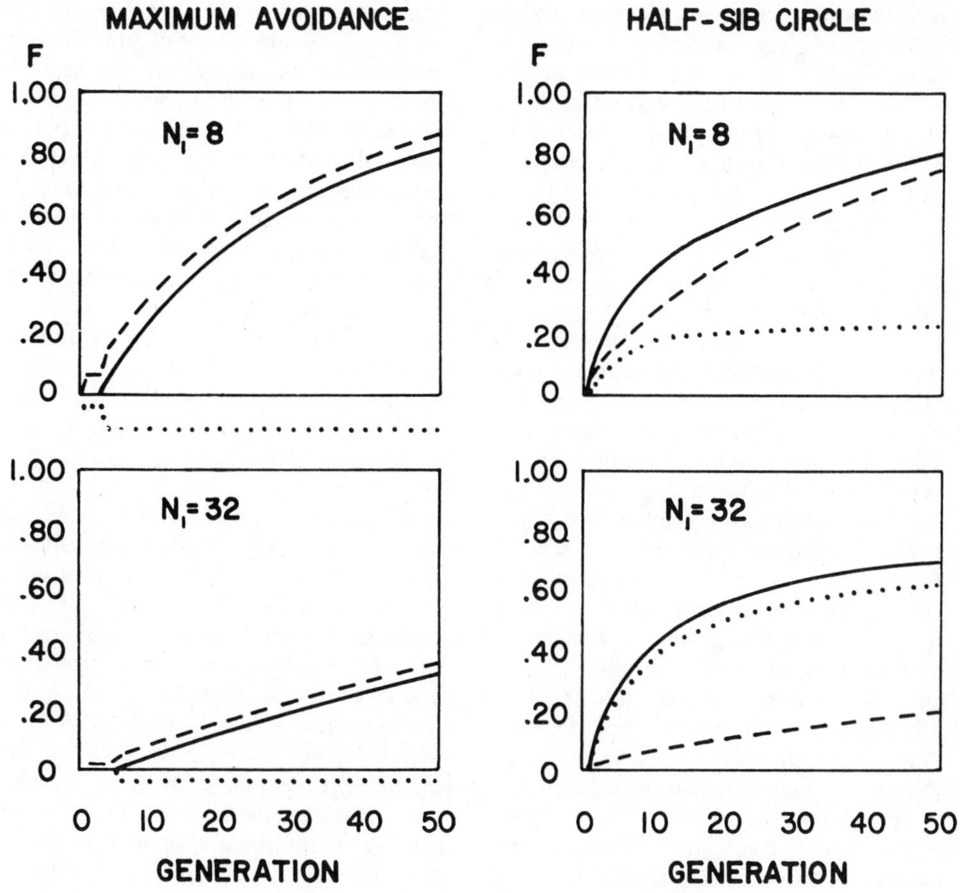


FIG. 8. Solid lines = $F_{IT}$; broken lines = $F_{ST}$; dotted lines = $F_{IS}$.

negative $F_{IS}$ if $N = 4$ (also maximum avoidance) to positive at some population size between 8 and 16. Even with circular half-sib mating $F_{IS}$ is negative if $N = 4$ but becomes positive between this and $N = 8$ and reaches the high value +0.73 if $N = 32$.

There is one property of groups of size $N$ that has been found to be the same whether mating of consanguineous individuals is avoided as much as possible, or the reverse, or is at random. It was shown in a previous paper (Wright, 1933b) that under maximum avoidance an array of lines that started from the mating $AB/ab$, where $A$ and $B$ show the proportion $c$ of recombination, ultimately arrive at correlation $r_\infty = (1-2c)/[1+2(2N-1)\,c]$ and a proportion of recombinant lines of $2X_\infty = \frac{1}{2}(1-r_\infty) = 2Nc/[1+2(2N-1)\,c]$ when fixation is complete. Thus under double first cousin mating ($N = 4$) $r_\infty = (1-2c)/(1+14c)$ and $2X_\infty = 8c/(1+14c)$, (the latter correcting an error in the paper as published). Kimura (1963), applying the probability concept arrived at exactly the same general formula for the proportion of recombinants in the case of circular half-sib mating as given above for maximum avoidance. I have confirmed this by path analysis in the special case $N = 4$.

In the case of random mating in groups of $N_m$ males and $N_f$ females, I obtained $r_\infty = (1-2c)/[1+2(N_e+1)\,c]$ and amount of recombination $(N_e+2)\,c/[1+2(N_e+1)\,c]$ where $N_e = 4N_mN_f/(N_m+N_f)$. This has been confirmed by Kimura by his method. The correlation approaches $(1-2c)/2Nc$ in large groups instead of $(1-2c)/4Nc$ as in the two preceding cases, but this is because the offspring generation was assumed to be drawn at random instead of just two from each parent. Using the formula of Kimura and Crow (1963a), $N_e = (4N-4)/(2+\sigma_k^2)$, $N_e$ above must be replaced by $2N_e - 2$ for the case in which $\sigma_k^2 = 0$ giving $r_\infty = (1-2c)/[1+2(2N_e-1)c]$ exactly as under the other two systems. It appears that the ultimate result of the race between recombination and fixation is exactly the same in these three cases. The differences in rates of fixation are ultimately exactly compensated for by differences in amounts of recombination.

Fig. 8 compares the ways in which $F_{IT}$, $F_{IS}$, and $F_{ST}$ change in the course of 50 generations after starting lines of 8 or 32 from random breeding stock under the two extreme systems. Under maximum avoidance, $F_{ST}$ (measuring the permanent inbreeding effect) is always higher than $F_{IT}$ (measuring the total inbreeding) because of negative $F_{IS}$. It is shown in Fig. 7 that in populations of eight, $F_{IT}$ is very slightly higher under maximum avoidance than under random mating. Thus $F_{ST}$ is still more in excess under the former since $F_{ST} = F_{IT}$ under random mating. These differences are, however, slight even in populations of eight and become less in larger populations as shown for $N = 32$.

The very different character of the progress of $F_{IT}$ under circular half-sib mating as compared with random mating in populations of eight was brought out in Fig. 7. It is evident from Fig. 8 that this is due to the rapid rise and approach to constancy of positive $F_{IS}$. Because of positive $F_{IS}$, $F_{ST}$ is always lower than $F_{IT}$.

With larger $N$, illustrated by the case of $N = 32$, there is an almost qualitative difference. $F_{IT}$ is almost wholly dominated at first by the large amount of current inbreeding measured by $F_{IS}$ while the more permanent differentiation of lines as wholes, measured by $F_{ST}$, builds up very slowly. The excess correlation between adjacent individuals tends to maintain different alleles in different regions around the circle. There is an approach to the situation in a population that is broken up into permanently distinct isogenic lines, which has been recognized as the best way to maintain the potentiality for maximum heterozygosis realizable by crossing, since G. H. Shull and D. F. Jones developed the theoretical basis for the enormously successful hybrid corn program. The basis for the exceedingly low rates of decrease of heterozygosis in large populations under circular systems of mating, demonstrated by Kimura and Crow, is obvious from Fig. 8.

Open systems of half-sib and first-cousin mating in infinite populations were studied as first approaches to the problem of isolation by distance. They were not very satisfactory models, however, and at the time the results were merely presented without discussion (Wright, 1921). When more satisfactory models were studied later (Wright, 1940, 1943a, 1946, 1951) it did not seem worthwhile to go back to these very artificial systems. It is, however, instructive in the present context to do so, first for the open system of half-sib mating and then for the corresponding closed circle of 32 individuals.

Fig. 9 shows the values of $F$ under half-sib mating in an infinite population, carried to generation 50 instead of merely to generation 15 as in 1921. This is compared with the progress of $F_{IT}$ under linear isolation by distance with neighborhoods of various sizes. Parent-offspring distances are assumed to be distributed normally, and hence variances of ancestral population rise linearly with number of generations but sizes of these populations rise only as the square roots of the latter. In the case of the half-sib system the parent-offspring distance is constant and equals 0.50 in terms of the distance between adjacent individuals. The grandparental distances have a 1 : 2 : 1 distribution about zero, the great grandparental a 1 : 3 : 3 : 1 distribu-

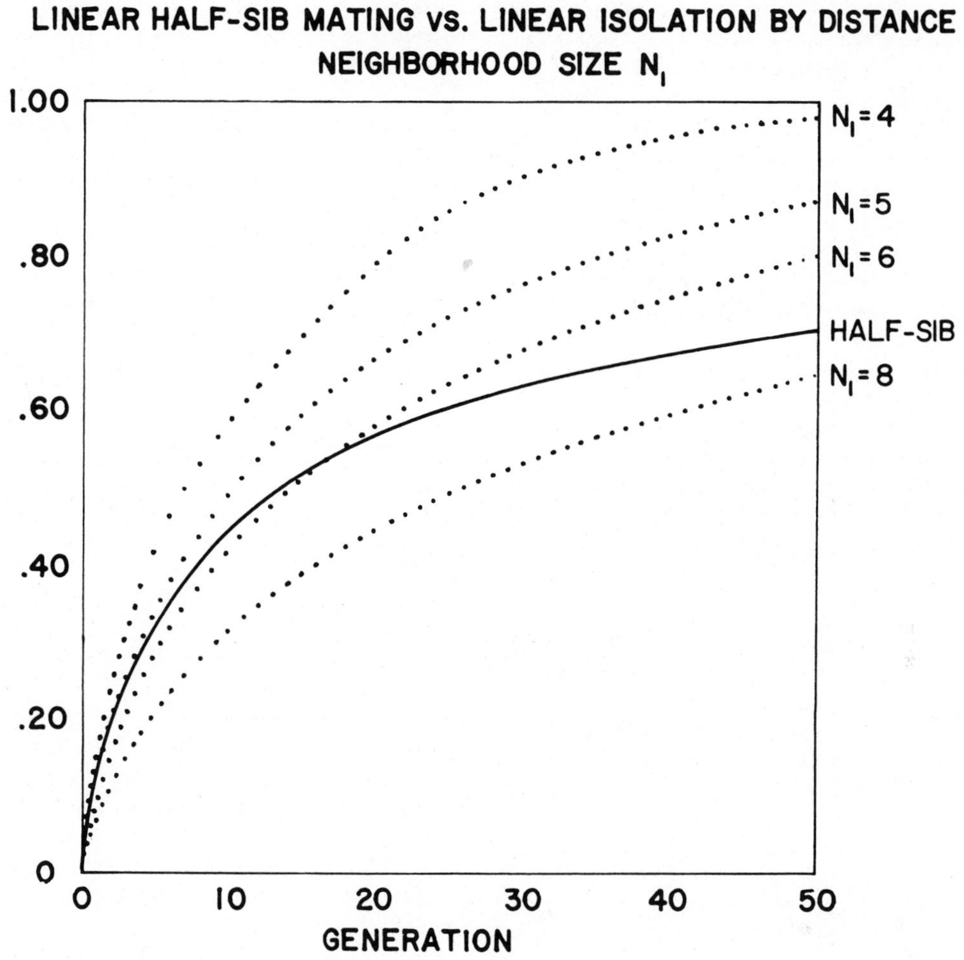


FIG. 9.

tion. the great great grandparental a 1 : 4 : 6 : 4 : 1 distribution. The variances rise linearly (0.25. 0.50. 0.75. 1.00. etc.) as in the model for linear isolation by distance. On the other hand the range also rises linearly (1. 2. 3. 4, etc.) as does the number of different ancestors (2, 3, 4, 5, etc.) in contrast with the sizes of ancestral populations in the model of linear isolation by distance that rise only as the square root of the number of ancestral generations. This difference is due to the fact that in the half-sib model the parental distribution is as platykurtic as possible and the ancestral distributions only gradually approach normality instead of all being normal.

At first sight it would appear that the size of "neighborhood" to be used in making a comparison should be two since each individual is produced by two adjacent individuals. This, however, does not allow for the differences between drawing at random from normally distributed parents and taking exactly two offspring from each parent. The effective number is thus expected to be considerably larger than two.

On calculating the values of $F$ by generations with $N = 4$, by the formula given earlier for the linear model, it turns out that these rise much too rapidly. With $N = 5$ the values start out rather similarly (actually very slightly too small) but soon are considerably in excess. With $N = 6$, the values are too small for 16 generations but are considerably too large at 50 generations. Even with $N = 8$, the course of $F$ under the linear model is clearly rising toward that under the half-sib model, though still below at 50 generations.

The difference in form of the course of $F$ under the two systems can be accounted for by the differences discussed above. In the half-sib system the effective number of ancestors in ancestral generation $X$ is somewhere between $\sqrt{X}$, expected if parental and all ancestral distributions were normal, and $X$. There is in consequence greater damping of progress than under the linear model in which this number varies with $\sqrt{X}$. It should be noted that in terms of the linear model the scale of generations of Fig. 9 should be transformed to one of square roots.

The increase in $F_{IT}$ in the circular half-sib system, $N = 32$, does not differ appreciably from that in the infinite population, up to 50 generations. It is of interest to make an analysis in terms of subdivisions of the total circle, similar to that discussed earlier for the models of isolation by distance. Since $S$ has been used for the whole circle in relation to an infinite array of such groups, it will be convenient to use $X$ for fractions of the circle, of length $X$ in the sense used earlier, but giving only half weight to the terminal individuals. The smallest such fraction ($X = 1$) includes two individuals with correlation between gametes of $r_0$ and $r_1$ in equal frequencies. The next to be considered ($X = 2$) includes four individuals with gametic correlations of $r_0$, $r_1$, and $r_2$ in frequencies 0.25, 0.50, and 0.25 respectively. The distance $X = 16$, includes all 32 individuals. Again it is the square roots of these "distances" that correspond to the distances on the linear model.

Since the total of all groups ($T$) is subdivided into groups of size $S = 32$ and these into fractions ($X$), the panmictic index for individuals relative to the total can be analyzed into three factors, $P_{IT} = P_{IX}P_{XS}P_{ST}$.

TABLE 6. *Analysis of system of circular half-sib mating, N = 32 with S pertaining to a single group of the 50th generation, T to the totality of all such groups and X (in $F_{XT}$, $F_{IX}$, and $F_{XS}$) fractions of groups terminating at "distance" x (in $r_x$) from individuals. Terminal distances have half weight.*

| Distance $X$ | $r_x$ | $F_{XT}$ | $F_{IX}$ | $F_{XS}$ |
|---|---|---|---|---|
| 0 | 0.8477 | 0.8477 | — | — |
| 1 = 31 | 0.6982 | 0.7730 | −0.3293 | 0.7169 |
| 2 = 30 | 0.5594 | 0.7009 | −0.0090 | 0.6270 |
| 3 = 29 | 0.4354 | 0.6331 | +0.1776 | 0.5424 |
| 4 = 28 | 0.3288 | 0.5703 | +0.2977 | 0.4641 |
| 5 = 27 | 0.2406 | 0.5132 | +0.3801 | 0.3929 |
| 6 = 26 | 0.1704 | 0.4619 | +0.4392 | 0.3289 |
| 7 = 25 | 0.1167 | 0.4164 | +0.4829 | 0.2722 |
| 8 = 24 | 0.0772 | 0.3765 | +0.5160 | 0.2224 |
| 9 = 23 | 0.0493 | 0.3417 | +0.5416 | 0.1790 |
| 10 = 22 | 0.0304 | 0.3115 | +0.5617 | 0.1414 |
| 11 = 21 | 0.0181 | 0.2854 | +0.5777 | 0.1088 |
| 12 = 20 | 0.0104 | 0.2628 | +0.5906 | 0.0806 |
| 13 = 19 | 0.0057 | 0.2432 | +0.6012 | 0.0562 |
| 14 = 18 | 0.0032 | 0.2261 | +0.6100 | 0.0349 |
| 15 = 17 | 0.0019 | 0.2113 | +0.6174 | 0.0163 |
| 16 | 0.0015 | 0.1981 | +0.6237 | 0 |

Table 6 shows the values of $r_X$ for all values of $X$ from 0 to 31 for the 50th generation based on calculations for all previous generations. $F_{IT}$ is 0.69822 and thus $P_{IT} = 0.30178$, $P_{ST}$ is 0.80187 and $P_{IS}$ $(= P_{IT}/P_{ST})$ is thus 0.37635. Thus $F_{ST}$ $(= 0.19813)$ is far below its limiting value while $F_{IS}$ $(= 0.62365)$ is not very far below its limiting value 0.72736 (see Table 5). The coefficients involving $X$ were found as follows for each $X$.

$$P_X = 1 - r_X$$

$$P_{XT} = \frac{1}{X}\left[\tfrac{1}{2}P_0 + \sum_{1}^{x-1} P_i + \tfrac{1}{2}P_X\right], \qquad F_{XT} = 1 - P_{XT}$$

$$P_{IX} = P_{IT}/P_{XT} = 0.30178\ P_{XT}, \qquad F_{IX} = 1 - P_{IX}$$

$$P_{XS} = P_{IS}/P_{IX} = 0.37635\ P_{IX}, \qquad F_{XS} = 1 - P_{XS}$$

The values of $F_{XT}$ (Table 6) are of little interest in themselves. The correlations $F_{IX}$ (also Table 6) between uniting gametes relative to the array of their own fractional group are, however, of interest. $F_{I1} = -0.329$ is the correlation between

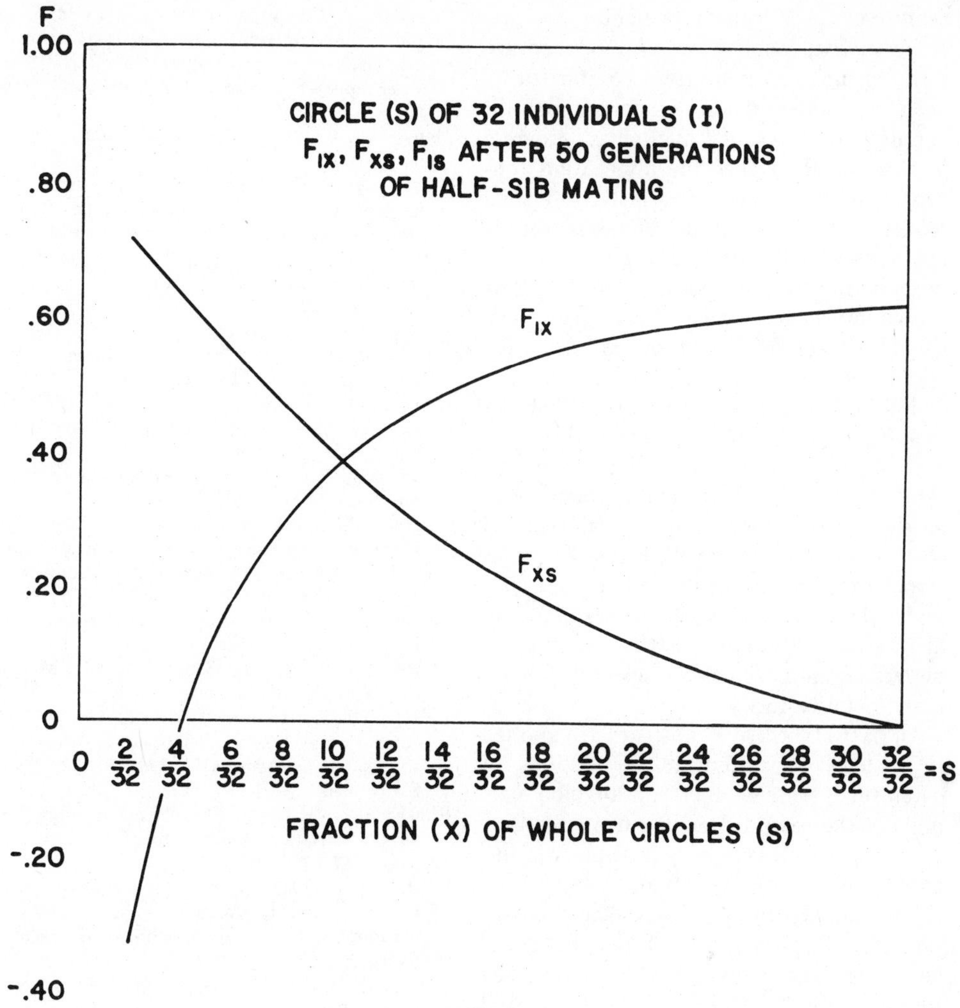


FIG. 10.

uniting gametes relative to the array of gametes from two neighboring individuals. $F_{I2} = -0.009$ is virtually zero, indicating that such groups of four individuals (effective number about six by Kimura and Crow's formula) are roughly equivalent to neighborhoods. Beyond this, $F_{IX}$ rises gradually as $X$ increases, toward the value $F_{IS} = 0.624$. $F_{XS}$ in the last column of Table 6 measures the amount of differentiation of the fractional groups relative to that of complete local fixation ($q_S(1-q_S)$). It falls from 0.717 for groups of two to 0.222 for groups of 16 (half of the circle) and to 0.016 for groups of 30. That for "neighborhoods" (groups of four) is approximately 0.627. These points are illustrated in Fig. 10, similarly to the case of isolation by distance in an area continuum in Fig. 4. This brings out in another way, the strong local differentiation within the circles that interferes with the progress of fixation of circles as wholes as measured by $F_{ST}$.

## SUMMARY

The general conclusion of part I is that the theoretical correlation between representatives of a locus in gametes, uniting or otherwise, relative to one or another array of such representatives ($F$-statistics), gives a broader basis for comparison of population structures, including progress in fixation, than does the alternative concept: the probability of identity of such representatives by origin. One reason is that correlations vary from −1 to +1 while probabilities vary only from 0 to +1. The probability concept gives, however, a very useful supplementary interpretation where applicable.

The relation of the basic set of $F$-statistics, $F_{IT}$, $F_{IS}$, $F_{ST}$, to variances within populations is discussed in part II and applications to diverse patterns of population structure are reviewed (the island model with or without selective differences, isolation by distance in continuous populations under balancing of local inbreeding and dispersion, uniformly distributed clusters under a similar balance, selective clines, breeds of livestock).

In part III, these $F$-statistics are applied to systems of mating in populations of given small size, in which consanguine mating is either avoided as much as possible, or pursued as much as is possible without any disruption of the group.

The apparently paradoxical result obtained by Kimura and Crow that heterozygosis declines more rapidly under the former than under the latter is discussed from the standpoint of these statistics.

These systems have been found to agree in one respect, the ultimate proportion of recombinant lines in the race between fixation and recombination among lines starting from double heterozygotes.

### Literature Cited


BARTLETT, M S., AND J. B. S. HALDANE. 1934. The theory of inbreeding in tetraploids. J. Genet., **29**: 175–180.

——. 1935. The theory of inbreeding with forced heterozygosis. J. Genet., **31**: 327–340.

BATEMAN, A. J. 1950. Is gene dispersion normal? Heredity, **4**: 353–363.

BERNSTEIN, F. 1930. Fortgesetzte Untersuchungen aus der Theorie der Blutgruppen. Z. indukt. Abstamm.—u Vererb. Lehre, **56**: 233–273.

COCKERHAM, C. CLARK. 1954. An extension of the concept for partitioning hereditary variance for analysis of covariances among relatives, when epistasis is present. Genetics, **39**: 859–882.

FISHER, R. A. 1949. The theory of inbreeding. Oliver and Boyd, Edinburgh, 120 pp.

KEMPTHORNE, O. 1954. The correlations between relatives in a random mating population. Proc. Royal Soc. London, **B143**: 103–113.

KIMURA, M. 1953. "Stepping stone" model of population. Ann. Rept. Nat. Inst. Genetics of Japan, **3**: 62–63.

——. 1963. A probability method for treating inbreeding systems, especially with linked genes. Biometrics, **19**: 1–17.

KIMURA, M., AND J. F. CROW. 1963a. The measurement of effective population number. Evolution, **17**: 279–288.

——. 1963b. On the maximum avoidance of inbreeding. Genet. Res. Cambridge, **4**: 399–415.

KIMURA, M., AND G. H. WEISS. 1964. The stepping stone model of population structure, and the decrease of genetic correlation with distance. Genetics, **49**: 561–576.

MALÉCOT, G. 1948. Les mathématiques de l'hérédité. Masson et Cie, Paris, 63 pp.

MCPHEE, H. C., AND S. WRIGHT. 1925. Mendelian analysis of the pure breeds of livestock III. The Shorthorns. J. Hered., **16**: 205–215.

——. 1926. Mendelian analysis of the pure breeds of livestock IV. The British Dairy Shorthorns. J. Hered., **17**: 397–401.

ROBERTSON, A. 1952. The effect of inbreeding on the variation due to recessive genes. Genetics, **37**: 189–207.

WAHLUND, S. 1928. Zuzammensetzung von Populationen und Korrelationserscheinungen vom Standpunkt der Vererbungslehre aus betrachtet. Hereditas, **11**: 65–106.

WRIGHT, S. 1921. Systems of mating. Genetics, **6**: 111–178.

——. 1922a. Coefficients of inbreeding and relationship. Amer. Nat., **56**: 330–338.

——. 1922b. The effects of inbreeding and crossbreeding on guinea pigs. Bull. U. S. Dept. Agric., **1121**: 1–59.

——. 1923a. Mendelian analysis of the pure breeds of livestock I. The measurement of inbreeding and relationship. J. Hered., **14**: 339–348.

——. 1923b. Mendelian analysis of the pure breeds of livestock II. The Duchess family of Shorthorns as bred by Thomas Bates. J. Hered., **14**: 405–422.

——. 1931. Evolution in Mendelian populations. Genetics, **16**: 97–159.

——. 1933a. Inbreeding and homozygosis. Proc. Nat. Acad. Sci., **19**: 411–420.

——. 1933b. Inbreeding and recombination. Proc. Nat. Acad. Sci., **19**: 420–433.

——. 1938a. The distribution of gene frequencies under irreversible mutation. Proc. Nat. Acad. Sci., **24**: 253–259.

——. 1938b. The distribution of gene frequencies of polyploids. Proc. Nat. Acad. Sci., **24**: 372–377.

——. 1938c. Size of population and breeding structure in relation to evolution. Science, **87**: 430–431.

——. 1939. Statistical genetics in relation to evolution. Exposés de biometrie et de statistique biologique. No. 802, 63 pp., Hermann and Cie, Editeurs.

——. 1940. Breeding structure of populations in relation to speciation. Amer. Nat., **74**: 232–248.

——. 1943a. Isolation by distance. Genetics, **28**: 114–138.

——. 1943b. An analysis of local variability of flower color in Linanthus Parryae. Genetics, **28**: 139–156.

——. 1946. Isolation by distance under diverse systems of mating. Genetics, **31**: 39–59.

——. 1951. The genetical structure of populations. Ann. Eugenics, **15**: 323–354.

——. 1952. The theoretical variance within and among subdivisions of a population that is in a steady state. Genetics, **37**: 312–321.

——. 1954. The interpretation of multivariate systems. Chapter 2, pp. 11–33, *in* Statistics and mathematics in biology, edited by O. Kempthorne, W. A. Bancroft, J. W. Gowen, and J. L. Lush.

——. 1963a. Discussion of "Systems of mating used in mammalian genetics" by E. L. Green and D. P. Doolittle. *In* Methodology in mammalian genetics, pp. 41–53, edited by W. J. Burdette. Holden-Day, Inc., San Francisco.

——. 1963b. Discussion of "Estimation of genetic variances" by C. C. Cockerham. *In* Statistical genetics and plant breeding, pp. 93–94, edited by W. D. Hanson and H. T. Robinson. Publ. No. 982, Nat. Acad. Sci. Nat. Res. Council, Washington, D. C.

WRIGHT, S., AND H. C. MCPHEE. 1925. An approximate method of calculating coefficients of inbreeding and relationship. J. Agric. Res., **31**: 377–383.

# Editor's Comments on Paper 19

19 **Cavalli-Sforza, Kimura, and Barrai:** The Probability of Consanguineous Marriages
*Genetics,* **54**, No. 1, Pt. 1, 37–60 (1966)

Early fundamentally important mathematical papers not in English are rare, but in studying the structure of population isolates one must begin with the German article by Wahlund (1928), which showed that subdivision of a larger population into smaller subpopulations can increase the expected amount of heterozygosity. A summary of the subsequent literature strongly emphasizing its relationship to other work of the present first author (Cavalli-Sforza) and to the techniques of Sewall Wright is by Yasuda (1968).

Principal papers to see in this history are those by Wright (1943, 1946), which used continuous process models to study the diffusion of genes in the reproductive neighborhood of individuals. However, these techniques assumed a too simplistic model, so that other ideas have had to be introduced. Kimura and Weiss (1964) provided a large improvement in the variety of techniques available by looking at a rectangular plane grid (stepping-stone model in two dimensions), while, for example, Maruyama (1970) studied anew the effects of a circular habitat (i.e., colonies that exchange migrants in a circle and of the geometrically similar linear habitat).

The importance of such work is found by noting that the proportion of individuals of particular genetic makeup can be strongly affected by both population size and mating structure. Critical papers are by Kimura and Crow (1964), Kimura and Maruyama (1971), Maruyama (1971), and Li and Nei (1972). Of these four papers the second and third emphasize the effect of geographic structure of a population on the occurrence of heterozygosity. (Maruyama, in particular, concluded that the geographic structure is not important in the total number of heterozygotes that occur on the average as a result of a particular mutation.) The first and fourth papers are more concerned with the potential medical (genetic load, and the number of affected individuals) effects of population size.

However, the principal reason for selection of this paper, mentioned in the introduction to this volume, is that it suggests that kin-based migration problems are treatable by matrix methods. The papers mentioned above (except that by Yasuda) in general do not consider the kin-based structure at all, or assume "random" mating within each isolate. Unfortunately, the immediate structure of a real population is quite important, so that in addition to the present paper, see also Hiorns et al. (1969) and Hajnal (1963) for further study of exchange between local groups and of the effect of structures within groups, respectively.

# 19

Reprinted from *Genetics*, **54**, No. 1, Pt. 1, 37–60 (1966)

# THE PROBABILITY OF CONSANGUINEOUS MARRIAGES

L. L. CAVALLI-SFORZA, M. KIMURA[1], AND I. BARRAI

*International Laboratory of Genetics and Biophysics, Pavia Section, Istituto di Genetica, University of Pavia, Pavia, Italy*

Received December 22, 1965

THE frequency with which a husband is related to his wife, which we call the probability of consanguineous marriage, is determined by a number of factors, among which the following seem to be the most important ones: (1) the abundance of relatives, which depends on the type of relationship and on population growth. (2) The availability of consanguineous individuals in the "mating range", (migration causes a dispersal of relatives whose effect increases as the relationship becomes more remote: the more migration, the less consanguineous marriage). (3) The availability of the consanguineous individuals in the right age groups (age effect). (4) Assortive mating for socio-economic conditions and physical traits may have to be considered because of the similarity between relatives. (5) Traditions for or against some types of consanguineous marriages may also be a factor of importance. (6) There may exist other factors of social or economic nature.

In an earlier paper (BARRAI, CAVALLI-SFORZA and MORONI 1962), we showed the influence of factors 2 and 3. Another effect was also found, belonging to group 6. In the present paper we will concentrate on evaluating the effect of the first four factors, with a view to estimating the probability of consanguineous marriage in a population for which some necessary demographic parameters are available.

The necessary parameters are essentially those specifying the distributions of the distance between birth places, as well as of the age differences: between sibs, between father and offspring, between mother and offspring, between husband and wife. As estimates of these parameters are not usually available, a sample survey was carried out in an area (in the Parma province) for which information on consanguineous marriages was already at hand (MAINARDI, CAVALLI-SFORZA and BARRAI 1962). Data from the sample survey will be used in this paper.

*The number of relatives:* We will consider the simplest, and commonest type of relationship represented by individuals who are related via two sibs, as in Figure 1. In that example, the chains of descent via two sibs lead to the two relatives A and B, who are second cousins once removed. We will call $i$ the number of ancestors between the common ancestors and the male relative A, $j$ the number between the common ancestors and the female relative, B. Thus, in Figure 1, $i = 3$, $j = 2$, and $n = i + j$ is the number of *intermediate* ancestors. Of these, $n_0$

[1] On leave from the National Institute of Genetics, Mishima, Sizuoka-Ken, Japan. Present address: The same institute. This paper also constitutes Contribution No. 606 from the National Institute of Genetics. Aided in part by a Grant-in-Aid for Fundamental Scientific Research from the Ministry of Education in Japan.

Genetics **54**: 37–60 July 1966.

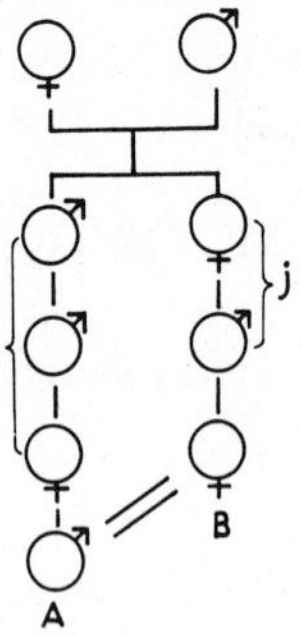


FIGURE 1.—Example of consanguineous mating: for definition of symbols, see text.

will be females and $n_1$ will be males, with $n_0 + n_1 = n$. In Figure 1, $n_0 = 2, n_1 = 3$.

If $s$ is the expected number of sibs per individual, $p$ the expected number of progeny per individual, and if there is no correlation in fertility, there are (ignoring sex) $2^2p^3s$ relatives of type A per B individual in the example illustrated, because B has $2^2$ grandparents each of which has $s$ sibs, each of which has $p^3$ great grandchildren. In general, there are $2^ip^js$ relatives of the wife in a consanguineous marriage which have the same consanguinity degree with her as her husband, and $2^jp^is$ relatives of the husband.

In a closed, stationary population in which everybody marries, the expected number of married progeny per couple is $p = 2$. In a stable population $p = 2a$, where $a$ is the factor of increase per generation. The average number of sibs $s$ depends on the distribution of progeny size. In fact, a family with progeny size $p$ ascertained through the progeny is counted $p$ times and each individual in the progeny has $(p-1)$ sibs. Then if $\phi\,(p)$ is the frequency of progeny size $p$, $p\,\phi\,(p)/\sum_{p=1}^{\infty} p\,\phi\,(p) \equiv \psi\,(p)$ is the frequency of $(p-1)$ sibs (FROTA-PESSOA 1957). The mean of the distribution given by $\psi$ is equal to $\sum_{p=1}^{\infty}(p-1)\,\psi\,(p) = (V + \bar{p}^2 - \bar{p})/\bar{p}$ where $\bar{p}$ and $V$ are the mean and the variance of the number of progeny. When this is distributed as in Poisson, the mean number of sibs is equal to the mean number of progeny $\bar{p}$. For distributions with variances higher than the mean, the mean number of sibs is higher than $\bar{p}$ (MAINARDI *et al.* 1962).

In many populations, however, it is so closely $s = p = 2$ that, if the variance and correlation for fertility can be ignored, the expected number of relatives $n_c$ is $2^n$ for even cousins, namely, for cousins having $i = j$, and $2^{n+1}$ for cousins once removed $(i = j \pm 1)$. But these are also the numbers of pedigrees which can be distinguished on the basis of the sex of intermediate ancestors (BARRAI *et al.* 1962) and therefore each individual has one expected relative for each pedigree type ignoring the sex of the relative. Ignoring the type of pedigree, each individual has four first cousins, four relatives with $F = 1/8$ (uncles, aunts, nieces, and nephews), 16 first cousins once removed, 16 second cousins, 64 second cousins once removed, and 64 third cousins. These classes of relatives will be here called *degrees* of relationship; while the relationship specified by a type of pedigree as determined by the arrangement of males and females among common ancestors will be called *type of relationship.*

*Migration effect:* 1. *The discontinuous case.* The study of migration demands a choice of the model of the geographic distribution of population. A first choice is that between continuous and discontinuous models. If we prefer a discontinuous distribution, for computing the probability of consanguineous marriage, we must have two matrices, $X$ and $M$, both of order $k$, where $k$ is the number of groups of people (villages, tribes, castes, etc) into which the population is clustered. Elements $x_{ij}$ of matrix $X$ specify the probability that for a given type of relationship a relative of an individual born in village i is born in village j. Row elements must add to unity in each row. In matrix $M$, elements $m_{ij}$ specify the frequencies that of all marriages in the area, one member is born in village i and the other is born in village j. This matrix is symmetric because we ignore sex, and the sum of its elements is unity. The probability of consanguineous marriages corresponding to a given type of relationship, will then be

$$P = n_c \sum \frac{x_{ij}\, m_{ij}}{n_i} \tag{1}$$

where the sum is extended to all combination of $i$ and $j$, $n_i$ is the size of group $i$ and $n_c$ is the number of relatives of that type.

Matrix $X$ is not easy to obtain from field data. It can be computed however, as a product of other matrices $S$, $A_0$, $A_1$, defined below, each representing one step in the path connecting one consanguineous individual to his consanguineous mate in the pedigree, on the assumption that migration in successive steps is independent and therefore can be treated as a Markov process. In fact, a correlation between migration steps may exist, especially because of stratification in socio-economic conditions, not accounted for by the grouping method employed, when this is for instance, a purely geographic one. We shall give the treatment for equal group size $n$, in a stationary population at equilibrium for migration, and discuss a possible generalization later.

The computation of matrix $X$ from matrices $S$, $A_0$, $A_1$ will be shown for simplicity using an example, namely, the pedigree of second cousins once removed shown in Figure 1. $A_1$ is a matrix of the transition probabilities for father-offspring migration. Its element $a_{ij}$ is the probability that a child of a father born in village i is born in village j, with row elements adding to unity in each row. Because of migration equilibrium and equal group size, also columns will add to 1. The $A_0$ matrix is the same transition matrix for mother-offspring. Matrix $S$ is a transition matrix for sib migration, whose element $s_{ij}$ is the probability that the sib of an individual born in village i is born in village j. Matrix $S$ is symmetric with rows and columns adding to 1.

Then, the matrix $X$ for the example of Figure 1 can be equated to the product

$$A'_0\, A'_1\, A'_1\, S\, A_0\, A_1 \tag{2}$$

where $A_0'$ is the transpose of $A_0$, etc. The above matrix product is obtained by following the path from A to B in Figure 1, via intermediate ancestors. Following the reverse path from B to A one obtains

$$A'_1 \, A'_0 \, S \, A_1 \, A_1 \, A_0 \tag{3}$$

which, by a well known theorem of matrix algebra, is shown to be the transpose of product (2). Note that $S$ is symmetric.

The probability of consanguineous marriage $P_c$ will then become

$$\frac{1}{N} \sum x_{ij} \, m_{ij} \tag{1'}$$

where $N$ is the size of the individual group, and $n_c$ is put equal to 1.

Since the $m_{ij}$'s are the elements of a symmetric matrix, it is immaterial if we use expression (2) or (3) for computing matrix $X$ whose elements appear in (1′).

The above treatment is based on the assumption of equal group size. On the other hand, the group size may be different in the actual case, but we can still apply the above theory by considering actual groups as collections of subgroups with approximately equal size.

Although matrices $A_0$, $A_1$, $S$ are not difficult to obtain, they are not usually available. It may therefore be convenient sometime, as a first rough approximation, to use the method suggested by Barrai *et al.* (1962) which is essentially the same as that followed by Hajnal (1963), of ignoring the possibility of marriage in the group outside the one in which the individual is born, as this probability is often small, and use an average of the probabilities that an individual will have a child born in the same group, as the one in which he was born. We shall see later to what formulas this method leads, and their shortcomings.

The probability for a given degree of relationship should be obtained by adding up the probabilities for the various types belonging to this degree, as these differ one from the other. It is only if there is no difference between male and female migration rates that the expected frequency of consanguinity types that form them is independent of the proportion of the sexes among intermediate ancestors.

2. *The continuous case.* The use of a discontinuous model may be unsatisfactory if the population distribution is nearer to the continuous one. Also, much detailed knowledge is necessary if we want to use the discontinuous model. An approximation by a continuous model may therefore be useful, as it requires somewhat less detailed statistical information.

In analogy to the study of isolation by distance, put forward by Sewall Wright (1963), we might consider two types of continuous population distributions, a one-dimensional and a two-dimensional type. It may be noted, however, that the first type is far less frequently encountered, at least in a pure form, in human populations, and represents in any case a simpler model than the two-dimensional type. We have therefore concentrated our attention on a model of two-dimensional isotropic migration. The one-dimensional case could be obtained fairly easily as a simplified treatment, following the lines that we will give here for the two dimensional model. For isotropic migration in a two-dimensional habitat given by coordinates $x$, $y$, the density function that the marriage of an individual born at the origin takes place with a mate born in $(x, y)$ will be

$$f(x, y) = \frac{f(r)}{2\pi r} \tag{4}$$

where $r = (x^2 + y^2)^{1/2}$, $f(r)$ is the probability density that individual A marries an individual born at distance $r$ from A's birthplace, and $-\infty < x, y < +\infty$.

Suppose that an individual, say a male A, is at the origin, and consider a small area $dS$ $(= dx \cdot dy)$ around a point $(x, y)$. The number of females in that area is $(D/2)dS$, where $D$ is the population density, and $D/2$ the population density of females.

If we know a function $M_c(x, y)$ giving the probability density that one relative of A with a given type of relationship is born at point $(x, y)$ and if there are expected to exist altogether $n_c$ relatives of that type, the expected number of A's female relatives living in area $dS$ will be

$$\frac{n_c}{2} M_c(x, y) dS \tag{5}$$

The probability of marriage between two individuals with given type of relationship will be

$$P = \iint_{-\infty}^{+\infty} \frac{n_c M_c(x, y)}{D} \frac{f(r)}{2\pi r} dx\, dy \tag{6}$$

Noting that $dxdy = rdrd\theta$ and integrating over all values of $\theta$ between 0 and $2\pi$ we have

$$P = \frac{n_c}{D} \int_0^\infty M_c(x, y) f(r) dr \tag{7}$$

The function $M_c(x, y)$ measures the dispersal of relatives and is the convolution of the following distributions: (1) The probability distribution of the distance between the birth places of the sibs which start the chains of relationship; (2) $n$ probability distributions, each representing one generation in the chain of relationship starting with the two sibs, where $n = i + j$ is the number of intermediate ancestors, i.e.; the ancestors between the common ancestors and the consanguineous mates.

Since it is important to distinguish male and female migration, the $n$ one-generation steps of migration will have to be subdivided into $n_0$ female, and $n_1$ male generations $(n_0 + n_1 = n)$.

In taking $M_c$ as the convolution of $n + 1$ distributions it is assumed that there is no correlation between migration and successive generations and, in the absence of information, this might be taken as a first approximation.

In order to give $M_c(x, y)$ one must know the elementary distributions of which $M_c$ is made. The contribution of sib-sib migration was neglected, because it is very modest with respect to the other components. Migration distributions of interest to genetics have been recently analyzed for European populations

(SUTTER and TRAN NIGOC TOAN 1957; LUU-MAU-THANH and J. SUTTER 1963; CAVALLI-SFORZA 1958, 1963) and are extremely skew. Perhaps the best fit was obtained with gamma distributions which, when fitted with respect to $r$, had exponents close to $-1$, and could not therefore permit convolutions to be obtained over two dimensions.

Accordingly, it was tried to fit distribution functions that would lend themselves more easily to obtain the $M_c$ function. Two such functions are the exponential distribution, and a two-dimensional normal distribution (with equal variances for $x$ and $y$). When expressed with respect to $r$, such distributions are given in (8) and (9):

"exponential" $$m_E(r) = k\,e^{-kr} \quad (8)$$

"normal" $$m_N(r) = \frac{r}{V} e^{-r^2/2V} \quad (9)$$

Neither of these functions seems to fit adequately the observed data. It is not unreasonable, however (considering the variety of means of transportation employed), to use sums of two or more of such functions. Fitting these distributions to the Parma data by numerical maximum likelihood, it was found that the sum of two exponentials

$$m_E(r) = phe^{-hr} + (1-p)ke^{-kr} \quad (10)$$

or the sum of three normals:

$$m_N(r) = \frac{pr}{V_S} e^{-r^2/2V_S} + \frac{qr}{V_M} e^{-r^2/2V_M} + \frac{(1-p-q)r}{V_L} e^{-r^2/2V_L} \quad (11)$$

fit the data reasonably well (Table 1).

TABLE 1

*Distributions of birth distances for father-offspring (F-O), mother-offspring (M-O), husband-wife (H-W) pairs. Observed frequencies are given from a sample of fam:lies living in 1958 in the Parma province*

| | F-O | | | M-O | | | H-W | | |
|---|---|---|---|---|---|---|---|---|---|
| Distance (km) | obs | trinorm* | biexp* | obs | trinorm | biexp | obs | trinorm | biexp |
| 0–1.56 | 340 | 339.7 | 338.4 | 293 | 289.2 | 291.7 | 133 | 132.1 | 132.2 |
| 1.56–4.06 | 11 | 11.0 | 11.1 | 18 | 17.8 | 18.1 | 10 | 9.92 | 10.0 |
| 4.06–7.81 | 8 | 7.3 | 5.7 | 21 | 18.5 | 13.9 | 6 | 7.0 | 8.4 |
| 7.81–12.81 | 10 | 9.8 | 6.9 | 16 | 24.0 | 15.4 | 12 | 11.5 | 9.7 |
| 12.81–19.06 | 6 | 7.2 | 7.5 | 16 | 15.5 | 15.1 | 14 | 11.7 | 10.0 |
| 19.06–26.56 | 4 | 4.4 | 7.5 | 7 | 6.4 | 13.5 | 9 | 8.7 | 9.7 |
| 26.56–35.31 | 4 | 4.7 | 7.2 | 5 | 5.0 | 11.1 | 2 | 7.4 | 8.7 |
| 35.31–45.31 | 7 | 5.7 | 6.5 | 6 | 5.9 | 8.5 | 10 | 7.7 | 7.3 |
| 45.31–∞ | 24 | 24.1 | 23.1 | 21 | 20.7 | 15.7 | 19 | 19.0 | 19.0 |
| Total | 414 | | | 403 | | | 215 | | |
| $\chi^2_{[6]}$ | | | 5.6 | | | 12.63 | | | 8.93 |
| $\chi^2_{[4]}$ | | 0.70 | | | 3.15 | | | 5.28 | |

* Theoretical distributions are given by equations (10) for the biexponential and (11) for the trinormal, and the method of fitting is given in text. Parameters of the theoretical distributions are given in Table 2.

3. *"Sum of Exponentials" for the migration distributions.* Using distribution (10) for the elementary migration step, integral (6), giving the probability of consanguineous marriage under consideration of the sole migration effect, requires numerical integration. If we use the cartesian coordinate system $(x,y)$ for isotropic migration in two dimensions, the density function for migration in one generation can be expressed as

$$m(x,y) = \frac{m_1(r)}{2\pi r} \quad \text{where } r = \sqrt{(x^2+y^2)} \tag{12}$$

If $C(u,v)$ is the characteristic function of $m(x,y)$ such that

$$C(u,v) = \iint_{-\infty}^{+\infty} e^{iux+ivy}\, m(x,y)\,dx\,dy\;, \quad (i=\sqrt{-1}), \tag{13}$$

we obtain, in terms of $r$,

$$C(u,v) = \int_0^\infty m_1(r) J_0(sr)\,dr \tag{14}$$

where $J_0$ is the Bessel function and

$$s = \sqrt{(u^2+v^2)} \tag{15}$$

In particular, when $m_1(r)$ is given by $m_E(r)$ of (10), the characteristic function reduces to

$$C(s) = \frac{ph}{(h^2+s^2)^{1/2}} + \frac{(1-p)k}{(k^2+s^2)^{1/2}} \tag{16}$$

If we consider the distribution of migration distances after $n_0$ female generations and $n_1$ male generations, with parameters $p_0$, $h_0$, $k_0$ and $p_1$, $h_1$, $k_1$ respectively, if the migrations in different generations are independent, the density function will be

$$M(x,y) = \frac{1}{(2\pi)^2} \iint_{-\infty}^{+\infty} e^{-i(xu+yv)}\, C_1^{n_1}(s) C_0^{n_0}(s)\,du\,dv$$

$$= \frac{1}{2\pi} \int_0^\infty C_1^{n_1}(s) C_0^{n_0}(s)\, s J_0(rs)\,ds \tag{17}$$

where $C_0(s)$ and $C_1(s)$ are the characteristic functions of the distributions of the female and male migrations over one generation. Figure 2 illustrates some of the results of convolution based on (17).

Equation (17) gives us the $M_c$ function desired for equation (6). We now have

$$\mathrm{P} = \frac{n_c}{2\pi D} \iint_0^\infty C_1^{n_1}(s) C_0^{n_0}(s)\, s J_0(rs) f(r)\,dr\,ds \tag{18}$$

If $f(r)$ is also given (see Table 1) as a sum of two exponentials with parameters $p_m$, $h_m$, $k_m$, then, noting that

$$\int_0^\infty e^{-h_m r} J_0(sr)\,dr = \frac{1}{(h_m^2+s^2)^{1/2}} \tag{19}$$

one obtains

$$P = \frac{n_c}{2\pi D} \left\{ \frac{p_m}{2} \int_0^\infty D_1^{n_1}(t) D_0^{n_0}(t) \frac{h_m dt}{(h_m^2+t)^{1/2}} + \frac{1-p_m}{2} \int_0^\infty D_1^{n_1}(t) D_0^{n_0}(t) \frac{k_m dt}{(k_m^2+t)^{1/2}} \right\} \tag{20}$$

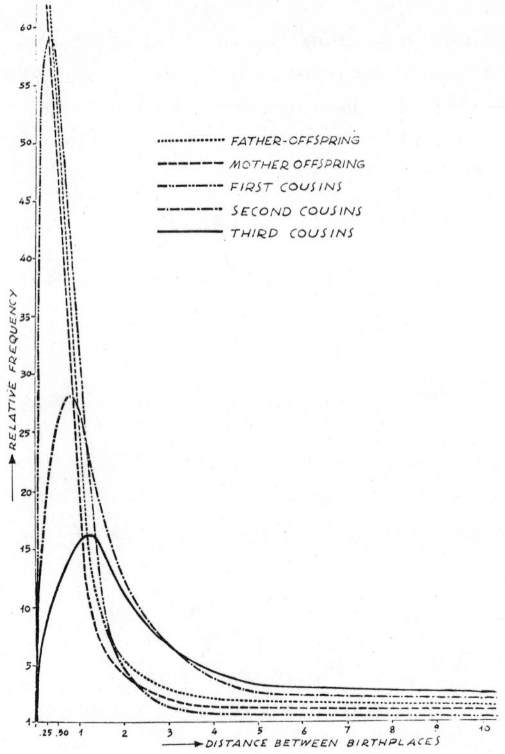


FIGURE 2.—Distributions based on equation (17).

in which

$$D_0(t) = \frac{p_0 h_0}{(h_0^2 + t)^{1/2}} + \frac{(1-p_0) k_0}{(k_0^2 + t)^{1/2}}$$

$$D_1(t) = \frac{p_1 h_1}{(h_1^2 + t)^{1/2}} + \frac{(1-p_1) k_1}{(k_1^2 + t)^{1/2}} \quad (21)$$

Calling

$$I_m = P\,D \quad (22)$$

and using for $p$, $h$, and $k$ parameters the estimated set of values for male, female, and mating range distributions, expressions (21) can be used for the numerical evaluation of $I_m$, and hence, $P$.

4. *Sum of normals for fitting migration distributions:* If the distribution of migration distance is expressed by the sum of normal distributions, our calculation becomes much easier. This is mainly because the sum of any number of independent random variables, each of which is distributed normally, is again distributed normally with the mean and variance given respectively by the sums of means and variances of component normal distributions. Furthermore, if the distribution of migration in two dimensional cartesian coordinate $(x,y)$ is given by normal distribution,

$$\frac{1}{2\pi V_i} e^{-(x^2+y^2)/2V_i} \tag{23}$$

and if the distribution of distance between mates is also given by another normal distribution,

$$\frac{1}{2\pi V_m} e^{-(x^2+y^2)/2V_m} \tag{24}$$

the integral of the product of the above two distributions can be expressed in the simple form:

$$I = \frac{1}{4\pi^2 V_i V_m} \iint_{-\infty}^{+\infty} e^{-(x^2+y^2)(V_i^{-1}+V_m^{-1})/2} dxdy = \frac{1}{2\pi(V_i+V_m)} \tag{25}$$

In what follows, we will assume that the individual migration is expressed by the sum of three normal distributions, corresponding to short range ($S$), medium range ($M$) and long range ($L$) components of migration. These distributions all have mean 0, but their variances are $V_S$, $V_M$ and $V_L$ respectively for short range, medium range and long range components. Thus the male migration in one generation may be expressed *symbolically*

$$p_1 S_1 + q_1 M_1 + (1-p_1-q_1) L_1 \tag{26}$$

where $p$ and $q$ are constants and $S_1$, $M_1$, and $L_1$ represent normal migration with variances, $V_{S1}$, $V_{M1}$ and $V_{L1}$ respectively. The corresponding expression for a female migration is

$$p_0 S_0 + q_0 M_0 + (1-p_0-q_0) L_0 \tag{27}$$

where $S_0$, $M_0$ and $L_0$ have variances $V_{S0}$, $V_{M0}$ and $V_{L0}$. We will also assume that the distribution of distance between the birth places of mates in two-dimensional cartesian coordinate is expressed by the sum of three normal distributions:

$$f(x,y) = \frac{p_m}{2\pi V_{Sm}} e^{-(x^2+y^2)/2V_{Sm}} + \frac{q_m}{2\pi V_{Mm}} e^{-(x^2+y^2)/2V_{Mm}} + \frac{(1-p_m-q_m)}{2\pi V_{Lm}} e^{-(x^2+y^2)/2V_{Lm}} \tag{28}$$

This distribution will be expressed symbolically by

$$p_m S_m + q_m M_m + (1-p_m-q_m) L_m \tag{29}$$

With these expressions, the probability of marriage between two mates connected by $n_1$ generations of male migrations and $n_0$ generations of female migrations in their ancestors is

$$\mathrm{P} = \frac{n_c}{D} \iint_{-\infty}^{\infty} M_c(x,y) f(x,y) dxdy \tag{30}$$

in which $M_c(x,y)$ can be expressed symbolically by

$$M_c = [p_1S_1+q_1M_1+(1-p_1-q_1)L_1]^{n_1}[p_0S_0+q_0M_0+(1-p_0-q_0)L_0]^{n_0} \tag{31}$$

which is a sum of normal bivariate distributions:

$$M_c(x,y) = \sum_j c_j\, e^{-(x^2+y^2)/2V_j}/2\pi V_j\ , \tag{32}$$

each term of this sum corresponding to a term of the expansion of (31). Each such term contains the symbolical product

$$S_1^a M_1^b L_1^{n_1-a-b} S_0^c M_0^d L_0^{n_0-c-d} \equiv T\ ; \tag{33}$$

which corresponds to a normal distribution whose variance is given by

$$V_j = aV_{S1} + bV_{M1} + (n_1 - a - b)V_{L1} + cV_{S0} + dV_{M0} + (n_0 - c - d)V_{L0} \quad (34)$$

while $c_j$ in (32) will be given by the numerical coefficient accompanying the symbolical product (33) in the expansion. The integration in (30) may then be carried out using formula (25). Thus, the integral $I_m$ in the right side of (30) may be expressed symbolically as follows:

$$I_m = M_c \cdot [p_m S_m + q_m M_m + (1 - p_m - q_m) L_m] \quad (35)$$

and will give rise to a quantity

$$I_m = \sum_j b_j/(2\pi W_j) \quad (36)$$

which can be computed by expanding the symbolical product (35). The symbolical part of each term of the expansion contains $T \cdot Y$, where $T$ is given by (33), and Y stands for either $S_m$, $M_m$ or $L_m$. Corresponding to each $TY$ there is a term in (36) with values $b_j = c_j p_m$ and $W_j = V_j + V_{Sm}$ if $Y = S_m$; $b_j = c_j q_m$ and $W_j = V_j + V_{Mm}$ if $Y = M_m$ and so on.

*Age effects:* Cousins of even degree are usually of similar age; cousins of uneven degree have usually some difference in age and therefore tend to marry less frequently. For an exact estimation of age effects, we need to know the probability distribution of age at marriage

$$d\,\phi_m(t,t') = \phi_m(t,t')dt\,dt' \quad (37)$$

where $t'$ and $t$ are the ages at marriage of male and female. We need also information on the distribution of the ages of consanguineous individuals.

$$\phi_c(t/t') \quad (38)$$

is the probability density of female (aged $t$) who are relatives of males aged $t'$. while

$$\phi_g(t) \quad (39)$$

is that of individuals from the general population, then

$$\psi_t = \phi_c/\phi_g \quad (40)$$

will express the frequency ratio of a certain age class among the relatives of an individual of age $t'$ to the same age class in the general population.

This frequency ratio can be averaged over all the marriages taking place in the population by computing the integral

$$I_a = \iint \phi_m(t,t')\psi_t\,dtdt' \quad (41)$$

extended to the whole range of ages at marriage for $t$, and $t'$.

The function $\phi_g$, namely the frequency of individuals of age $t$ in the general population is given, to a first approximation, by a rectangular distribution which is constant over the range of ages of persons eligible for marriage. It may, however, decrease with increasing $t$ in populations with high mortality in the reproductive period, or with decreasing birth rate, or it may also fluctuate as a consequence of irregularities of birth and death rates. If, however, population by ages were rectangular between 0 and $\omega$,

$$\phi_g(t) = 1/\omega \quad (42)$$

where $\omega$ is the upper limit of the age distribution and

$$I_a = \omega \iint \phi_m(t,t') \phi_c(t,t') dt dt' = \omega J \tag{43}$$

where $J$ represents the double integral.

The choice of males or females as the starting point in equation 38 is arbitrary and the procedure should be repeated after reversing the sexes, averaging the results. Since $\phi_c(t/t') = \phi_c(t'/t)$, it is enough to take as $1/\omega$ the average frequency per year of age, of individuals of either sex in the population, averaging over reproductive years.

In order to obtain the function (38) it is useful to obtain the distribution of age difference $\Delta$ between two relatives of different type of relationship. If there is no correlation between the age at marriage in various generations, the distribution of age difference $\Delta$ between relatives is easily computed. One needs, to this aim, information on: (1) the distribution of the age difference between sibs; as the order of birth of sibs is immaterial, this distribution has expected mean 0, and variance $\sigma_s^2$; (2) the distribution of generation times, namely the distribution of the age of the parent at birth of an offspring. These distributions are usually different for males and females and their means and variances will be given by the symbols $\tau_m$, $\sigma_m^2$ for males and $\tau_f$,$\sigma_f^2$ for females.

If we refer to Figure 1, we shall see that the computation of the expected age difference $\Delta$ between individual A and his relative B involves the difference between the sum of as many generation times $\tau$ as there are intermediate ancestors in the branch leading to B, minus the sum of as many generation times as there are intermediate ancestors in the branch leading to A. $\tau_m$ or $\tau_f$, namely male or female generation times, must be taken each time depending on the sex of the intermediate ancestors. The age difference between sibs does not contribute to the expected value of $\Delta$, because the order of birth of sibs is not taken into consideration, but it does contribute to the variance of $\Delta$. Therefore, if $m_i$ is the number of males among the common ancestors in the branch of the tree leading to A (the husband) and having $i$ generations, and $m_j$ is that in the other branch with $j$ generations, (where $m_i + m_j = n_1$) the expected age difference is

$$\bar{\Delta} = E(\Delta) = (m_j - m_i)\tau_m + (j - i + m_i - m_j)\tau_f \tag{44}$$

and the variance of $\Delta$ will be given by

$$\sigma_\Delta^2 = \sigma_s^2 + n_1\sigma_m^2 + (n - n_1)\sigma_f^2 \tag{45}$$

Formulas (44) and (45) were obtained by us (see BRAGLIA 1962) and also independently by HAJNAL (1963).

In order to provide material necessary for this type of evaluation, a sample of the population of the Parma province was subjected to analysis by questionnaire in 1958. The numerical results thus obtained will be used in the later part of this paper. Among other things, distributions of generation times and of the difference in age between sibs were derived. These distributions are somewhat asymmetric, but as $\Delta$ is the sum of several of them, it tends to normality rapidly with increasing $(i + j)$. A direct check of data available between first cousins showed good agreement with the expectation of normality (MAINARDI *et al.* 1966).

We can therefore use the following expression for the desired distribution given in (38) above:

$$\phi_c(t/t') = \frac{1}{\sigma_\Delta \sqrt{2\pi}} \exp[-(t'-t-\bar{\Delta})^2/2\sigma_\Delta^{\ 2}] \tag{46}$$

If we want to proceed to obtain integral $J$ given in (43) above, we need to specify the bivariate distribution of age at marriage, $\phi_m$. This distribution is certainly not normal in most populations in which the data are available. We found it could be rather well represented by a normal correlated surface after transforming ages to $\log(t-t_{min})$ where $t_{min}$ is the minimum age at marriage. However, the evaluation of $j$ demanded in this case numerical integration.

When compared with the more exact treatment just mentioned, the results obtained by using the normal approximation to function $\phi_m$ were also satisfactory except for the uncle-niece or aunt-nephew case.

If we then consider $\phi_m$ to be a normal bivariate function with means $\mu_h,\mu_w$ for the ages at marriage of husband and wife (where $\mu_h-\mu_w \doteq \tau_m-\tau_f$) respective variances $\sigma^2{}_h$, $\sigma^2{}_w$, and a correlation coefficient $\rho_{hw}$, then the integral $J$ reduces to

$$J = \frac{1}{S\sqrt{2\pi}} e^{-M^2/2S^2} \tag{47}$$

where

$$M = \bar{\Delta} - (\mu_h - \mu_w) \doteq (m_j - m_i - 1)\tau_m + (j - i + m_i - m_j + 1)\tau_f \tag{48}$$

$$S^2 = \sigma_\Delta^{\ 2} + \sigma_h^{\ 2} - 2\rho_{hw}\,\sigma_h\,\sigma_w + \sigma_w^{\ 2} = \sigma_0^{\ 2} + n_1\,\sigma_m^{\ 2} + (n-n_1)\sigma_f^{\ 2}$$

where

$$\sigma_0^{\ 2} = \sigma_s^{\ 2} + \sigma_h^{\ 2} - 2\rho\sigma_h\,\sigma_w + \sigma_w^{\ 2}$$

a result comparable to that obtained by HAJNAL (1963).

It is interesting to compare the results given in (47) with observations in the paper by BARRAI *et al.* (1962) on the problem of age effects. The analysis summarized in Figures 1 and 2 of that paper indicated an approximately parabolic relationship between the logarithm of the frequency of a given type of consanguineous marriage and a function indicated in the abscissa which is a linear transformation of the quantity $M$ given in (48) above. It will be noted that formula (47) gives an exactly parabolic relationship between $\log J$ and $M$ if there is no difference between the variance of male and female generation times ($\sigma_m{}^2$ and $\sigma_f{}^2$). It is very probable, therefore, that the deviation from a parabola observed in Figures 1 and 2 of BARRAI *et al.* (1962) is due to the fact that the variances of male and female generation times are unequal.

The computation of the quantity $I_a$ from formula (43) thus supplies a correction factor for the expected frequency of a given pedigree type of consanguineous marriage which can be applied to the expected frequency computed on the basis of migration alone, provided that between migration and age there is no important correlation. The corrected probability will be:

$$P_c = P\,I_a \tag{49}$$

*Assortative mating for heritable traits:* Assortative mating for socio-economic conditions or other traits which are inherited via biological or social mechanisms may also affect the frequency of consanguineous matings because of the higher resemblance between relatives. In order to assess its influence, one needs knowledge of the correlation between husband and wife ($\rho$) and that between relatives ($r_c$) for the trait or traits responsible for assortative mating. We will assume that the trait is measured in such a scale that $x$, $y$, the male and female values respectively, are normal with mean 0 and standard deviation $\sigma$ in the general population. The expected value $y_c$ of the trait, in female relatives of individuals of trait $x$, will then be $y_c = r_c x$, with variance $\sigma^2(1-r^2_c)$, and therefore the frequency of relatives with trait $y_c$ of individuals with trait $x$ will be:

$$\phi_c(y_c|x) = \frac{\exp\left[-(y_c - r_c x)^2/2\sigma^2(1-r_c^2)\right]}{\sigma\sqrt{2\pi(1-r_c^2)}} \tag{50}$$

while the frequency of individuals with trait $y$ in the general population will be

$$\phi(y) = \frac{e^{-y^2/2\sigma^2}}{\sigma\sqrt{2\pi}} \tag{51}$$

The ratio between the frequency of individuals with trait $y$ among the relatives of an individual with trait $x$, and the same frequency among individuals from the general population will be

$$\psi(y|x) = \phi_c(y|x)/\phi(y) \tag{52}$$

If $\phi$ $(x, y)$ is the bivariate correlated distribution of the trait, with correlation $\rho$ between husband and wife

$$\phi(x,y) = \frac{\exp\left[-(x^2 - 2\rho xy + y^2)/2(1-\rho^2)\right]}{2\pi\sigma^2\sqrt{1-\rho^2}} \tag{53}$$

and

$$\phi(x,y) = \phi(x)\,\phi(y|x) \tag{54}$$

the integral

$$I_s = \iint_{-\infty}^{+\infty} \phi(x,y)\psi(y|x)\,dxdy \tag{55}$$

will give the average over all marriages of the ratio $\psi$. On integration this is found to be

$$I_s = \frac{1}{1-\rho r_c} \tag{56}$$

Factor $I_s$ can be used to multiply the probability of consanguineous marriages computed on the basis of migration and age (if the trait in question is independent of both migration and age), in order to correct the probability for the effect of assortative mating.

It is likely that the postulated independence does not exist for socio-economic conditions and migration, so that this method of evaluation may be valid only as a first approximation.

In any case, it would seem, from what little knowledge is available that the correction factor is not likely to be large. From data collected in the Parma region $\rho$ is of the order of .5, $r_c$ is not known, but must be low. $r_c$ could be estimated from knowledge of $\rho$ and of the correlation between parent and offspring for the trait.

If $r_{PO}$, $r_{MO}$ and $r_s$ are the correlation coefficients between father and offspring, mother and offspring, sib and sib respectively, the correlation between relatives with $n_0$ and $n_1$ female or male intermediate ancestors will be

$$r_c = r_s \, r_{PO}^{n_1} \, r_{MO}^{n_0} \tag{57}$$

Even assuming that $\rho = r_{PO} = r_{MO} = 0.8$ the correlation between first cousins would be $r_c = .18$, that between second cousins $r_c = .116$, leading to correction factors of 1.20 and 1.11.

For the traits determined entirely by additive genes $r_{PO} = r_{MO} = r_s = 0.5$, and if $\rho = .25$ the correction factor would be 1.03. It would take a great many independent heritable traits to make the correction factor important. It therefore seems likely that one can neglect assortative mating effects at the first approximation.

*Agreement between theory and observation:* A complete test of the theory just given would require demographic knowledge which is not available in the literature. Material which has been collected in the Parma province contains such information but the work of analysis is not complete, especially for the part regarding demographic data of the past two centuries (BARRAI, CAVALLI-SFORZA and MORONI 1964). Changes of demographic patterns are known to have taken place especially in the last and the present century, and the study of consanguineous marriages requires demographic knowledge valid for ancestral generations. Therefore, until data for earlier times than now available is at hand, no satisfactory test will be possible.

At the moment, the source of data coming nearest to the requirements is that from the Upper Parma River Valley. Here a sample of almost 500 families coming from various villages of the area was investigated by questionnaire (MAINARDI *et al.* 1962). Some of this material is already published (CAVALLI-SFORZA 1963; CAVALLI-SFORZA *et al.* 1964). This is, at the moment, the only source of information on distribution of distances between birth places, and age differences, in which we are interested, but it comes from a contemporary population living in the same area in which we have collected consanguineous marriages.

1. *Migration.* The migration data utilized in Table 1 are from the source just cited. Biexponential and trinormal distributions were fitted by a fully numerical version of maximum likelihood estimation, computing the likelihood of a set of trial values of the parameters, then obtaning first derivatives by recomputing likelihoods with small increments added to each parameter in turn. From this the information matrix could be calculated and corrections of the trial values obtained. The procedure was iterated until the increase in likelihood was negligible. In some cases there were difficulties in obtaining a good fit using maximum likelihood and another method was employed in which chi-square is minimized as follows. One defines plausible intervals for each parameter and on the basis of chi-square decides which half-interval to use as trial estimate. One then proceeds by taking the better half-interval for each parameter in a new cycle of computation, progressing in this way until the required precision is reached.

Maximum likelihood estimates for biexponential and for trinormal migration

TABLE 2

*Parameters of the fitted distribution of Table 1*

| | F-O | M-O | H-W |
|---|---|---|---|
| Trinormal | | | |
| $p$ | .839 | .740 | .646 |
| $q$ | .061 | .163 | .144 |
| $1-p-q$ | .100 | .097 | .210 |
| $V_S$ | .32 | .36 | .42 |
| $V_M$ | 60.0 | 58.7 | 89.9 |
| $V_L$ | 1890. | 1608. | 1182. |
| Biexponential | | | |
| $p$ | .828 | .725 | .614 |
| $h$ | 2.493 | 2.330 | 2.266 |
| $k$ | .0248 | .0425 | .0325 |

curves to the distribution of birth place distances of the pairs father-offspring, mother-offspring, and husband-wife are given in Table 2.

The distribution of distances between birthplaces of sibs showed such a high concentration in the zero class that this distribution was neglected throughout. This has the only disadvantage that, when using biexponentials, the integral of equation (20) does not converge for uncle-niece or aunt-nephew, while it would if the convolution had included the sib-sib migration. Therefore we are unable to give expected values for this class of relatives under the hypothesis of biexponential migration.

Table 3 shows the probabilities of consanguineous marriage, $P$ computed from equation (30) for trinormal migration and from equation (20) for biexponential migration. The numerical integration necessary in the latter case was carried out by computer. The migration parameters are those given in Table 2 and the population density employed in the calculation is that valid on average for the Upper Parma River Valley, years 1860–1962, and is $D = 45.7$ inhabitants per square kilometer. All probabilities are given for $n_c = 1$. All consanguineous marriages from uncle-niece or aunt-nephew to third cousin are considered, but only pedigrees that have different probabilities are distinguished, namely those pedigrees in which the number of males $n_1$ among intermediate ancestors ($n$) varies from 0 to $n$.

It will be noted that there is a marked discrepancy between the trinormal and the biexponential, which is more serious the nearer the relationship. The cause of this is believed to be the difference in behavior of the two functions at the origin. In fact, even if the two functions are fitted to the same observed distributions, the class at the origin, which represents on the average some 75% of the observations, is fitted in the biexponential by a monotonically decreasing function which cuts the ordinate at a value different from 0, while for the trinormal the function used is zero at the origin and goes to a peak thereafter. It seems therefore that with data such as the present ones, in which there is an accumulation of most of the observations in the class at the origin, the choice of one or the other

TABLE 3

*The probability of consanguineous marriage under assumption of a trinormal and of a biexponential migration distribution as given in Tables 1 and 2. Only migration effect is considered*

| Degree of relationship | $n$ | $n_1$ | P values trinormal | P values biexponential |
|---|---|---|---|---|
| Uncle niece, aunt nephew | 1 | 0 | .002150 | ...... |
| | | 1 | .002558 | ...... |
| First cousins | 2 | 0 | .001088 | .003886 |
| | | 1 | .001278 | .004542 |
| | | 2 | .001499 | .005311 |
| 1½ cousins | 3 | 0 | .000613 | .001635 |
| | | 1 | .000714 | .001889 |
| | | 2 | .000830 | .002182 |
| | | 3 | .000961 | .002522 |
| Second cousins | 4 | 0 | .000366 | .000843 |
| | | 1 | .000425 | .000966 |
| | | 2 | .000432 | .001107 |
| | | 3 | .000570 | .001270 |
| | | 4 | .000654 | .001457 |
| 2½ cousins | 5 | 0 | .000227 | .000490 |
| | | 1 | .000263 | .000559 |
| | | 2 | .000304 | .000628 |
| | | 3 | .000350 | .000716 |
| | | 4 | .000404 | .000817 |
| | | 5 | .000462 | .000933 |
| Third cousins | 6 | 0 | .000145 | .000301 |
| | | 1 | .000168 | .000342 |
| | | 2 | .000193 | .000388 |
| | | 3 | .000222 | .000441 |
| | | 4 | .000255 | .000501 |
| | | 5 | .000294 | .000570 |
| | | 6 | .000335 | .000649 |

$n$ is the number of intermediate ancestors and $n_1$ is the number of males among them.

continuous model is not an easy one. The choice cannot be done on the only evidence of the goodness of fit of the theoretical distribution. As the type of the distribution is critical in determining the expectations, it is believed that it will be preferable in cases like this to adopt, when adequate demographic data are available, the discontinuous model, using migration matrices which do not assume any specific form of migration distribution.

Other evidence shows, in any case, that the demographic data available from the present day populations show only qualitative agreement with the consanguinity data. When the values of probabilities of consanguineous marriages of Table 3 are plotted on a graph (Figure 3) it will be seen that $P$ rises almost exponentially with the number of males for a given $n$. The slope of increase of log $P$ is about log 1.14 from Figure 3 data. This slope should correspond to the logarithm of the quantity called $c$ and estimated from actual data coming from

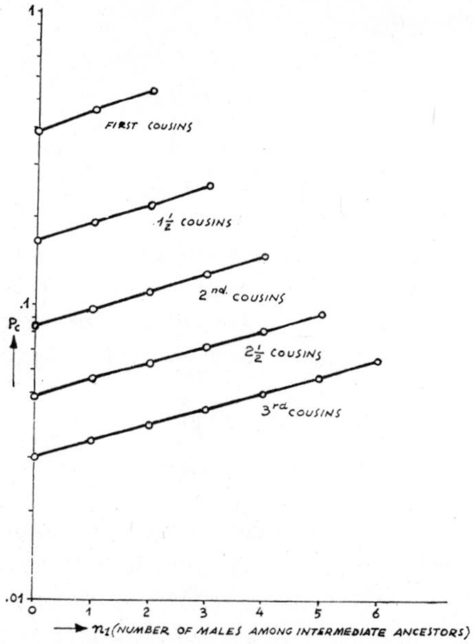


FIGURE 3.—Probabilities of consanguineous marriages, based on values in Table 3.

a comparable area in a former paper (BARRAI *et al.* 1962). There, values of $c$ were, for the mountainous region of Parma, 1.389 for second cousins, 1.366 for third cousins and lower than 1 (.918) for first cousins, the last value being influenced (as shown in that paper) by a sociological factor which should account for the aberration. The agreement is thus only a qualitative one in the sense that the difference between male and female migration observed in a modern population, although in the right direction, is not as high as would be expected from the $c$ value determined from second and third cousin marriage frequencies in the same area.

*A priori* the approach by the "sum of exponentials" is probably more satisfactory than that by the "sum of normals". In fact, the biexponential seems to represent a little more accurately the clustering of people in villages by its having the mode at the origin.

It is, however, of interest to follow further the "sum of normals" approach, because it does not require any numerical integration and is therefore more directly available for the calculation of expected frequencies of consanguinity. Although the migration integral $I_m$ requires, with a trinormal distribution, the computation of $3^{n+1}$ terms, whose sum constitutes $I_m$, only one term is important in the present circumstances. This depends on the fact that the short range migration is so much more important in terms of frequency than the other two. It therefore happens that for the values given in Table 2, the term

$$\frac{p_1^{\,n_1}\, p_0^{\,n_0}\, p_m}{n_0 V_{s0} + n_1 V_{s1} + V_{sm}} \tag{58}$$

accounts for some 95% of the value of $I_m$. Since, approximately, $V_{s0} = V_{s1} =$

$V_{sm} = V_s$, formula (58) simplifies to

$$P \doteq \frac{1}{2\pi D} \frac{p_1^{n_1} p_0^{n_0} p_m}{V_s(n_1+n_0+1)} \tag{59}$$

It is interesting to compare this value with that obtained by the simplified discontinuous treatment, in which only marriages inside the village where both mates are born are considered. In the latter case let us take as $P_1$ the proportion of children born in the same village as their fathers, $P_0$ the proportion born in the same village as their mothers, $P_m$ the proportion of husbands and wives who were born in the same village, and $N$ the village size; then, always considering $n_c = 1$ and neglecting age and other corrections, for a pedigree with $n_0$ and $n_1$ female and male common ancestors,

$$P \doteq \frac{P_1^{n_1} P_0^{n_0} P_m}{N} \tag{60}$$

With the numerical values so far used (namely $V_S$ approximately $= 0.4$, and $D = 45.7$) the two formulas from simplified normal migration and the simplified discontinuous treatment give approximately equal results only for first cousins, ($n = n_1 + n_0 = 2$) as the average village size ($N$) in the corresponding region is not far from 300 and as $P_0 = .821$, $P_1 = .727$, $P_m = .645$, are not far from the corresponding values $p_0$, $p_1$, $p_m$ of Table 2. For other types of relationships these two estimates inevitably diverge.

2. *Age effects:* A direct comparison between the $P$ values given in Table 3 and the observed ones is not possible until we correct $P$ values for age according to (49). To this aim, we need to know the required age distributions. From the 1958 Parma sample (MAINARDI *et al.* 1962) the values given in Table 4 were computed. Here also, however, we find a discrepancy between data from the contemporary population and those suited for the analysis of the consanguineous marriages, but again there is at least qualitative agreement. A test would be provided by the fitting of parabolas to the logarithms of the frequencies of consanguineous matings of various degrees versus the expected age difference between mates, as was done in Figures 1 and 2 of the paper by BARRAI *et al.* (1962), but a more direct test was used in Table 2 of the paper by CAVALLI *et al.* (1964), in which the demographic data given here in Table 4 were used to fit the frequencies of 29 classes of consaguineous matings from the Upper Parma Valley. Consanguineous marriages 1850 to 1950 were considered ranging from uncle-niece to second cousins, and grouped according to $n_1$, and $n$, which gives rise to the 29 classes shown in that table. Expectations of each class were computed as proportional to

$$P_m^{n_0} P_f^{n_1} e^{-M^2/2V} \tag{61}$$

where $M$, $V$ are as defined by equations 48. When, however, the values of the contemporary sample, given in Table 4, were inserted in this equation, the fit to the observed consanguinity frequencies was not satisfactory. We therefore tried to fit $P_m$, $P_f$, $\tau_m$, $\tau_f$, $\sigma^2{}_m$, $\sigma^2{}_f$, $(\sigma^2{}_s + \sigma^2_\Delta)$ values to the data by numerical maximum likelihood. The expectations improved considerably and the only major dis-

TABLE 4

*Estimates of parameters necessary for age corrections, obtained from a contemporary population in the Upper Parma River Valley (Mainardi* et al. 1966)

| | | | | |
|---|---|---|---|---|
| Generation time: | | | | |
| | Males | mean | $\tau_m$ | 33.24 ± 0.19 |
| | | variance | $\sigma^2_m$ | 46.60 |
| | Females | mean | $\tau_f$ | 28.73 ± 0.16 |
| | | variance | $\sigma^2_p$ | 33.78 |
| | Variance of age difference between sibs | | $\sigma_s^2$ | 25.78 |
| | Variance of age difference between mates | | $\sigma_0^2$ | 33.31 |

crepancy left was between classes of first cousins, where other factors of sociological nature might be involved (Cavalli-Sforza *et al.* 1964). The demographic parameters thus estimated are given also here, in the heading of Table 6.

It will be noted that these estimates indicate a later average age at reproduction than in the modern population (by 2 or 3 years) as well as an increased variance of generation times. The consanguineous marriages from which these parameters were estimated extend from 1850 to 1950, and therefore, their intermediate ancestors lived mostly in the 19th Century or even earlier. The difference between the fitted values, which are estimated for earlier generations, and the values obtained for the present generation is probably in the right direction. Thus, removing geographical heterogeneity was not sufficient to give an agreement, other than qualitative, between our present estimates of the demographic parameters and those necessary for a good fit. It should be noted, in addition that in the period 1850 to 1950 some changes in the consanguineous frequencies were observed, and a more recent research (Moroni, in preparation) shows that very extensive changes in consanguineous frequencies took place in the 19th Century. Unfortunately, breaking down the figures further by periods of time would reduce them too much for a meaningful comparison with expected values to be possible. We shall have to be content at the moment with an approximate agreement, as is possible with the present data.

It may also be argued that the normal approximation to the distribution of ages at mating or at reproduction may be inadequate. However, it has been seen (Mainardi *et al.* 1966) that for first cousins, age differences are normally distributed, in agreement with expectation, and the effect of the non-normality of the distribution of age at mating is noticeable only slightly for the extreme case of uncle-niece or aunt-nephew. It was found that ages at mating are well fitted by log ($t$-$t_{min}$) where $t_{min}$ is the minimum legal age at marriage (Hald 1952). We have found that also the correlation surface between male and female ages at marriage is reasonably normal using the above transformation, and have used it to compute $I_a$, although it requires numerical integration. Table 5 shows that

TABLE 5

*Age effect*

| Type of relationship | $i$ | $j$ | $m_i$ | $m_j$ | J values: Transformation log $(t\text{-}t_{min})$ | J values: Untransformed ages | Difference % |
|---|---|---|---|---|---|---|---|
| Uncle-niece | 0 | 1 | 0 | 0 | .001688 | .001649 | 2.4 |
| | 0 | 1 | 0 | 1 | .000776 | .000688 | 13.0 |
| Aunt-nephew | 1 | 0 | 0 | 0 | .000146 | .000133 | 10.0 |
| | 1 | 0 | 1 | 0 | .000057 | .000050 | 14.0 |
| First cousin | 1 | 1 | 0 | 0 | .033564 | .033006 | 1.7 |
| | 1 | 1 | 0 | 1 | .034278 | .034000 | .8 |
| | 1 | 1 | 1 | 0 | .026568 | .026188 | 1.4 |
| | 1 | 1 | 1 | 1 | .031437 | .031030 | 1.3 |

Discrepancy between numerical integration using a normal bivariate distribution fitted to ages of mates transformed according to log $(t\text{-}t_{min})$ and direct integration using a normal bivariate distribution fitted to untransformed ages (1954 marriages, Italy)

the divergence between the two methods of computation is rather small, even in the most critical cases.

The quantity $I_a$ from formula (43) gives the correction factor by which $P$ values obtained from migration should be multiplied to obtain $P_c$. In order to obtain it, we need $J$ from formula (47) and $\omega$. The value $\omega$ will be computed here as the reciprocal of the mean frequency per time unit (years in this case) of individuals of either sex in the population, averaging over reproductive years. In practice, we have simply taken ages between 15 and 50 years and averaged the corresponding frequencies (unweighted for simplicity), using data from the 1901 Italian census, a time which corresponds to the middle of the period examined before. The value of $\omega$ thus obtained is 70.36 years.

Using as demographic values those given at the head of Table 6, the quantities $I_a = \omega J$ given in the body of Table 6 were obtained.

3. *Probabilities of consanguineous marriage.* It is now possible to obtain the $P_c$ values corrected for age effects, multiplying each $P_c$ value times the appropriate $I_a$ value. The correspondence between $P$ and $I_a$ values is easily established, keeping in mind that $m_i$, $m_j$ $(m_i + m_j = n_1)$ are the parameters specifying the number of males among intermediate ancestors in the branches leading respectively to the consanguineous husband and wife, and that $i$, $j$ $(i+j=n)$ are the total numbers of intermediate ancestors in the two branches. An example of of calculations will be found in Table 7.

The result (always keeping $D = 45.7$ as in Table 3) in the calculation of $P_c$ is given in Table 8, where both the biexponential and the trinormal models of migration are retained and compared with observations, after adding up for all pedigrees belonging to the same degree of relationship. When doing this, it should be remembered that some pedigrees are represented more than once, and that if $P_c(i,j,m_i,m_j)$ is the expected frequency of a pedigree with given $i$, $j$, $m_i$, and

TABLE 6

*Correction factors for age* ($I_a$) *computed from formulas (43), (47), and (48) using demographic values* estimated from frequencies of consanguineous marriages in the paper by Cavalli* et al. *(1964)*

| | $m_i$† | $m_j$ | $I_a$ |
|---|---|---|---|
| Uncle-niece | 0 | 0 | 0.08851 |
| | 0 | 1 | 0.03018 |
| Aunt-nephew | 0 | 0 | 0.00316 |
| | 1 | 0 | 0.00091 |
| First cousins | 0 | 0 | 2.1688 |
| | 0 | 1 | 2.29962 |
| | 1 | 0 | 1.61077 |
| | 1 | 1 | 2.04407 |
| First cousins once removed: husband older generation | 0 | 0 | 0.33139 |
| | 0 | 1 | 0.16646 |
| | 0 | 2 | 0.07878 |
| | 1 | 0 | 0.66639 |
| | 1 | 1 | 0.38086 |
| | 1 | 2 | 0.20306 |
| Second cousins once removed | 0 | 0 | 0.14860 |
| | 0 | 1 | 0.29895 |
| | 0 | 2 | 0.52017 |
| | 1 | 0 | 0.07800 |
| | 1 | 1 | 0.17162 |
| | 1 | 2 | 0.32619 |
| | 2 | 0 | 0.03915 |
| | 2 | 1 | 0.09375 |
| | 2 | 2 | 0.19378 |
| | 3 | 0 | 0.01887 |
| | 3 | 1 | 0.04898 |
| | 3 | 2 | 0.10966 |
| Third cousins | 0 | 0 | 1.53060 |
| | 0 | 1 | 1.57073 |
| | 0 | 2 | 1.48415 |
| | 0 | 3 | 1.30132 |
| | 1 | 0 | 1.33034 |
| | 1 | 1 | 1.48415 |
| | 1 | 2 | 1.52047 |
| | 1 | 3 | 1.44166 |
| | 2 | 0 | 1.07619 |
| | 2 | 1 | 1.30132 |
| | 2 | 2 | 1.44166 |
| | 2 | 3 | 1.47475 |
| | 3 | 0 | 0.81582 |
| | 3 | 1 | 1.06610 |
| | 3 | 2 | 1.27388 |
| | 3 | 3 | 1.40262 |

* $\tau_m = 36.10$, $\tau_f = 30.95$, $\sigma^2_m = 53.32$, $\sigma^2_f = 42.59$, $\sigma^2_o = 53.08$, $\omega = 70.36$.

† As before, $i,j$ are the numbers of intermediate ancestors in the branches of the pedigree leading to the husband and to the wife (see Figure 1) and $m_i$, $m_j$ are the numbers of males among them.

TABLE 7

*Computation of probabilities of consanguineous marriages* $P_c$ *in percent. Example for* $1\frac{1}{2}$ *cousins, husband in shorter branch* ($i = 1, j = 2$)

| Pedigree types | $m_i$ | $m_j$ | $n'$ | $I_m$ Trinormal | $I_m$ Biexp | $I_a$ | $P_c$ percent Trinormal | $P_c$ percent Biexp |
|---|---|---|---|---|---|---|---|---|
| 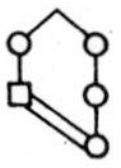 | 0 | 0 | 1 | .000613 | .001635 | .33139 | .0203 | .0541 |
| 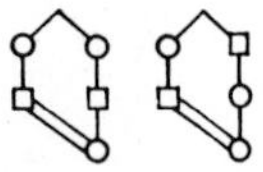 | 0 | 1 | 2 | .00714 | .001889 | .16646 | .0119 | .0314 |
| 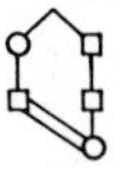 | 0 | 2 | 1 | .000830 | .002182 | .07878 | .0065 | .0172 |
| 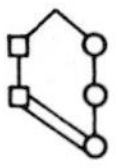 | 1 | 0 | 1 | .000714 | .001889 | .66639 | .0476 | .1259 |
| 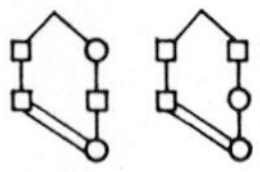 | 1 | 1 | 2 | .000830 | .002182 | .38086 | .0316 | .0831 |
| 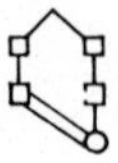 | 1 | 2 | 1 | .001961 | .002522 | .20306 | .0195 | .0512 |
| | | | | | | $\Sigma n'P_c =$ | .1809 | .4774 |

$P$ data from Table 3 and $I_a$ data from Table 6. $m_i$, $m_j$ are the number of males among intermediate ancestors in the branches of the tree leading to the husband and wife respectively; $n'$ is the number of pedigrees. $P_c$ values are products $I_m \times I_a$ and are given in percent. In pedigrees, squares are males, circles females.

$m_j$, the sum of the frequencies of all pedigree types for a given pedigree $n = i + j$ will be given by:

$$P_c(n) = \sum^{i}\,(m_i)\overset{j}{(m_j)}P_c\,(i, j, m_i, m_j) \tag{62}$$

where the $\Sigma$ is extended to all pedigree types with different $i$, $j$, $m_i$, $m_j$ values belonging to the same type of relationship. The expectations given as probabilities of consanguineous marriages in Table 8 are then obtained, corresponding to the two continuous models.

It will be noted that there is only an approximate agreement, but the migration estimates here employed are probably too large. This can account for the discrepancy becoming higher with less close relationship. In any case, the biexponential function gives a better fit. Also, the vagaries of demographic values and of consanguineous matings with time and geography are very probably responsible for a part of the discrepancy. As better estimates of the demographic parameters will be forthcoming in the not too distant future, we hope to have better opportunities for a more satisfactory test of the theory.

TABLE 8

*Comparisons between probabilities of consanguineous marriages corrected for age (as in Table 7) and observed frequencies for upper Parma River Valley (communes of Corniglio, Monchio, Tizzano, Palanzano)*

| | | | Probability $P_c$ | | Observed marriages | |
|---|---|---|---|---|---|---|
| | $m_i$ | $m_j$ | Trinormal | Biexp | Number | Percent |
| Uncle-niece | 0 | 0 | .0190 | .... | 5 | .037 |
| | 0 | 1 | .0077 | .... | 4 | .030 |
| Aunt-nephew | 0 | 0 | .0007 | .... | 1 | .007 |
| | 0 | 1 | .0002 | .... | 0 | 0 |
| Total | | | .0276 | | | .074 |
| First cousins | 0 | 0 | .2360 | .8384 | 109 | .8066 |
| | 0 | 1 | .2939 | 1.0445 | 157 | 1.1618 |
| | 1 | 0 | .2059 | .7316 | 99 | .7326 |
| | 1 | 1 | .3052 | 1.0856 | 96 | .7104 |
| Total | | | 1.0410 | 3.7001 | | 3.4114 |
| 1½ cousins, in shorter branch | 0 | 0 | .1809 | .4774 | 153 | 1.322 |
| 1½ cousins, in shorter branch | 0 | 0 | .0399 | .0919 | 45 | .3330 |
| Second cousins | 0 | 0 | 1.2870 | 2.8953 | 938 | 7.2520 |

It is also possible that some of the assumptions here made: independence of age and migration, lack of parent-offspring correlation in fertility, and in mobility, may limit the usefulness of the simple model here described, to an extent that further research may show.

This work has been supported by grants from the U.S. Atomic Energy Commission and by EURATOM-CNR-CNEN Contract No. 012–61–12 BIAI. We wish to thank Dr. Raymond Appleyard for reading the manuscript.

## SUMMARY

Theories were developed to predict the frequencies of various types of consanguineous marriages based on demographic data of migration patterns, age distributions, and similarity of mates in the general population. The effect of migration was formulated both with discrete and continuous models. In the former, the entire population is subdivided into discrete groups (villages etc.) and migration and marriage are treated using transition and matrimonial migration matrices. It was then shown that the method of matrix algebra leads to simple expressions of the results. However, information is not at the moment sufficient to construct numerically the migration and marriage matrices to treat the actual cases, but may become available in the future. On the other hand, using continuous models, fitting of either biexponential or trinormal distribution to migration distances allows us to predict the frequencies observed in an actual case (the Parma Valley area) when age effect on marriage is also taken into account.—The agreement between observed and expected results for Parma is only fair. In part, at least, this seems to be the consequence of the inadequacy of the demographic information now available and that should be improved by future research.—As an indicator for the breeding structure of populations, the probabilities of consanguineous marriages should have an important bearing especially for human population genetics.

## LITERATURE CITED

Barrai, I., L. L. Cavalli-Sforza, and A. Moroni, 1962 Frequencies of pedigrees of consanguineous marriages and mating structure of the population. Ann. Hum. Genet. **25**: 347–377.

Barrai, I., L. L. Cavalli-Sforza, and A. Moroni, 1964 Record linkage from parish books. pp. 51–60. *Mathematics and Computer Science in Biology and Medicine*, H.M.S.O., London.

Braglia, G. L., 1962 Frequenze di matrimoni consanguinei. Tesi di Laurea in Fisica, Università di Parma.

Cavalli-Sforza, L. L., 1957 Some notes on the breeding patterns of human populations. Acta Genet. Statist. Med. **6**: 395–399. —— 1963 The distribution of migration distances: models, and applications to genetics. pp. 139–158. *Entretien de Monaco en Sciences Humanines; Les déplacements humains.* Edited by Jean Sutter.

Cavalli-Sforza, L. L., I. Barrai, and A. W. F. Edwards, 1964 Analysis of human evolution and random genetic drift. Cold Spring Harbor Symp. Quant. Biol., **23**: 10–20.

Frota-Pessoa, O., 1957 The estimation of the size of isolates based on census data. Am. J. Hum. Genet. **2**: 9–16.

Hajnal, J., 1963 Random mating and the frequency of consanguineous marriages. Proc. Royal Soc. London B **159**: 125–177.

Hald, A., 1952 *Statistical Theory with Engineering Applications.* Wiley, London.

Luu-mau-Thanh, and J. Sutter, 1963 Contribution à l'étude de la répartition des distances séparant les domiciles des époux dans un département francais. Influence de la consanguinité. Entretien de Monaco en sciences humaines; Les déplacements humains, 123–137.

Mainardi, M., L. L. Cavalli-Sforza, and I. Barrai, 1962 The distribution of the number of collateral relatives. Atti Ass. Genet. Ital. **7**: 123–130. —— 1966 Some demographic estimates of genetic interest. (in preparation).

Morton, N. E., 1955 Non-randomness in consanguineous marriage. Ann. Hum. Genet., **20**: 116–124.

Moroni, A., Inbreeding explosion in the 19th Century in a Catholic country. (In preparation).

Sutter, J., and Tran Ngoc Toan, 1957 The problem of the structure of isolates and of their evolution among human populations. Cold Spring Harbor Symp. Quant. Biol., **22**: 379–383.

Wright, S., 1943 Isolation by distance. Genetics **28**: 114–138.

# Editor's Comments on Paper 20

20 **Jacquard:** Genetic Information Given by a Relative
*Biometrics,* **28**, 1101–1114 (Dec. 1972)

Because of the preponderance of English-speaking theorists in the early days of population genetic theory, developments in languages other than English have not always been given proper attention, either for historical or current research purposes. This error should be corrected. The present article and associated references are selected as examples of this imperative.

Although most readers will have been aware of the work of Malécot (written in French in 1948, but translated only in 1969), few are aware of other works in French. For example, the papers of M. Gillois (1965a, 1965b, 1966) are original and fundamental contributions to axiomatic foundations of population genetics. His mathematical techniques are at least more modern and in many ways easier to follow than those of Cotterman (this comment may only reflect the editor's bias toward algebraic and set theoretic methods). A very useful study would be to integrate these methods with those of Lyubich.

Although the English may have ignored the French, the reverse is certainly not true, and it is arguable that recent developments in French language structural genetic theory have been due to awareness by French theorists of structural literature in English. The undoubtedly most comprehensive treatment of genetic algebra as presented by Etherington was done by Bertrand (1966) and most certainly needs to be translated. Bibliographic references in that work are almost entirely to I. M. H. Etherington or to A. A. Albert.

For a general survey of the work of French (and other) geneticists, the excellent text of Jacquard will luckily appear in English (1970, 1973).

# 20



Reprinted from *Biometrics*, **28**, 1101–1114 (Dec. 1972)

# GENETIC INFORMATION GIVEN BY A RELATIVE

ALBERT JACQUARD

*Institut National d'Etudes Démographiques, 27, rue du Commandeur, Paris 14*

## SUMMARY

What genetic information concerning an individual can be provided by the knowledge of the genetic make-up of one of his relatives? In general, the well-known "coefficient of kinship" is not sufficient to solve the problem. It is necessary to use a more complete measure of kinship, the set of nine "condensed identity coefficients." This article reviews these coefficients and their use to establish equations which give the "genotype structure" (i.e. the set of probabilities of the various genotypes) of an individual, given the genotype of one of his relatives and the gene structure of the population. When the ancestry network is simple enough, these equations may be put in a form in which only the coefficient of kinship appears. But the conditions for such a simplification are restrictive: not only must both relatives not be inbred, but also the kinship between them must be unilineal.

## 1. INTRODUCTION

Genetics teaches us what the biological link between relatives is: at some loci they have genes which are the replicas of the same gene of a common ancestor. Malécot [1948; 1966] calls such genes, which are transmitted from one generation to the next without alteration, *identical*. The kinship between two individuals may thus be characterized by the possibility of finding such identical genes in their genetic make-ups.

This concept is the basis of fundamental reasonings which have proven useful in solving many problems of population genetics: the rate of genetic drift, genetic structure of the progeny of related mates, effect of systems of mating, etc. Since this concept has sometimes been misunderstood, it seems necessary to stress one particular aspect: that two specified genes are actually identical is beyond our knowledge. We cannot state "these two genes are identical"; all we can say is "the probability that these genes are identical is equal to $\cdots$". Hence, in this paper all the statements dealing with identity will refer to probabilities, and all the measures based on the possibility of identity will be expressed in terms of probabilities. In particular, the kinship of two individuals $X$ and $Y$ may be measured by the coefficient of kinship $\phi(XY)$ defined as: $\phi(XY)$ is the probability that a gene randomly selected from $X$ and a gene randomly selected at the same locus from $Y$ are identical.

Thus the kinship of two individuals is not a characteristic of these individuals themselves but a characteristic of the knowledge we have about their genealogies.

For many problems (notably the determination of the progeny of related mates) this measure of kinship is sufficient, but not for all cases. In particular it does not permit us to treat the most fundamental question: what can we infer about the genetic make-up of an individual $Y$ when we know the genetic make-up of his relative $X$? The purpose of this paper is to review more sophisticated measures of kinship and to show how they make it possible to solve this last problem.

## 2. MEASURE OF KINSHIP

The kinship of two individuals results from the fact that among the ancestors of one, we find one or more ancestors of the other, or the other himself. However intricate the network of common ancestors, the genetic information that an individual $X$ gives about an individual $Y$ is contained in the probability that one of $X$'s genes and one of $Y$'s homologous (taken at the same locus) genes should be replicas of the same ancestral gene, i.e. "identical by descent" or "identical".

### 2.1 *Detailed identity coefficients*

Let us now consider the genes possessed at an autosomal locus by $X$ and $Y$: $G_X$ transmitted to $X$ by his father, $G_X^*$ transmitted to $X$ by his mother, $G_Y$ transmitted to $Y$ by his father, and $G_Y^*$ transmitted to $Y$ by his mother. These four genes may be identical or non-identical by pairs, but when, for example, $G_X$ is identical to $G_X^*$ and $G_X^*$ is identical to $G_Y$, then $G_X$ is identical to $G_Y$. In other words, the relation "identity" is transitive. Taking this transitivity into consideration, it is possible to find 15 cases, first defined by M. Gillois [1964] as "15 identity states $S$". The symbols assigned to the various states are given in Table 1, where identical genes are linked by a line.

When the network of ancestry of two individuals $X$ and $Y$ is known, it is possible to determine the probability $\delta_i$ of each of the identity states $S_i$ : these probabilities are the "detailed identity coefficients" of $X$ and $Y$. This set of 15 coefficients supplies a summarized description of the kinship of $X$ and $Y$ without any loss of genetic information. Every kinship is thus characterized by the values of the 15 coefficients; for example, if $X$ and $Y$ have no common ancestor and if their fathers and mothers are without parental links, we have $\delta_{15} = 1$, $\delta_i = 0$ for $i \neq 15$; if $X$ and $Y$ are monozygotic twins born from non-related parents, each gene of $X$ is identical to the gene of the same source of $Y$, a condition which corresponds to state $S_9$, hence $\delta_9 = 1$, $\delta_i = 0$ for $i \neq 9$.

### 2.2. *Condensed identity coefficients*

For most problems the paternal and maternal origin of a gene is irrelevant; the genotype $a_i a_j$ is the same whether $a_i$ comes from the father and $a_j$ from the mother or vice versa. States such as $S_{10}$, $S_{11}$, $S_{13}$, and $S_{14}$, defined above, are then equivalent and it is possible to condense them into a single

TABLE 1
DETAILED IDENTITY STATES

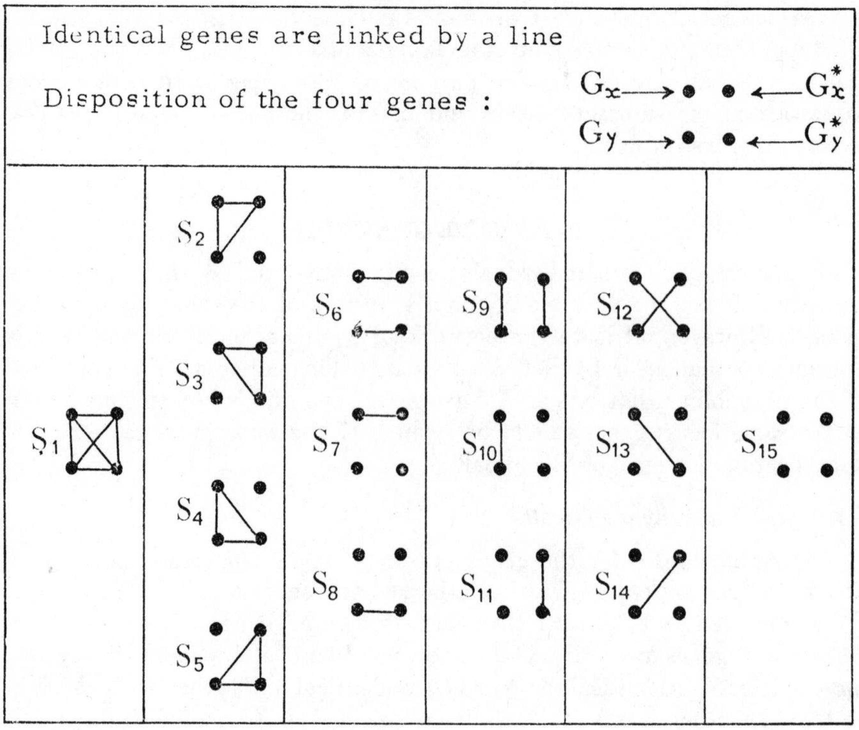


state. Thus, to characterize the kinship of $X$ and $Y$, only nine states, denoted as "condensed identity states", are necessary. They correspond to the various arrangements of the following events:

(a) the two genes of $X$ are identical (an event described by $C_X$) or not ($C_{\bar{X}}$);
(b) the two genes of $Y$ are identical ($C_Y$) or not ($C_{\bar{Y}}$); and
(c) the number $N$ of cases of identity between a gene of $X$ and a gene of $Y$ is 0, 1, 2, or 4.

The nine condensed identity states are defined by the following list and by Table 2:

| | | |
|---|---|---|
| $\Sigma_1 : C_X , C_Y , N = 4$ | $\Sigma_4 : C_X , C_{\bar{Y}} , N = 0$ | $\Sigma_7 : C_{\bar{X}} , C_{\bar{Y}} , N = 2$ |
| $\Sigma_2 : C_X , C_Y , N = 0$ | $\Sigma_5 : C_{\bar{X}} , C_Y , N = 2$ | $\Sigma_8 : C_{\bar{X}} , C_{\bar{Y}} , N = 1$ |
| $\Sigma_3 : C_X , C_{\bar{Y}} , N = 2$ | $\Sigma_6 : C_{\bar{X}} , C_Y , N = 0$ | $\Sigma_9 : C_{\bar{X}} , C_{\bar{Y}} , N = 0.$ |

Each of the states $\Sigma_i$ is equivalent to one or more states $S_i$ ; for example, $\Sigma_8$ is equivalent to $S_{10}$ , $S_{11}$ , $S_{13}$ , and $S_{14}$ .

When the network of the ancestry of two individuals $X$ and $Y$ is known, it is possible to determine the probability $\Delta_i$ of each of the condensed identity

TABLE 2
CONDENSED IDENTITY STATES

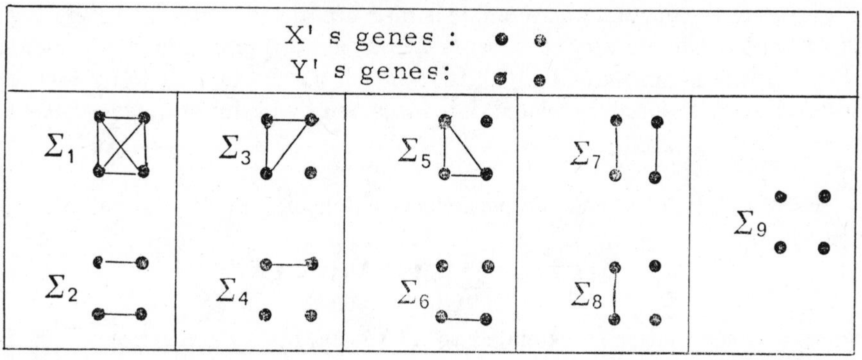


states $\Sigma_i$ . These probabilities are the 'condensed identity coefficients' of $X$ and $Y$. It is easy to see that the coefficients $\Delta_i$ and the coefficients $\delta_i$ satisfy the relations:

$$\begin{array}{lll} \Delta_1 = \delta_1 & \Delta_4 = \delta_7 & \Delta_7 = \delta_9 + \delta_{12} \\ \Delta_2 = \delta_2 & \Delta_5 = \delta_4 + \delta_5 & \Delta_8 = \delta_{10} + \delta_{11} + \delta_{13} + \delta_{14} \\ \Delta_3 = \delta_2 + \delta_3 & \Delta_6 = \delta_8 & \Delta_9 = \delta_{15}. \end{array} \quad (1)$$

## 2.3. *Coefficient of kinship and inbreeding coefficient*

For some problems the genotypes of the individuals are irrelevant; it is only useful to know the genetic make-up of their gametes. In such cases the only effect of kinship is to make possible the identity of a gene in a gamete of $X$ and a gene in a gamete of $Y$; states such as $\Sigma_3$ , $\Sigma_5$ , and $\Sigma_7$ are then equivalent. In this way, we can characterize the kinship of $X$ and $Y$ by a single coefficient the so-called "coefficient of kinship" defined, as we have seen above, as the probability $\phi(XY)$ that a gene taken at random in $X$ and a gene taken at random at the same locus in $Y$ will be identical. It immediately follows that the coefficients $\phi$, $\Delta_i$ , and $\delta_i$ satisfy the relations:

$$\begin{array}{l} \phi = \delta_1 + \frac{1}{2}(\delta_2 + \delta_3 + \delta_4 + \delta_5 + \delta_9 + \delta_{12}) + \frac{1}{4}(\delta_{10} + \delta_{11} + \delta_{13} + \delta_{14}) \\ \phi = \Delta_1 + \frac{1}{2}(\Delta_3 + \Delta_5 + \Delta_7) + \frac{1}{4}\Delta_8 . \end{array} \quad (2)$$

Finally, it is sometimes useful to define another coefficient, the "inbreeding coefficient" $f$ of an individual, as being equal to the probability that the two genes he possesses at a locus are identical. Obviously, according to these definitions, the inbreeding coefficient of an individual is equal to the coefficient of kinship of his parents. It follows that the inbreeding coefficients $f_X$ and $f_Y$ are related to the identity coefficients by the relations

$$\begin{array}{l} f_X = \delta_1 + \delta_2 + \delta_3 + \delta_6 + \delta_7 = \Delta_1 + \Delta_2 + \Delta_3 + \Delta_4 \\ f_Y = \delta_1 + \delta_4 + \delta_5 + \delta_6 + \delta_8 = \Delta_1 + \Delta_2 + \Delta_5 + \Delta_6 . \end{array} \quad (3)$$

*Note:* It should be observed that according to these definitions (a) kinship is a characteristic of two individuals $X$ and $Y$; it results from the fact that $X$ and $Y$ have common ancestors; it is measured by their kinship coefficient $\phi(XY)$ or by their identity coefficients $\delta_i$ or $\Delta_i$ and (b) inbreeding is a personal characteristic of an individual $X$; it results from the fact that $X$'s mother and father are themselves related; it is measured by his inbreeding coefficient.

### 2.4. *Computation of identity coefficients*

It is well known that the coefficient of kinship can be computed from the equation

$$\phi(XY) = \sum_i (\tfrac{1}{2})^{n_i+m_i+1}(1 + f_i), \tag{4}$$

where the summation is extended to all the pathways going from $X$ to $Y$ through a common ancestor $i$, $f_i$ is the inbreeding coefficient of the ancestor $i$, $n_i$ is the number of generations between $X$ and $i$, and $m_i$ is the number of generations between $Y$ and $i$ (Malécot [1948], to [1954]).

In most cases equation (4) leads to straightforward calculations. On the other hand, the computation of the identity coefficients $\delta_i$ or $\Delta_i$ is quite involved and tedious, and as soon as ancestry networks become slightly more complex the use of computers becomes necessary. A computer program developed by Nadot and Vayssex [1972] is now available. To compute the identity coefficients of two individuals, when their ancestry network is known for five generations, an IBM 360-50 requires 3 or 4 seconds. Table 3 gives the coefficients $\Delta_i$ and $\phi$ for the kinships which are usually considered.

## 3. CONDITIONAL GENETIC STRUCTURES OF RELATIVES

### 3.1. *Definition of the probabilistic genotype structure and of the probabilistic gene structure of an individual*

Consider the genes possessed by an individual $Z$ at an autosomal locus, and suppose that for this locus there exist $n$ alleles $a_1, \cdots, a_n$ whose frequencies in the population are $p_1, \cdots, p_n$ (with, of course, $p_1 + \cdots + p_n = 1$). In many cases, we do not know exactly which genes are possessed by $Z$, but, according to the information we do have, we can assign probabilities $P_{ij}(Z)$ to the various possible genotypes $(a_i a_j)$. In what follows we shall always have $i, j = 1, 2, \cdots, n, i \leq j$.

By definition the set of the $\frac{1}{2}n(n + 1)$ probabilities $P_{ij}(Z)$ is the "probabilistic genotype structure" of $Z$; we note this structure $S(Z)$. For example, if the only information we have concerning $Z$ is the fact that he belongs to the population, and if we can assume that this population has reached Hardy–Weinberg equilibrium for this locus, we can write $S(Z) = S_p$, where $S_p$, called "equivalent panmictic structure" of the population, is a set of $\frac{1}{2}n(n + 1)$ probabilities defined by

$$S_p\begin{cases}\text{for all } i & P_{ii} = p_i^2\\ \text{for all } i \text{ and } j \neq i & P_{ij} = 2p_ip_j\,.\end{cases}$$

TABLE 3

| Kinship | $\Delta_1$ | $\Delta_2$ | $\Delta_3$ | $\Delta_4$ | $\Delta_5$ | $\Delta_6$ | $\Delta_7$ | $\Delta_8$ | $\Delta_9$ | $\phi$ |
|---|---|---|---|---|---|---|---|---|---|---|
| Father - son | 0 | 0 | 0 | 0 | 0 | 0 | 0 | 1 | 0 | 1/4 |
| Half - sibs | 0 | 0 | 0 | 0 | 0 | 0 | 0 | 1/2 | 1/2 | 1/8 |
| Sibs | 0 | 0 | 0 | 0 | 0 | 0 | 1/4 | 1/2 | 1/4 | 1/4 |
| First-cousins | 0 | 0 | 0 | 0 | 0 | 0 | 0 | 1/4 | 3/4 | 1/16 |
| Double first-cousins | 0 | 0 | 0 | 0 | 0 | 0 | 1/16 | 6/16 | 9/16 | 1/8 |
| Second-cousins | 0 | 0 | 0 | 0 | 0 | 0 | 0 | 1/16 | 15/16 | 1/64 |
| Uncle - nephew | 0 | 0 | 0 | 0 | 0 | 0 | 0 | 1/2 | 1/2 | 1/8 |
| Sibs whose parents are sibs | 1/16 | 1/32 | 1/8 | 1/32 | 1/8 | 1/32 | 7/32 | 5/16 | 1/16 | 3/8 |

Similarly, it is useful for some problems to define the probabilistic gene structure $s(Z)$ of an individual $Z$ as the set of the $n$ probabilities, one for each gene, that a gamete randomly selected among the gametes issued from $Z$ carries that gene. For example, if as above the only information we have concerning $Z$ is the fact that he belongs to the population, we can write $s(Z) = s$, where $s$, gene structure of the population, is the set of the $n$ frequencies $p_i$.

### 3.2. *Conditional genotype structures*

The problem now is: what is the probabilistic genotype structure of an individual $Y$, given

(a) the gene structure $s$ of the population (assumed to be in Hardy–Weinberg equilibrium for the locus) and

(b) the genotype of an individual $X$, related to $Y$ by a parental network characterized by the nine identity coefficients $\Delta_1 \cdots \Delta_9$?

We should first note that the information given by the knowledge of the genotype of $X$ is not the same if this genotype is homozygous $a_k a_k$ or if it is heterozygous $a_k a_l$. When it is heterozygous we know that the two genes of $X$ are not identical, and consequently the identity states which imply this identity, $\Sigma_1$, $\Sigma_2$, $\Sigma_3$, and $\Sigma_4$, have not materialized; conversely, when $X$ is homozygous all the identity states have to be considered. Thus these two cases must be analyzed separately.

*Heterozygous genotype $a_k a_l$ of the relative X*

The only identity states to be considered are $\Sigma_5$, $\Sigma_6$, $\Sigma_7$, $\Sigma_8$, and $\Sigma_9$, the total probability of which is, according to (3), $\Delta_5 + \Delta_6 + \Delta_7 + \Delta_8 + \Delta_9 = 1 - f_X$, where $f_X$ is the inbreeding coefficient of $X$. Let us consider each of these five states of identity between the genes of $X$ and the genes of $Y$ and search for the possible genotypes of $Y$.

State $\Sigma_5$. The probability of this state is $\Delta_5/(1 - f_X)$. In this case the two genes of $Y$ are identical to each other and they are identical either to one gene of $X$ or to the other. Therefore, the possible genotypes of $Y$ are $a_k a_k$ and $a_l a_l$, each of them being equiprobable. We can write

$$S_{Y|\Sigma_5} = \tfrac{1}{2}S_{kk} + \tfrac{1}{2}S_{ll}$$

where $S_{kk}$ is a set of probabilities defined by

$$S_{kk}\begin{cases}\text{if} \quad i = j = k & P_{ij} = 1\\ \text{otherwise} & P_{ij} = 0.\end{cases}$$

State $\Sigma_6$. The probability of this state is $\Delta_6/(1 - f_X)$. In this case the two genes of $Y$ are identical to each other but have no link with the genes of $X$. Therefore $Y$ may have all homozygous genotypes $a_i a_i$, the probability of each being equal to the frequency $p_i$ of the gene $a_i$ in the population. We can write $S_{Y|\Sigma_6} = S_H$, where $S_H$ (called "equivalent homozygous structure of the population") is a set of probabilities $P_{ij}$ defined by

$$S_H\begin{cases}\text{for all } i & P_{ii} = p_i\\ \text{for all } i \text{ and } j \neq i & P_{ij} = 0.\end{cases}$$

State $\Sigma_7$. In this case, the probability of which is $\Delta_7/(1 - f_X)$, each gene of $Y$ is identical to a gene of $X$. The two individuals have, therefore, the same genotype $a_k a_l$. We can then write $S_{Y|\Sigma_7} = S_{kl}$, where $S_{kl}$ is a set of probabilities $P_{ij}$ defined by

$$S_{kl}\begin{cases}\text{if} \quad i = k, j = l & P_{ij} = 1\\ \text{otherwise} & P_{ij} = 0.\end{cases}$$

State $\Sigma_8$. The probability of this state is $\Delta_8/(1 - f_X)$. In this case one of the genes of $Y$ is identical to one of the genes of $X$; it is then either $a_k$ or $a_l$. The other gene of $Y$ has no link with the genes of $X$; it can be any allele $a_j$ with probability $p_j$. $Y$ may have genotypes such as $a_k a_j$ or $a_l a_j$ for any $j$, each with probability $\tfrac{1}{2}p_j$. We can write $S_{Y|\Sigma_8} = S_{F(kl)}$, where $S_{F(kl)}$, called "filial structure of $a_k a_l$", is a set of probabilities $P_{ij}$ defined by

$$S_{F(kl)}\begin{cases}\text{if} \quad i = k, j = l & P_{ij} = \tfrac{1}{2}(p_k + p_l)\\ \text{if} \quad i = k, j \neq l & P_{ij} = \tfrac{1}{2}p_j\\ \text{if} \quad i \neq k, j = l & P_{ij} = \tfrac{1}{2}p_i\\ \text{otherwise } P_{ij} = 0. \end{cases}$$

State $\Sigma_9$. The probability of this state is $\Delta_9/(1 - f_X)$. In this case there is no identity between the genes of $X$ and $Y$. $Y$ may have every genotype $a_i a_j$ with a probability equal to the frequency of this genotype in the population. We can write $S_{Y|\Sigma_9} = S_p$, where $S_p$ is the so-called "equivalent panmictic structure" defined as above.

Finally we can sum up all possible states and write the probabilistic genotype structure of $Y$, given that his relative $X$ has the heterozygous genotype $a_k a_l$ :

$$S_{Y|a_k a_l} = \frac{1}{1 - f_X} (\tfrac{1}{2}\Delta_5(S_{kk} + S_{ll}) + \Delta_7 S_{kl} + \Delta_8 S_{F(kl)} + \Delta_6 S_H + \Delta_9 S_p). \quad (5)$$

*Homozygous genotype $a_k a_k$ of the relative X*

When $Y$'s relative $X$ has a homozygous genotype all the identity states are possible. It is necessary to examine each of these and to determine the probabilities of the possible genotypes of $Y$ in each case. A reasoning similar to that presented in the previous section leads to the equation:

$$S_{Y|a_k a_k} = (\Delta_1 + \Delta_5 + \Delta_7)S_{kk} + (\Delta_3 + \Delta_8)S_{F(kk)} + (\Delta_2 + \Delta_6)S_H + (\Delta_4 + \Delta_9)S_p \quad (6)$$

where $S_{kk}$, $S_H$, and $S_p$ are defined as above and $S_{F(kk)}$ is a set of probabilities defined by

$$S_{F(kk)} \begin{cases} \text{if } i = k & P_{ij} = p_j \\ \text{if } j = k & P_{ij} = p_i \, . \\ \text{otherwise} & P_{ij} = 0 \end{cases}$$

Using the two relations (5) and (6) we can determine the probabilities of the various genotypes of $Y$ regardless of the complexity of the ancestry network by which he is related to the relative $X$ whose genotype is known. These equations may appear sophisticated, but it is impossible to avoid this complication unless the kinship between $X$ and $Y$ is simple.

3.3. *Some particular cases of kinship*

Some of the identity coefficients $\Delta_i$ are nil if the kinship of $X$ and $Y$ has particular characteristics:

(a) $X$ is non-inbred (i.e. his parents are not related). Then $\Delta_1 = \Delta_2 = \Delta_3 = \Delta_4 = f_X = 0$; the relations (5) and (6) may be combined into a single equation

$$S_{Y|a_k a_l} = \tfrac{1}{2}\Delta_5(S_{kk} + S_{ll}) + \Delta_7 S_{kl} + \Delta_8 S_{F(kl)} + \Delta_6 S_H + \Delta_9 S_p , \quad (7)$$

which is valid whether $k = l$ or $k \neq l$.

(b) $X$ and $Y$ are non-inbred. Then $\Delta_1 = \Delta_2 = \Delta_3 = \Delta_4 = \Delta_5 = \Delta_6 = f_X = f_Y = 0$; thus both relations (5) and (6) can be written in the simplified form

$$S_{Y|a_k a_l} = \Delta_7 S_{kl} + \Delta_8 S_{F(kl)} + \Delta_9 S_p , \quad (8)$$

where $\Delta_7$ is the probability that each of the two genes of $Y$ is identical to one of the two genes of $X$; $\Delta_8$ is the probability that only one of the two genes of $Y$ is identical to a gene of $X$; and $\Delta_9$ is the probability that no gene of $Y$ is identical to a gene of $X$.

Equation (8) is fundamentally the same as the one given by Li and Sacks [1954] using "*ITO* matrices"; but we see here that this method is applicable only when both individuals concerned are not inbred (which fortunately is most often the case).

(c) $X$ and $Y$ are not inbred and their kinship is unilineal (i.e. one of the four parents of $X$ and $Y$ is not related to the three others). In this case, one of the four genes of $X$ and $Y$ is necessarily not identical to any of the three others. In addition to the nil coefficients given for (b) above, state $\Sigma_7$ is impossible and $\Delta_7 = 0$; the relation (8) becomes $S_{Y|a_ka_l} = \Delta_8 S_{F(kl)} + \Delta_9 S_p$. But going back to relation (2), we see that in this case $\Delta_8 = 4\phi$, $\Delta_9 = 1 - 4\phi$, and thus the above relation may be put in the form

$$S_{Y|a_ka_l} = 4\phi S_{F(kl)} + (1 - 4\phi)S_p . \tag{9}$$

In this case the relation between the genotypes of two relatives $X$ and $Y$ may be expressed by using their coefficient of kinship $\phi$, but we see how restrictive the required conditions are: not only must $X$ and $Y$ not be inbred but also their kinship must be unilineal. The relation (9) is therefore valid for kinships such as first-cousins or uncle-nephew pairs, but not for sibs or double-cousins.

### 3.4. *Conditional gene structures*

As we have seen, to solve some problems it is sufficient to know the probabilistic gene structure of an individual $Z$, i.e. the set of $n$ probabilities $p_i(Z)$ that a gamete randomly selected among the gametes issued from $Z$ carries a gene $a_i$ ; this set will be denoted $s(Z)$. Knowledge of the genotype structure obviously implies knowledge of the gene structure. If $s$ is the gene structure of the population, i.e. the set of the $n$ frequencies of the alleles $a_i$ ; and $s_k$ is the gene structure of a gamete carrying a gene $a_k$ , i.e. a set of $n$ frequencies $p_i$ such that if $i = k$, $p_i = 1$ and $i \neq k$, $p_i = 0$, we can write the correspondence between the various genotype structures previously defined and the gene structures:

$$S_{kk} \rightarrow s_k \qquad\qquad S_H \rightarrow s$$

$$S_{kl} \rightarrow \tfrac{1}{2}s_k + \tfrac{1}{2}s_l \qquad\qquad S_p \rightarrow s$$

$$S_{F(kl)} \rightarrow \tfrac{1}{4}s_k + \tfrac{1}{4}s_l + \tfrac{1}{2}s \qquad\qquad S_{F(kk)} = \tfrac{1}{2}s_k + \tfrac{1}{2}s.$$

Referring to equations (5) and (6), we see that the gene structure of an individual $Y$, given the genotype of his relative $X$, is

(a) if $X$ is heterozygous $a_ka_l$

$$s_{Y|a_ka_l} = 1/(1 - f_X)[(\tfrac{1}{2}\Delta_5 + \tfrac{1}{2}\Delta_7 + \tfrac{1}{4}\Delta_8)(s_k + s_l) + (\Delta_6 + \tfrac{1}{2}\Delta_8 + \Delta_9)s];$$

(b) if $X$ is homozygous

$$s_{Y|a_ka_k} = (\Delta_1 + \tfrac{1}{2}\Delta_3 + \Delta_5 + \Delta_7 + \tfrac{1}{2}\Delta_8)s_k + (\Delta_2 + \tfrac{1}{2}\Delta_3 + \Delta_4 + \Delta_6 + \tfrac{1}{2}\Delta_8 + \Delta_9)s.$$

These two relations may be put in the same form when the relative $X$ is non-inbred. Indeed, in this particular case, we have $f_X = 0$, $\Delta_1 = \Delta_2 = \Delta_3 = \Delta_4 = 0$, and these relations become

$$s_{Y|a_ka_l} = (\Delta_5 + \Delta_7 + \tfrac{1}{2}\Delta_8)\tfrac{1}{2}(s_k + s_l) + (\Delta_6 + \tfrac{1}{2}\Delta_8 + \Delta_9)s \qquad (10)$$

and

$$s_{Y|a_ka_k} = (\Delta_5 + \Delta_2 + \tfrac{1}{2}\Delta_8)s_k + (\Delta_6 + \tfrac{1}{2}\Delta_8 + \Delta_9)s. \qquad (11)$$

But, in (10) $\frac{1}{2}(s_k + s_l)$ is simply the probabilistic gene structure $s_X$ of the heterozygote $X$, and in (11) $s_k$ is the probabilistic gene structure $s_X$ of the homozygote $X$. Furthermore, referring to relations (2), we see that $\Delta_5 + \Delta_7 + \frac{1}{2}\Delta_8 = 2\phi$ and $\Delta_6 + \frac{1}{2}\Delta_8 + \Delta_9 = 1 - 2\phi$, since we assume that $\Delta_1 = \Delta_2 = \Delta_3 = \Delta_4 = 0$. Finally (10) and (11) may be written

$$s_{Y|X} = 2\phi s_X + (1 - 2\phi)s. \qquad (12)$$

This relation, which may be demonstrated directly without considering the genotype structures, shows that the coefficient of kinship is sufficient to determine the gene structure of an individual if we know the gene structure of one of his relatives, but this is true only when this relative is not himself inbred. Otherwise, it is necessary to use the set of identity coefficients $\Delta_i$ and to refer to equations (10) and (11).

## 4. SOME APPLICATIONS

### 4.1. *Notation*

For specific applications the gene structures $s$ can be easily represented by a row-vector with $n$ components. On the other hand, the genotype structures $S$ have $\frac{1}{2}n(n + 1)$ components, and it is convenient to represent such a set of probabilities by means of a triangular array (trimat) of $n$ columns and $n$ rows. This is the lower triangular part of a square matrix in which the probability corresponding to the homozygous genotype $a_ia_i$ is on the hypotenuse on the $i$th row and column, and the probability corresponding to the heterozygous genotype is in the $i$th column and $j$th row, $i > j$. For example, in the case of the ABO blood system, the probabilities of the six genotypes can be presented in the following order:

$$S \equiv \begin{array}{|lll} P_{\text{AA}} & & \\ P_{\text{AB}} & P_{\text{BB}} & \\ P_{\text{AO}} & P_{\text{BO}} & P_{\text{OO}} \\ \hline \end{array},$$

while the set of probabilities of the three genes is the row-vector $s \equiv$

$(p_A \, q_B \, r_O)$. Using this notation, the equivalent panmictic and homozygous structures of a population, whose gene structure is $s = (p\ q\ r)$, are

$$S_p = \left| \begin{array}{lll} p^2 & & \\ 2pq & q^2 & \\ 2pr & 2qr & r^2 \end{array} \right. \quad \text{and} \quad S_H = \left| \begin{array}{lll} p & & \\ 0 & q & \\ 0 & 0 & r \end{array} \right. ,$$

and the "filial" structures of an individual with, for example, AA or AB genotypes are

$$S_{F(AA)} = \left| \begin{array}{lll} p & & \\ q & 0 & \\ r & 0 & 0 \end{array} \right. \quad \text{and} \quad S_{F(AB)} = \left| \begin{array}{lll} \frac{1}{2}p & & \\ \frac{1}{2}(p+q) & \frac{1}{2}q & \\ \frac{1}{2}r & \frac{1}{2}r & 0 \end{array} \right. .$$

4.2 *Relatives of an individual whose blood group is* AB

Let us assume that the only information we have concerning an individual $Y$ is that he belongs to a population of gene structure $s = (p\ q\ r)$ for the blood group ABO, and the fact that one of his relatives $X$ has the blood group AB. The kinship between $X$ and $Y$ is measured by a set of nine identity coefficients $\Delta_i$. Using the trimat notation, equation (5) gives

$$\left| \begin{array}{lll} P_{AA} & & \\ P_{AB} & P_{BB} & \\ P_{AO} & P_{BO} & P_{OO} \end{array} \right. = \frac{1}{1-f_X} \left\{ \tfrac{1}{2}\Delta_5 \left| \begin{array}{lll} 1 & & \\ 0 & 1 & \\ 0 & 0 & 0 \end{array} \right. + \Delta_7 \left| \begin{array}{lll} 0 & & \\ 1 & 0 & \\ 0 & 0 & 0 \end{array} \right. \right.$$

$$\left. + \tfrac{1}{2}\Delta_8 \left| \begin{array}{lll} p & & \\ p+q & q & \\ r & r & 0 \end{array} \right. + \Delta_6 \left| \begin{array}{lll} p & & \\ 0 & q & \\ 0 & 0 & r \end{array} \right. + \Delta_9 \left| \begin{array}{lll} p^2 & & \\ 2pq & q^2 & \\ 2pr & 2qr & r^2 \end{array} \right. \right\}.$$

For example, in the case where $X$ and $Y$ are sibs issued from sibs, we have $\Delta_5 = \frac{1}{8}$, $\Delta_6 = \frac{1}{32}$, $\Delta_7 = \frac{7}{32}$, $\Delta_8 = \frac{5}{16}$, $\Delta_9 = \frac{1}{16}$, $f_X = \frac{1}{4}$ from which we see

$$S_{Y|AB} = \frac{4}{3} \left\{ \frac{1}{16} \left| \begin{array}{lll} 1 & & \\ 0 & 1 & \\ 0 & 0 & 0 \end{array} \right. + \frac{7}{32} \left| \begin{array}{lll} 0 & & \\ 1 & 0 & \\ 0 & 0 & 0 \end{array} \right. + \frac{5}{32} \left| \begin{array}{lll} p & & \\ p+q & q & \\ r & r & 0 \end{array} \right. \right.$$

$$\left. + \frac{1}{32} \left| \begin{array}{lll} p & & \\ 0 & q & \\ 0 & 0 & r \end{array} \right. + \frac{1}{16} \left| \begin{array}{lll} p^2 & & \\ 2pq & q^2 & \\ 2pr & 2qr & r^2 \end{array} \right. \right\}.$$

The probability that $Y$'s genotype is AB is thus $\Pr\{Y(AB) \mid X(AB)\} = \frac{1}{24}(7 + 5p + 5q + 4pq)$.

### 4.3. *Offspring of relatives for whom one genotype is known*

Let us assume that $F$ and $M$ are double-cousins, and that $F$ has for the ABO blood system the genotype AB. What are the probabilities of the various possible genotypes of their child $C$?

The gene structure of $F$ is $s_F = (\frac{1}{2}, \frac{1}{2}, 0)$. The gene structure of $M$, knowing that the kinship coefficient between $F$ and $M$ can be written $\phi = \frac{1}{8}$ (double first cousins) is, according to equation (12),

$$s_{M\mid F} = \tfrac{1}{4}(\tfrac{1}{2}\ \tfrac{1}{2}\ 0) + \tfrac{3}{4}(p\ q\ r) = \tfrac{1}{8}(1 + 6p\ \ 1 + 6q\ \ 6r).$$

It is then easy to compute the probabilistic genotype structure of the child $C$:

$$S_C = \frac{1}{16}\left|\begin{array}{ccc} 1 + 6p & & \\ 2 + 6(p + q) & 1 + 6q & \\ 6r & 6r & 0 \end{array}\right. .$$

### 4.4. *Probability of being a carrier of a recessive gene*

Let us consider a deleterious recessive autosomal gene $d$, whose frequency in the population is $q$; normal gene $D$ has the frequency $(1 - q)$. We know that an individual $X$ suffers from the condition caused by this gene; $X$'s genotype is then $dd$. What is the probability for a relative $Y$ of $X$, who does not suffer from this condition, to be a carrier of gene $d$?

Let $\Delta_i$ be the identity coefficients between $Y$ and $X$; equation (6) written in trimat notation gives the probabilities for $Y$ to have the genotypes $DD$, $Dd$, and $dd$:

$$S_Y = (\Delta_1 + \Delta_5 + \Delta_7)\left|\begin{array}{cc} 0 & \\ 0 & 1 \end{array}\right. + (\Delta_3 + \Delta_8)\left|\begin{array}{cc} 0 & \\ 1 - q & q \end{array}\right.$$

$$+ (\Delta_2 + \Delta_6)\left|\begin{array}{cc} 1 - q & \\ 0 & q \end{array}\right. + (\Delta_4 + \Delta_9)\left|\begin{array}{cc} (1 - q)^2 & \\ 2q(1 - q) & q^2 \end{array}\right. .$$

Now we know that $Y$ does not have the genotype $dd$; thus the probability that he is a $d$-carrier is:

$$\Pr\{Y(Dd)\} = P_{Dd}/(P_{Dd} + P_{DD})$$

$$= \frac{\Delta_3 + \Delta_8 + 2(\Delta_4 + \Delta_9)q}{\Delta_2 + \Delta_3 + \Delta_6 + \Delta_8 + (\Delta_4 + \Delta_9)(1 + q)} .$$

As an example let us consider the pedigree of two Indians $A$ and $B$ of the Jicaques tribe studied by Chapman *et al.* [1971]. They are sibs and their parents have many common ancestors as shown in Figure 1.

The identity coefficients of $A$ and $B$ are calculated to be:

$$\Delta_1 = 0.0657 \qquad \Delta_4 = \Delta_6 = 0.0238$$

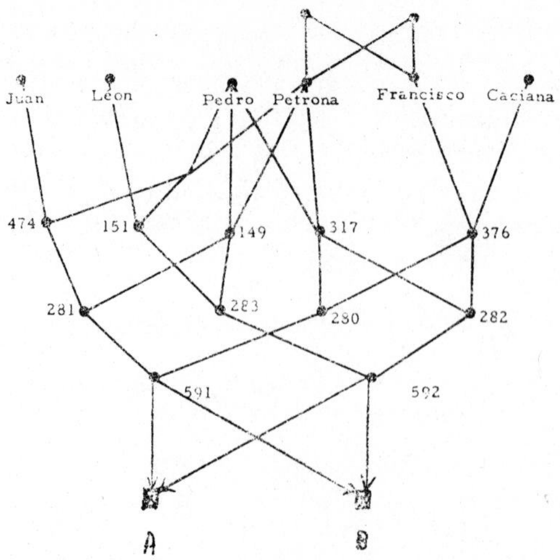


FIGURE 1

$$\Delta_2 = 0.0073 \qquad \Delta_7 = 0.2544 \qquad \Delta_9 = 0.0859.$$

$$\Delta_3 = \Delta_5 = 0.0867 \qquad \Delta_8 = 0.3657$$

If we know that $A$ has a disease caused by a rare recessive gene $dd$ whose frequency is $q \simeq 0$ the probabilistic genotype structure of $B$ is

$$S_{B|dd} = 0.41\left|\begin{array}{ll} 0 & \\ 0 & 1 \end{array}\right. + 0.45\left|\begin{array}{ll} 0 & \\ p & q \end{array}\right. + 0.03\left|\begin{array}{ll} p & \\ 0 & q \end{array}\right. + 0.11\left|\begin{array}{ll} p^2 & \\ 2pq & q^2 \end{array}\right. \simeq \left|\begin{array}{ll} 0.14 & \\ 0.45 & 0.41 \end{array}\right. .$$

If $B$ were a child not yet born, the probability that he would be homozygous $dd$ is 0.41, instead of 0.25 as expected for sibs whose parents are not related; if $B$ is a living child who does not suffer from this disease, the probability that he is a carrier is equal to $0.45/0.59 = 0.76$, instead of 0.67 as expected for two ordinary sibs.

## RESUME

### INFORMATION GENETIQUE FOURNIE PAR UN APPARENTE

Le propos de cet article est de répondre à la question: quelle information génétique peut-on obtenir au sujet d'un individu à partir de la connaissance du génôme d'un de ses apparentés? Le 'coefficient de parenté est en général insuffisant pour résoudre ce problème. Il faut avoir recours à une mesure plus fine de l'apparentement: l'ensemble des 9 coefficients d'identité contractés.

Cet article passe en revue ces coefficients et leur emploi pour établir des relations donnant la structure génotypique (ensemble des probabilités des divers génotypes) d'un individu, connaissant le génotype d'un apparenté, et l'ensemble des fréquences géniques de la population. Lorsque le réseau parental est simple les relations peuvent être mises sous une forme nécessitant seulement la connaissance du coefficient de parenté. Mais les conditions pour qu'une telle simplification soit possible sont assez restrictives; il faut non seulement que les deux individus considérés soient non consanguins mais que leur apparentement soit unilinéal.

## REFERENCES


Chapman, A. and Jacquard, A. [1971]. Un isolat d'Amérique Centrale: les indiens Jicaques de Honduras. *Génétique et Population*, P. U. F. Paris

Gillois, M. [1964]. La relation d'identité en génétique. Thèse Faculté des Sciences de Paris.

Li, C. C. and Sacks, L. [1954]. The derivation of joint distribution and correlation between relatives by the use of stochastic matrices. *Biometrics 10*, 347–60.

Li, C. C. [1954]. *Population Genetics.* University of Chicago Press, Chicago.

Malécot, G. [1948]. *Les mathématiques de l'hérédité.* Masson et Cie, Paris.

Malécot, G. [1966]. *Probabilités et hérédité.* Presses Universitaires de France, Paris.

Nadot, R. and Vayssex, G. [1972]. Computation of identity coefficients. Submitted to *Biometrics.*



*Received September* 1971, *Revised April* 1972

*Key Words:* Genetic structures; Identity coefficients between relatives; Measure of kinship; Population genetics.

# Editor's Comments on Paper 21

**21 Schull and Neel:** The Interpretation of the Effects of Inbreeding
*The Effects on Inbreeding of Japanese Children,* Harper & Row, Inc., New York, 1965, pp. 328–351

A large number of genetically significant problems are definitely related or relatable to the system of mating of a natural population. Empirically, one must ask what the particular conditions are or were, but theoretically, one must ask what they could be under specified conditions. In the introduction and elsewhere in this volume I have loosely grouped many such questions as statistical, hence not of primary concern in a structural study. I have included this paper precisely because it reviews these problems from a considerably more classical view than is taken by the present volume while embedding its discussion in our historic context.

The text from which this selection is borrowed, as well as numerous other studies by Schull and Neel and their associates, should certainly be studied by those interested in the *empirical* side of the present problems. Schull, in particular, has pioneered in using social theoretical concepts to orient genetic studies (Schull, 1958, was a candidate for inclusion in this volume, omitted because of its size). His work should be considered along with the referenced papers of Haldane for a complete study of genetics and social structure from an empirical perspective. In addition, Yasuda and Morton (1967) and Morton (1968) provide reasonably comprehensive and recent surveys of the problems and literatures of population biology of human natural populations. The complete issue of *Eugenics Quarterly* (1968) should also be reviewed for surveys of topics not considered here, as should Spuhler (1972).

# 21



Reprinted from *The Effects of Inbreeding on Japanese Children*, by W. J. Schull and J. V. Neel, 328–351 (1965) with permission of Harper & Row, Inc.

# *The Interpretation of the Effects of Inbreeding*

W. J. SCHULL AND J. V. NEEL

Attempts to delineate the risks associated with inbreeding and to understand their implications with respect to human evolution are not new. The nineteenth century saw a substantial effort in this direction. Unfortunately, these earlier investigations were often undertaken with the aim of demonstrating the deleterious nature of consanguineous matings, on the one hand, or their harmlessness, on the other. Seldom were they pursued in an unfettered spirit. An even greater handicap to their success than the special interests which motivated so many was the absence of a theory of inheritance which afforded some direction to the investigative efforts. The rediscovery of Mendel's work and the spread of a particulate notion of heredity provided this missing, and needed, frame of reference. Since 1900, interest in inbreeding has centered to a large extent on the changes in gene, genotype, and phenotype frequencies to be expected in populations following specified systems of inbreeding. While Bateson (1902) appears to have been the first to recognize, at least in a general way, the effect of consanguineous matings upon the manifestation of rare genes, numerous other individuals soon made substantial contributions to our understanding of the effects of inbreeding on the genetic composition of a population (e.g., Pearl, 1913, 1914; Fish, 1914; Jennings, 1914, 1916, 1917; Robbins, 1917, 1918; East and Jones, 1919; and Wright, 1921). Our purpose in this chapter, however, is not to review in detail the contributions of these and other subsequent investigators but rather, having provided a brief theoretical framework for the effects of inbreeding, to examine certain recent efforts to analyze population structure through the use of inbreeding data, particularly data on man, and to consider the bearing of the present data on these efforts.

**15.1. Inbreeding and Homozygosity.** As early as 1908, G. H. Hardy and W. Weinberg had established that in an infinitely large, randomly mating population, gene and genotype frequencies were invariant in the absence of mutation and selection. This finding has been and is a central principle of population genetics. It was soon patent, however, that the Hardy-Weinberg law, as it is now generally designated, is not and cannot be applicable to all populations; in some there is inbreeding or assortative mating, and in most mutation and selection are occurring. Clearly, then, it was of more than heuristic interest to know the effect of departures from random mating on gene and genotype frequencies in the absence as well as in the presence of mutation and selection. Of particular concern is the departure from randomness which obtains when mates are more closely related biologically than they would be if they were chosen at random from the population. The effect of consanguineous matings on the genetic composition of the population can be most readily comprehended by a consideration of the most extreme form of inbreeding, namely, self-fertilization.

Suppose there existed a randomly mating population in which self-fertilization suddenly became obligatory. It is clear that homozygous individuals in this population would give rise only to other homozygous individuals. Heterozygous individuals, however, would be expected to produce homozygous and heterozygous progeny with equal frequency. Thus, if all individuals were equally fertile, it is obvious that the proportion of heterozygosity in the population must diminish by one-half with each generation of self-fertilization. It follows, intuitively, that the only effect that inbreeding can have, per se, is to increase the frequency of homozygotes at the expense of the heterozygotes, and hence the increase in uniformity within a single line, the usual decline in vigor in various respects, and the other consequences generally associated with inbreeding in plants and animals must largely be consequences of the increase in homozygosis.

While the rate of decay in heterozygosis is a function of the particular system of inbreeding, it is important to bear in mind that in an infinitely large population complete homozygosis is not a necessary consequence of inbreeding. Certain systems of mating will lead to genetic equilibria short of complete homozygosis (cf. Wright, 1921). Examples of such are obligatory second-cousin marriage or obligatory half-first-cousin marriage, and so forth. These remarks are not true for populations of finite size, however, where complete homozygosis can occur even in a randomly mating population. It should also be noted that since inbreeding merely alters genotype frequencies without altering gene frequencies (in the absence of selection), it follows that where gene action is additive, inbreeding will have no effect on the average of a population but will increase the variability between families while diminishing that within families.

**15.2. The Notion of a "Genetic Load."** With this brief background, let us now direct our attention to matters of more immediate concern, namely, the insight which inbreeding data may provide into the maintenance of genetic variability in populations. Evolution may be defined as the systematic changes in gene frequencies which occur in populations. More precisely, the study of evolution involves the study of those factors which determine the persistence and spread of a mutant gene (or genes) in a population. Important in this regard are two classes of forces, namely, one whose effects are "directional" and one whose effects lack direction and are, in a sense, merely "dispersive." Among the former are the mating formula of the population, mutation, migration, and selection. Among the latter are the random fluctuations in gene frequency which stem from the size of the population as well as random variations in mutation rates, selection intensities, and so forth. Despite the existence of directional forces which tend to the fixation of genes, almost without exception all populations adequately studied to date have shown a degree of genetic variability which requires explanation. Two simple models have been proposed to account for this persistence of variability (intermediates are, of course, possible if the genetic variability rests on the occurrence of more than two genes at the relevant locus or upon genes at more than one locus). Experience tells us that most new mutations are deleterious and that, accordingly, there must exist an equilibrium between the frequency of their origin and their elimination. Alternatively stated, we recognize that at some genetic loci the allelic polymorphism which exists represents in the main an equilibrium between selection and recurrent mutation. When this obtains—that is, when the variability persists because of an equilibration of selection with mutation—we describe the locus in question as being "classical." Not all genetic variability is of this kind; there exist instances wherein the polymorphism apparently stems from an equilibrium between selective forces, with mutation playing a very minor role. Loci of this kind have been termed "balanced," or heterotic. In either event, it is obvious that the maintenance of the genetic variability involves a disproportionate failure of certain genotypes to survive and/or reproduce. This failure, loss, burden, or, if you will, "load" is the debt paid for the existence of genetic variability.

Inasmuch as the manner in which genetic variability is maintained is perhaps the central problem of population genetics, and inasmuch as "classical" and "balanced" loci respond rather differently to changes in mutation rates or selection intensities, increasingly interest has converged on the relative role of these two kinds of loci in the maintenance of the "load" of a population. It seems clear, however, that if these roles are to be evaluated, an explicit definition of the "load" of a population is needed. Crow (1958) has defined the genetic load of a population as "the extent to which it (the population) is impaired by the fact that not all individuals are of the optimum type." Optimality in this context implies fitness; the usual metric for the latter has been derived from the expected number of children of an individual, counted at the same age as that individual. We shall adopt a metric of that type in what follows, while recognizing that, especially in the case of man, it is in some respects an oversimplification of a complex issue (cf. Huxley, 1962). But even with this relatively simple definition of fitness, it is, in practice, not always an easy task to estimate the mean number of progeny associated with individuals of a given genotype (or phenotype for that matter). We shall ignore these problems in the pages to follow and assume that fitness can, in fact, be measured. If we are concerned only with gene frequencies, as is frequently true, then the absolute fitnesses of an array of genotypes can all be multiplied by a common constant without altering the relationship of these genotypes one to another. We speak of such fitnesses as relative, and they represent the ratio of the mean number of children produced by some genotype relative to a standard. The choice of the standard is, as we shall see, a matter of disagreement.

In the evaluation of the relative roles of "classical" as contrasted to "balanced" loci in maintaining the genetic load of a population, we are primarily concerned with four questions. These have been succinctly formulated by Crow (personal communication) as follows:

1. Does mutation result in mutants of the "balanced" or segregational type at the majority (or at least a substantial minority) of loci?

2. Do the majority (or at least a substantial minority) of loci in an equilibrium population have two or more alleles of the balanced kind maintained in the population?

3. Is the concealed (apparent upon inbreeding) variability in a population mainly due to genes associated with balanced loci?

4. Is the expressed variability in a population mainly due to genes associated with "balanced" loci?

Questions 1 and 2, it will be noted, involve estimates of the number of loci of the two kinds, whereas Questions 3 and 4 involve estimates of the magnitude of the effect to be ascribed to these different loci. Our primary concern in this chapter is with the amount of information that studies on inbreeding can contribute to answers to these questions. To determine this requires some further consideration of the genetic load and, especially, its components.

**15.3. Some Components of the Load.** It is apparent upon reflection that the load, in the sense defined by Crow, can be partitioned into a number of discrete components. A convenient classification of the more important of these follows:

1. *Mutational load.* This is the extent to which the fitness of a population is diminished by recurrent mutation. In principle, the mutational load can be measured by a comparison of the fitness of a specified population with that which would obtain in that population if the mutation rate were zero and the gene frequencies in equilibrium.

2. *Segregational load.* As we have previously stated, in some instances the individuals in a population with maximum fitness may not be homozygous. When this obtains, these individuals may be expected to produce persons less fit than themselves as a consequence of segregation. The segregational load of a population is the extent to which these segregants reduce the fitness of the population.

3. *Recombinational load.* If two or more linked loci are considered simultaneously, then one can define a *recombinational load*. By analogy with the segregational load, the recombinational load is the loss in population fitness occasioned by the occurrence of recombinants less fit than the optimum.

4. *Substitutional load.* It is conceivable that a change in the environment may give rise to a situation wherein the genotype favored prior to the change is not favored after it. The burden imposed upon the population by virtue of the fact that the newly favored genotypes are rare and the poorer ones common has been termed the substitutional load. But if the environmental change is for the better and accompanied by a rapid expansion of the species, it is by no means clear that the load here involved can be treated in the same fashion as the load of recurrent mutation.

5. *Incompatibility load.* We recognize that certain genotypes have reduced fitness with certain parents, for example, the Rh-positive child of the Rh-negative mother. In this instance, the fitness associated with a given genotype does not stem solely from some inherent property of that genotype but varies as a function of another's genotype. As yet, the only genetic systems known wherein an incompatibility load seems to exist involve serotypes.

6. *Migrational load.* The fitness of a population may be lowered through the immigration of less favored genotypes or through the emigration of favored ones; conversely, the fitness of a population may be raised by the immigration of favored genotypes or the emigration of the less favored. As in the case of the mutational load, the load to be ascribed to migration can, in theory at least, be evaluated by a comparison of the fitness of a particular population with that which would obtain in that population if the migration rate were zero and the gene frequencies in equilibrium. Again, as in the case of the substitutional load, if migration introduces a fitter genetic complement and the population expands, the load due to the fact that all individuals are not of the favored type would seem to differ importantly from that imposed by recurrent mutation.

There are still other types of loads which can be identified, but at this writing the above enumerated appear to be those of greatest importance to man. Intuitively one recognizes that the evaluation of one or more of these components must vary with the genetic nature of the polymorphism under scrutiny. There are, however, a limited number of genetic systems of sufficient prevalence to warrant consideration, and it is to a brief account of these that we now turn.

**15.4. The Types of Genetic Systems for Which Inbreeding Effects Must Be Considered.** Basic to the belief that inbreeding data can be used to investigate population structure is the assumption that different genetic systems will respond to inbreeding in different manners. To ascertain whether this is, in fact, true one must examine singly the types of genetic systems which might underlie a given polymorphism. Conceivably, any one of the following might obtain:

1. The most fit genotype might be a homozygote (or homozygotes if more than two kinds exist). This represents, of course, the "classical locus" to which reference has already been made and where, it will be recalled, there is an equilibrium between selection and recurrent mutation.

2. The most fit genotype might be a heterozygote (or heterozygotes if more than one such type exists). We have previously referred to this case as the "balanced locus." It will be recalled that "balanced" implies that the equilibrium is primarily maintained by a balance of opposing selective forces.

3. The most fit genotype may vary with time or environments. Thus, for example, at one time or in one environment the fitness order may be $AA > Aa > aa$ whereas at another time or in another environment the order may be $AA < Aa < aa$. It will be noted that at no time or in no environment is the heterozygote favored, but over a long period of time (or numerous environments) its mean fitness may be greater than either homozygote. This example serves to illustrate that the short-term optimum need not be the long-term one, and thus the "true" optimum may be operationally impossible to define save for a few special characteristics such as survival.

4. In a system of multiple alleles, more than one genotype may have optimum fitness. If this obtains, it is conceivable that the favored genotypes will involve only homozygous individuals, only heterozygous individuals, or some combination of the two. The number and genetic nature of the genotypes with optimum fitness will determine whether an equilibrium, if one exists, involves a balance of (1) selection and recurrent mutation, as is the case if one homozygous class is superior to all others; (2) opposing selective forces, as in the case where all heterozygotes are optimally fit; or (3) both. It is in this latter possibility, where some genes may be maintained by a balance of opposing selective forces and other genes by a balance of selection and recurrent mutation—the so-called "mixed locus"—that we are here interested. As a simple but by no means the only illustration of a mixed locus consider a population wherein three genes, say, $A_1$, $A_2$, and $A_3$, occur with frequencies $p$, $q$, and $r$. Let us assume that mating is at random, but not all genotypes are equally fit. Let the fitness be $A_1A_1 = A_1A_2 = A_2A_1 = A_2A_2 = 0$; $A_1A_3 = A_3A_1 = A_3A_3 = 1 - s$; $A_2A_3 = A_3A_2 = 1$. Suppose that $A_3$ mutates to $A_1$ with frequency $\mu$ but all other mutation rates are negligible.

It can readily be shown that under these circumstances the gene frequencies at equilibrium will be, to a close approximation,

$$p \sim \frac{\mu(1+s)}{2s}$$

$$q \sim \frac{s}{1+s}$$

$$r \sim \frac{1}{1+s}$$

These results, it will be noted, are intuitively satisfying since the relationship of $A_1$ to $A_3$ is that of a "classical" pair of alleles whereas the relationship of $A_2$ to $A_3$ is that of a balanced pair of alleles and the equilibrium values are a mixture of the simpler two-allele cases.

5. The four systems thus far enumerated have in common the fact that the polymorphism is presumably the result of genetic variation at only one locus. More complex, multifactorial systems are obviously possible, in which selection either favors individuals whose phenotype rests upon being simultaneously homozygous at multiple loci or simultaneously heterozygous at multiple loci. These two situations may be viewed as the multifactorial counterparts to the "classical" and "balanced" locus hypotheses.

6. Finally, and again with respect to multifactorial systems, selection might favor individuals who were simultaneously heterozygous at more than one locus, with the phenotype against which selection is directed resulting from a number of different homozygous states. This possibility has been championed by Lerner (1954) and forms the keystone of the notion of "genetic homeostasis."

As we have previously stated, if inbreeding data are to be of value in the analysis of population structure, then clearly the frame of reference within which such data are to be used must permit discrimination between some combination (and ideally all) of the genetic possibilities just enumerated. Does there now exist a frame of reference or a theory capable of providing this discrimination?

**15.5. The Morton-Crow-Muller Attempt to Develop a Theoretical System for Utilizing Inbreeding Data to Analyze Population Structure.** The earliest efforts to utilize inbreeding data from man (Muller, 1948, 1950; Russell, 1952; Reed, 1954; Slatis, 1954; Penrose, 1957) were, with the exception of Muller's work, primarily directed toward a single problem, namely, the estimation of the average number of deleterious genes possessed by the average individual, without reference to the question of how these genes were maintained in the population. The application of the various formulations was so handicapped by the limited data available that the results are largely of historic interest. It was not until 1956 that Morton, Crow, and Muller advanced a method with considerably broader objectives; the method was enlarged upon in a number of subsequent publications, notably by Crow (1958) and Morton (1960, 1961). The argument proceeds somewhat as follows:

Consider the $k$th genetic locus ($k = 1, 2, \ldots, N$) at which any one of $n$ alleles may occur. Let $q_i$ be the frequency of the $i$th allele, say, $A_i$ ($i = 1, 2, \ldots, n$). Then for a diploid population inbreeding at some fixed rate, $F$, we may write

| | | | |
|---|---|---|---|
| GENOTYPE: | $A_iA_i$ | $A_iA_j$ | $A_jA_j$ |
| FREQUENCY: | $q_i^2(1-F)+q_iF$ | $2q_iq_j(1-F)$ | $q_j^2(1-F)+q_jF$ |
| FITNESS: | $m_{ii}$ | $m_{ij}$ | $m_{jj}$ |

where $m_{ii}$, $m_{ij}$, and $m_{jj}$ are the mean number of offspring of individuals with the genotypes $A_iA_i$, $A_iA_j$, and $A_jA_j$, respectively. We presume that these values have been determined at that stage in the offspring generation equivalent to the stage in the parental generation for which the assigned frequencies are appropriate, with each actual child scored as 0.5 children. The expected number of children of an individual randomly selected from this population is, of course,

$$E(m) = \sum_{i=1}^{n} m_{ii}[q_i^2(1 - F) + q_i F] + 2 \sum_{\substack{i,j=1 \\ i<j}}^{n} m_{ij} q_i q_j (1 - F)$$

The population will be decreasing, stable, or increasing depending upon whether $E(m)$ is less than, equal to, or greater than 1.

Since we are here concerned only with gene frequencies, we may replace the absolute fitnesses, the $m$, with relative fitnesses. This is tantamount to multiplying each of the $m$'s by a common constant. The relative fitnesses we might designate as $r_{ii}$, $r_{ij}$, and $r_{jj}$, respectively. They represent the ratios of the expected number of offspring of individuals of genotypes $A_iA_i$, $A_iA_j$, and $A_jA_j$ to the expected number of offspring of some standard genotype. Morton, Crow, and Muller (1956) select as the standard the optimum genotype, i.e., the one with the largest absolute fitness.

Now it will be recalled that the load of a population is defined as the loss in fitness attributable to the fact that not all individuals in the population are of the optimum genotype. Thus, in a stable population, the load at the $k$th locus, say ,$L_F^{(k)}$, is

$$L_F^{(k)} = \sum_{i=1}^{n} (1 - r_{ii})[q_i^2(1 - F) + q_i F] + 2 \sum_{\substack{i,j=1 \\ i<j}}^{n} (1 - r_{ij})[q_i q_j (1 - F)]$$

In most treatments it is convenient to contrast the load in two specific populations, namely, where random mating occurs—that is, where $F = 0$—and where complete inbreeding occurs—that is, where $F = 1$. Evaluation of the more general case at these two points reveals

$$L_0^{(k)} = \sum_{i=1}^{n} (1 - r_{ii})q_i^2 + 2 \sum_{\substack{i,j=1 \\ i<j}}^{n} (1 - r_{ij})q_i q_j$$

and

$$L_1^{(k)} = \sum_{i=1}^{n} (1 - r_{ii})q_i$$

In both instances, we assume $q_i$ and $q_j$ to be the equilibrium frequencies. The metric by which Morton, Crow, and Muller propose to evaluate the contributions to the load of classical as contrasted with heterotic loci is the ratio of loads, $L_0^{(k)}$ and $L_1^{(k)}$. This we observe to be

$$\frac{L_1^{(k)}}{L_0^{(k)}} = \frac{\sum_{i=1}^{n} (1 - r_{ii})q_i}{\sum_{i=1}^{n} (1 - r_{ii})q_i^2 + 2 \sum_{\substack{i,j=1 \\ i<j}}^{n} (1 - r_{ij})q_i q_j}$$

We turn now, to their demonstration that in theory, at least, this ratio discriminates between classical and balanced loci provided certain assumptions are fulfilled. For convenience, consider the two-allele case. As we have previously pointed out, Morton, Crow, and Muller use the optimal genotype as the standard to which the individual fitnesses are made relative. If we are concerned with a classical locus, this implies, of course, that the optimum is a homozygote, say, the class $AA$. Thus, we may take as the relative fitnesses associated with the genotypes $AA$, $Aa$, and $aa$ the values $1$, $1 - hs$, and $1 - s$ where $s$ is the selective disadvantage of the $aa$ homozygote and $h$ may be viewed as a measure of dominance and $0 \leqslant h \leqslant 1$, $0 \leqslant s \leqslant 1$. It can be easily shown that the equilibrium frequency of the allele $a$—say, $q_a$—is the real, positive root of the equation

$$sq^2(1 + \mu)(1 - 2h) + qhs(1 + 2\mu) - \mu = 0$$

where $\mu$ is the rate of mutation of $A$ to $a$. We note that if $h = 0$ (the case where $aa$ is the recessive phenotype) then

$$q_a \sim \sqrt{\frac{\mu}{s}}$$

If $h = 1$ (the case where $aa$ is the dominant phenotype), then

$$q_a \sim \frac{\mu}{s}$$

whereas if $h = 1/2$ (the case of no dominance),

$$q_a \sim \frac{2\mu}{s}$$

If any one of these values is substituted in the ratio given above, we find that to a close approximation

$$\frac{L_1}{L_0} \sim \frac{1}{2h}$$

Since we presume $h$ to be small (the data from Drosophila suggest a value of the order of 0.02), the above ratio will be large.

Consider, now, the two-allele case when selection favors the heterozygote. The relative fitnesses are, in this instance, $1 - t$, 1, and $1 - \alpha t$ for $AA$, $Aa$, and $aa$, respectively, where $t$ is the selective disadvantage of the $AA$ genotype and $\alpha$ a constant. The equilibrium frequencies we know to be

$$q_a = \frac{1}{1 + \alpha}$$

and

$$q_A = 1 - q_a$$

Substitution of these values in the ratio yields

$$\frac{L_1}{L_0} = 2$$

More generally, it can be shown that at a locus with $n$-alleles if one homozygote is superior to all other classes, then

$$\frac{L_1^{(k)}}{L_0^{(k)}} \sim \frac{1}{2\bar{h}}$$

where $\bar{h}$ is the average degree of dominance.

If the favored classes are the heterozygotes, all of which are equally fit, then

$$\frac{L_1^{(k)}}{L_0^{(k)}} = n$$

It appears to have been generally assumed that these two cases represent the limiting values for an $n$-allele locus. That is to say, other genetic systems, e.g., the "mixed locus," are presumed to lead to ratios intermediate between the number of alleles, $n$, and $1/2\bar{h}$. There is no general proof of this nor does it seem likely that such is possible in the light of the following case: We have seen that in a three-allele system subject to the conditions set forth in Section 15.4, the equilibrium frequencies for the three alleles are, to a close approximation, $\mu(1+s)/2s$, $s/(1+s)$, and $1/(1+s)$. If these values and the two selective disadvantages, namely, $s$ and 1, are inserted into $L_0$ and $L_1$ we find

$$\frac{L_1}{L_0} = 2 + \mu(1 - 6s - 7s^2)/2s^2$$

which is less than the number of alleles, and hence the ratio is outside the interval previously described. Whether this holds for other cases is not immediately obvious. It is worth noting that the random load in the case under discussion is the sum of the mutational load for a classical locus, $2\mu$, and the segregational load for a locus with two alleles, namely, $s/(1+s)$. Moreover, in general, the segregational and mutational components of the load are approximately additive, especially for the random load.

Clearly, if the formulation of the preceding paragraphs is to be given substance, we must be able to utilize inbreeding data from populations not completely inbred. Here, again, Morton, Crow, and Muller have advanced an ingenious argument, which proceeds somewhat as follows: We have stated that the load associated with a locus with $n$-alleles in a population inbreeding at rate $F$ is

$$L_F^{(k)} = \sum_{i=1}^{n}(1 - r_{ii})[q_i^2(1-F) + q_iF] + 2\sum_{\substack{i,j=1\\i<j}}^{n}(1 - r_{ij})q_iq_j(1-F)$$

Alternatively, we could have written

$$1 - L_F^{(k)} = \sum_{i=1}^{n} r_{ii}q_i^2 + 2\sum_{\substack{i,j=1\\i<j}}^{n} r_{ij}q_iq_j - F\left[\sum_{i=1}^{n} r_{ii}q_i^2 + 2\sum_{\substack{i,j=1\\i<j}}^{n} r_{ij}q_iq_j - \sum_{i=1}^{n} r_{ii}q_i\right]$$

and this, it will be noted, is merely

$$1 - L_F^{(k)} = 1 - L_0^{(k)} - F[L_1^{(k)} - L_0^{(k)}]$$

If, now, the load at a particular locus is measured in terms of mortality, then the quantity $1 - L_F^{(k)}$ may be viewed as the probability of not dying of genetic causes ascribable to genes at the $k$th locus. If there were $N$ such loci, *and all independent of one another* (in terms both of gene action and segregation), then the probability of failing to die from genetic causes associated with any one of these loci is merely the product of the individual probabilities, or

$$\prod_{k=1}^{N}[1 - L_F^{(k)}]$$

Suppose that in addition to the causes of death associated with the $N$ loci there are $c$ independent (in the probability sense) environmental causes of death where $c = 1, 2, \ldots, v$ (which in practice cannot be clearly distinguished from genetic causes of death and may, in fact, overlap with them). Let $x_c$ be the probability of death due to the $c$th cause, then the probability $P(S)$ of surviving environmental as well as genetic causes of death, if we assume these to be independent of one another, is

$$P(S) = \prod_{c=1}^{v}\prod_{k=1}^{N}[1 - x_c][1 - L_F^{(k)}]$$

or

$$\log P(S) = \sum_{c=1}^{v} \log(1 - x_c) + \sum_{k=1}^{N} \log(1 - L_F^{(k)})$$

$$= \sum_{c=1}^{v} \left[ -x_c - \frac{x_c^2}{2} - \frac{x_c^3}{3} - \cdots \right] + \sum_{k=1}^{N} \left[ -L_F^{(k)} - \frac{(L_F^{(k)})^2}{2} - \frac{(L_F^{(k)})^3}{3} - \cdots \right]$$

If the individual $x_c$ and $L_F^{(k)}$ are all small, then to a close approximation

$$\log P(S) = -\left[ \sum_{c=1}^{v} x_c + \sum_{k=1}^{N} L_F^{(k)} \right]$$

or, in terms of $L_0$ and $L_1$,

$$\log P(S) = -\left[ \sum_{c=1}^{v} x_c + \sum_{k=1}^{N} L_0^{(k)} + F \sum_{k=1}^{N} [L_1^{(k)} - L_0^{(k)}] \right]$$

If we view this relationship in more traditional regression notation, we have

$$-\log P(S) = A + BF$$

where

$$A = \sum_{c=1}^{v} x_c + \sum_{k=1}^{N} L_0^{(k)}$$

and

$$B = \sum_{k=1}^{N} [L_1^{(k)} - L_0^{(k)}]$$

If, now,

$$\sum_{c=1}^{v} x_c \geqslant 0, \qquad \sum_{k=1}^{N} L_0^{(k)} \geqslant 0, \quad \text{and} \quad \sum_{k=1}^{N} L_1^{(k)} \geqslant \sum_{k=1}^{N} L_0^{(k)}$$

as they must be, then clearly the ratio $B/A$, which in the present state of our knowledge is what we are forced to employ in this treatment as our best estimate of $L_1/L_0$, is in the presence of any environmental causes of death always less than the true $L_1/L_0$ and *thus necessarily underestimates the ratio of interest to us.* It should be noted, however, that if

$$\sum_{c=1}^{v} x_c = 0$$

that is, if there are no environmental causes of death, the ratio $(A + B)/A$ is an unbiased estimate of $L_1/L_0$.

The first attempts to apply the foregoing argument to human data involved mortality data (Morton, Crow, and Muller, 1956). Subsequently (Morton, 1960), it was shown that the argument could be extended to "detrimental" traits without loss of generality. Presumably the best estimate of load comes from a summing of both mortality and morbidity (although the relevance of the latter as currently measured in human populations to reproduction is not entirely clear). In summary, the foregoing argument states that if classical loci are the major contributors to the load of mortality and morbidity *revealed by inbreeding*, then the ratio $L_1/L_0$ will be large, whereas if balanced loci are the major contributors to the load *revealed by inbreeding*, then the ratio $L_1/L_0$ will be small, of the order of the average number of alleles at such loci. However, in the $B/A$ ratio which is in practice the approach to estimating $L_1/L_0$, environmental causes of death are included in the $A$ term. The difficulties this creates will be discussed later.

**15.6. Theoretical Objections and Alternatives to the Morton-Crow-Muller Formulation.** As might be expected from the implications and importance of this treatment, it has been subjected to close scrutiny. Without question, this has been one of the stimulating formulations of recent genetic history. However, serious objections have now been raised to several of the assumptions of this treatment. These objections are essentially of two kinds. First, there are objections (Li, 1963*a*, *b*; Sanghvi, 1963) to the fact that in both systems the same fitness value (1.0) is assigned to the optimum genotype, that is, the one with the largest absolute fitness. It is pointed out that this results in the same genotype having a different fitness on the assumption that the locus is of the classical kind from what it would have on the assumption that the locus is of the balanced kind. Specifically, in the classical system the genotype $AA$ is assigned a fitness value of 1, whereas in the balanced system it is $Aa$ that receives this value. Under these circumstances, fitness becomes, not a property of the genotype but of the assumptions with respect to the genetic model which maintains the polymorphism. This has led some to advance other definitions for the "genetic load." Unfortunately while these definitions may have a high degree of validity, the alternates proposed thus far do not seem to lead to formulations capable of differentiating between mutational and segregational loads.

Insertion of specific fitness values into the Morton-Crow-Muller formulation leads to a ready demonstration that at a given gene frequency for $A_j$, the mean population fitness is lower in the balanced than in the classical system, i.e., with this formulation the load of maintaining a polymorphism is greater than of maintaining a classical locus. If, however, as Sanghvi and Li insist is logically more correct, we assign the genotype $A_iA_i$ the same value in both systems, then not only does the greater load of maintaining a polymorphism disappear, but so does the difference in the inbreeding effect. For a full appreciation of the nuances of the argument, the reader is referred to the original papers.

It seems important to emphasize that the validity of the manner in which Morton, Crow, and Muller develop their argument is not at issue but, rather, the nature of some of the assumptions. Crow (1963), in rejoinder to the comments of Li and Sanghvi, has pointed out, quite properly, that if in a complex formulation one alters definitions, one can of course expect different conclusions from those based on the original definitions. This fact does not necessarily imply that the original formulation was incorrect. A pressing problem is to develop a rational (rather than intuitive) set of definitions and frame of reference which will provide a point of departure for all those interested in such formulations. Incidentally, there may, in fact, be a number of tenable intermediate positions between these opposing viewpoints. For instance, one may argue that a balanced polymorphism tends to arise only when, for some reason, the $A_iA_i$ genotype is not functioning at an optimum level. In this case, the load of a polymorphism and the inbreeding effect would both be at a level intermediate between the two opposing formulations.

Secondly, and still at the theoretical level, the legitimacy of the assumption of independent gene action has been questioned (Levene, 1963). It is asserted that epistasis is an important factor in evolution and that the existence of epistatic effects further complicates the interpretation of the $L_1/L_0$ ratio. By way of example, consider the case of two loci at each of which two alleles occur and where the matrices of fitnesses and genotype frequencies are as follows:

| | *Fitness* | | | *Frequency* | | |
|---|---|---|---|---|---|---|
| | *BB* | *Bb* | *bb* | *BB* | *Bb* | *bb* |
| *AA* | $1-f_{11}$ | $1-f_{12}$ | $1-f_{13}$ | $p_{11}$ | $p_{12}$ | $p_{13}$ |
| *Aa* | $1-f_{21}$ | $1-f_{22}$ | $1-f_{23}$ | $p_{21}$ | $p_{22}$ | $p_{23}$ |
| *aa* | $1-f_{31}$ | $1-f_{32}$ | $1-f_{33}$ | $p_{31}$ | $p_{32}$ | $p_{33}$ |

Following Levene (1963), we consider only the symmetrical case where

$$f_{22} = 0$$

$$f_{12} = f_{21} = f_{23} = f_{32} = c + e_1$$

$$f_{11} = f_{13} = f_{31} = f_{33} = 2c + e_2$$

and the equilibrium frequencies for the genes $A$ and $B$ are 1/2. Under these circumstances, it can be shown that

$$\frac{L_1}{L_0} = \frac{8c + 4e_2}{4c + 2e_1 + e_2}$$

This ratio will have a maximum when $c = e_1 = 0$, at which point the ratio is 4, or $2^2$. More generally, we can show that if $s$ loci are involved, and interrelated in the form here assumed, the ratio $L_1/L_0$ will be

$$\frac{L_1}{L_0} = \frac{sc + e_s}{\frac{s}{2}c + 2^{-s}\sum_{s}\binom{s}{j}e_j}$$

The maximum of this ratio is $2^s$ when $c = e_1 = e_2 = \cdots = e_{s-1} = 0$. Clearly this could be a large number and indistinguishable from the ratio to be expected of the classical locus, despite the fact that the individual with maximum fitness is the individual heterozygous at all $s$ loci. Within this same system, if $e_1 = e_2 = \cdots = e_s$, the ratio is, of course, two; whereas if $e_1 > 0$ or $e_2 > 0$ or ... or $e_{s-1} > 0$, the ratio would be less than two.

This illustration is admittedly contrived, but it serves nonetheless to underscore the importance of the assumption of non-synergistic gene action if load ratios are, in man, to have the relevance to questions 3 and 4 in Section 15.2 that is frequently imputed to them. Two lines of evidence have been offered in support of the legitimacy of the assumption of independent gene action in man insofar as studies of mortality are concerned. First, in Drosophila more than half of the viability load is due to monogenic lethals which cannot, of course, exhibit interaction. In the absence of data to the contrary, it seems reasonable to presume that a similar situation may obtain in man. Secondly, it is argued that even if interactions do occur, they become an important complicating factor only at levels of inbreeding well above those which occur in man. Alternatively stated, at low levels of inbreeding such as those which prevail in man, one would expect only a small fraction of genes to become homozygous in any one individual, and thus few opportunities for interactions arise.

In a somewhat similar vein, it can be shown that if a particular polymorphism is related to $k$ loci and if the following assumptions obtain: (1) at each locus two alleles occur, (2) at every locus $A$ mutates to $a$ with frequency $\mu$, and (3) the fitness of a particular genotype is $1 - xhs - ys$ where $x$ is the number of loci at which heterozygosity occurs and $y$ is the number of loci at which homozygosity for the "unfavorable" genes occur and $h$ is the degree of dominance, then the "load" ratio is

$$\frac{L_1}{L_0} = \frac{1}{2ph + q}$$

where $p$ and $q$ are the equilibrium frequencies of the favorable and unfavorable genes at a specific locus. It can be shown, and is indeed obvious, from the conditions of symmetry which hold, that the equilibrium frequency at the $i$th locus—say, $p_i$—equals $p$ for every $i$. The ratio above is independent of $s$, as was to be expected, but it is also independent of $k$, the number of loci. The implications of this finding with respect to questions 1 and 2 (see Sec. 15.2) should be clear.

One interesting consequence of the Morton-Crow-Muller formulation is that given a balanced system, once the heterozygote advantage has been estimated, one can calculate the cost to the population, in terms of mortality or impaired fertility, of maintaining the polymorphism. Thus, it has been estimated that maintenance of the $ABO$ polymorphism through selection would require the death of 2% of all zygotes in Caucasian populations (Chung and Morton, 1961) and 3% in Japanese (Chung, Matsunaga, and Morton, 1960), and for the $MN$ locus, 3% in both Caucasian and Japanese populations (Morton and Chung, 1959*a*; Chung, Matsunaga, and Morton, 1961). There are now more than a dozen other genetic polymorphisms known, and the tempo of discovery shows no signs of abating. These persist in a variety of seemingly small, isolated, and inbred populations, some at remarkably similar gene frequencies, with—barring more migration than now seems likely—the implication that they too are subject to rather strong, stabilizing selective forces (cf. Neel et al., 1964). If each locus is independent of the others in its response to selection, then one is led to an intuitive concern that the "species tolerable load space" is rapidly being exhausted (Neel, 1965). This concern has, in fact, been used, and tellingly, to defend the point of view that the bulk of concealed variability revealed by inbreeding is of the classical kind. It is argued that the "cost" to a population of maintaining many heterotic genes is prohibitive. Thus, for example, to maintain 100 independent, heterotic loci where each leads to the loss of three zygotes in a hundred, as seems to be the case for the *ABO* and *MN* loci, would result in the killing of approximately 95.3 percent of the population. However, this excessive "cost" disappears if the Sanghvi-Li objection is met and also is materially lessened, if the genes concerned are linked so that simultaneous elimination can occur (see Lewontin, 1964). But if linkage is to be important in this respect, epistasis must obtain. Lewontin reasons that epistasis must, in fact, be sufficiently widespread to make linkage an important factor in the evaluation of a population's load.

**15.7. Practical Difficulties in the Interpretation of Inbreeding Results.** In addition to these objections to the assumptions and theory of the Morton-Crow-Muller formulation, a variety of what we may term "practical" impediments to the application of this argument to a set of actual data is becoming clear. The following, while not an exhaustive list, will serve to indicate the nature of these objections.

1. *The many interpretations of "gross" B/A ratios.* The foregoing treatment of inbreeding effects has been couched in terms of a single locus. In fact, in practice we measure the average effect resulting from segregation at many loci. In our own data, the $B/A$ ratio for mortality plus morbidity is 4.6 for Hiroshima and 4.5 for Nagasaki. Let us, in order to develop our argument, assume that approximately half of the deaths in the outbred population are "environmental" and unrelated to the genotype of the individual concerned—possibly an overgenerous allowance. Then the corrected $L_1/L_0$ ratio for these two cities might be as high as 10. Now, for a number of phenotypes whose frequencies and genetic behavior suggest they are for the most part due to recessive genes maintained primarily by mutation pressure, the following $B/A$ ratios have been computed for Caucasian populations (Morton, 1961): low-grade mental deficiency, 46.1; recessive muscular dystrophy, 225.9; recessive deaf-mutism, 442.0. In Table 15.1, modified from an earlier publication (Neel and Schull, 1962*a*), we have given a number of illustrations of how an observed $B/A$ ratio of 10 can be met by combining a variable proportion of loci of the segregational variety where $L_1/L_0$ values range between 2 and 10 with a variable proportion of loci of the mutational kind where $\bar{h}$, the average dominance, ranges from 0.035 to 0.050. It will be recalled that for loci of the latter type, $L_1/L_0$ is merely $1/2\,\bar{h}$. Which particular solution one favors depends at present to a large degree on one's intuition concerning the average degree of dominance and the average number of alleles at the average locus. Whatever one's choice may be, it is clear that the contribution of the mutational load to the load expressed upon inbreeding may be large but the contribution to the random load small. Alternatively stated, knowledge of the contribution of loci of the mutational variety to the inbred load tells us little about their contribution to the random load. Knowledge of the genetic component in the latter is of tremendous import to most, if not all, human societies, and the Morton-Crow-Muller formulation contributes little to the elucidation of these components.

2. *The uneven distribution of inbreeding and specific deleterious genes within populations.* The derivation of a $B/A$ ratio, be it for a specific trait or for nonspecific indicators such as total mortality or morbidity prior to age of reproduction, assumes a random or "even" distribution of both inbreeding and the responsible genes within the area from which the data are drawn. In fact, it has long been apparent that consanguineous marriages are much more frequent in some regions than others. More recently, with the development of methods permitting a rapid survey for the carriers of the rare recessive gene responsible for acatalasemia, it has become possible to demonstrate on a large scale a fact which in genetic theory comes as no surprise: even in long-settled populations,

rare recessive genes are not uniformly distributed throughout the country. In the particular case under discussion, there were tenfold differences between regions (Hamilton et al., 1961). It could further be demonstrated that on the basis of reasonable assumptions, unequal distributions of consanguineous marriage and gene frequencies could combine in such a manner as to produce

**Table 15.1**

*Illustration of the various interpretations possible of an $L_1/L_0$ ratio of 10 where one assumes two components in the ratio, one a "segregational" component and one a "mutational" component. These calculations assume that at loci with more than two alleles either one homozygote is superior to all other classes or, if the favored classes are heterozygotes, then all of the latter are equally fit. Recall that under these conditions the $L_1/L_0$ ratio at a segregational locus is merely the number of genes at that locus, whereas $L_1/L_0$ ratio at a mutational locus will be $1/2\bar{h}$.*

| Number of genes at a locus | Proportion of random load | | Proportion of inbred load | | Implied $\bar{h}$ |
|---|---|---|---|---|---|
| | Mutational | Segregational | Mutational | Segregational | |
| 2 | 0.10 | 0.90 | 0.82 | 0.18 | 0.043 |
| 2 | 0.20 | 0.80 | 0.84 | 0.16 | 0.043 |
| 2 | 0.50 | 0.50 | 0.90 | 0.10 | 0.046 |
| 3 | 0.10 | 0.90 | 0.73 | 0.27 | 0.040 |
| 3 | 0.20 | 0.80 | 0.76 | 0.24 | 0.041 |
| 3 | 0.50 | 0.50 | 0.85 | 0.15 | 0.045 |
| 4 | 0.10 | 0.90 | 0.64 | 0.36 | 0.037 |
| 4 | 0.20 | 0.80 | 0.68 | 0.32 | 0.039 |
| 4 | 0.50 | 0.50 | 0.80 | 0.20 | 0.043 |
| 5 | 0.10 | 0.90 | 0.55 | 0.45 | 0.035 |
| 5 | 0.20 | 0.80 | 0.60 | 0.40 | 0.038 |
| 5 | 0.50 | 0.50 | 0.75 | 0.25 | 0.043 |
| 10 | 0.10 | 0.90 | 0.10 | 0.90 | 0.050 |
| 10 | 0.20 | 0.80 | 0.20 | 0.80 | 0.050 |
| 10 | 0.50 | 0.50 | 0.50 | 0.50 | 0.050 |

$B/A$ ratios differing by a factor of 5. Whether this is an important source of inaccuracy in $B/A$ ratios involving gross mortality and morbidity statistics or whether the sources of error at individual loci average out is, of course, unknown.

3. *The limited picture afforded by present data.* Both Crow (1963) and Levene (1963) have dwelt at some length on the inadequacies of efforts to utilize $B/A$ ratios when the data are based on only a limited portion of the life cycle. Difficult and controversial though the measurements to date have been, they still have been mostly concerned with the easiest type of data to collect—early mortality and morbidity. How much more difficult to acquire will be the equally important data on fertility! It was, apparently, especially this thought that led Levene (1963, p. 591) to conclude: "It is unfortunate that the difficulties discussed in the present paper make it unlikely that the $(B + A)/A$ ratio will answer the question in practice."

4. *The role of other kinds of loads in obscuring the contributions of classical and balanced systems.* Earlier we listed six different kinds of genetic "loads" to be encountered in a population. The incompatibility load is especially deserving of further discussion, since limited evidence bearing on the manner of its action is at hand. It is quite clear that the incompatibility load, as discussed above, would obscure the effects of the homozygosity resulting from inbreeding and so introduce a further complication into the interpretation of inbreeding results. This load is decreased in an inbred child but increased in the offspring of an inbred mother. With respect to the inbred child, Crow and Morton (1960) have expressed the incompatibility load as

$$I = D(1 - E)$$

where

$$D = \sum d_i p_i(1 - p_i)(1 - p_i + p_i F_m)$$

and

$$E = 2F_c - \text{Prob.}(a \equiv b \equiv c)$$

with

$p_i$ = frequency of allele $A_i$
$d_i$ = decrease in fitness due to incompatibility for the antigen resulting from allele $A_i$
$F_m$ = coefficient of inbreeding for mother
$F_c$ = coefficient of inbreeding for child

and

Prob. $(a \equiv b \equiv c)$ = the probability that the particular allele contributed to the child by the father is identical by virtue of common origin to a pair of alleles of common origin possessed by the mother. For human marriages, this probability is usually zero and can be disregarded.

The actual evaluation of a specific incompatibility load involves a number of assumptions, such as to make an evaluation an approximation. Be this as it may, Crow and Morton (1960) and Chung and Morton (1961) have estimated the incompatibility load of the *ABO* locus (the $D$ of the above equation) as 0.066, from which it follows that the load would fall from 0.066 in outbred children to 0.058 in the children of first cousins, i.e., a decrease of 8 losses per 1000 pregnancies. Chung and Morton (1961) favor the view that this load becomes manifest during embryonic life. If this is an accurate appraisal of the incompatibility load, if a significant proportion of this embryonic loss is perinatal, and if there are many such loci with effects approximating those of the *ABO* locus, then the (lessened) incompatibility load in consanguineous marriages could assume a magnitude which would seriously distort our appraisal of the effects of homozygosity in man. Moreover, the increased survival in consanguineous marriages of compatible fetuses (who, if not homozygous, would have a 2/3 chance of being carriers for any genes for which both parents were carriers) would tend to offset the predicted rate of elimination of recessive genes as a result of consanguinity. That this possibility may yet prove of considerable significance is suggested by the data of Schull, Yanase, and Nemoto (1962) to be discussed in Section 15.9. Were there no inbreeding effect, the incompatibility load might lead to negative $B/A$ ratios. A few such ratios have been reported, but at this writing it seems probable they are due more to sampling errors than to incompatibility loads (cf. Neel, 1962*a*).

5. *The assumption of equilibrium.* It seems quite clear that genetic equilibrium is not a characteristic of many—if any—human populations. The environment is changing far too fast. Beyond that is the remarkable mobility of human populations, with the resulting migrational load. In groups characterized by considerable racial crossing in relatively recent times, such as the American Negro, the "Nordestino" of Brazil, or the populations of many Caribbean islands, the magnitude of this migrational load may be so great—and at the very least is so indeterminate—that conclusions concerning inbreeding effects based on such populations must forever be suspect. Even within "new" countries such as the United States or Australia, where the admixture has involved persons predominantly of one race, the extent of the migrational load remains a moot question. However, this much seems clear: if, as seems likely, there are racial and regional differences in the frequencies of some rare recessive genes (e.g., amaurotic idiocy, acatalasemia, cystic fibrosis of the pancreas), hybridization can only render these genes relatively less common than in the strain which supplies them—and so increase the significance of inbreeding in the appearance of mutant phenotypes.

6. *The sampling variance associated with the inbreeding parameters A and B.* The estimation of $B$, the load at full homozygosity, involves (until someone studies the uncle-niece marriages of

India) an extrapolation to values at $F = 1.0$ from observation on values of $F$ ranging from 0 to 0.06. Under these circumstances, unless the sampling variance of the regression term is small, the error in $B$ may be enormous. It is perfectly obvious from theoretical considerations, as well as from the size of many of the series which have been published, that the sampling variance must be large. Add to that the varying role of socioeconomic factors in the different series, and it is difficult to place great reliance in any particular estimate of $B$. Similar considerations apply to $A$. In Chapter 5 we pointed out that despite the widely varying numbers and standards of individual studies if one averages the values of the individual studies on mortality effects to date, there is a remarkable similarity between the findings in the three principal racial groups. Should this agreement persist in future studies, then it will begin to appear that inbreeding effects are no different in Negroes, Caucasians, and Mongolians—with the inference that the wide differences between individual studies illustrate just how great the sampling variances in these studies may be.

7. *The environmental contribution to A.* The degree to which the environmental contribution to $A$ may distort the estimate of the $L_1/L_0$ ratio will vary from population to population and in any event for the present is moot for any population. Crow (1963) has recently emphasized that—granted the assumptions of independence of gene action, an equilibrium population, and so forth—a large $B/A$ ratio could arise only if the concealed load is indeed mainly of the classical type but that a small $B/A$ ratio is uninterpretable since a small ratio could arise not only if the concealed load was balanced in type, but also if the concealed load was of the classical type and the environmental contribution to the variable under study was large. With the advances to be anticipated in both the understanding and control of disease processes, substantial progress may be expected in dispelling this source of ambiguity—in some populations, many of the environmental causes of death will have disappeared, and those remaining can be identified with some accuracy.

8. *The identification of the optimal genotype.* Finally, it should be noted that in practice it will frequently be difficult to identify, out of an array of genotypes, the one which is optimum. The term "optimum" implies evaluation under a specific set of conditions. For traits of commercial importance, such as those listed in Table 14.4 with reference to livestock, the frame of reference is obviously, for the moment at least, man made. What response of the species in a given environment is optimal in the evolutionary sense—especially where quantitative traits are concerned—will be difficult to determine.

**15.8. Phenodeviants.** Mentioned in Section 15.4, but not yet discussed in any detail, was a complex type of genetic system to which Lerner (1954) in particular has directed attention. We will now consider this system at some length because of the manner in which it illustrates the full range of current difficulties in the interpretation of inbreeding results. We have seen that the very limited data available do not lead us to attribute more than 20–30% of the increase in morbidity encountered in the children of first cousins to simple recessive inheritance. By inference, the balance of the inbreeding effect is due to homozygosity in irregularly expressed monomeric recessive systems or multifactorial systems. In all the systems thus far considered—whether maintained by mutation pressure or as a balanced polymorphism—it was assumed that the correspondence between genotype and phenotype was quite specific. By contrast, in this type of multifactorial system a given phenotype may arise from a number of different (albeit related) genotypes characterized by a high degree of homozygosity, these genotypes being maintained either by mutation pressure or by balanced polymorphisms.

In presenting a summary of the data from Hiroshima and Nagasaki regarding congenital defect, Neel (1958, p. 435) wrote as follows:

> It must be constantly kept in mind that the genetic phenomena responsible for the observed facts are undoubtedly mixed in nature. Thus facts (3), (5), and (6)[1] can to a considerable extent be "explained" by a judicious admixture of two simple genetic mechanisms, namely, simple recessive inheritance, and simple irregular dominant inheritance. When, however, the total picture is considered, a somewhat more complex

[1] Cf. list of facts in first paragraph of the quotation from Morton (1960), below.

possibility comes to mind, namely, that many congenital malformations of various types find a partial explanation in the existence in man of genetic systems of the type discussed in such penetrating and provocative detail by Lerner (1954; see also Dobzhansky, 1955), the malformations ("phenodeviants") being caused "by the intrinsic properties of multigenic Mendelian inheritance, due to which a certain percentage of individuals of every generation falls below the threshold of the obligate proportion of loci needed in a heterozygous state to ensure normal development."

For the detailed statement of this hypothesis and its application to man, the reader who is interested in this question is urged to refer to the original publications. We would like at this time to reiterate that suggestion, with particular reference to the possible role of phenodeviants in the consanguinity effect.

Morton (1960, p. 358) has strongly criticized this suggestion on a variety of grounds. Because of the important but somewhat unusual nature of these criticisms, as well as the manner in which they are affected by the contents of this monograph, they seem worthy of quotation in some detail (the Roman numerals have been inserted by the authors for the purpose of subsequent reference):

Recently Neel (1959) [*sic*] has suggested that many sporadic malformations in man may be phenodeviants, or "segregants resulting from the existence and functioning of complex (multilocal) genetic homeostatic systems, of the type particularly discussed by Lerner (1954)." However, it is not at all apparent that the heritable defects which characterize highly inbred lines (phenodeviants in the sense of Lerner) have any real similarity to sporadic defects in randomly mating populations (phenodeviants in the sense of Neel). By Lerner's criteria of increased incidence under inbreeding and under unfavorable environments, most of the malformations studied by Neel can hardly be considered phenodeviants, since they are at the same frequencies in the more inbred Japanese and less inbred Caucasian populations and are perhaps least frequent under the poor environment of American Negroes [I]. Neel cites the following facts in support of his interpretation.

1. Total malformation rates are substantially the same in different populations.
2. Specific malformations have different frequencies among populations.
3. Type-specific malformation rates are slightly increased in sibs of probands.
4. Monozygotic concordance is low.
5. Malformation rates in the U.S. are intermediate between England, Switzerland, and Sweden.
6. Malformation rates increase with inbreeding.
7. Specific malformation rates depend on sex, parental age, and associated defects.

It is difficult to see how these observations bear on the phenodeviant hypothesis. Fact (6) indicates that some malformations depend on homozygosity, fact (4) indicates that most malformations depend on special environmental circumstances, and the other points can be interpreted in a great variety of ways, and are therefore irrelevant [II].

To support Neel's hypothesis that a large proportion of sporadic malformations are phenodeviants [III], it will be necessary to show that

1. They are genetic.
2. They are polygenic.
3. They depend on homozygosity.
4. The genes involved are maintained through heterozygote advantage [IV].

Neel offers no critical evidence for any of these points, and it would seem that his hypothesis must be stated more clearly before a definite test is possible. Does he assert that *all* the genes which produce phenodeviants are maintained by heterozygote advantage, or that only a "significant fraction" are [IV]? In the latter case, if the significant fraction is sufficiently small, the hypothesis is neither disputable nor heuristic. Would a gene maintained by mutation pressure, with modifiers maintained by heterozygote advantage, satisfy his concept of a phenodeviant or not? And does he espouse Lerner's mystical thesis that phenodeviants cannot in principle be referred to any specifiable set of loci, but represent the effect of too high a level of homozygosity *per se* (an hypothesis that would require that *all* types of phenodeviants be increased in affected individuals and their sibs) [V]? This leads to the following dilemma. Suppose phenodeviant $A$ occurs when more than $n_A$ loci become homozygous, and phenodeviant $B$ when more than $n_B$ loci are homozygous. Then if $n_A$ is greater than $n_B$, defect $B$ should always accompany $A$, while if $n_B$ is greater than $n_A$, defect $A$ should always accompany $B$. Extending this to all defects, the one which requires the highest level of homozygosity, and therefore is most sensitive to inbreeding, must be accompanied by all the others which are less sensitive to inbreeding. Clearly this prediction is not true. Thus there is not sufficient information in the level of homozygosity per se to account for the specificity and distribution of defects.

In view of the extremely low penetrance of phenodeviants which must be postulated to fit the twin discordances, Neel's vague and complex hypothesis can neither be rigorously supported nor disproved at the present stage of human genetics. However, it does seem important (and critical evidence that the load represented by inbreeding is not a segregation load) that the inbreeding coefficient for isolated cases of limb-girdle muscular dystrophy, deaf-mutism, and low-grade mental defect agrees quantitatively with the assumption that sporadic cases are not associated with inbreeding [VI]. In fact, the estimate of the proportion of sporadic cases based on this assumption and the inbreeding coefficients of familial and isolated cases is slightly, but not significantly, higher than the estimate from SEGRAN, whereas just the opposite would be expected if sporadic cases were associated with inbreeding . . . . Since sporadic cases are not demonstrably related to homozygosis, the hypothesis of phenodeviants in man must be dismissed as either false or operationally undefined. Most sporadic cases are probably due to maternal and environmental factors and to a variety of nonrecessive genetic mechanisms, including polygenes, heterozygotes with low penetrance, dominant mutations, deletions, and aneuploidy [VII].

There are some seven aspects of this discussion which seem worthy of comment, identified by Roman numerals in the foregoing quotation and in the following commentary:

I. As explicitly pointed out by Neel (see fact 2 on Morton's list!), although the total impact of congenital defect is very similar in Caucasian and Japanese populations, there were significant differences in the frequency of five of the six specific defects for which accurate racial comparisons were felt to be possible. Indeed, it was this constancy in the total impact of congenital defect in the face of differences in subtype, despite the great racial differences in diet and disease experience, that stimulated thought concerning genetic regulatory mechanisms susceptible to selection. Not only is this quotation inaccurate, but one might draw from it the erroneous inference that there is no increase in malformation frequency with inbreeding.

II. With respect to the relevance of these seven facts to the phenodeviant hypothesis, it was, to begin with, clearly indicated in the original paper (p. 435) that these facts were amenable to many interpretations. However, it does not follow from the fact that a series of findings can be interpreted in a variety of ways that they are irrelevant to an issue. There is a difference between relevant evidence and conclusive evidence. These findings, as brought out in the original discussion, are all consistent with the manner in which one might visualize the expression of multifactorial systems whose phenotypic manifestations are related to environmental variables and wherein certain segregants behave as phenodeviants.

III. Neel's hypothesis is not that "a large proportion of sporadic malformations are phenodeviants." Nothing was said about sporadicity in the original presentation, and, indeed, it is clear from genetic theory that phenodeviants may be familial. This misinterpretation on Morton's part undoubtedly stems from his interest in "segregation analysis" (cf. Morton, 1959). A basic flaw in this approach which does not seem to have been previously pointed out is that whenever, in segregation analysis, two or more affected siblings are encountered without affected close relatives, these are categorized as due to recessive inheritance, and on the basis of such categorization an approach is then developed to estimating the proportion of a given phenotype due to simple recessive inheritance. However, there are, in fact, other obvious genetic mechanisms resulting in affected siblings with normal parents and near relatives (especially with the limited pedigrees often available for humans), namely, dominant genes of low penetrance, sectorial mutations, or multifactorial inheritance. Segregation analysis, as presented thus far, would not distinguish between familial phenodeviants and simple recessive inheritance—which perhaps accounts for the inadvertent identification of phenodeviants with sporadic defect. It would, in this connection, be helpful to have a set of criteria justifying the application of segregation analysis to a set of data. More specifically, what is the basis for applying segregation analysis to data on deaf-mutism but not to data on anencephaly?

IV. Twice in rapid succession, the concept of phenodeviants is mistakenly identified with heterozygote advantage despite the fact that the original article was quite explicit on this point, as follows (p. 435):

Two subpossibilities must be explored, if the possibility is to be considered that malformations to some extent are segregants from multilocal systems. One can, on the one hand, regard these systems, whatever their role in the organism, as in large part a function of mutation pressure. This point of view makes no assumptions concerning the role of the postulated loci in the economy of the species. One can, on the other hand, consider the possibility that the loci involved contribute to a balanced polymorphic or homeostatic system, with the inference that the genes are in fact playing an important role in the genetic stability of the species.

It should be made explicitly clear that the concept of phenodeviants has nothing to do with the concept of heterotic loci or heterozygote advantage in the sense that this term is usually employed but is a statement of the *relative disadvantage of some* (*but not necessarily all*) *homozygotes*. These homozygotes, as stated above, might result from simple mutation pressure and not, as the term "heterozygote advantage" usually implies, be segregants from balanced polymorphic systems. Thus, the fourth line of evidence required by Morton for support of the phenodeviant hypothesis would seem to be irrelevant. Since the frequency of phenodeviants is increased by inbreeding, but yet clearly the explanation does not rest with simple recessive inheritance, then a substantial step would seem to have been taken towards the first three lines of evidence required by Morton. However, in point of fact Morton's four criteria are really not sufficiently stringent—what is ultimately needed is not general evidence relating homozygosity to phenodeviants, such as comes from inbreeding effects, but very specific evidence involving known genetic systems in which homozygote and heterozygote can be distinguished.

V. Concerning the manner in which Lerner's hypothesis is misinterpreted at this point—and the resulting product then labeled "mystical"—Lerner (1961) himself has made such an instructive reply that no further comment is necessary.

VI. Morton has repeatedly emphasized that the results of segregation analysis of such traits as limb-girdle muscular dystrophy, deaf-mutism, and low-grade mental defect strongly suggest that these traits are maintained by mutation pressure, this in a context which suggests that this discredits the concepts of phenodeviants and "balanced" loci. Thus, in addition to the above quoted passage, we read (Morton, 1960, p. 355; see also Morton, 1961, p. 285) as follows:

> . . . about one gamete in four carries a gene which, if homozygous, would produce limb-girdle muscular dystrophy, deaf-mutism, or low-grade mental defect. According to one hypothesis, this is a segregation load maintained by heterozygote advantage. In its extreme form, the theory postulates that, at each relevant locus, all homozygotes produce the trait in question, while heterozygotes are normal. This obviously requires a large number of mutually heterotic alleles, at least $B/A$ alleles per locus by Crow's theorem (1958), or more than 200 for each of these conditions. This is much larger than $n$, the total number of contributory loci or complementary alleles. Since local human populations in the past were not of greater order than this, it is incredible that selection against rare homozygotes could maintain such a high number of alleles at each locus. Moreover, the average allele frequency would have to be less than $B/(B/A)$, or $A$, which is of the same magnitude as the mutation pressure, and so small that nearly every affected person would be expected to come from close consanguineous marriages. For all these reasons, the hypothesis of a segregational load in which all homozygotes are similarly affected is clearly inadmissible.

Now, although we have yet to identify the proponent of this extreme application of the hypothesis upon whose demolition Morton is so intent, we must agree with Morton's conclusion—even though Fraser (1962) has pointed out some errors in the logic whereby this conclusion was reached—but fail to find that conclusion either particularly relevant or surprising in view of the nature of the material under consideration. The demonstration that *all* the genes producing these *rare* defects which often exhibit *simple recessive* inheritance cannot be maintained by heterozygote advantage does not invalidate the possibility that there may be *some* genes responsible for more *common* traits *not following simple genetic patterns* which are maintained by heterozygote advantage. In addition, again we see the mistaken identification of phenodeviants with isolated cases, and we remind the reader that phenodeviants can be familial and that the practice in segregation analysis of assigning such familial cases (which under most multifactorial hypotheses would be expected to have a higher mean coefficient of inbreeding than do isolated cases) to simple recessive inheritance becomes an exercise in circuity.

Incidentally, even in the data on limb-girdle muscular dystrophy, the estimated $F$ for isolated cases is $119 \times 10^{-5}$ in a population where the mean coefficient of inbreeding is estimated at $62 \times 10^{-5}$ (Morton and Chung, 1959*b*, p. 363). Similarly, from the data of Stevenson and Cheeseman (1956) on deaf-mutism, Chung, Robison, and Morton (1959, p. 358) compute the coefficient of inbreeding for isolated cases as $278 \times 10^{-5}$ and for the general population as $39 \times 10^{-5}$. If one *assumes* that the higher coefficient of inbreeding for isolated affected individuals than for the general population is entirely due, in the isolated cases, to recessive inheritance with a single affected child in the sibship, and that for all instances of two or more affected in a sibship the cause is recessive inheritance, one can conclude, as Morton does, that "sporadic cases are not associated with inbreeding." Without these assumptions, one cannot reach this conclusion. While the assumption is relatively safe in these instances, it would obviously lead to difficulties if segregation analysis were applied to traits more consistent with Lerner's definition of a phenodeviant, such as harelip and cleft palate, congenital heart defects, clubfoot with associated CNS defects, or complex malformations.

VII. Finally, elsewhere in this same critique, Morton (1960, p. 357) has pointed out that with some formulations of the phenodeviant hypothesis the relationship between $F$ and frequency of defect should be nonlinear and stated that "these expectations are not borne out by the data." Since, as we have been at pains to show, over the limited range of $F$ for which human data are available the findings can be fitted almost equally well by a logistic or exponit type of curve, it is difficult to perceive a firm basis for that statement.

Lest there be misinterpretation, we emphasize that the foregoing does not constitute a statement of belief that a *major* portion of congenital defect and/or the inbreeding effect is due to phenodeviants. It does, however, reiterate the suggestion that in man one genetic mechanism playing a role in the etiology of congenital defect and the inbreeding effect may be multifactorial systems of the type envisioned by Lerner (1954). Furthermore, the foregoing expresses the conviction that the validity of this suggestion has not been challenged by the criticisms which have been raised to date.

At no place in the foregoing has an attempt been made to specify the percent contribution of phenodeviants to the total inbreeding effect. If the range of permissible interpretations regarding the relative contributions of "simple" classical and "simple" balanced polymorphic systems is as great as indicated in Table 15.1, then how premature it would be to attempt to set an exact figure on the contribution of phenodeviants to either the outbred or inbred loads. Since the mathematical treatment of the systems contributing to inbreeding effects thus far rests at a level of simplicity patently far below the complexity of real-life populations, it is clear that no general mathematical formulation of the systems resulting in phenodeviants is to be expected for some time. The best to be anticipated in the immediate future is population simulation on large capacity computers, out of which probable limits on interpretations may emerge.

**15.9. Current Estimates of the Number of Lethal Equivalents in Man.** Thus far in our consideration of the use of inbreeding data in analyzing population structure, attention has been directed primarily toward certain theoretical and practical difficulties attendant upon efforts to partition the genetic burden of a population, and especially that exposed by inbreeding, into a mutational and a segregational component. It seems appropriate now to consider briefly the bearing of data presently available upon these efforts. Attention will be restricted to observations on mortality, the least controversial area of application of the Morton-Crow-Muller formulation.

In Chapter 5 most of the recent studies of the effect of inbreeding upon mortality have been briefly described (see Tables 5.9, 5.11, and 5.13). These studies, it was pointed out, are conspicuous more for their dissimilarities than their similarities. They range from observations on a few hundred children to observations on thousands, from studies of isolated rural areas to investigations of sizable industrial cities, from observations made by trained observers to information obtained by a self-administered questionnaire. They further differ in the manner whereby the related and unrelated groups were ascertained, the number of years at risk of death, and so forth. Thus, rigorous

**Table 15.2**

*Summary of lethal equivalents as estimated for various stages of manifestation in contemporary populations. The column headed "Total" represents the sum of lethal equivalents manifesting themselves as stillbirths or deaths between parturition and maturity. For further explanation of the entries, see text.*

| Country and/or population | Lethal equivalents | | | | | |
|---|---|---|---|---|---|---|
| | Miscarriage | Stillbirth | Neonatal death | Infantile and juvenile death | Total | $B/A$ |
| Brazil: | | | | | | |
| Indians | | | | 1.37[a] | 1.37 | 0.71 |
| Negroes | 3.20 | 1.24 | | 1.38[a] | 2.62 | 9.22 |
| Whites | −0.59 | 0.88 | | −0.20[a] | 0.68 | 0.18 |
| France: | | | | | | |
| Finistère | | 0.38 | 0.68 | 2.44 | 3.50 | |
| Loir-et-Cher | | 0.29 | 0.48 | 1.93 | 2.70 | 18.53 |
| Morbihan | | 1.07 | 1.15 | 2.18 | 4.40 | |
| Germany | | | | | 0.86[b] | 1.77 |
| Italy | | | 1.03[c] | 2.06 | 3.09 | 8.37 |
| Japan: | | | | | | |
| Fukushima | | | | | 0.96[b] | 5.85 |
| Hiroshima | | 0.14 | 0.29 | 0.63 | 1.06 | 6.08 |
| Hoshino | | 0.15 | 0.67 | | 0.82 | 10.57 |
| Mishima | | 0.26 | −0.29 | 0.55 | 0.52 | 1.58 |
| Nagasaki | | −0.02 | 0.20 | 0.15 | 0.33 | 1.08 |
| Shizuoka | | | | | 1.11[b] | 5.74 |
| Tanganyika | | | | | −0.70[b] | −1.26 |
| Sweden | −2.60 | −1.45 | 0.90 | 0.82 | | −3.52 |
| U.S.A. | 0.52 | 0.08 | 0.04 | 1.83 | 1.95 | 7.52 |

[a] Includes neonatal deaths.
[b] Inclusion of stillbirths uncertain.
[c] Includes stillbirths.

comparisons seem ill-advised, if not wholly unjustified. It is of interest nonetheless to ask whether these studies, collectively, point toward certain conclusions.

Table 15.2 sets out the $B/A$ ratios associated with some 17 areas and/or populations of the world, as well as estimates of the number of lethal equivalents and the stage of development at which manifestation of the latter appears to occur. For convenience, the estimates of the number of lethal equivalents have been derived from comparisons restricted to data on the children of unrelated parents and of first cousins (for an illustration of the method of computation, see Slatis, Reis, and Hoene, 1958). Perusal of this table reveals certain obvious inconsistencies—negative $B/A$ ratios, to cite one. But there is an over-all element of agreement which, in respect of the aforementioned dissimilarities in the studies and the marked differences in medical practice, nutrition, and so on in the areas for which the data are reported, might be viewed as unexpected. Thus, for example, we note that with one exception, namely, the observations from France, the $B/A$ ratios are small—10 or less. It will be recalled that under the Morton-Crow-Muller formulation the ratio of the inbred to the random load provides crucial evidence with respect to the contributions of mutation and selection to the concealed variability in a population only if this ratio is large, and then only if gene action is, in general, non-synergistic. These studies do not reveal ratios which may be construed as unambiguously supporting the view that the load revealed by inbreeding is mainly of the mutational variety. The observed values are, in fact, consistent with a substantial portion of the load being segregational in origin.

Again, with singular exceptions, there is some agreement in the probable number of lethal equivalents, particularly at those stages of development, such as stillbirths and neonatal deaths, for which the studies are most comparable. We note, for example, that relatively fewer lethal genes appear to manifest themselves as stillbirths or deaths in the first month of life than at subsequent ages. It is also of interest that the variation in the estimates derived from a single area in, for example, France or Japan, nearly equals the variation from one area of the world to another. Alternatively stated, there seems to be as much intra-area variation as inter-areas variation. With regard to miscarriages and deaths after the neonatal period, interpretations of the data must be more guarded. Thus, it is difficult to ascertain whether a major portion of lethal genes are or are not manifest in early zygotic mortality because of the apparent heterogeneity between the four studies (estimates range from −2.60 to 3.20) and the admitted difficulty in obtaining reliable information on losses at this stage of life. Some investigators, notably Slatis (1963), have nonetheless interpreted these fragmentary data to imply that relatively more lethal equivalents will be expressed as miscarriages than as stillbirths or neonatal deaths. As to infantile and juvenile deaths, the differences which exist between the studies in number of years at risk of death preclude sweeping generalizations. This fact notwithstanding, it seems unlikely that the mean number of lethal equivalents acting after parturition, and more especially after the first month of life, is apt to exceed 3. The biological, psychological, and sociological impact of these genes upon a community or population is, as we have seen, presently beyond assessment.

**15.10. The Future of Studies of Inbreeding Effects in Man.** As brought out in Chapter 1, this study first began to take shape in 1955. At that time, the primary objective was to utilize an unusual opportunity to collect data on consanguinity effects and to reason therefrom concerning man's concealed genetic variabilty. With the Morton-Crow-Muller formulation of 1956, an added objective was an attempt to utilize these data to reach a decision concerning the relative role in human populations of mutational and segregational systems. As should be apparent from the foregoing, the modifications of, and objections to, the Morton-Crow-Muller formulation—which has played as stimulating a role as any in recent genetics—are such that it is clear this added objective will not be achieved at this time. It seems to us important, however, that the realization of what has been a recurrent theme of modern biology—that living systems are more complex than appeared at first evaluation—should not result in a wave of disillusion with the value of studies on consanguinity effects, for much of value remains to be done. Specifically, the immediate future of studies of the effect of inbreeding in human populations would seem to rest primarily with four types of investigations.

1. *Further Studies on Inbreeding Effects.* First, we need to extend our knowledge of consanguinity effects, especially through studies of the fertility of consanguineous marriages, the fertility of the inbred individual, and the effect of parental inbreeding on offspring characteristics. Such studies will at the very least broaden our empiric knowledge of consanguinity effects. In addition, these additional studies should provide a body of data suitable for comprehensive comparisons of inbreeding effects in man with those in other animals more susceptible to experimental manipulation. In this connection we note the possibility of improvement in both the type and analysis of data available from other mammals. Thus, if and when as a result of animal experimentation we gain a better insight into some of the genetic systems responsible for inbreeding effects, a body of data from man will be available for cautious comparisons and extrapolations. Further, should there occur unexpected mathematical insights into the interpretations of gross inbreeding effects, the necessary data for the application of these insights would be at hand. Since with every passing year consanguineous marriages are apt to be increasingly atypical of the population from which they are drawn and fertility potentials more subject to human regulation, it is important these data be collected now.

It is obvious that in a situation where the mortality of the children of first cousins is only a few percent greater than that of control children, as in these data, a slight increase in the fertility of consanguineous marriages would result in net family size being the same for consanguineous and non-consanguineous marriages. The only pertinent data from Japan on this specific point appear to be those of Schull, Yanase, and Nemoto (1962), based on a total census of three *buraku* of the small island of Kuroshima, off the western shore of Kyushu. For the 175 marriages contracted in these three *buraku* between 1920 and 1939, the mean number of children ever born to all couples surveyed was 5.71, a figure that suggests that a reasonable test of natural fertility is involved. The findings of that investigation are reproduced in Table 15.3. While the differences recorded therein are not

**Table 15.3**

*Summary of the reproductive performances of Buddhists and Catholics, related and unrelated, married in the years 1920–1939, living on the island of Kuroshima, Japan.* (*After Schull, Yanase, and Nemoto, 1962*)

| | Buddhists | | | Catholics | | | |
|---|---|---|---|---|---|---|---|
| | Related | Not related | Total | Related | Not related | Total | Grand total |
| Number of families | 14 | 21 | 35 | 22 | 118 | 140 | 175 |
| Number of childless couples | 1 | 2 | 3 | 0 | 8 | 8 | 11 |
| Number of children ever born | 74 | 101 | 175 | 149 | 676 | 825 | 1,000 |
| Number of children dying before 20 years of age | 10 | 11 | 21 | 20 | 61 | 81 | 102 |
| Percent childless couples | 7.14 | 9.52 | 8.57 | 0.00 | 6.78 | 5.71 | 6.28 |
| Percent mortality | 13.51 | 10.89 | 12.00 | 13.42 | 9.02 | 9.82 | 10.20 |
| Mean number of children ever born | 5.29 | 4.81 | 5.00 | 6.77 | 5.73 | 5.89 | 5.71 |
| Variance of children ever born | 7.60 | 9.56 | 8.59 | 9.33 | 8.93 | 9.07 | 9.06 |
| Mean number of children surviving to 20 years of age | 4.57 | 4.29 | 4.40 | 5.86 | 5.21 | 5.31 | 5.13 |

at the level of statistical significance, they do raise the possibility that the net reproductive performance of consanguineous marriage equals or even exceeds that of a non-consanguineous marriage. Whether the implied "compensation" foı the increased mortality and morbidity of a consanguineous marriage is conscious or a natural biological phenomenon, due to such factors as the lessened frequency of maternal-fetal antigenic incompatibility in consanguineous marriages, will become a very interesting question should this finding be confirmed. Be this as it may, it is apparent that important aspects of the "consanguinity effect" in man have thus far been very inadequately explored and that at this stage of knowledge we cannot exclude the possibility that in at least some areas the consanguineous marriage may have a greater net reproductive potential than the non-consanguineous.

A word should be said about the question of the most desirable controls for these future studies. Our approach has been to utilize a random population sample for controls and then attempt to identify differences between these controls and the study group and to correct insofar as possible for the biases introduced by such differences as may be identified. Two other types of controls must be considered. Sibling controls have much to offer in theory, since they offer a partial "standardization" of the genetic background against which inbreeding and outbreeding effects are being contrasted. Since, however, important socioeconomic differences may exist between siblings, the investigator still is not relieved of the responsibility of collecting data on such variables. Furthermore, in populations with high mobility, such as that of the United States, and, increasingly, Europe and Japan, difficulty may be experienced in arranging for the examinations of the children of sibling controls under standardized conditions. A second type of control is a properly stratified or matched population sample. While in theory this is an excellent approach, in practice it poses many problems.

Thus, before one can match the consanguineous sample, one must either have conducted (simultaneously) a sufficiently extensive survey to provide matching material, which implies a very large-scale study, or else, at a somewhat lower level of precision, employ a "nearest neighbor" approach or some other arbitrary matching procedure. The problems which can arise in the nearest neighbor approach are well illustrated by the experience of Böök (1957), in which a significant age difference existed between consanguineous couples and their neighbors. In our opinion, the difficulties of matching samples in population studies involving man are such that the approach employed in this study, of identifying and making allowance for biases, is equally valid.

2. *Studies on the Biological Effects of Racial Crossing.* In the genetic sense, racial crossing in man is the antithesis of inbreeding. The effects of race-crossing on the frequency of congenital defect and other causes of morbidity have yet to be adequately studied. Neel (1958) (see also Frazier, 1960) drew attention to the apparently low frequency of congenital defect (aside from polydactyly) in the American Negro (who is approximately one-third Caucasian in ancestry) and emphasized the possible significance of this for hypotheses which attempted to relate congenital defect to genetic homozygosity. At that time, no data on "pure" Negro populations could be found; since then, Simpkiss and Lowe (1961) have published the only series known to us which, while small (2,068 births), appears otherwise satisfactory. In this series, without autopsies of stillbirths or neonatal deaths, there were 26 children with what would seem to be major defect (plus 28 with polydactyly). This 1.26% frequency agrees with the data on Caucasian and Japanese populations and keeps alive the possibility of lower rates in the American Negro. On the other hand, Saldanha (1964) reports no notable difference between malformation rates in births to whites, Negroes, and mulattoes in Brazil; yet the same author (Saldanha, 1961) reports lower malformation rates in Brazilian-Italian hybrids than in the Brazilian or Italian ethnic groups in Brazil.

Recently Morton (1962; see also Chung and Morton, 1963) has published a brief account of an extensive study of birth and death records in Hawaii, with particular reference to the results of race crossing. The previously mentioned facts of a similarity in total malformation frequency but differences in the frequency of specific types among different racial groups are confirmed. No effect of race-crossing on malformation frequency was detected, a fact interpreted as militating strongly against an important role for phenodeviants in the etiology of congenital defect. Finally, the convergence of spina bifida rates in Hawaiians of Japanese ancestry and of Caucasian ancestry leads to the conclusion "that this, the most striking difference in specific malformations between the two races, is of environmental origin" (Morton, 1962, p. 24). An extended consideration of these conclusions must await the presentation of the full material. While environmental factors undoubtedly strongly influence the expression of whatever genetic systems are involved in congenital defect, the sweeping emphasis on such factors in this report is to considerable extent offset by Miller's recent finding (1964) that in British Columbia both the descendants of Japanese immigrants and, of greater significance, of American Indians show the high frequency of harelip and cleft palate and the low frequency of spina bifida encountered in native-born Japanese, despite the great differences in diet and disease patterns in the two regions, despite the long separation of the Japanese and American Indian ancestry and despite the fact that the Caucasian frequencies in British Columbia agree closely with figures on American and European Caucasian populations. With further respect to the "convergence" of the spina bifida rates for Caucasians and Japanese, attention is directed to Hewitt's demonstration (1963) of a striking correlation between longitude and mortality from spina bifida in the United States and to his suggestion that the lower rates in the American West might find a partial explanation in a greater degree of marriage between persons of different geographical origins (hybridization) in the West. We are at present confronted with conflicting evidence, a conflict which must certainly be resolved before it is possible to assess the meaning of the evidence.

3. *Machine Simulation of Populations.* The third line of investigation which appears especially hopeful at this point is the machine simulation of the effects of inbreeding in populations with a variety of genetic structures. Thus, whereas it may be extremely difficult to develop general mathematical

formulations covering models having the complexity already known to exist in human populations, it should be much less difficult to program computers to simulate situations now known or readily envisioned. These simulations, while obviously far less desirable than general formulations, will nevertheless constitute an important step towards replacing intuition with demonstrable logic.

4. *Inbreeding Effects on Specific Systems.* The fourth line of investigation which will certainly receive attention in the future is the analysis of inbreeding effects on specific genetic systems in which a high proportion—preferably all—of the genotypes can be identified. As has been brought out, attempts to derive significant genetic parameters from gross inbreeding results draw heavily on the concept of equilibrium. Both because of a changing environment and increasing population mobility, this assumption is at obvious variance with the facts. The errors it introduces are often indeterminate. Studies involving specific genetic systems may go far to reveal just how unrealistic the assumption is.

The systems to come under scrutiny should involve not only the obvious genetic polymorphisms, but, with the burgeoning development of methods for detecting carrier states, the systems identified by less common, "recessively" inherited phenotypes. It is recognized that the necessary studies must be laborious, comprehensive, and at times a little boring because of the repetition implied by the magnitude they must assume if one is to detect small but significant differences in mortality and fertility between genotypes. They must also be repeated in populations living under diverse conditions. There seems no escape from a great deal of work if these and other studies of the genetics of human populations are to reveal their potential. If the issues have the importance and implications which we believe them to have, the work is justified.

# References


Albert, A. A. Non-associative algebras: I. Fundamental concepts and isotopy. *Ann. Math.*, **43** (4):685–707, Oct. 1942a.

———Non-associative algebras: II. New simple algebras. *Ann. Math.*, **43**(4):708–723, Oct. 1942b.

———Algebras derived by non-associative matrix multiplication. *Amer. J. Math.*, **66**:30–40, 1944.

Atkins, J. R. The cardinality of a kin relationship. *Philadelphia Anthropol. Soc. Bull.*, **13**(1), 1959.

———On the fundamental consanguineal numbers and their structural basis. *Amer. Ethnol.*, 1974.

Ballonoff, P. A. Stability properties of marriage systems: Theory of minimal structures. IX IUAES (in press) 1973.

———(ed). *Mathematical Models of Social and Cognitive Structures.* Urbana, Ill.: University of Illinois Press, 1974a.

———*Elementary Theory of Minimal Structures: Mathematical Foundations of Social Anthropology.* The Hague: Mouton and Co. (in press) 1974b.

Bartlett, M. S., and J. B. S. Haldane. The theory of inbreeding with forced heterzygosis. *J. Genet.*, **30**(3):327–340, 1935.

Bertrand, M. Algèbres non-associatives et algèbres génétiques. *Mém. Sci. Math.*, **162**:1–103, 1966.

Bruck, R. H. Some results in the theory of linear non-associative algebras. *Trans. Amer. Math. Soc.*, **56**:141–199, Sept. 1944.

Chapple, E. D., and C. S. Coon. The family. In: *Principles of Anthropology.* New York: Holt, Rinehart and Winston, Inc., 1942.

Clark, R. W. *JBS: The Life and Work of J. B. S. Haldane.* New York: Coward, McCann, & Georghegan, Inc., 1969.

Cockerham, C. C. Avoidance and rate of inbreeding. In: *Mathematical Topics in Population Genetics, No. 1,* New York: Springer-Verlag New York, Inc., 1970.

Cotterman, C. W. A note on the detection of interchanged children. *Amer. J. Human Genet.*, **3**(4):362–375, Dec. 1951.

———Estimation of gene frequencies in nonexperimental populations. In: Kempthorn et al. (eds). *Statistics and Mathematics in Biology.* Ames, Iowa: Iowa State University Press, 1954.

———Factor-union phenotype systems. In: Morton (ed). *Computer Applications in Genetics.* Honolulu: University Press of Hawaii, 1969.

Crawley, E. *The Mystic Rose: A Study of Primitive Marriage and of Primitive Thought in Its Bearing on Marriage.* New York: Boni and Liveright, Publishers, 1927.

Crow, J. F. Breeding structure of populations: II. Effective population number. In: Kempthorn et al. (eds). *Statistical Mathematics in Biology.* Ames, Iowa: Iowa State University Press, 1954.

———and N. E. Morton. Measurement of gene frequency drift in small populations. *Evolution,* **9**:202–214, June 1955.

Dahlberg, G. Inbreeding in man. *Genetics,* **14**:421–454, 1929.
———*Mathematical Methods for Population Genetics.* New York: John Wiley & Sons, Inc. (Interscience Division), 1948.
Dumont, L. Descent or intermarriage? A relational view of Australian section systems. *Southwestern J. Anthropol.,* **55**:231–250, 1966.
Durkheim, E., and M. Mauss. *Primitive Classification.* Chicago: The University of Chicago Press, 1963.
Elkin, A. P. The complexity of social organization in Arnhemland. *Southwestern J. Anthropol.* **6**(1):1–20, 1950.
Etherington, I. M. H. Commutative train algebras of ranks 2 and 3. *J. London Math. Soc.,* **15**:136–149, 1940.
———Special train algebras. *Quart. J. Math.,* **12**(45):1–7, Mar. 1941.
*Eugenics Quarterly.* Fourth Princeton conference, population genetics and demography. **15** (2), 1968.
Fish, H. D. On the progressive increase of homozygosis brother–sister matings. *Amer. Nat.,* **48**(576):759–761, 1914.
Fisher, R. A. The correlation between relatives on the supposition of Mendelian inheritance. *Trans. Roy. Soc. Edinburgh,* **56**(2):399–433 (1918).
———*The Theory of Inbreeding.* New York: Academic Press, Inc., 1949; 2nd ed., 1965.
Gillois, M. Corrélation génétique dans le cas de dominance. *Ann. Inst. Henri Poincaré,* **2**(1): 38–94, 1965a.
———Relation d'identité en génétique: I. Postulats et axiomes Mendeliens. *Ann. Inst. Henri Poincaré,* **2**(1):1–94, 1965b.
———La Relation de dependance en génétique. *Ann. Inst. Henri Poincaré,* **2**(3):261–278, 1966.
Gonshor, H. Special train algebras arising in genetics. *Proc. Edinburgh Math. Soc.,* **12**:41–53, 1960.
———Contributions to genetic algebras. *Proc. Edinburgh Math. Soc.,* **17**(4):289–298, Dec. 1971.
Greechie, R. J., and M. Ottenheimer. An introduction to a mathematical approach to the study of kinship (unpublished manuscript).
Hajnal, J. Concepts of random mating and the frequency of consanguineous marriages. *Proc. Roy. Soc. (London),* **B159**(974):125–177, Dec. 1963.
Haldane, J. B. S. *The Causes of Evolution.* Ithaca, N.Y.: Cornell University Press, 1931.
———The formal genetics of man. *Proc. Roy. Soc. (London),* **B135,** 1948.
———and P. Moshinsky. Inbreeding in Mendelian populations with special reference to human cousin marriage. *Ann. Eugen,* **9**:321–340, 1939.
Halsey, A. H. Social Mobility. In: Harrison and Boyce (eds).*The Structure of Human Populations.* New York: Oxford University Press, 1972.
Hamilton, W. D. Altruism and related phenomena, mainly in social insects. In: Johnson (ed). *Annual Review of Ecology and Systematics,* Vol. 3. Palo Alto, Calif.: Annual Reviews, Inc., 1972.
Heuch, I. Sequences in genetic algebras for overlapping generations. *Proc. Edinburgh Math. Soc.,* **18**(1):19–29, June 1972.
Hilden, J. GENEX—An algebraic approach to pedigree probability calculus. *Clin. Genet.,* **1**:319–348, 1970.
Hiorns, R. W., G. A. Harrison, A. J. Boyce, and C. F. Kuchemann. A mathematical analysis of the effects of movement on the relatedness between populations. *Ann. Human Genet.,* **32**:237, 1969.
Holgate, P. Genetic algebras associated with sex linkage. *Proc. Edinburgh Math. Soc.,* **17**(2): 114–120, Dec. 1970.
Imaizumi, Y., and M. Nei. Variability and heritability of human fertility. *Ann. Human Genet.,* **33**:251, 1970.
Jacquard, A. *Structures génétiques des populations.* Paris: Masson et Cie, 1970.
———*Genetic Structures of Populations.* New York: Springer-Verlag New York, Inc., 1973.
Jennings, H. S. Formulae for the results of inbreeding. *Amer. Nat.,* **48**(575):693–696, 1914.

———The numerical results of diverse systems of breeding. *Genetics* **1**:53–89, 1916.
Karlin, S. *Equilibrium Behavior of Population Genetic Models with Non-random Mating.* New York: Gordon and Breach, Science Publishers, 1968.
Kay, P. Taxonomy and semantic contrast. *Language,* **47**(4):866–887, 1971.
Kimura, M. *Diffusion models in population genetics.* London: Methuen & Co. Ltd., 1964.
———and J. F. Crow. The number of alleles that can be maintained in a finite population. *Genetics,* **49**:725–738, Apr. 1964.
———and T. Maruyama. Pattern of neutral polymorphism in a geographically structured population. *Genet. Res.,* **18**:125–131, 1971.
———and Ohta, T. *Theoretical Aspects of Population Genetics.* Princeton, N.J.: Princeton University Press, 1971.
———and G. H. Weiss. The stepping stone model of population structure and the decrease of genetic correlation with distance. *Genetics,* **49**:561–576, Apr. 1964.
Lévi-Strauss, C. *The Elementary Structures of Kinship.* Boston: Beacon Press, Inc., 1969.
Li, W. H., and M. Nei. Total number of individuals affected by a single deleterious mutation in a finite population. *Amer. J. Human Genet.,* **24**(6):667–679, 1972.
Lilleståøl, J. Pedigree probability calculus by means of linear operators. *Theoret. Population Biol.,* **2**:328–338, 1971.
Liu, P.-H. Murngin: A mathematical solution. *Current Anthropol.* **14**(1–2): 103–110, Feb.–Apr. 1973.
Lorrain, F. Manuscript in press. Paris: Hermann & Cie, 1974.
Lyubich, Y. I. Basic concepts and theorems of the evolutionary genetics of free populations. *Russ. Math. Surv.,* **26**(5):51–123, Sept.–Oct. 1971.
Maine, H. S. *Ancient Law.* London, 1861.
Malécot, G. *The Mathematics of Heredity.* San Francisco: W. H. Freeman and Company Publishers, 1969.
Maruyama, T. The rate of decrease of heterozygosity in a population occupying a circular or a linear habitat. *Genetics,* **67**:437–454, Mar. 1970.
———An invariant property of a structured population. *Genet. Res.,* **18**:81–84, 1971.
Mauss, M. *The Gift: Forms and Functions of Exchange in Archaic Societies.* New York: W. W. Norton and Co., Inc., 1967.
Meggitt, M. J. "Marriage classes" and demography in central Australia. In: Lee and DeVore (eds). *Man the Hunter.* Chicago: Aldine-Atherton, Inc., 1968.
Monod, J. *Chance and Necessity.* New York: Alfred A. Knopf, Inc., 1971.
Moran, P. A. P. The rate of approach to homozygosity. *Human Genet.,* **23**(1):1–5, Nov. 1958.
Morgan, L. H. Systems of consanguinity and affinity of the human family. Washington, 1871.
———*Ancient Society: Or, Researches in the Lines of Human Progress from Savagery, Through Barbarism to Civilization.* London, 1877.
Morton, N. E. Problems and methods in the genetics of primitive groups. *Amer. J. Phys. Anthropol.,* **28**:191–202, 1968.
Mukherjee, R. Concepts and methods for the secondary analysis of variations in family structures. *Current Anthropol.,* **13**(3–4):417–444, June–Oct. 1972.
Mycielski, J., and S. M. Ulam. On the pairing process and the notion of genealogical distance. *J. Combinatorial Theory,* **6**:227–234, 1969.
Needham, R. (ed). *Rethinking Kinship and Marriage.* New York: Tavistock Publishers, 1971.
Nei, M., and Y. Imaizumi. Random mating and frequency of consanguineous marriages. *Ann. Rep. Nat. Inst. Radiol. Sci., Sci. Technol. Agency,* Japan, 1962.
———Y. Imaizumi, and T. Furusho. Variability and heritability of human fertility. *Ann. Human Genet.,* **33**:251–259, 1970.
Pearl, R. A contribution towards an analysis of the problem of inbreeding. *Amer. Nat.,* **47**(562): 577–614, Oct. 1913.
———On the results of inbreeding a Mendelian population: A correction and extension of previous conclusions. *Amer. Nat.,* **48**(565):57–62, 1914c.
———Studies on inbreeding:IV. *Amer. Nat.,* **48**(572):491–494, 1914a.

———Studies on inbreeding: V. *Amer. Nat.,* **48**(573):513–523, 1914b.

———Studies on inbreeding: VI. Some further considerations regarding the measurement and numerical expression of degrees of kinship. *Amer. Nat.,* **51**(609):545–559, 1917a.

———Studies on inbreeding: VII. A single numerical measure of the total amount of inbreeding. *Amer. Nat.,* **51**(611):636–639, Nov. 1917b.

———, and J. R. Miner. Tables for calculating coefficients of inbreeding. *Statist. Bull.,* **215**: 123–138, 1913.

Provine, W. B. *The Origins of Theoretical Population Genetics.* Chicago: The University of Chicago Press, 1971.

Raffin. R. Axiomatisation des algèbres génétiques. *Acad. Roy. Belg. Bull. Classe Sci.,* **37**:359–366, 1951.

Robbins, R. B. Some applications of mathematics to breeding problems. *Genetics,* **2**:498–504, 1917.

———Applications of mathematics to breeding problems: II. *Genetics,* **3**:73–92, 1918a.

———Some applications of mathematics to breeding problems: III. *Genetics,* **3**:375–389, 1918b.

Robertson, A. The effect of non-random mating within inbred lines on the rate of inbreeding. *Genet. Res,* **5**:164–167, 1964.

Ruheman, B. A method for analyzing classificatory relationship systems. *Southwestern J. Anthropol.,* **1**:531–576, 1945.

Schafer, R. D. Structure of genetic algebras. *Amer. J. Math.,* **71**:121–135, 1949.

Schull, W. J. Empirical risks in consanguineous marriages: Sex ratio, malformation, and viability. *Amer. J. Human Genet.,* **10**(3):294–343, 1958.

Spuhler, J. N. Behavior and mating patterns in human populations. In: Harrison and Boyce (eds). *The Structure of Human Populations.* New York: Oxford University Press, 1972.

Tyler, S. A. The myth of P: Epistemology and formal analysis. *Amer. Anthrbpol.,* **71**(1):71–79, Feb. 1969.

Wahlund, S. Zusammensetzung von Populationen und Korrelationsersheinungen vom Standpunkt der Vererbungslehr aus betrachtet. *Hereditas,* **11**:65–106, 1928.

Wake, C. S. *The Development of Marriage and Kinship.* Chicago: The University of Chicago Press, 1967.

Watterson, G. A. Non-random mating, and its effect on the rate of approach to homozygosity. *Human Genet.,* **23**(3):204–232, July 1959.

White, H. C. *An Anatomy of Kinship.* Englewood Cliffs, N.J.: Prentice-Hall, Inc., 1963.

Woodger, J. H. *The Axiomatic Method in Biology.* New York: Cambridge University Press, 1937.

———*Biology and Language.* New York: Cambridge University Press, 1952.

Wright, S. Systems of Mating: I, II, III, IV, V. *Genetics,* **6**:111–178, 1921.

———Coefficients of inbreeding and relationship. *Amer. Nat.,* **56**(645):330–338, 1922.

———Isolation by distance. *Genetics,* **28**:114–138, Mar. 1943.

———Isolation by distance under diverse systems of mating. *Genetics,* **31**:39–59, Jan. 1946.

Yasuda, N. An extension of Wahlund's principle to evaluate mating type frequency. *Amer. J. Human Genet.,* **20**(1):1–23, Jan. 1968.

———and N. E. Morton. Studies on human population structure. In: Crow and Neel (eds). *Proceedings III International Congress of Human Genetics.* Baltimore: The Johns Hopkins Press, 1967.

Yengoyan, A. A. Demographic and ecological influences on Aboriginal Australian marriage sections. In: Lee and DeVore (eds). *Man the Hunter.* Chicago: Aldine-Atherton, Inc., 1968.

# Author Citation Index

# Subject Index